Computational Aspects of Heat Transfer in Structures

Howard M. Adelman, *Compiler*
Langley Research Center
Hampton, Virginia

Proceedings of a joint NASA/George Washington
University/Old Dominion University symposium
held at Langley Research Center
Hampton, Virginia
November 3-5, 1981

National Aeronautics
and Space Administration

**Scientific and Technical
Information Branch**

1982

Use of trade names or names of manufacturers in this report does not constitute an official endorsement of such products or manufacturers, either expressed or implied, by the National Aeronautics and Space Administration.

PREFACE

This document contains the proceedings of the Symposium on Computational Aspects of Heat Transfer in Structures held at the Langley Research Center, November 3-5, 1981. The symposium was jointly sponsored by the NASA Langley Research Center, George Washington University, and Old Dominion University.

The main theme of the symposium was recent progress in improving national capability to efficiently calculate the temperature and thermal stress in complex structures with an emphasis on reentry flight-vehicle structures. Speakers were drawn from NASA, various other government scientific organizations, universities, aerospace companies, and computer software organizations.

The symposium was organized in the following six sessions:

I. Computer Programs for Thermal Analysis

II. Advancements in Analysis Techniques

III. Thermal Analysis of Large Space Structures

IV. Thermal Analysis of High-Speed Vehicles

V. Impact of New Computer Systems

IV. Concerns, Issues, and Future Directions

Additionally, a special paper entitled "Historical Perspectives on Thermo-structural Research at the NACA Langley Aeronautical Laboratory From 1948 to 1958" was presented by Richard R. Heldenfels. The sixth session was a panel discussion conducted on the final day of the symposium. A transcript of the panelists' remarks is included in this document.

Papers and the authors are grouped by session and identified in the Contents. The order of papers is the actual order of speaker appearance at the symposium. The papers contained in this compilation were submitted as camera-ready copy. A list of attendees is included at the end of this document.

I would like to express appreciation to the session chairmen, speakers, and panelists whose efforts contributed to the technical excellence of the symposium. The session chairmen were: Allan Wieting (Langley Research Center), Earl Thornton (Old Dominion University), Edwin Kruszewski (Old Dominion University), James Robinson (Langley), and Jules Lambiotte, Jr. (Langley). The moderator of the panel discussion was Sidney Dixon of Langley.

Howard M. Adelman
Symposium Chairman

CONTENTS

PREFACE iii

SESSION I - COMPUTER PROGRAMS FOR THERMAL ANALYSIS
Chairman: Allan R. Wieting

1. NASTRAN THERMAL ANALYZER IN A UNIFIED FINITE-ELEMENT TREATMENT OF THERMO-STRUCTURAL ANALYSES 1
   Hwa-Ping Lee

2. THERMAL CAPABILITIES AND GRAPHICAL OUTPUT OF PAFEC 21
   J. E. Akin

3. THE SPAR THERMAL ANALYZER - PRESENT AND FUTURE 35
   M. B. Marlowe, W. D. Whetstone, and J. C. Robinson

4. A COMPARISON OF THE FINITE DIFFERENCE AND FINITE ELEMENT METHODS FOR HEAT TRANSFER CALCULATIONS 51
   A. F. Emery and H. R. Mortazavi

5. A METHOD TO MODEL LATENT HEAT FOR TRANSIENT ANALYSIS USING NASTRAN 83
   Robert L. Harder

SESSION II - ADVANCEMENTS IN ANALYSIS TECHNIQUES
Chairman: Earl A. Thornton

6. SOME ASPECTS OF ALGORITHM PERFORMANCE AND MODELING IN TRANSIENT THERMAL ANALYSIS OF STRUCTURES 91
   Howard M. Adelman, Raphael T. Haftka, and James C. Robinson

7. ALGORITHMIC ASPECTS OF TRANSIENT HEAT TRANSFER PROBLEMS IN STRUCTURES 99
   Raphael T. Haftka and M. Hassan Kadivar

8. EVALUATION OF AN IMPROVED FINITE-ELEMENT THERMAL STRESS CALCULATION TECHNIQUE 115
   Charles J. Camarda

9. STATUS REPORT ON DEVELOPMENT OF A REDUCED BASIS TECHNIQUE FOR TRANSIENT THERMAL ANALYSIS 133
   Charles P. Shore

10. APPLICATIONS OF PERTURBATION TECHNIQUES TO HEAT-TRANSFER PROBLEMS 147
    Osama A. Kandil

11. DEVELOPMENT OF MIXED TIME PARTITION PROCEDURES FOR THERMAL ANALYSIS OF STRUCTURES 161
    Wing Kam Liu

SESSION III - THERMAL ANALYSIS OF LARGE SPACE STRUCTURES
Chairman: Edwin Kruszewski

12. INTEGRATED THERMAL-STRUCTURAL ANALYIS OF LARGE SPACE STRUCTURES 179
    Jack Mahaney, Earl A. Thornton, and Pramote Dechaumphai

13. INTERACTIVE MODELING, DESIGN AND ANALYSIS OF LARGE SPACECRAFT 199
    L. Bernard Garrett

14. INTERACTIVE COMPUTATION OF RADIATION VIEW FACTORS 221
    A. F. Emery, H. R. Mortazavi, and C. J. Kippenhan

15. RECENT DEVELOPMENTS IN THERMAL RADIATION SYSTEM ANALYZER (TRASYS) 243
    Robert A. Vogt

16. ROLE OF IAC IN LARGE SPACE SYSTEMS THERMAL ANALYSIS 253
    G. K. Jones, J. T. Skladany, and J. P. Young

SESSION IV - THERMAL ANALYSIS OF HIGH-SPEED VEHICLES
Chairman: James C. Robinson

17. REENTRY HEATING ANALYSIS OF SPACE SHUTTLE WITH COMPARISON OF
    FLIGHT DATA 271
    Leslie Gong, Robert D. Quinn, and William L. Ko

18. REENTRY HEAT TRANSFER ANALYSIS OF THE SPACE SHUTTLE ORBITER 295
    William L. Ko, Robert D. Quinn, Leslie Gong, Lawrence S. Schuster,
    and David Gonzales

19. SPACE SHUTTLE ORBITER ENTRY HEATING AND TPS RESPONSE:
    STS-1 PREDICTIONS AND FLIGHT DATA 327
    Robert C. Ried, Winston D. Goodrich, Chien P. Li, Carl D. Scott,
    Stephen M. Derry, and Robert J. Maraia

20. TRANSIENT THERMAL ANALYSIS OF A TITANIUM MULTIWALL THERMAL
    PROTECTION SYSTEM 349
    M. L. Blosser

21. HISTORICAL PERSPECTIVES ON THERMOSTRUCTURAL RESEARCH AT THE NACA
    LANGLEY AERONAUTICAL LABORATORY FROM 1948 TO 1958 363
    Richard R. Heldenfels

SESSION V - IMPACT OF NEW COMPUTER SYSTEMS
Chairman: Jules J. Lambiotte, Jr.

22. GUIDELINES FOR DEVELOPING VECTORIZABLE COMPUTER PROGRAMS 393
    E. W. Miner

23. EVALUATION OF THE SPAR THERMAL ANALYZER ON THE CYBER-203 COMPUTER 405
    J. C. Robinson, K. M. Riley, and R. T. Haftka

24. DEVELOPMENT OF A CRAY 1 VERSION OF THE SINDA PROGRAM 425
    Susan M. Juba and Peter E. Fogerson

25. AN EVALUATION OF SUPERMINICOMPUTERS FOR THERMAL ANALYSIS 437
    Olaf O. Storaasli, James B. Vidal, and Gary K. Jones

SESSION VI - CONCERNS, ISSUES, AND FUTURE DIRECTIONS
Moderator: Sidney C. Dixon

26. PANEL DISCUSSION - CONCERNS, ISSUES, AND FUTURE DIRECTIONS 453
    Sidney Dixon, Moderator

27. SURVEY OF COMPUTER PROGRAMS FOR HEAT TRANSFER ANALYSIS
    (Not presented at conference) 487
    Ahmed K. Noor

ATTENDEES 563

NASTRAN THERMAL ANALYZER IN A UNIFIED FINITE-ELEMENT TREATMENT OF THERMO-STRUCTURAL ANALYSES

Hwa-Ping Lee
Goddard Space Flight Center
Greenbelt, Maryland

ABSTRACT

For solution accuracy, modeling efficiency and cost effectiveness, the NASTRAN Thermal Analyzer (NTA) is suited to treat large-scale unified thermo-structural analyses with the NASTRAN (NAsa STRuctural ANalysis) computer program. The mathematical similitude between these two distinct disciplines of thermal and structure is examined. It serves as the theoretical basis upon which the implementation of the thermal capability in NASTRAN was accomplished. The program structure, the functional flow, the solution algorithms, the organization of an input data deck and the solution capabilities of NTA are summarized. Emphasis is placed on the interface of the unified approach in thermo-structural analyses where stresses, deflections, vibrations and bucklings induced by the effect of temperature change are of concern. Attentions are also directed to the pre-processor and post processors. As a specially designed pre-processor, the VIEW program is capable of generating exchange factors which can be output, at user's option, in formats compatible with that required by NTA. Two post processors that serve specific objectives are included. They are the thermal variance analysis and the graphical displaying capability of temperatures in color or B&W.

INTRODUCTION

The NASTRAN Thermal Analyzer is a general-purpose thermal analysis computer program that has been developed and integrated in the NASTRAN System (Refs. 1, 2). This thermal analysis capability was implemented using applicable functional modules of NASTRAN which had been developed for structural analysis originally. However, a number of new modules were developed and added to satisfy unique requirements for thermal analysis. They comprised new elements and new solution algorithms. The feasibility of utilizing the structural elements and functional modules in thermal application lies with the mathematical similitude that exists in the two distinct disciplines. The *intrinsic modular structure of NASTRAN permits a direct abstraction of its matrix functional modules to be arranged in proper solution sequences for thermal analysis.* As a consequence, the NTA is unique in that its thermal model is fully compatible with the structural NASTRAN model at the grid point and element level. This feature is invaluable in unified treatment of thermo-structural analyses especially for problems of large size and complex configuration, where stresses, deflections, vibrations and bucklings induced by the temperature effect are of concern.

This paper starts with an examination of the mathematical similitude between the two disciplines of thermal and structure after both governing equations are cast in the matrix form following the finite element methodologies. Equivalence of terms between the two physical systems will be identified as they are essential to use structural functional modules or terminologies in thermal analysis. The program structure which conforms to that of the NASTRAN will be presented. The most frequently employed bulk data cards will be listed. They are thermal conduction elements, boundary surface elements, material properties, thermal loadings, etc. The three solution algorithms spanning the whole spectrum of interested thermal problems will be included.

In aerospace applications, thermal radiation plays an important role in heat transporting process. Geometric view factors are required in thermal analysis when radiative exchanges prevail. The pre-processor VIEW (Refs. 3, 4) was designed to compute the exchange factors. This stand-alone software program is in full compatibility with the NTA. At the input end, the VIEW uses the same boundary surface elements CHBDY* of the NTA model to define radiatively active surfaces. At the other end of output, an option to output the exchange factors in formatted cards directly useable in an input data deck of the NTA is available to users. The functional structure and the program organization of VIEW together with the unique data card $VIEW will be presented. A partial listing of a typical data deck to generate view factors, results and an NTA input data deck embracing the exchange factors in the formatted card forms of RADLST and RADMTX will be illustrated.

Regarding post processors, a few relevant programs that are operational at Goddard Space Flight Center will be included. They serve specific objectives. The thermal variance analysis (Ref. 5) and the visual display of temperatures or temperature gradients (Ref. 6) will be described.

THEORETICAL BASIS FOR PROGRAM IMPLEMENTATION

Theoretical finite element treatment of thermal conduction analysis and its extensions to include radiative exchanges can be cited in Refs. 7-9. The NTA was implemented in accordance with the general heat equation in the matrix form as follows

$$[C]\{\dot{T}\} + [K]\{T\} = \{Q^\ell\} + \{Q^n\} \qquad (1)$$

where

$\{T\}$ = a vector of temperatures at grid points

$\{\dot{T}\}$ = a vector of rate-change-of-temperatures at grid points

---

*Names of actual NTA cards are capitalized and underlined.

[C] = a symmetric matrix of heat capacitance

[K] = a symmetric matrix of thermal conductance

$\{Q^{\ell}\}$ = a vector of applied thermal loads (constant or time-dependent)

$\{Q^n\}$ = a vector of nonlinear thermal loads.

The preceding expression implies three classes of problems that require separate solution algorithms. As stands, Eq. (1) represents an unsteady-state heat equation. However, it represents a steady-state case when the term $\{\dot{T}\}$ vanishes, and the linear and nonlinear steady-state cases are treated differently.

The NTA is a component in the NASTRAN system and mathematical similitude can be drawn between the structural and thermal systems (Fig. 1). A number of elements, modules, and the input and output parameters of NASTRAN were used in "borrowed" forms for the NTA as a consequence. They include the input file processor, the geometry processor, constraining, partitioning of matrices, etc. together with all structural finite elements except the excessive DOF's (degrees-of-freedom) at each vertex of an element being constrained properly.

PROGRAM STRUCTURE AND FUNCTIONS

The NTA also shares the same program structure as the structural counterpart. Its input and output formats were so designed that they are fully compatible with those of NASTRAN. A complete NTA input data deck consists of three parts:

(1) Executive Control Deck
(2) Case Control Deck
(3) Bulk Data Deck

Their functions are listed in Fig. 2. The functional flow of the bulk data cards relative to the definition, constraints and thermal loadings of a thermal model is shown in Fig. 3. The Bulk Data Deck constitutes the main body of a complete input data deck. The frequently used bulk data cards appear in Fig. 4.

In a model, the heat conducting structure is formed by heat conduction elements that are interconnected at grid points. Various heat conduction elements are provided. They consist of 1-D rods, 2-D triangular and quadrilateral plates and axisymmetric rings of triangular as well as trapezoidal cross-sections, and 3-D solids such as wedges, tetrahedra and hexahedra, as presented in Fig. 5. Also included are scalar heat conduction elements (CELASi) which may serve as linear thermal conductors connecting pairs of grid points with specified thermal conductances. The CHBDY is a special boundary surface element which serves as a medium in exchanging heat from external environment to the overlaid conduction element through the attached grid points.

Thermal loads of constant and time-varying quantities may be applied directly to grid points, or via the boundary surface elements. The types of thermal load included in this program are the concentrated load applied to a grid point, the internally generated heat within an element, and the uniform heat flux as well as the directional thermal radiant source applied to the surface of an element.

Various constraints can be applied to grid points. The single-point constraint is used in the steady-state case to specify prescribed temperature at a grid point. The multipoint constraint is used to specify a linear relationship of temperatures at selected grid points. Omitted points are constrained to reduce the number of unknown temperatures in the transient thermal analysis.

The NTA has been provided with three specialized solution algorithms that are able to yield accurate, efficient and stable solutions. They are:

(1) Linear steady-state case: This solution is of a matrix inversion process,

(2) Nonlinear steady-state case: This solution employs an iterative process,

(3) Transient thermal analysis including both linear and nonlinear boundary conditions: This integration algorithm uses the modified Newmark-β method (Ref. 10), which allows a user to select a value for the parameter β in the range of $0 \leq \beta \leq 1$. This expression together with special cases is given in Fig. 6.

To solve a specific type of physical problem, the NASTRAN has a formatted and permanently stored sequence of macro-instructions to execute mathematical modules, and it is called a Rigid Format. Therefore, three Rigid Formats have been formed for NTA. Other NASTRAN features such as the DMAP (direct matrix abstraction program) and ALTER (a similar user-oriented program modification but to a lesser degree) are also available to the NTA users.

With NASTRAN, the NTA is especially suited to treat large-scale unified thermo-structural problems. The only limitations on the problem size are those imposed by practical considerations of the execution time and by the ultimate capacity of auxiliary storage devices. There is no dimension statement in the program.

In unified thermo-structural analyses, the grid point temperature data, as required by NASTRAN to analyze thermally induced structural responses, are provided directly by the NTA through the TEMP data cards. This can be achieved by using a pair of compatible models. Specifically, the input data decks for these two distinct disciplines share the same basic model of the finite element discretization. Two separate input card decks, however, are still required. Either the NASTRAN structural model or the NTA thermal model can be the first to become available in the back-to-back analyses. Alterations of cards in the input deck for the second model need be made only from the one first in existence to accommodate constraints, loadings, material properties, parameters, etc. in accordance with the problem description. While remaining useable for those cards defining grid points and connection cards (element descriptions), they generally constitute the main body of a bulky input data deck, which would be the most labor-intensive and time-consuming effort to model and prepare independently. When thermo-structural analyses are performed in tandem, the structural model, satisfying mechanical requirements and design criteria, is usually the first to be created. The modification or transformation of model is, therefore, from a NASTRAN structural model to an NTA thermal model.

A PREPROCESSOR – VIEW

The VIEW program (Ref. 3) was specifically designed to yield the view factors, F_{ij}, and then the exchange factors, $A_i F_{ij}$, that are required in thermal analysis with surfaces that are active in radiative exchanges. The VIEW was originally designed to run on an IBM System/360 operating under OS (Operating System), with a minimum region size of 110 K bytes. This computer program takes into account the presence of any intermediate surfaces. It computes these view factors either by the contour integration or by the double area summation method. The former is known to be more accurate but less efficient. Either method may be selected or a criterion may be specified which causes the program to select the best method based upon the geometry of the problem.

As a preprocessor to NTA, the following compatibility requirements are featured:

(1) Accept GRID and CHBDY from the NTA model as the input to VIEW for surface definition,

(2) Produce the output RADLST and RADMIX in the formats acceptable to the NTA model.

Additional features of this program include:

(3) A restart capability, which protects a user against having to rerun an entire problem should a computer failure occur.

(4) The ability to dynamically allocate available core space, thus allowing the user to request the amount of space in the computer required for ones problem by using the region parameter on the job card. There is no maximum number of elements to which the user is limited, except that the computer's capacity may not be exceeded.

(5) The ability to accept one or a combination of two input formats. The VIEW program can accept both data formats of the RAVFAC-type and the NASTRAN-type inputs.

(6) The ability to run several problems in sequence in one job submission. Each problem run is referred to as a "case."

Five basic element shapes may be described by using the NTA data card <u>CHBDY</u>. Each of these shapes has a given name as shown in the table:

Element Shape	Name
Circular plate	POINT element
Rectangular plate	LINE element
Conical or cylindrical shell	REV element
Triangular plate	AREA3 element
Quadrilateral plate	AREA4 element

Describing the dimension and location of these five elements is accomplished by using grid points.

The functional flow of the VIEW program is given in Fig. 7, and the organization of an input data deck is shown in Fig. 8. The unique input data card <u>$VIEW</u> is used to define element characteristics such as the specifications of sub-element mesh sizes in x and y directions for surface integrations, the shading flags, etc.

The listing of a typical input data deck of VIEW and a typical output are reproduced in Fig. 9. The exchange factors in the force directly admissible to an NTA data deck is illustrated in Fig. 10.

POST PROCESSORS

(1) Thermal Variance Analysis

This solution capability developed and integrated in NTA is capable of assessing the sensitivity of temperature variance resulting from uncertainties inherent in input parameters, which may include geometry, material properties, applied thermal loads, etc. The computational process is to modify the input data, to calculate partial derivatives of the output temperatures and to compute the variances of the output quantities.

Two new data cards /VARY and /PARM were introduced for modifying the input bulk data. A module VARIAN was added to compute variance of any output quantity ϕ, and it is based on the relationship of the form

$$\text{Variance}(\phi) = [\Sigma \, (\frac{\partial \phi}{\partial S_k} \cdot \Delta S_k)^2]^{1/2}$$

(2) Visual Display of Temperatures

The capability of visual display of temperatures or temperature gradients in color or B&W has been installed at GSFC. The Grinnell GMR-275 Image Display System together with a software package MOVIE·BYU have served as a post processor to NTA to show temperature results graphically. The basic capability of the Grinnell GMR-275 includes the following:

(A) Color or black and white display of up to 512 × 512 pixel images with 8-bits of data at each pixel.

(B) Software-controlled hardware-implemented zoom and pan.

(C) Three 8 × 10 look-up tables to control false color displays.

(D) Vector drawing.

(E) Split-screen.

The software MOVIE · BYU (Brigham Young University) contains several components which are Fortran programs for the display and manipulation of data representing mathematical, topological or architectural models where geometry may be described in terms of polygonal elements or contour line definitions. The source of the polygonal element data can be a finite-element analysis. The program has hidden line, contour, animation, shading, and full color capabilities in addition to many others. For a detailed description of the system, consult Ref. 6.

Applied thermal loads or temperature results at grid points can be displayed in color using the described hardware and software system. The input for MOVIE·BYU can be prepared by preprocessing the NTA Bulk Data Deck and thermal load or temperature card decks to produce Geometry and Function files. The displays which were produced in color (five slides shown at the presentation only) associated blue with cold temperatures and red with warm temperatures.

CONCLUDING REMARKS

Features and capabilities of the NTA are summarized in the following:

- A general-purpose heat transfer analysis computer program using finite-element method
- Linear and nonlinear transient and steady-state cases
- Conduction in discretized elements with temperature-variable (DOF) output at grid points
- Boundary conditions:
  (1) Specified temperatures at grid points
  (2) Thermal loadings with
     (A) Internal (volumetric) heat generation
     (B) External heat flux
        (A) Constant
        (B) Directional
        (C) Time-dependent
  (3) Convective boundary with
     (A) Constant convective film coefficient
     (B) Temperature-dependent convective film coefficient
  (4) Radiative boundary with
     (A) Diffuse-grey surfaces
     (B) Specular surfaces (Ref. 9)
- Arbitrary initial temperatures prescribed at grid points

- Material properties:
  (1) Isotropic and anisotropic thermal conductivity properties
  (2) Temperature-dependent thermal conductivity or convective film coefficient in nonlinear steady-state case
  (3) Temperature-dependent emissivity and absorptivity in transient-state case (Ref. 11)
  (4) Temperature-dependent convective film coefficient and heat capacitance in transient-state case (Ref. 12)
- Provision of user selected β-value for stability in transient solution algorithm
- Graphical displaying capabilities
  (1) Conduction elements
  (2) Boundary surface elements
  (3) On-line printer plot of temperature vs. time and $\partial T/\partial t$ vs. time at grid points
  (4) Isothermal contour plot
- Miscellaneous
  (1) DMAP, ALTER
  (2) Restart, punch-card or tape output, etc.
  (3) Direct matrix input to $[c]$ or $[k]$
  (4) Ability to be used as a conventional lumped-mass thermal network
- Preprocessor
  The VIEW program
- Post processors
  (1) Thermal variance analysis
  (2) Visual displays of temperature and temperature-gradient in color or b&w

The advantage of using NTA in thermo-structural analyses over other combinations is clearly shown in Fig. 11. The convenience and useability of NTA is further enhanced with the addition of the post processors. These features provide a flexibility far beyond that available in other known software systems in the public domain. This fact, combined with the NTA's proven reliability, has made it a valuable tool in the analytical arsenal suitable for unified thermo-structural analyses.

REFERENCES

1. Lee, H. P., "NASTRAN Thermal Analyzer — Theory and Application Including a Guide to Modeling Engineering Problems, Vol. 1," NASA TM X-3503, 1977.

2. "The NASTRAN User's Manual (Level 17.0)," NASA SP-222(04), Dec. 1977, Washington, D. C.

3. Puccinelli, E. F., "View Factor Computer Program (Program VIEW) User's Manual," NASA TM X-70538, 1973.

4. Puccinelli, E. F., and Jackson, C. E. Jr., "VIEW — A Modification of the RAVFAC View Factor Program for Use with the NASTRAN Thermal Analyzer," NASTRAN: Users' Experiences, NASA TM X-2637, 1972, pp. 455-463.

5. Harder, R. L., "NASTRAN Variance Analysis and Plotting of HBDY Elements," NASA CR-139007, The MacNeal-Schwendler Corp., May 1974.

6. Christiansen, H., and Stephenson, M., "Movie • BYU," Civil Engineering Dept., Brigham Young University, Provo, Utah.

7. Zienkiewicz, O. C., "The Finite Element Method," 3rd ed., McGraw-Hill Book Co., 1977.

8. Lee, H. P., "Application of Finite-Element Method in the Computation of Temperature with Emphasis on Radiative Exchanges," AIAA Progress in Astronautics and Aeronautics: Vol. 31 — Thermal Control and Radiation. The MIT Press, Cambridge, Mass., 1973, pp. 491-520.

9. Lee, H. P., and Jackson, C. E. Jr., "Finite-Element Solution for Radiative-Conductive Analyses with Mixed Diffuse-Specular Surfaces," AIAA Progress in Astronautics and Aeronautics: Vol. 49 — Radiative Transfer and Thermal Control, 1976, pp. 15-46.

10. Newmark, N. M., "A Method of Computation for Structural Dynamics," Proc. ASCE, J. of Eng'g. Mech. Div., EM-3, July 1959, pp. 67-94.

11. Lee, H. P., and Harder, R. L., "The GSFC NASTRAN Thermal Analyzer New Capabilities." NASTRAN: Users' Experiences, NASA TM X-3428, 1976, pp. 119-126.

12. Lee, H. P., "Simulation of Nonlinear Thermal Boundaries in Finite Element Analysis Using NASTRAN Thermal Analyzer." Presented at Second National Symposium on Numerical Methods in Heat Transfer, American Institute of Numerical Heat Transfer, held at University of Maryland, College Park, Md., Sept. 28-30, 1981.

$$[M]\{\ddot{x}\} + [C]\{\dot{x}\} + \underbrace{\underbrace{[K]\{x\} = \{F(t)\}}_{\text{STEADY-STATE HEAT EQUIATION}}}_{\text{TRANSIENT-STATE HEAT EQUATION}}$$

Symbol	Structural System		Thermal System
$\{x\}$	Displacement		Temperature
$\{\dot{x}\}$	Velocity		Rate change of temperature
$\{F\}$	Applied Load	(Equivalence) →	Heat source or Thermal load
$[K]$	Stiffness		Thermal conductance
$[C]$	Damping		Heat capacitance
$[M]$	Mass		None

Figure 1. Mathematical Similitude between Structural and Thermal Systems

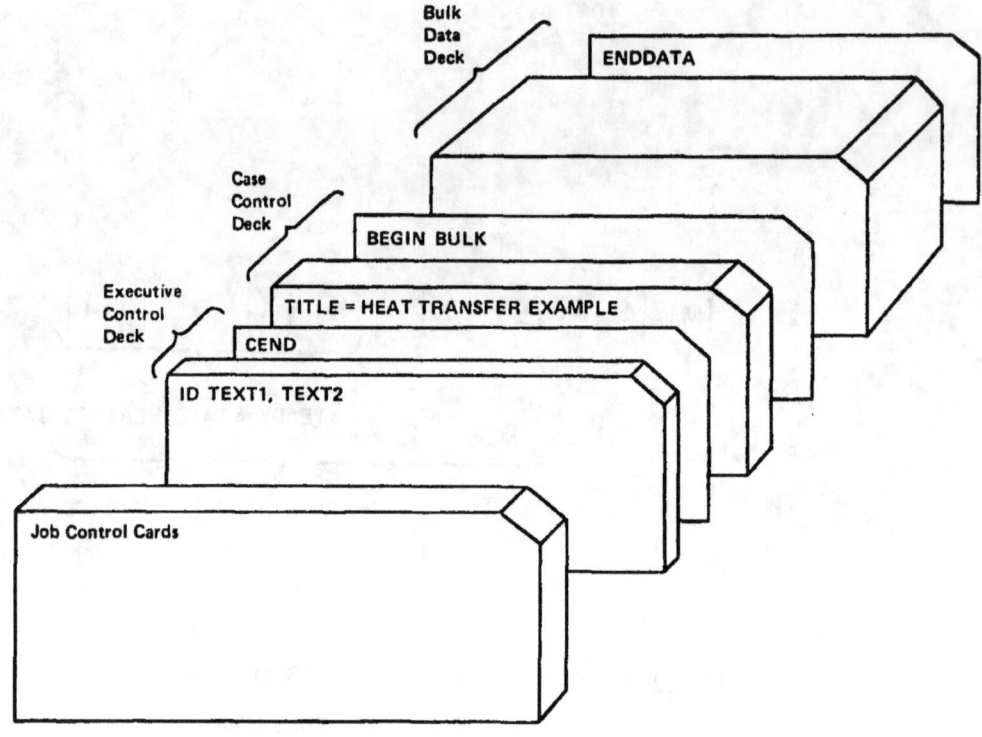

1. Executive Control Deck
    (a) To identify the job
    (b) To select solution type
    (c) To limit maximum CPU time in min.
    (d) To select types of execution
    (e) To embrace DMAP sequence if used

2. Case Control Deck
    (a) To select loading and boundary condition sets
    (b) To request output

3. Bulk Data Deck
    (a) To define the finite-element model
    (b) To specify loading and boundary conditions

Figure 2. NASTRAN Thermal Analyzer Program Structure

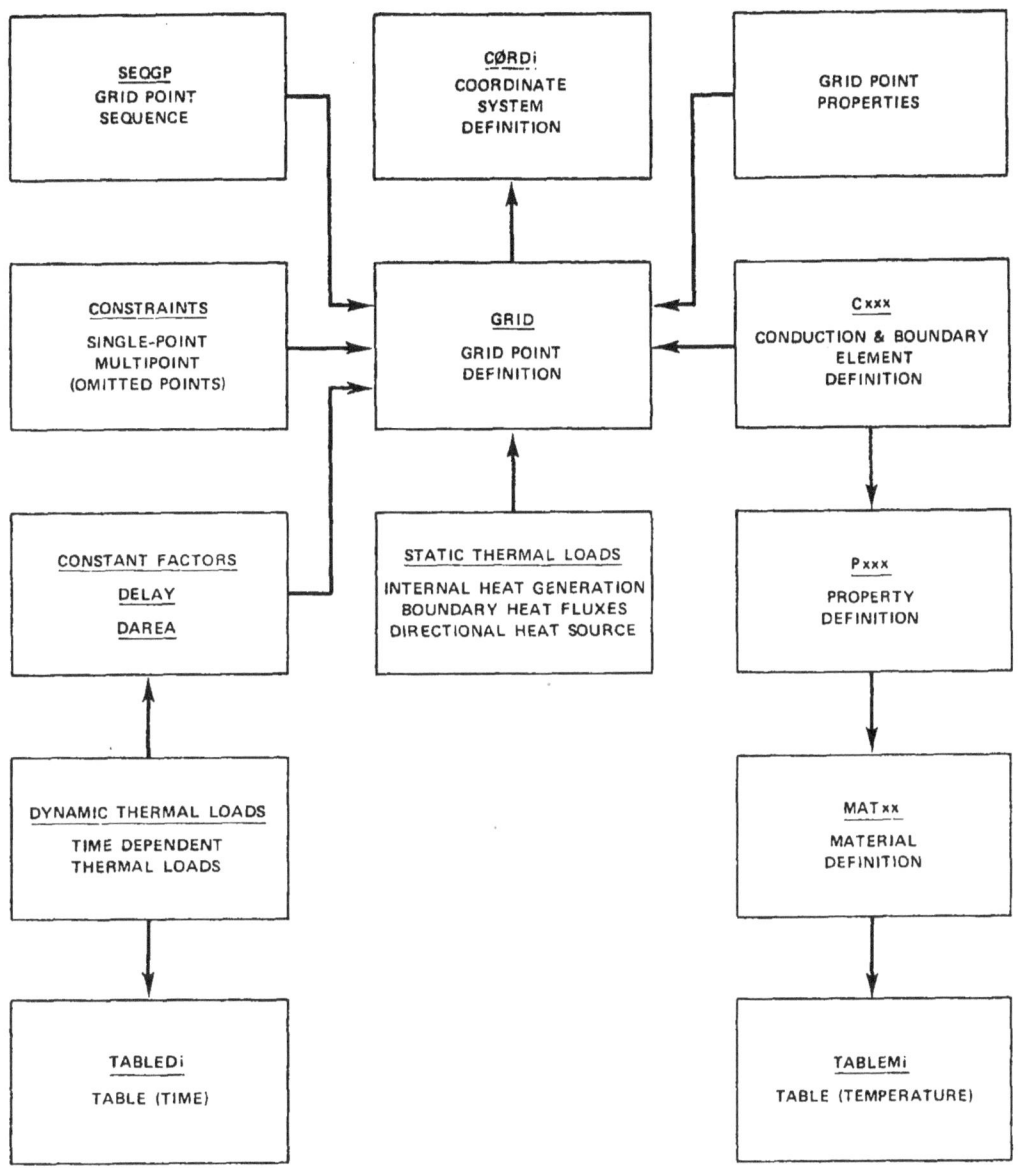

Figure 3. Functional Diagram of the NTA Bulk Data Cards in a Thermal Model

GEOMETRIC DEFINITION:

    CORDi¡
    GRID, GRDSET, SEQGP
    EPØINT
    SPØINT

ELEMENTS

  HEAT CONDUCTION:

    CBAR, CRØD
    CELASi
    CDAMPi
    CQUADi, CTRIAi
    CHEXAi, CTETRA, CWEDGE

  BOUNDARY SURFACE:

    CHBDY (POINT, LINE, REV, AREA3, AREA4, ELECYL)

ELEMENT PROPERTIES

    PBAR, PRØD
    PQUADi
    PTRIAi
    PHBDY

THERMOPHYSICAL PROPERTIES

    MAT4, MAT5
    MATT4, MATT5
    TABLEMi

CONSTRAINT AND PARTITIONS:

    SPC, SPC1
    MPC
    ASET, ØMIT, ØMIT1

THERMAL LOADINGS:

    DAREA
    DELAY
    DLØAD
    NOLINi
    QBDY1, QBDY2
    QHBDY
    QVECT
    QVØL
    TABLEDi
    TLØADi

RADIATIVE EXCHANGE DESCRIPTION:

    RADLST
    RADMTX

MISC:

    DMIG
    NFTUBE
    PARAM
    TEMP, TEMPD
    TF
    TSTEP
    $

Figure 4. Bulk Data Cards Frequently used in Thermal Analysis

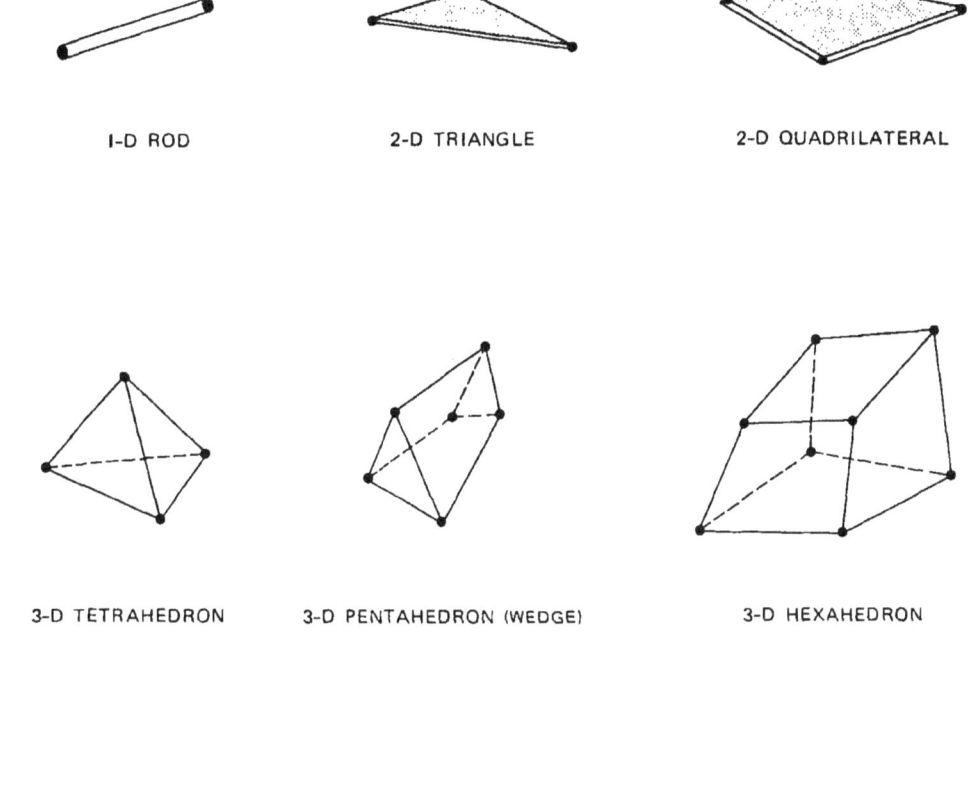

Figure 5. Representative Heat Conduction Elements

$$[K]\{\beta T_{n+1} + (1-\beta)T_n\} + \frac{1}{\Delta t}[C]\{T_{n+1} - T_n\}$$

$$= \{\beta Q_{n+1}^I + (1-\beta)\{Q_n^I\}\} + (1+\beta)\{Q_n^n\} - \beta\{Q_{n-1}^n\}$$

$$\left[\frac{1}{\Delta t}[C] + \beta[K]\right]\{T_{n+1}\}$$

$$= \left[\frac{1}{\Delta t}[C] - (1-\beta)[K]\right]\{T_n\} + \{\beta Q_{n+1}^I + (1-\beta)Q_n^I\} + (1+\beta)\{Q_n^n\} - \beta\{Q_{n-1}^n\}$$

$\beta = 0$ EULER INTEGRATION
$\beta = \frac{1}{2}$ CENTRAL DIFFERENCES
$\beta = 1$ BACKWARD DIFFERENCES

Figure 6. The Modified Newmark-Beta Method

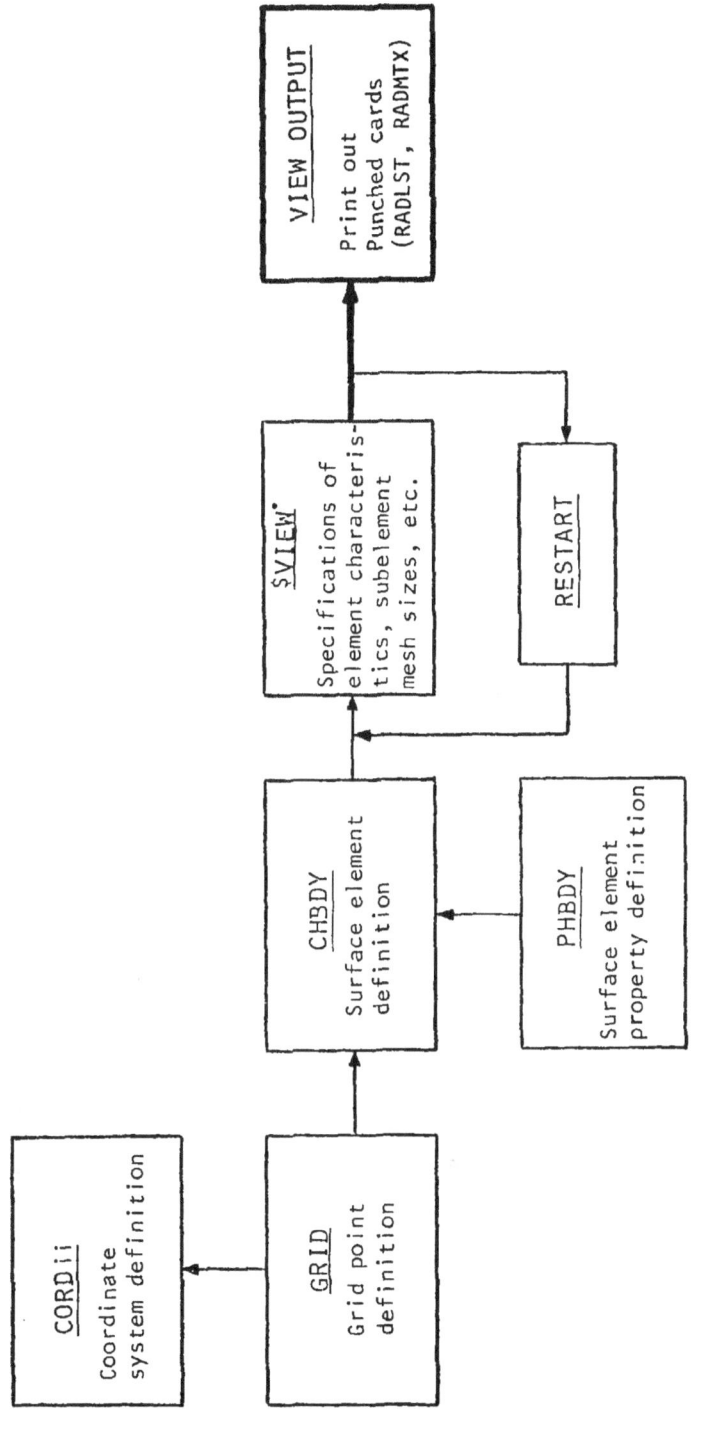

Figure 7. Functional Diagram of VIEW to Generate View Factors

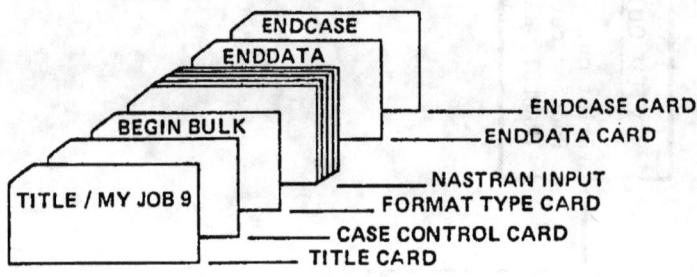

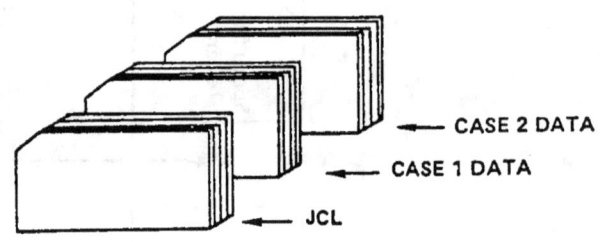

1	2	3	4	5	6	7	8	9	10
$VIEW	IVIEW	KSHD	KBSHD	NB	NG	DISLIN			
$VIEW	14	0	1	5	7	-.25			

Figure 8. Organization of a VIEW Input Data Deck and the Unique Input Data Card of $VIEW

```
//B9EFPT01 JOB (        1831C,T,B00081,001H00),YYY,MSGLEVEL=(2,0)
// EXEC VIEW,REGION=175K
TITLE    *** SAMPLE PROBLEM NO. 1 ***
CASE=      1
BEGIN BULK
CHBDY     10        AREA4         10      11      12      13      50
CHBDY     20        REV           20      30                              55
CHBDY     30        AREA4        100     103     102     101              60     +C1
CHBDY     40        POINT         50                                      65     +C2
+C1
CHBDY     50   10   POINT         40       0.0     0.0     1.            65
+C2                                        0.0     0.0    -1.0                  +C2
SVIEW     50     0    1
SVIEW     55     1    0
SVIEW     60     0    1           0.0     15
SVIEW     65     1    1           0.0     20
PHBDY     10          .786                         5
GRID      10                2.5    -5.0    0.0
GRID      11                2.5     5.0    0.0
GRID      12               -2.5     5.0    0.0
GRID      13               -2.5    -5.0    0.0
GRID     100                2.5    -2.5   10.0
GRID     101                2.5     2.5   10.0
GRID     102               -2.5     2.5   10.0
GRID     103               -2.5    -2.5   10.0
GRID      20                              10.0
GRID      30          .5
GRID      40          .5
GRID      50                              10.0
ENDDATA
ENDCASE
```

VIEW FACTORS

ELEMENT	AREA	ELEM TO ELEM = VIEW FACTOR	ELEM TO ELEM = VIEW FACTOR	VF SUM FROM ELEMENT
10	0.50000E 02	10 - 20 =0.673263E-01	10 - 10 =0.107153E 00	
		10 - 30 =0.503031E-01	10 - 10 =0.100606E 00	0.11763E 00
20	0.31416E 02	20 - 30 =0.728946E-01	30 - 20 =0.916021E-01	0.18005E 00
30	0.25000E 02			0.19221E 00
40	0.78600E 00			0.0
50	0.78600E 00			0.0

Figure 9. A Typical Program Input to Generate View Factors and the Output

```
GRID       29          105.0     105.0     25.98
GRID       41          30.       30.0      30.0
GRID       43          60.       60.0      30.0
GRID       45          90.0      60.0      30.0
GRID       47          105.0     60.0      30.0
GRID       49          3.9       105.0     25.98
MAT4       302
PARAM      SIGMA       .167E-12
PARAM      TABS        273.16
PHBDY      401
QVECT      321         302                            .85
QVECT      751         .14       .866      .5         -0.5
RADLST     41          42        43        44         45        47        48
RADMTX     1 0.320E 01 .0        .0        .0         .399E 01  .0        .0
65000000                                                                          COVS
RADMTX     2           .0        .776E 01 .129E 02    .0        .0        .181E 02 .813E 01 65000000
RADMTX     3           .00       .320E 01 .813E 01    .181E 02  .0
RADMTX     4           .00       .0       .0          .0
RADMTX     5           .00       .0       .0          .0
RADMTX     6           .00       .776E 02 .129E 02   .399E 01
RADMTX     7           .0
TEMP       901         41        500.0     500.0      29        47
TEMP       901         49        280.0     280.0      29        47        280.0
TEMP       951         1         500.0     500.0      21        280.0
TEMP       951         3         500.0     500.0      21
TEMP       901         49        280.0     280.0
TEMPD      901         100.0
TEMPD      951         500.0
TLOAD2     601         0         0.0       1.65       0.0       0.0
QTL2       888                   10.0      1                                      QTL2
ENDDATA
```

Figure 10. A Typical Input Data Deck Containing RADLST and RADMTX

APPROACH	CONVENTIONAL METHOD	UNIFIED FINITE-ELEMENT METHOD
VIEW FACTOR GENERATION	(SURFACES)	(FINITE ELEMENTS)
THERMAL ANALYSIS	(NODAL-NETWORK)	
STRUCTURAL ANALYSIS (NASTRAN)	(FINITE ELEMENTS)	
TOTAL	3 DIFFERENT MODELS	1 BASIC MODEL

Figure 11. Number of Models Required for Thermo-Structural Analysis Including Radiative Exchanges

THERMAL CAPABILITIES AND GRAPHICAL OUTPUT OF PAFEC

J. E. Akin
University of Tennessee
and
PAFEC Engineering Consultants, Inc.
Knoxville, Tennessee

SUMMARY

The computation of heat transfer in structures is enhanced by the utilization of passive and interactive graphics. These capabilities of the PAFEC system are presented and future developments are outlined. This finite element system is shown to have significant thermal capabilities in support of its general structures.

INTRODUCTION

The Program for Automatic Finite Element Calculations - PAFEC - is a powerful, general purpose code designed for easy use in thermal and structural analysis. Data preparation for PAFEC is perhaps simplest and fastest of all such finite element codes. PAFEC employs free format input with engineering key words, powerful mesh general facilities, and extensive plotting options. PAFEC has restart capabilities and control options for maximum versatility and economy. The PAFEC modules are particularly easy to learn and remember, making the use of PAFEC attractive for the beginner or occasional use.

PAFEC Interactive Graphics Suite - PIGS - interactively views and modifies the PAFEC data base. PIGS employs a graphics terminal with cursor control and optionally, a digitizing tablet. PIGS can be used for data generation and to study the pre- and post-solution results.

PAFEC (Level 4) offers the following types of analysis:

Interactive graphics, free format input
User defined program control steps
Steady state, transient heat transfer
Boundary element methods
Linear static, stress and displacements
Modes and frequencies calculations
Direct dynamic time integration
Frequency response analysis
Elastohydrodynamic lubrication
Large deflection analysis, buckling
Creep and plasticity analysis
Substructures, cyclic symmetry

Comparisons of the general PAFEC (Level 3) capabilities with those of other major finite element codes, like NASTRAN and ANSYS, can be found in reference [1]. A more specific tabular summary of its heat transfer features is included in reference [2]. In this paper the emphasis will be on the thermal analysis capabilities of PAFEC, its built-in pre- and post-processing features, and user control options. Future enhancements will also be outlined.

PAFEC is released in Levels that are always upward compatible. The current, Level 3, version was released in 1979. Some of the new features to be presented here are in the Level 4 release that is scheduled for January 1982. This FORTRAN software is installed on most computer hardware ranging from small computers, e.g. Prime 250, through the super mainframes, e.g. CRAY 1.

ANALYSIS WITH PAFEC

The theoretical basis for a finite element thermal formulation has been well established and is given in typical texts, such as reference [3]. Commonly utilized computational procedures for applied thermal analysis are presented in reference [4], and elsewhere. Most of the large finite element codes offer similar capabilities. The differences that are most noticeable center around the ease of use and the model building aids.

PAFEC uses English keywords (which can be shortened to four characters) and completely free format data. Data input in PAFEC is done in modular form, where only those modules actually needed for a job are used. The modules can be in any order and repeated as desired. Thus, the typical names of the modules in Table 1 imply information about the types of data available in the PAFEC thermal analysis option.

The advantages of the modular data system are many. From the standpoint of the user, the data are clear and intelligible without decoding them from the documentation. Copying data from files produced by other finite element programs or using input data prepared for other finite element programs can be done with far less editing. The numerical values are in free format with commas or spaces as separators between data items. The computer program thrives on the modular data, too. A scan is made of the module headers. This information indicates which phases to run and which subroutines and libraries are to be loaded in each phase.

The module names are known to the program by only the first four characters. Hence, anything that follows the first four characters can be thought of as merely comment material. The modular data construction also allows for constant properties to be inserted just after the header card with the form 'property' = numerical values. In the program input, the modules may be in any order. However it is recommended that modules serving a like function, e.g. mesh generation, be grouped together to aid in the user's grasp of data structure.

The keywords used to identify typical data modules employed in a thermal analysis indirectly indicate the capabilities of the PAFEC system. To be more specific the thermal analysis options include both transient and steady state solutions. Temperature dependent properties can be utilized for nonlinear thermal problems. The available boundary conditions include time dependent nodal temperatures or heat flux. Before considering the specific thermal options some of the general model building aids will be discussed.

MODEL BUILDING AIDS

PAFEC offers extensive mesh generation and data supplementation options. Eight default coordinates systems are available as user-defined axes. The generation of nodes on lines and arcs is included. The isoparametric generation of meshes for 2-D elements, surfaces and 3-D solids is included in the use of PAFBLOCKS. There are transitions for mesh refinement, and user defined spacing ratios and holes. PAFBLOCKS permit the independent creation of a continuous geometric model of curved blocks. Additional data control the separate subdivision of the blocks into finite elements. Thus once the geometric model is established the subdivision data are easily changed. The use of PAFBLOCKS is available in both interactive and batch modes. Other features allow for the generation of repetitive mesh segments by translation, rotation, or scaling. Powerful boundary condition generators are also included in PAFEC. PAFEC offers extensive warning and error messages in the data validation and geometry checks. Several mesh plotting options are included in PAFEC. Most of these plots allow user selected windows for more detail.

For nodal coordinates generation it is most direct to use the NODES module, but other modules are also used in describing the nodal coordinates. These are:

NODES   gives coordinates of nodes in any axis set.

AXES   describes the axis sets used for giving coordinates.

SIMILAR.NODES   Once a group of nodes has been described, this module may be used to locate other nodes which happen to be similar to them.

LINE.NODES   are used to force any number of nodes to lie on the
ARC.NODES   same line or circular arc, respectively.

PAFBLOCKS   are used in conjunction with each other to cause blocks
MESH   of both nodes and elements to be generated.

Several facilities are used in description of element topology, types, material properties, and thickness. Elements may optionally be referred to by GROUP.NUMBER to facilitate assignment of properties, making drawings, and other uses.

ELEMENTS   describes the properties, element type, group number and topology for individual elements.

GROUP.OF. SIMILAR. ELEMENTS	copies selected elements into new element groups. This facility may be used to describe repeated geometry.
REFERENCE.IN. PAFBLOCK	permits treating exceptions to elements and nodes generated with PAFBLOCKS.
PLATES.AND. SHELLS	sets element thickness, material number, and other information.

There are ten types of PAFBLOCKS available for model building. The most common, Type 1, block is shown in figure 1. It can be utilized to generate surfaces, solids, or sets of diaphragms. In figure 1, the quantities N1, N2, N3, N4, N5 refer to spacing ratios which are given in the MESH module; nodal point numbers specify the order in which the topology is listed in the PAFBLOCKS module. Dotted lines show resulting meshes with triangular or wedge elements. Numerous element types can be generated using PAFBLOCKS.

The PAFEC element library contains over 80 elements. Most involve linear to cubic isoparametric interpolation but hybrid and semi-loof elements are also employed. For thermal analysis options both the linear and quadratic surface and solid elements are available.

PIGS can be used interactively to generate and modify PAFEC data. When generating data, curved and straight boundaries may be defined and, if required, the individual points and node numbers may be specified. However, it is more common to generate the minimum number of nodes required to define a two or three dimensional structure and mesh the structure using an interactive PAFBLOCK. Where appropriate, any part of the mesh may be replicated, scaled, rotated and translated to form new parts of the structure with a minimum of effort. Information can be input from cursor, keyboard or digitizer and the interactive plot can be manipulated in many ways to ensure that the user sees as clearly as possible those sections of the structure which are of most interest.

It is also possible to have PAFEC automatically create an interface file for the MOVIE.BYU [5] program. This allows interactive hidden line plots of perspective or isometric views to be generated. A post-processing MOVIE interface is also available. Since the PAFEC model building capabilities are so powerful and popular at least one PAFEC site has developed an interface to create NASTRAN data. Often one page of PAFEC data would create ten to fifteen pages of NASTRAN input.

THERMAL MATERIAL DATA

In many types of analysis, the materials are isotropic with constant properties. For user convenience, ten standard material types are built into PAFEC, and include a typical steel, stainless steel, cast iron, aluminum alloys, titanium, glass, and concrete. The standard data may be overridden, or other materials may be added with the MATERIALS module.

Occasionally, nonuniform or anisotropic material descriptions are needed. Some PAFEC modules for this include:

ORTHOTROPIC. MATERIAL
: permits specification of the nine components of the compliance matrix, and orthotropic thermal properties.

LAMINATES
: Layered ply material may be described here.

VARIABLE.MATERIAL TABLES
: describes the temperature dependence of user selected properties. Linear interpolation is used between tabulated values.

The experienced user can also supply subroutines to replace or expand these options. Thus specific nonlinear material responses or alternate constitutive relations can be introduced.

Convection coefficients are described in the standard MATERIALS modules. However, the forced mass flow convection conditions are not standard input. This type of property would also have to be defined by a user supplied subroutine.

THERMAL ANALYSIS OPTIONS

PAFEC may be used for finding temperature distributions in order that thermal distortions and stresses can be found, or for other applications requiring the solution of heat transfer.

There are two main types of thermal calculations: steady state and transient. In transient problems there is usually a thermal shock and it is required to know how the temperature varies with time. At any point in time it may be supposed that the temperature distribution is known completely; a finite element solution is needed to determine how the temperature will vary during a short interval of time. The transient temperature solution involves marching forward in time. For the process to begin, temperatures are required at a start time which is conveniently taken at time, $t = 0$. Initially, all temperatures may be known and input as data for the problem, or, alternatively PAFEC may have to carry out a steady state calculation as a prelude to the transient analysis.

During the steady state solution at each node either the temperature or the heat flux entering the structure is an unknown. For most nodes in the structure there will probably be an unknown temperature and the heat flux entering the structure from external sources will be zero. The following modules are used to describe the thermal boundary conditions:

TEMPERATURE
: gives the nodal temperatures. Nodes not mentioned are assumed to be unknown.

FLUX
: gives the heat inputs at nodes. For any node at which neither the flux nor the temperature is specified it is assumed that the flux input is zero.

FACTOR.LOADS   can be used to sum various thermal load cases together. This can also be used to simulate unsymmetrical cases by combining results of symmetrical and antisymmetrical cases.

For a transient thermal analysis it is assumed that the initial temperature field is completely specified. If this is not the case then a steady state solution should usually be run first.

The following modules are used in a transient calculation:

TEMPERATURE   gives the initial temperature distribution if a steady state solution was not run. Any node not mentioned is assumed to be at zero temperature.

THERMAL.SHOCK   describes the variations with time of any nodal temperatures which are prescribed.

NODAL.FLUX.SHOCK   gives the variation with time of prescribed heat fluxes at nodes.

UNSTEADY.THERMAL.TIMES   is used to define the time step selected and the time at which the solution is to end. It also controls the times for printed and graphical output.

Since the thermal analysis is not the default operation the program must be told to execute the desired option. This is accomplished by placing the commands CALC.STEADY.TEMPS and/or CALC.TRANS.TEMPS in the control module. If both are present then the steady state solution is automatically used as the starting condition for the transient analysis. In that case the TEMPERATURE module only describes boundary conditions on the steady state solution. If the calculated temperatures are to also be utilized in a structural analysis then the control command SAVE.TEMPS will cause the required files to automatically be created and named. As will be discussed later, the results of a thermal analysis can be displayed graphically in both passive and interactive formats.

The treatment of a radiation boundary condition is not currently easily included in a PAFEC analysis. This nonlinear boundary condition can be introduced by special user supplied subroutines. The program does not calculate any radiation view factors. However, the interactive program described at this conference, reference [6], could aid an analyst in obtaining the necessary view factors.

ADDITIONAL BOUNDARY CONDITIONS

To apply constraints at certain nodes, the following modules are used:

RESTRAINTS	can be used for temperatures that are known. It can describe the constraints at one node, or all the nodes on a line or a whole plane can be constrained at once.
REPEATED.FREEDOMS	is used to constrain two or more nodes to have unknown temperatures that are the same.
GENERALIZED. CONSTRAINTS	permits writing arbitrary linear functions relating temperatures among any number of nodes.

PASSIVE GRAPHICS

Extensive passive graphics options are available. It is also possible to interactively preview these plots before transmitting them to a plotter. Numerous options are available to display the results of the model building. These include standard mesh plots, exploded mesh views, boundary plots, etc. The orientation of the plots can be specified but the program will select a default view for each plot.

Various items of information can be displayed on the plots. These include node, element, and material numbers; active DOF and restraints; wave front position; and all axis sets. For a thermal analysis the output plot options include steady state and transient temperature contours. Specific element groups can be selected for display. The ability to plot temperatures along a user defined nodal path is another useful feature. Modules for the selection of PAFEC graphical output are listed here:

IN.DRAW	is used for drawings of the structure and the constraints. This controls the information contained on the drawing, and the groups of elements to be drawn.
OUT.DRAW	produces drawings of the temperature after solution.
SELECT.DRAW	supplements IN.DRAW and OUT.DRAW by allowing a spatial selection or selection by element type for drawing some portion of the structure.
GRAPH	gives plots of temperature as ordinates with selected nodes as abscissas.

INTERACTIVE GRAPHICS

In the post-processing mode, PIGS is used for displaying deformed shapes, temperatures, stresses, and mode shapes. Any number of load cases or modes may be stored and retrieved. The temperature at individual nodes can be displayed simply by hitting that node with the cursor. Menu options are chosen in the same way.

Three different contouring options are available and individual faces of structure may be drawn selectively to classify the output. Any PIGS drawing may be reproduced on a suitably interfaced plotting device, and, conversely, plot files created in PAFEC may be viewed on an interactive graphics terminal.

All PIGS facilities are selected by the user from a menu of available options. Only one menu is displayed at any time, occupying a column on the left-hand side of the screen. Options are displayed in alphabetical order. Whenever the selected option is hit it will be underlined. There are four different menus, the ROOT menu, the ANALYSIS branch menu, the VIEW branch menu, and the DIGITIZE menu. The VIEW menu not only offers finite element mesh viewing but also data modification and generation facilities which constitute the most useful options in PIGS. The DIGITIZE menu allows the direct input of nodal coordinates and element or PAFBLOCK topology. It utilizes special hardware in the form of a digitizing tablet. The ANALYSIS menu, for post-solution interactive graphics, permits the displaced shapes, temperatures, stresses, and mode shapes to be displayed. Most of the many options in these menus employ cursor input. However, the analyst can request typed input and prompting assistance.

Figure 2 shows a selected segment of a three-dimensional mesh that has been rotated and displayed by PIGS. Upon request the temperature contours on these elements can be displayed as illustrated in figure 3. If the analyst desires more specific information then individual nodes can be selected with the screen cursor. When this is done the node number and computed temperature is added to the display. Figure 4 shows that display format. These and other features in PIGS makes it a very user friendly system.

Another interactive option is available for very inexperienced users. That is the Automatic data Preparation and Edit Systems - APES. It aids in digitizing the model, prompts the user for material and boundary condition data, etc. While this is useful for a beginner an experienced finite element user would quickly outgrow the need for such an option.

DISCUSSION

A new feature of PAFEC is the option to utilize Boundary Element Formulations in conjunction with the standard finite element thermal solutions. This is well suited to semi-infinite regions and other specialized treatments of the Poisson and bi-harmonic equations. Thermal results using this capability will be reported in the near future.

The PAFEC and PIGS systems provide a powerful thermal and structural analysis capability. It is a well documented and easy to use system. However, it currently has a weakness in the thermal area, that is, the lack of a user friendly treatment of radiation problems. Such nonlinear applications have been solved with PAFEC. But this is usually done by way of user supplied subroutines. The system is designed to easily accept such routines via the CONTROL module. However an experienced user is usually needed for a radiation analysis. Hopefully this current shortcoming will be overcome in the near future. In conclusion, the PAFEC system provides another useful tool for the computation of heat transfers in structures.

REFERENCES

1. Akin, J. E., and Dewey, B. R.: PAFEC Compared with Other Major Finite Element Codes. Structural Mechanics Software Series, Vol. IV. Perrone, N., and Pilkey, W. D. (Eds), Univ. Virginia Press, 1981.

2. Noor, Ahmed K.: Survey of Computer Programs for Heat Transfer Analysis. Computational Aspects of Heat Transfer in Structures. NASA CP-2216, 1982. (Paper 27 of this compilation.)

3. Huebner, K. H.: The Finite Element Method for Engineers. John Wiley & Sons, 1975.

4. Akin, J. E.: Application and Implementations of Finite Element Methods. Academic Press, London, 1982.

5. Christiansen, H.: MOVIE.BYU, A General Purpose Computer Graphics Display System. Brigham Young Univ., 1980.

6. Emery, A. F.; Mortazavi, H. R.; and Kippenhan, C. J.: Interactive Computation of Radiation View Factors. Computational Aspects of Heat Transfer in Structures. NASA CP-2216, 1982. (Paper 14 of this compilation.)

TABLE 1   TYPICAL DATA MODULES

A) Model Building

      AXES
      ARC.NODES
      ELEMENTS
      GROUP.OF.SIMILAR.ELEMS
      LINE.NODES
      MESH
      NODES
      PAFBLOCKS
      REFER.TO.PAFBLOCK
      SIMILAR.NODES

B) Thermal Material Properties

      LAMINATES
      MATERIALS
      ORTHOTROPIC.MATERIAL
      TABLE.OF.PROPERTIES
      VARIABLE.MATERIAL

C) Thermal Boundary Conditions

      FACTOR.LOADS
      FLUX
      GENERALIZED.CONSTRAINT
      NODAL.FLUX.SHOCK
      OMIT.FROM.FRONT
      REPEATED.FREEDOMS
      RESTRAINTS
      TEMPERATURE
      THERMAL.SHOCK
      UNSTEADY.SOLUTION

D) Passive Graphics

      GRAPHS
      IN.DRAW
      OUT.DRAW
      SELECT.DRAW

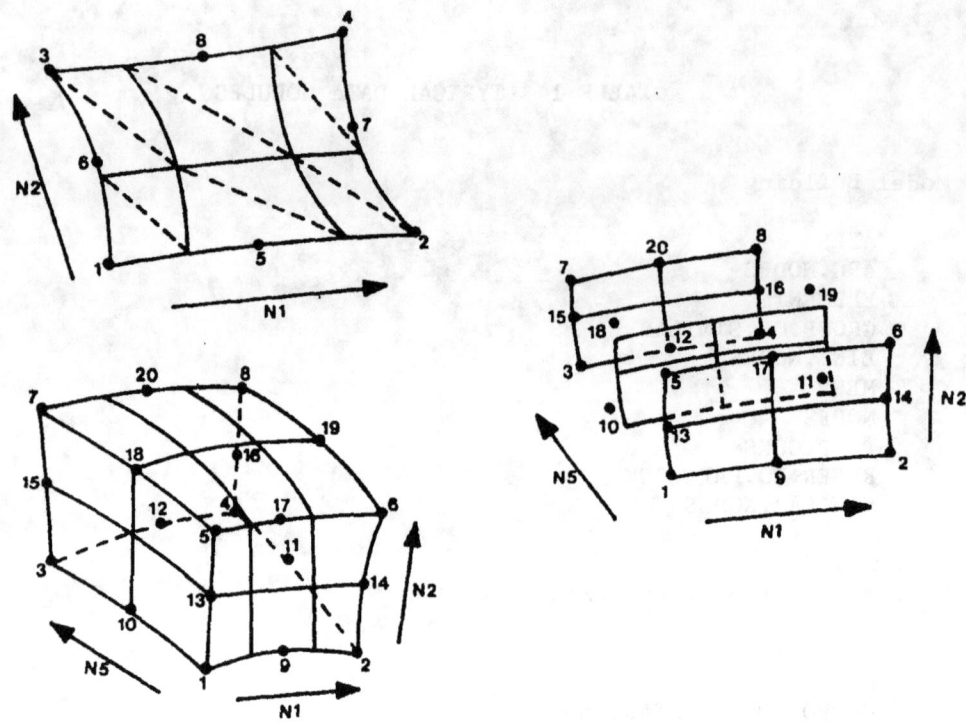

Figure 1.- A typical PAFBLOCK available for model building.

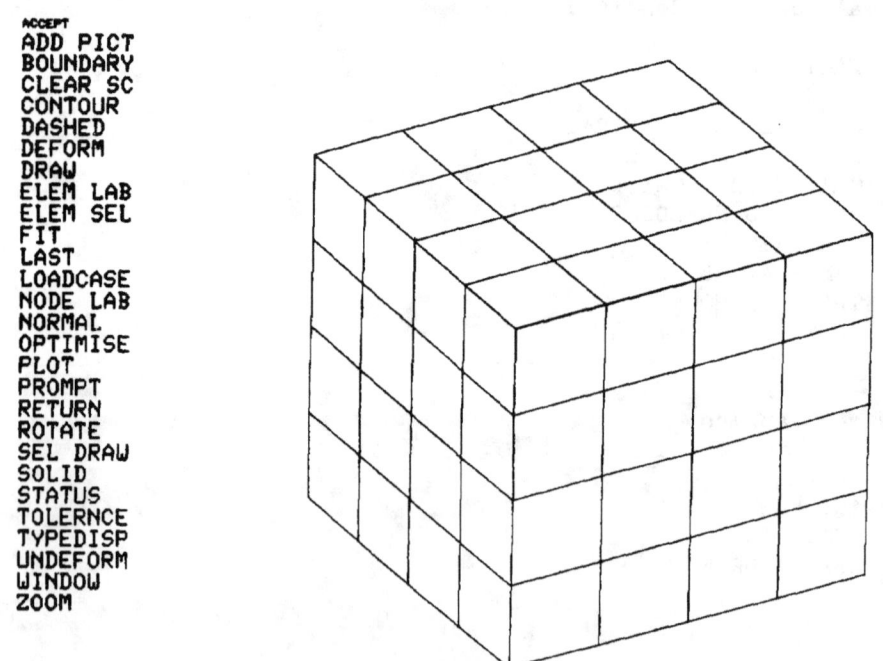

Figure 2.- Interactive display of 3-D mesh segment.

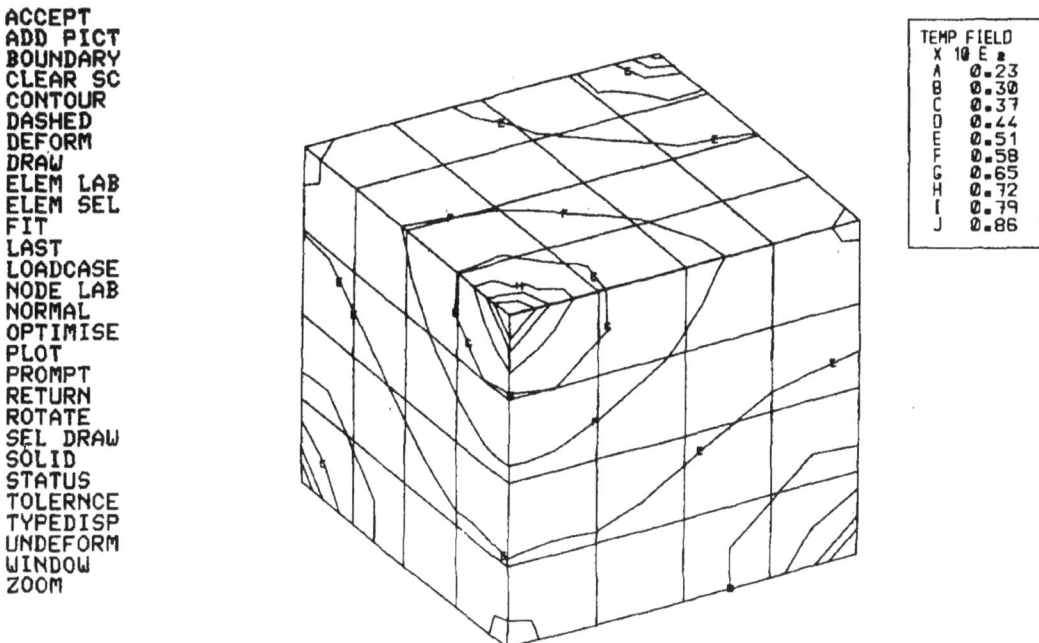

Figure 3.- Temperature contours added to mesh.

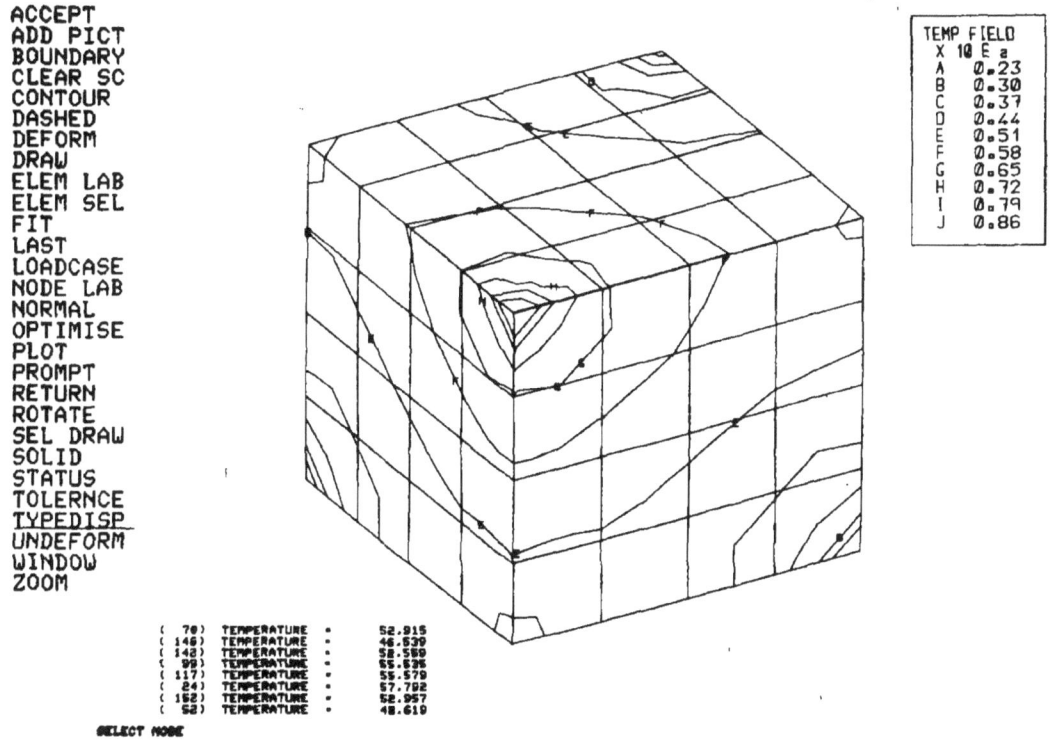

Figure 4.- Temperature values displayed during interactive node selection.

THE SPAR THERMAL ANALYZER - PRESENT AND FUTURE

M. B. Marlowe and W. D. Whetstone
Engineering Information Systems, Inc.
San Jose, California

J. C. Robinson
Langley Research Center
Hampton, Virginia

SPAR THERMAL ANALYZER

To provide a general in-house integrated thermal-structural analysis capability the Langley Research Center is having the SPAR Thermal Analyzer (fig. 1) developed under contract by Engineering Information Systems, Inc. The SPAR Thermal Analyzer is a system of finite-element processors for performing steady-state and transient thermal analyses. The processors communicate with each other through the SPAR random access data base. As each processor is executed, all pertinent source data is extracted from the data base and results are stored in the data base.

The tabular input (TAB), element definition (ELD) and arithmetic utility system (AUS) processors are used to describe the finite element model. The data base utility (DCU) processor operates on the data base. The plotting processors (PLTA, PLTB) provide the capability to plot the finite element model for model verification but do not directly plot temperatures. The thermal geometry (TGEO) processor performs geometry checking of the thermal elements and total model. The thermal processors for steady state analysis (SSTA) and transient analysis (TRTA, TRTB and TRTG) are described in References 1 and 2. In addition there are several processors not shown in the figure for extraction of thermal fluxes, system matrices and system operating characteristics.

On a scalar computer the processors may be executed interactively or in a batch mode. A typical analysis is usually performed as a sequence of interactive and batch operations where model development and verification is performed interactively and actual thermal calculations performed in batch mode. The program operates on UNIVAC, CDC, PRIME and VAX computers.

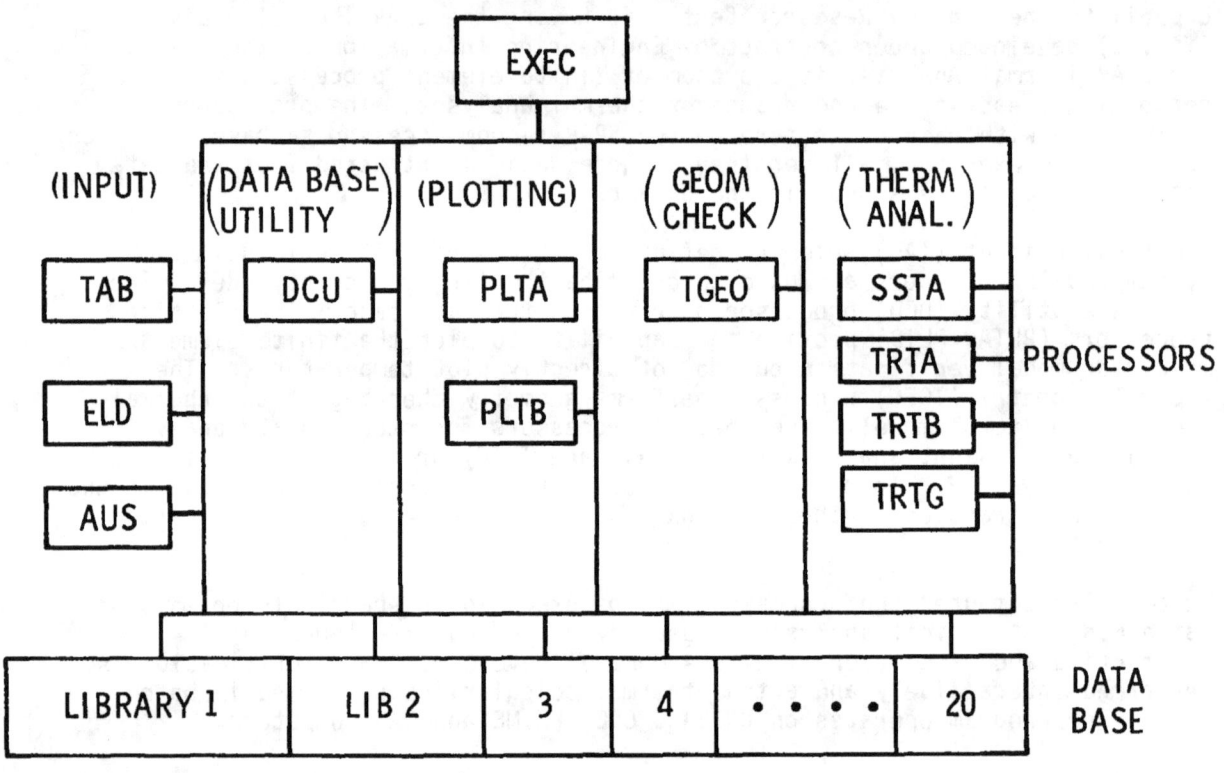

Figure 1

DATA BASE

The SPAR data base (fig. 2) stores all the program data and provides for data transfer between processors. The DCU processor permits the user to access data in the data base. The table of contents (TOC) function produces a twelve word record for each data entry that includes date and time of creation, four word data name, type, length, block size and error code. Additional data documentation is available through the use of text data sets which may be used to describe the problem solution. Restart capability is provided by saving one or more libraries on a disk file or a restart tape. The program can be restarted after any processor execution. In addition, DCU permits the transfer of data to or from a file external to the SPAR data base and allows condensing several libraries into one library for storage. The maximum size of the data base on CDC equipment is a maximum of 10 libraries of 2048 data sets per library. Use of libraries containing a large number of data sets increases I/O costs.

In addition to the DCU function described above, AUS permits the definition of arbitrary data sets for entry into the data base. In the SPAR Thermal Analyzer, AUS is used to define material properties, thermal loads and element properties. AUS also allows the transfer of all or part of one data set to another as well as performing arithmetic and other functional operations on applicable data in the data base.

- **STORES ALL DATA**

- **DATA TRANSFER BETWEEN PROCESSORS**

- **DOCUMENTS CONTENTS**

   TABLE OF CONTENTS (DATE, TIME, TYPE, SIZE, NAME)
   TEXT DATA SETS

- **RESTART CAPABILITY (FILE OR TAPE)**

- **TRANSFER DATA TO/FROM EXTERNAL FILE**

- **ALL DATA ACCESSIBLE**

- **MAX SIZE (CDC), 10 LIB, 2048 REC/LIB**

- **LARGE NO. ENTRIES PER LIB. INCREASES I/O COST**

Figure 2

ELEMENT REPERTOIRE

Several element types are included in the SPAR Thermal Analyzer to model the different heat transfer functions and the storage of heat (fig. 3). The K21, K31, K41, K61 and K81 elements model conduction and heat capacity. These are isoparametric elements which have a consistent capacitance matrix and are bounded by 1 or 2 nodes for one dimensional elements, 3 or 4 nodes for two dimensional elements and 6 or 8 nodes for three dimensional elements. A "zero-length" form of the 2-D element is used with a default length of unity to provide a lumped capacitance. All temperature dependent element properties are based on the average element temperature.

Two types of convective elements are included. The C21, C31 and C41 elements model convective heat transfer between a surface and a medium at a known temperature which may be time dependent. Convective exchange between a surface and a medium at an unknown temperature is modeled by the C32 (2 surface nodes (SN), 1 fluid node (FN)), C42 (2SN, 2FN) and C62 (4SN, 2FN) elements.

A two node mass-transport element with fluid conduction (MT21 element) is available in either a conventional or variably up-winded formulation. Combined mass-transport, convective-exchange elements are the MT42 (2SN, 2FN) and the MT62 (4SN, 2FN) elements which model typical pipe and plate-fin flow passage configurations.

The radiation-exchange elements are the R21 (1 or 2 SN, 1-D), R31 (3SN, 2-D) and R41 (4SN, 2-D) elements. In addition there is an experimental element capability that provides the user with the capability to include an element having from 2 to 32 nodes. This requires the user to insert the coding for the element in the proper subroutines, compile those subroutines and load the program. This capability was provided to allow check-out of new element formulations, but it can also be used to include capabilities not present in the program.

FUNCTION	NODES	COMMENTS
• CONDUCTION-CAPACITY	1, 2, 3, 4, 6, 8	ISOPARAMETRIC * CONSISTENT CAPACITANCE
• FORCED CONVECTION	1, 2, 3, 4	PRESCRIBED CONVECTIVE TEMP
• FLUID SURFACE CONVECTION	3, 4, 6	UNKNOWN CONVECTIVE TEMP
• MASS TRANSPORT	2	CONVENTIONAL OR UPWINDED FORMULATION
• COMBINED MASS-TRANSPORT AND CONVECTION	4, 6	
• RADIATION	1, 2, 3, 4	USER SUPPLIED SHAPE FACTORS
• EXPERIMENTAL	2 - 32	ELEMENT CHECK OUT, USER FORMULATION

* ALL ELEMENT PROPERTIES BASED ON AVERAGE ELEMENT TEMPERATURE

Figure 3

MATERIAL PROPERTIES

Thermal conductivity may be anisotropic with the specification of six terms to describe conductivity. Thermal conductivity, density, specific heat and convection coefficients may be temperature, time and pressure dependent. The properties affecting radiation, emissivity, reflectivity and transmissivity, may be temperature and time dependent. (See fig. 4.)

- CONDUCTIVITY

  ANISOTROPIC
  TEMPERATURE DEPENDENT
  TIME DEPENDENT
  PRESSURE DEPENDENT

- DENSITY
  SPECIFIC HEAT
  CONVECTION COEFFICIENTS

  TEMPERATURE DEPENDENT
  TIME DEPENDENT
  PRESSURE DEPENDENT

- EMISSIVITY
  REFLECTIVITY
  TRANSMISSIVITY

  TEMPERATURE DEPENDENT
  TIME DEPENDENT

Figure 4

BOUNDARY CONDITIONS AND THERMAL EXCITATION

The boundary conditions that may be specified include time-dependent nodal temperatures, convective exchange temperatures and mass transport rates (fig. 5). The imposed thermal excitation can consist of time and temperature dependent surface heat fluxes, volumetric heat generation and time dependent incident radiative heat fluxes. "Perfect" conductors are available to force two separate nodes to have the same temperature.

- PRESCRIBED, TIME-DEPENDENT TEMPERATURES
- CONVECTIVE EXCHANGE TEMPERATURES - TIME DEP.
- "PERFECT" CONDUCTORS - ENFORCED TEMPERATURES
- SURFACE HEAT FLUX - TIME OR TEMPERATURE DEP.
- VOLUMETRIC HEAT GEN. - TIME OR TEMPERATURE DEP.
- PRESCRIBED, TIME DEPENDENT RADIATION FLUX
- PRESCRIBED, TIME DEPENDENT MASS TRANSPORT RATE

Figure 5

RADIATION EXCHANGE

The radiation heat transfer model assumes that the radiating element is at a uniform temperature with uniform emissive power and incident heat flux (fig. 6). Surfaces emit and absorb diffusely and reflect diffusely and/or specularly. Complete radiation exchange factors or script F factors may be used and the complete matrix must be input since the program does not calculate the exchange factors or assume symmetry of the factor matrix. There is no specific limit on the number of radiating surfaces in a problem but the size of the exchange factor matrix may become so large that several data sets are required. The radiation contribution to the total "conductance" matrix is a diagonal term. When complete exchange factors are used, the radiation load vector is calculated by an iterative solution of the incident heat equation which usually converges in 2 to 10 iterations.

- RADIATION ELEMENTS - RADIATING SURFACE WITH UNIFORM EMISSIVE POWER AND INCIDENT HEAT FLUX

- SURFACES EMIT AND ABSORB DIFFUSELY, REFLECT DIFFUSELY AND/OR SPECULARLY

- COMPLETE EXCHANGE OR SCRIPT F FACTORS MAY BE USED

- NO LIMIT ON NUMBER OF RADIATING SURFACES

- RADIATION K MATRIX IS DIAGONAL

- RADIATION LOAD VECTOR - COMPLETE EXCHANGE FACTOR
    ITERATIVE SOLUTION TO INCIDENT HEAT EQUATION
    USUALLY CONVERGES IN 2 TO 10 ITERATIONS

Figure 6

SSTA

The SSTA processor calculates steady state nodal temperatures (fig. 7). The governing equation is

$$(K_k + K_h + K_r + K_m)T = Q + H + R$$

where K_k, K_h and K_r are the symmetric conduction, convection and radiation matrices and K_m is the asymmetric mass transport matrix, T is the nodal temperature vector and Q, H and R are the source, convection and radiation load vectors. A direct linear solution is performed when there are no radiation or temperature dependent properties. If there are temperature dependent material properties or radiation present a nonlinear solution is performed using a modified Newton-Raphson method. A good procedure is to perform a linear analysis to obtain a starting estimate of the temperature for the nonlinear solution. Through processor RESET controls the user can control factoring of the total matrix and the number of iterations performed between factorings for nonlinear analysis.

- LINEAR AND NONLINEAR STEADY STATE ANALYSIS OF
  $(K_k + K_h + K_r + K_m)T = Q + H + R$

- LINEAR SOLUTION

    NO RADIATION, NO TEMP DEPENDENCY

- NONLINEAR SOLUTION

    MODIFIED NEWTON-RAPHSON
    USER CONTROL - FACTORING AND ITERATIONS

Figure 7

TRTA

The TRTA processor (fig. 8) calculates transient temperature distributions in structures using an explicit algorithm based on a Taylor series expansion solution of the governing equation

$$(K_k + K_n + K_r + K_m)T + C\dot{T} = Q + H + R.$$

In this equation, C is a diagonal (lumped) heat capacitance matrix, $\dot{T}$ is a vector of the first derivative of nodal temperature with respect to time and the remaining terms are as defined for the previous figure. The time step, DT, used in the solution process is calculated automatically to assure stability of the solution. The user specifies the times at which nonlinear effects such as temperature dependent material properties and the radiation contribution, R, are recalculated. Recalculation of the conduction matrices for the 4, 6 and 8 node elements required when the material properties are temperature dependent is performed by scaling when the material properties are isotropic. "Arithmetic" node (nodes with negligible capacitance) capability is available in TRTA. The solution process may be easily restarted from any point in time for which a temperature distribution is available.

- EXPLICIT TRANSIENT THERMAL ANALYSIS BASED ON TAYLOR SERIES EXPANSION SOLUTION OF
  $(K_k + K_n + K_r + K_m)T + C\dot{T} = Q + H + R$
- AUTO CALCULATION OF DT
- ELEMENT PROP RECOMPUTED AS SPECIFIED BY USER
- CONDUCTION MATRIX FOR 4, 6, 8 NODE ELEMENTS SCALED IF ISOTROPIC
- "ARITHMETIC" NODES (ZERO CAPACITANCE)
- LUMPED CAPACITANCE MATRIX
- RESTART FROM ANY POINT IN TIME

Figure 8

TRTB AND TRTG

These two processors calculate transient temperature distributions using implicit solution algorithms (fig. 9). TRTB uses a Galerkin method with a variable weighted residual parameter, β, which may be set by the user. Different values of β correspond to the following algorithms

β = 1/2, Crank-Nicholson
β = 2/3, Galerkin
β = 1, Backward differences

The user must select the value of DT and the recalculation times for temperature-dependent material properties and radiation load vector. TRTB uses a diagonal (lumped) heat capacitance matrix.

The TRTG processor uses the Lawrence Livermore Laboratories packaged GEARIB algorithms based on the development by W. C. Gear of the University of Illinois. The user may select either of two variable order solution algorithms. The backward difference algorithm is usually chosen for stiff problems and the Adams-Moulton for non-stiff problems. The TRTG processor automatically determines DT and property recalculation times and uses either a diagonal or consistent heat capacitance matrix.

- IMPLICIT TRANSIENT ANALYSIS

- TRTB - GALERKIN WITH VARIABLE PARAMETER
    - USER SELECTS DT AND K MATRIX RECOMPUTATION TIME
    - LUMPED CAPACITANCE MATRIX

- TRTG - BASED ON LLL GEARIB PACKAGE
    - VARIABLE ORDER ADAMS-MOULTON OR BACKWARD DIFFERENCE ALGORITHM
    - AUTOMATICALLY CALCULATES DT AND K MATRIX
    - CONSISTENT OR LUMPED CAPACITANCE MATRIX

Figure 9

TYPICAL TRANSIENT ANALYSIS DATA FLOW

The flow of data in a typical transient thermal analyses is shown in figure 10. While the data base is not shown in the figure, the data input by the user, with the exception of individual processor control commands, and all data created by the processors, with the exception of the actual plot vector file, reside in the data base.

The processors used in the model definition and checking phase and their data outputs are:

   TAB - node (joint) locations
   AUS - thermal and element section properties
   ELD - element definition
   PLTA, PLTB - model plots
   TGEO - degree of freedom list compiled in checking element geometry

The processors used in the transient-thermal analysis include AUS and one of the transient analyses processors (TRTA, TRTB or TRTG). AUS is used to input the thermal excitation, boundary conditions and define arithmetic nodes if used. The transient analysis processor produces structural temperature distributions for times specified by the user. Printing of the temperatures is accomplished by the DCU processor (not shown).

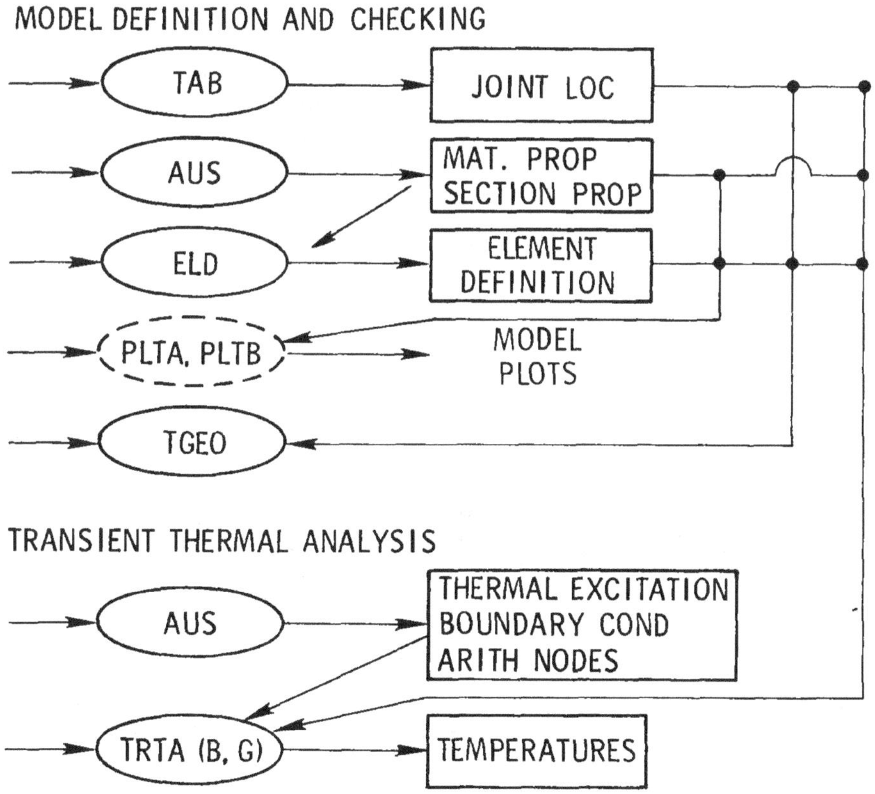

Figure 10

NATIONAL TRANSONIC FACILITY DOWNSTREAM NACELLE

A recent application of the SPAR Thermal Analyzer was part of the certification of the Langley National Transonic Facility (NTF). The NTF is a cryogenic, transonic wind tunnel using nitrogen as a test medium. Since the test medium can vary in temperature from 88 K (-300°F) to 340 K (150°F), transient temperatures of the nacelle and its supporting strut were required to determine thermal stresses.

Figure 11 shows the finite element model used to determine transient temperatures in the downstream nacelle of the NTF at Langley. The model has 2300 nodes and 6100 elements; 2000 time steps (DT) were used in the analysis.

The conduction elements of the thermal finite element model were the same as the structural elements in the structural finite element model. In addition, convection elements were added to the outside of the horizontal strut and outside and inside of the nacelle.

THERMAL MODEL

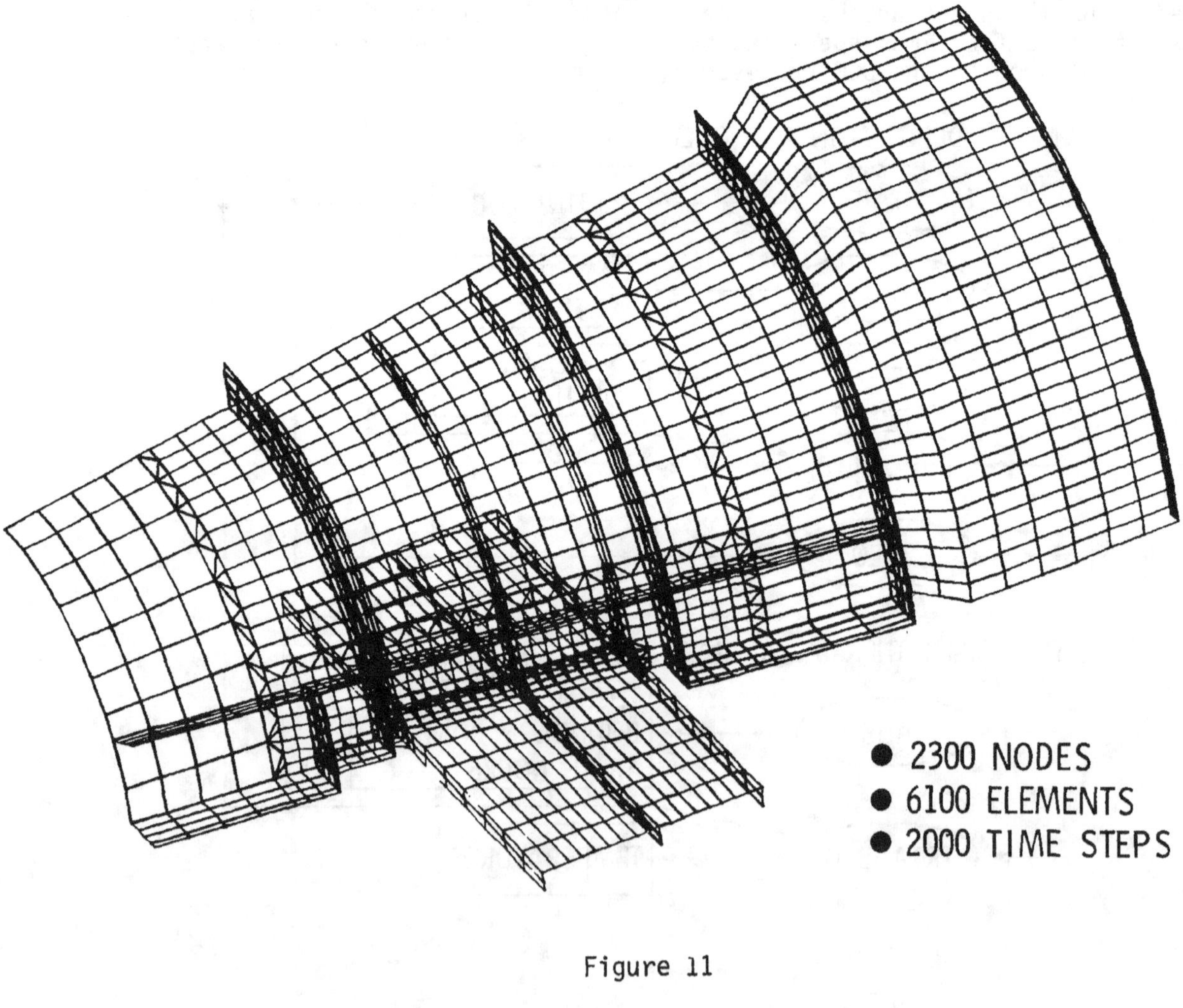

- 2300 NODES
- 6100 ELEMENTS
- 2000 TIME STEPS

Figure 11

THERMAL-STRUCTURAL ANALYSIS METHODOLOGY

The strategy for improving thermal-structural analysis capability is shown in the chain of boxes in the upper part of figure 12. From the left, there is the identification of a need and conception of an idea to fill that need. This leads to development or identification of a better methodology which is the complete definition of the method to apply the original idea. This method is then evaluated in a study code which allows numerical experimentation with the least amount of effort and greatest opportunity for experimentation. Those methods that prove worthwhile are then incorporated in a "production" code and evaluated on large scale problems.

Some of the ideas presently in varying stages of the development process for possible incorporation into SPAR are shown in the lower portion of the figure. Ideas that are being reported elsewhere in this symposium are identified with the authors name. These include radiation view factor calculation, improved nonlinear equation solution methods, use of a vector computer, switching in the GEAR method, reduced basis technique and integrated thermal-structural elements.

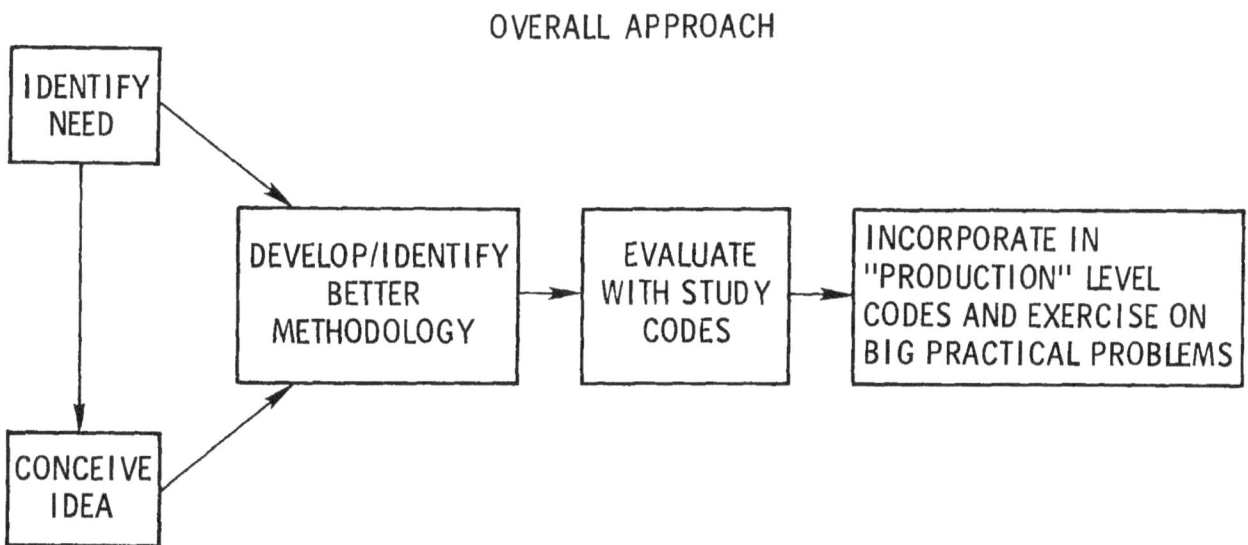

OVERALL APPROACH

RADIATION VIEW FACTORS (EMERY)
IMPROVED NONLINEAR EQUATION SOLUTION METHODS (HAFTKA AND KADIVAR)
VECTOR COMPUTER EVALUATION (ROBINSON, RILEY AND HAFTKA)
SWITCHING IN GEAR PACKAGE
REDUCED BASIS (SHORE)
INTEGRATED THERMAL-STRUCTURAL ELEMENTS (THORNTON)

Figure 12

POSSIBLE NEW CAPABILITY

The possible near-term new capabilities to be added to the SPAR Thermal Analyzer include radiation view factor calculation, solution logic capability and thermal-structural model data transfer (fig. 13). The radiation view factor computation method to be placed in SPAR is discussed by Emery in a another paper in the symposium.

Solution logic capability permits the analyst to supply Fortran code that modifies or can modify the solution process at each time step. Similar capability presently exists to modify the solution at each time that the conductance matrix is recalculated through the use of the experimental element capability. Solution logic capability and use of experimental elements requires recompilation of parts of the program and reloading.

When the thermal and structural finite-element models use the same node numbers and locations, transfer of temperature data to the structural model simply requires renaming a data set. Quite often, however, disciplinary requirements cause the two models to be different. In this case, transfer of temperature data to the structural model is much more difficult and automation of the process is highly desirable. When the thermal model elements are larger than the structural model elements, the temperature distribution functions used to formulate the element conductivity matrix may be used. When the thermal model is more refined than the structural model, the structural-node temperature is approximated as a weighted average of the temperatures of the surrounding thermal nodes.

- RADIATION VIEW FACTOR CALCULATION

- SOLUTION LOGIC CAPABILITY

- THERMAL-STRUCTURAL MODEL DATA TRANSFER

Figure 13

SUMMARY

The SPAR Thermal Analyzer is a modular, interactive program for general heat-transfer analysis (fig. 14). It analyzes problems having conduction, convection, radiation and mass-transport heat transfer with time, temperature and pressure dependent properties. Steady-state temperature distributions are determined by a direct solution method for linear problems and a modified Newton-Raphson method for nonlinear problems. An explicit and several implicit methods are available for the solution of transient heat-transfer problems.

Finite-element plotting capability is available for model checkout and verification. Temperature plotting is currently not directly available. The SPAR system uses a data base for all data transfer between processors and allows recovery of almost all data and a good capability to access all data.

At the present time, the SPAR Thermal Analyzer is the main software focal point for research and technology efforts in structural heat transfer at the Langley Research Center. The SPAR Thermal Analyzer will be available through COSMIC in the near future.

SPAR THERMAL ANALYZER

- MODULAR, INTERACTIVE PROGRAM

- CONDUCTION, CONVECTION, RADIATION AND MASS-TRANSPORT HEAT TRANSFER

- STEADY STATE AND TRANSIENT ANALYSIS
    EXPLICIT AND IMPLICIT SOLUTION METHODS

- PLOTING CAPABILITY

- EXCELLENT DATA ACCESS

- WILL BE AVAILABLE FROM COSMIC

- MAIN SOFTWARE FOCAL POINT OF LaRC R & T EFFORTS IN STRUCTURAL HEAT TRANSFER

Figure 14

REFERENCES

1. Marlowe, M. B.; Moore, R. A.; and Whetstone, W. D.: SPAR Thermal Analysis Processors Reference Manual, System Level 16. NASA CR-159162, 1979.

2. Hindmarsh, A. C.: A Collection of Software for Ordinary Differential Equations. Rept. No. UCRL-82091, Jan. 1979, Lawrence Livermore Laboratory.

A COMPARISON OF THE FINITE DIFFERENCE AND FINITE ELEMENT
METHODS FOR HEAT TRANSFER CALCULATIONS*

A. F. EMERY
H. R. MORTAZAVI
UNIVERSITY OF WASHINGTON
SEATTLE, WASHINGTON

INTRODUCTION

Of the many approximate or numerical methods used to solve heat transfer problems, the finite difference and finite element approaches have become the most widely known and used. The finite difference method (FDM) is the older, being introduced in 1928[1], and until recently, probably the better known and more extensively used for both research and production studies. The finite element method (FEM) is newer[2] and has not been used for heat transfer studies to a comparable degree, although it is the dominant method for structural analyses.

Because the FEM has a short history of use for heat transfer, there is considerable confusion about the relative values of the two methods and under which conditions one method is to be preferred. Both methods are based upon minimizing the weighted error in satisfying the first law of thermodynamics. Two common FDM's are: the heat balance FDM (BFDM) [3], based upon minimizing, over a nodal volume with unit weight; the mathematical FDM (MFDM) [4,5], based upon collocating at discrete points, usually regularly spaced upon coordinate lines. The FEM is based on the Galerkin method [6] in which the error is minimized over an elemental volume with a weight function equal to the temperature.

The purpose of this paper is to compare the two methods by describing their bases and their application to some common heat transfer problems. It will be no surprise that neither method is clearly superior or that, in many instances, the choice is quite arbitrary and depends more upon the codes available and upon the personal preference of the analyst than upon any well defined advantages of one method.

ANALYTICAL PROCEDURES

The numerical solution of a heat transfer problem by either FDM or FEM, is usually done in the following steps as shown in table 1:

*This work was supported by NASA grant NAG-1-41

TABLE 1
ORDER OF ANALYSES
1. Subdivision of the region into nodes or elements
2. Definition of material properties, initial temperatures and boundary conditions
3. Evaluation of the thermal conductance between nodes and the capacitance of each node
4. Formation of the global conductance and capacitance matrices
5. Imposition of the boundary conditions
6. Solution for the nodal point temperatures and heat fluxes
7. Display of the results

Both methods tend to use these same steps, although the division between the steps may be more or less clearly defined depending upon the method used and upon the organization of the computer code.

A. The Mesh

The process of subdividing the region into elements or nodal volumes, commonly called 'meshing', is usually accomplished by a separate program. Although the FDM mesh was generally created by hand, the FEM programs relied, almost from their inception, upon numerical mesh generators. This difference in approach was probably due to:

1. FDMs usually utilize nodal points which are oriented along coordinate lines. Thus irregular regions are difficult to model automatically and use is often made of hand meshing with special imaginary nodal points. If the region is susceptible to treatment by a different coordinate system, the mathematical problem is usually recast. For example, elliptical or oblique coordinate systems are often used.
2. FEM codes were first used in structural problems where an organized nodal point mesh was clearly evident and an arbitrary mesh could not be used (e.g., the frame and stringer structures of aircraft[7]). By the time the FEM was applied to continuum problems, it had been observed that the solution of the Laplacian gave rise to an acceptable mesh. Because the FEM could be applied to such a wide range of geometries and because the number of elements tended to increase quickly, a need for automatic mesh generation became obvious. As a consequence an emphasis was placed upon automatic mesh generation for the FEM.

Several symposia[8] have been organized soley to discuss sophisticated methods for the FEM and for the FDM used in computational fluid dynamics. Unfortunately, few of the thermal FDM use other than the simplest mesh generators. Lately we find that most codes, whether FDM or FEM tend to use the same mesh generators. Since this mesh generation is also used to identify the regions occupied by the different materials, to define intial temperatures, as well as to prescribe the boundary conditions, the entire mesh generation process is usually referred to as 'PRE-PROCESSING'.

B. Thermal Conductance

The primary difference between the FDM and the FEM lies in the methods used to construct the thermal conductance between the nodes.

1. FDM

In the FDM, the conductance is formed in one of two ways. The simplest approach is illustrated in figure 1 and is based upon the heat balance equation

$$\int_V \rho c \frac{\partial T}{\partial t} dv = \int_A q_n da + \int_V q' dv \qquad (1)$$

where the energy conducted through the surface is approximated by the linear relationship

$$q_n(i,j) = q_{ij} = k_{ij} A_{ij} (T_j - T_i)/\ell_{ij} \qquad (2)$$

The quantity A_{ij}/ℓ_{ij} is a geometric value which, in two dimensions, can be related to the angles of the triangle formed from the nodes[9]. Equivalently it can be found as illustrated in figure 1 by constructing the area A_{ij} which is perpendicular to the line joining the nodes and located at the mid point of the line. Although this method is difficult to implement, it can be done automatically and also provides a technique for determining the set of closest nodes which surround a node. Unless the nodes are regularly spaced, in which case the BFDM becomes the MFDM, the BFDM is of low accuracy, $O(\Delta x)$.

The second method, the mathematical FDM (MFDM), is based upon expressing the differential equation

$$\rho c \frac{\partial T}{\partial t} = \frac{\partial}{\partial x}\left(k_x \frac{\partial T}{\partial x}\right) + \frac{\partial}{\partial y}\left(k_y \frac{\partial T}{\partial y}\right) + q' \qquad (3)$$

by finite differences using a variety of different numerical approximations to achieve a desired level of accuracy. The method consists of establishing an interpolating function in space, usually a polynomial, and performing the needed differentiation. Another method is to express T_j in terms of T_i through a Taylor series expansion. By combining the expressions for the temperatures at different nodal points, i.e., establishing different stencils, figure 2, it is possible to achieve a very high order of accuracy. For example the 5 point stencil is accurate to $O(\Delta x^2)$, while the 9 point is of $O(\Delta x^4)$. However, the 13 point is accurate not only to $O(\Delta x^4)$ but ensures that $\nabla^4 T$ as well as $\nabla^2 T=0$ is satisfied[10].

For a non-linear problem it is usual to express $\frac{\partial}{\partial x}\left(k_x \frac{\partial T}{\partial x}\right)$ as

$$\frac{\partial}{\partial x}\left(k_x \frac{\partial T}{\partial x}\right) = k_x \frac{\partial^2 T}{\partial x^2} + \frac{\partial k_x}{\partial x} \frac{\partial T}{\partial x} \qquad (4)$$

and to evaluate each term separately. Unfortunately this approach does not give a conservative set of equations (in P. D. Lax's sense) and energy is not conserved in steady state problems. As a consequence, the non-linear FDM is usually confined to the 5 point stencil for regular regions and the heat balance method for irregular regions.

2. FEM

The finite element method is based upon the Galerkin approach in which a stationary value of the integral

$$I = \int_V \frac{1}{2}\left[k_x \left(\frac{\partial T}{\partial x}\right)^2 + k_y \left(\frac{\partial T}{\partial y}\right)^2 - \rho c \frac{\partial T^2}{\partial t} + 2q\hat{\ }T\right] dv + \int_A q_n T da \qquad (5)$$

is sought[11]. For steady state problems, the Galerkin and the variational approaches are equivalent and the method, in conjunction with the dual formulation (i.e., the adjoint problem), can be used to establish bounds upon the solution. For transient problems, there is no variational formulation and no bounds can be found.

The procedure consists of approximating the temperature within the element by

$$T(x,y) = \sum_{\ell=1}^{I} n_\ell(x,y) a_\ell = <n> \{a\} \qquad (6)$$

where the interpolating functions, $n(x,y)$ can be chosen at will to satisfy any desired criteria, subject only to the restriction that there be a sufficient number of nodes in the element to determine the coefficients, a_ℓ. Since each element is treated as an entity, it is common to map the element into a simpler shape to expedite the integration, figure 3. Consider that a local set of coordinates, ξ, η, are defined and that the mapping

$$x = \sum_{\ell=1}^{J} \bar{n}_\ell(\xi,\eta) \, b_\ell = <\bar{n}> \{b\} \qquad (7)$$

is used where the number of terms J need not equal the number of terms I used in the temperature interpolation. We then have

$$\{\hat{T}\} = [A]\{a\}, \quad T(x,y) = <n> [A]^{-1}\{\hat{T}\} = <N> \{\hat{T}\} \qquad (8a)$$

$$\{\hat{x}\} = [B]\{b\} \quad x = <\bar{n}> [B]^{-1}\{\hat{x}\} = <\bar{N}> \{\hat{x}\} \qquad (8b)$$

Thus for one element the conduction term is of the form

$$\frac{\partial T}{\partial x} = <\frac{\partial N}{\partial x}> \{\hat{T}\} = \left[<\frac{\partial N}{\partial \xi}\frac{\partial \xi}{\partial x}> + <\frac{\partial N}{\partial \eta}\frac{\partial \eta}{\partial x}>\right] \{\hat{T}\} \qquad (9a)$$

$$\frac{\partial I}{\partial \{\hat{T}\}} = \left[\quad k_x \left\{\frac{\partial N}{\partial x}\right\} <\frac{\partial N}{\partial x}> |J| d\xi d\eta \right] \{\hat{T}\} \qquad (9b)$$

and

$$\begin{Bmatrix} \frac{\partial N}{\partial x} \\ \frac{\partial N}{\partial y} \end{Bmatrix} = [J]^{-1} \begin{Bmatrix} \frac{\partial N}{\partial \xi} \\ \frac{\partial N}{\partial \eta} \end{Bmatrix} \qquad (10)$$

where J is the Jacobian of the mapping.

Although it is obvious that the spatial interpolation function, $\overline{N}$, must ensure compatability of elements (i.e., that an edge common to two elements must be the same when viewed from either element) it is not necessary that the temperatures be compatible. In structural mechanics FEM, incompatible formulations are sometimes used, but almost all thermal problems use compatible temperature functions. When the same number of functions are used to express the mapping and the temperature, the element is termed 'iso-parametric'. Although the isoparametric element is the usual, subparametric elements are often used for heat source problems or when an unusual temperature distribution is desired.

The most common 2-D elements are the 3 and 6 node triangles and the 4 and 8 node quadrilaterals, figure 4. Much of the research in the FEM has been devoted to developing new interpolating functions (i.e., new elements) and identifying their characteristics, particularly with respect to the numerical integration required.

3. Differences

Probably the most striking difference between the FDM and the FEM is the usual lack of continuity of the heat flux in the FEM. Each element is treated separately and if only compatibility of temperature is imposed, the heat flux at the edge common to two elements is not continuous. By contrast, the FDM has a continuous heat flux. Although the effect of lack of continuity of the heat flux is usually not important, it can show up in the form of an oscillating temperature profile where an overshoot in one node is compensated for by an undershoot at a neighboring node in order to minimize the integral over the entire region.

C. Capacitance Matrix

In the FDM, the capacity term, $\rho c \, \partial T / \partial t$ is represented by the simple lumped term

$$(\rho_i c_i V_i) \frac{\partial T_i}{\partial t} \quad (11)$$

where V_i is the volume surrounding the node i. In the FEM, the use of the interpolating function gives rise to the consistent capacitance matrix

$$[C]\left\{\frac{\partial \hat{T}}{\partial t}\right\} = \begin{bmatrix} c_{11} & c_{12} \cdots c_{1n} \\ c_{21} & \vdots \\ \vdots & \\ c_{n1} & \cdots \cdots c_{nn} \end{bmatrix} \{\hat{T}\} \quad (12)$$

in which the storage at each node is related to that at every other node in the element. This difference in the two capacitance matrices is responsible for much of the difference in the solutions to transient problems and is discussed further in the EXAMPLE section. For 3 node triangular elements with regular node spacing, the FDM and FEM have identical conductance and capacitance matrices. Other FEM elements give different matrices, usually as indicated by the sign of the off-diagonal terms. In the FEM the integration is usually accomplished through Gaussian integration and the need to evaluate the properties and the Jacobian at each point contributes to the longer execution times of the FEM.

D. Assemblage of the Global Matrix

1. FDM

In the FDM, each node is treated in turn, and the conductances are entered into the global matrix as they are determined. The process is simple and straightforward and requires very little computer time. Unless an FEM preprocessor mesh is used, FDM codes generally make use of a separate matrix which identifies the nodes which interact, the corresponding value of the term A_{ij}/ℓ_{ij} and the material numbers to permit rapid determination of the average conductance $(k_i+k_j)/2$.

2. FEM

In the FEM, the element matrix is first evaluated, then assembled into the global matrix. If the global matrix cannot be stored in core, it is necessary to search through the elements to treat only those containing the nodes whose equations are currently in core. This assembly process tends to be reasonably time consuming and may contribute to the reduced speed of the FEM.

3. Comparison

Table 2 lists some typical execution times for FDM and FEM solutions. In general, the FEM assembly costs rise quickly as the number of nodes per element is increased. Figure 5[12] shows some comparable results for a structural problem and the extra expense of the assembling is easily seen. (The figure also shows that for some variables the FDM is superior while for others the FEM is. Because of this, it is difficult to define one method as being the best for all parameters of a problem.)

TABLE 2
RELATIVE EXECUTION TIMES FOR THE TRANSIENT PROBLEM
Equal Time Step Sizes

n	FDM			FEM			
	L-W	Explicit	C-N	Linear	Quadratic	Cubic	Special Cubic
3	.8		1	1	1.8		
5	1.4	1.5	2	5.4	6.5		4.4
7						5.5	
9	3.7	3.5	6	32	36		24
13						36	
17	17	12	24	218	249		183
19						125	
31						540	
33	107	64	117	1624	1880		1500
61						4350	
65	760	407	1230	12200	14700		

n = number of elements   L-W = Lax Wendroff
C-N = Crank Nicolson

E. Boundary Conditions and Irregular Meshes

Although the greatest fundamental difference between the FDM and the FEM lies in the formulation of the conductances and capacitances, the greatest practical difference lies in their treatment of boundary conditions and irregularly shaped regions.

1. FDM

Specified temperature nodes are treated by ignoring the equation for the node. Boundary flux conditions are difficult to treat for irregular meshes unless one uses the simple heat balance formulation (BFDM) with its low order of accuracy. Irregular meshes have been studied for years with emphasis on using triangular surface elements[13]. Since many analysts tend to define their nodes along coordinate lines, it is most common to make use of imaginary nodes, figure 6, special differencing equations to achieve the desired accuracy[14], arbitrary grids[15], or even no nodes, but only cells[16]. In general, the advanced mesh generators used in fluid computations and the automatic high order boundary gradient differencing methods are not used in production FD methods. Consequently they suffer from reduced accuracy at irregular boundaries. On the other hand, research FDM's almost always maintain high accuracy at all types of boundaries.

2. FEM

Irregular regions are handled in the same manner as any other region. The accuracy is limited only by the ability of the mesh generator to represent the boundaries. This ability is probably the greatest strength of the FEM.

Specified temperature nodes are handled in two ways. In the first, the equation is deleted from the matrix and the right hand sides of all affected equations are modified. Another way is to replace the diagonal coefficient by a very large number, L, and the right hand side by L*T[17]. Although faster and simpler than the first method, the accuracy of the computer may limit the effectiveness of this approach.

Prescribed heat flux boundary conditions are handled better by FDM than by FEM. Because in the FEM each element is treated separately, the accuracy is limited by the interpolating accuracy of the element shape functions. By contrast, the FDM analyst can easily modify a difference equation to achieve the desired accuracy. The FEM analyst must call for a boundary element which is different from the interior element, figure 4, and may have some difficulty in merging this higher order element with the rest of the mesh unless a special mesh generator or slide lines are used.

F. Solving the Equations - Steady State Problems

Although sparse matrix solvers have become popular recently[18], most FDM and FEM codes use some form of elimination (usually Gauss or Cholesky) and rely upon a banded matrix or sky line approach to achieve fast solution times[19]. For highly non-linear problems, iterative solutions may be used, especially by FDM codes.

1. FDM

FDM codes normally use direct reduction or iteration. Because of the way in which the global matrix is assembled, iteration is a common procedure and, when used, the global matrix is not created. If reduction is used, only enough equations to treat the band are kept in core and they are shifted immediately after each nodal equation is reduced (the wave front method). Although iteration is an appropriate technique, convergence is difficult to determine and because iterative acceleration factors are often unknown, many users are reluctant to employ iterative techniques. Fortunately most heat transfer problems can be treated by using the SOR method and values of the over relaxation factor can be quickly approximated. For problems which are not symmetric or do not have Young's property A[20], the SSOR technique[21] is easy to implement and convergence is very rapid. Non-linear problems are ideally suited to the iterative method and rarely increase the solution time by an appreciable amount. Considerable research is still being conducted into improving the rate of convergence and treating highly non-linear problems[22].

2. FEM

Most FEM codes use a reduction process with a subsequent back substitution. Even with a skyline procedure, care should be taken to minimize the band width. To do this, most FEM codes will use a band width minimizing program after the mesh generation[23], although this may complicate the identification of the nodal point positions within the code. Non-linear problems are treated in two ways:

1. The matrices are reformed and solved for each iteration.

$$K^{(n)} T^{(n)} = R^{(n)} \tag{13a}$$

2. A residual is found, and an increment, ΔT determined from

$$\{\varepsilon\}^{(n)} = \{R\}^{(n)} - [K]\{T\}^{(n-1)}$$
$$\Delta T^{(n)} = [K]^{-1}\{\varepsilon\}^{(n)} \tag{13b}$$

It is common to use a mixture of these two methods by applying the boundary loads in increments, calculating a new K for each increment, but iterating within the increment using a constant K until the residual is sufficiently small[24]. Method 1 requires excessive computer time and may not converge. Method 2 requires that the original conductance matrix be stored in order to calculate the residual. Either method is expensive, and for highly non-linear problems, the iterative FDM appears to be preferred. Suendermann [25] suggests that the ratio of the execution times for hydrodynamic problems is 100 to $27b^2/2+18b$ for implicit methods and 100 to $23bi$ for explicit methods where b=band width and i the number of FEM iterations.

In both methods it is often desireable to know the amount of heat added to a constant temperature node. In the FDM, the calculation of Q_i is simply done by calculating the conductances and evaluating the heat balance. In the FEM, this may be done by using the original K matrix if it is stored. A simpler way is to permit a specified temperature node, T_i to float by connecting it to a constant temperature, T_c, through a very high conductance, K_H, and evaluating

$$Q_i = K_H (T_c - T_i) \qquad (14)$$

Although much simpler, the size of K_H required may be so large that some computers may not accurately calculate the correct inverse of the global conductance matrix. In this case the energy balance will have to be computed for the node under question.

G. Solving the Equations - Transient Problems

In general, both the FDM and the FEM simply step along in time, re-evaluating K and C at each time step, if needed. If the time steps are large or the problem is highly non-linear, then iterations may be needed within each time step[26]. If the non-linearities may be expressed analytically a Runge-Kutta technique may be applied to the resulting non-linear equations[27]. For very non-linear problems, Gear's method may be used [28]. Regardless of why the matrices must be re-evaluated (i.e., changing boundary conditions or non-linear properties) Young's method [29] can be used to modify the matrix and the solution with a minimum increase in time, although the code structure may have to be changed considerably.

H. Graphical Display

Besides printed output, the most common output is a spatial plot of a single variable or contour plots. Such plots of the temperature are relatively easy to generate for either the FDM or the FEM if a mesh generator was used. If the FDM was based upon the heat balance approach, then it is necessary to create a set of minimum size triangular elements to perform the contour plotting [30], especially if contour smoothing is to be used [31]. Table 3 illustrates the times necessary to compute a set of triangular elements connecting 418 nodes. If the region is complete, the element calculation is rapid. If there is one or more internal voids, then the establishment of the mesh may be very expensive, as indicated by the third table entry.

TABLE 3
Contour Plotting Times

1.	Temperature contours	5.6 seconds
2.	Temperature contours-establishing a mesh	30 seconds
3.	Temperature contours-establishing a mesh with an internal void	120 seconds

If the contours are to be based upon other than linear interpolation or if the heat fluxes between elements are to be plotted, the FEM is much simpler to use since the element interpolation functions may be used, both to give a higher order curve fit and to interpolate to establish the values of q_x and q_y at the nodal points.

In general the FEM output is much better suited than the FDM output to graphical display, particularly for curved surfaces. To permit rapid detection of input errors and for rapid analyis, both a plane and an isometric (with hidden line capability) contour plotter should be available.

EXAMPLES

We present some typical examples of the use of the two methods.

A. Distributed Heat Sources

Consider a one dimensional problem, $0 \leq x \leq 1$, with a distributed source strength of $q' = x^m$. The problem was treated with an FDM using two different boundary conditions and the FEM source term and the FEM utilizing 4 different elements[32]. Figure 7 illustrates the error in the temperature at the insulated surface. The solid line is for both the MFDM with a virtual point and for the BFDM. The dashed line is for the MFDM using the zero heat flux boundary condition and shows the underprediction of the temperature because of the lack of a source at the last node. The FEM results are substantially better for two reasons. Firstly, the source is not lumped at a single node, but is distributed between nodes, thus permitting a better representation of the spatially varying heat source (this is clearly shown by the equivalence of the results for the FDM using the FEM source term and the FEM.) Secondly, the higher accuracy of the FEM temperature interpolation permits consideration of temperature variations higher than the linear. The special cubic element of Tocher[33] provides for continuity of the temperature and the heat flux at the interface between elements. Over a range of source functions, $0 \leq m \leq 5$, this special cubic was not found to be any better than the quadratic or the normal cubic element, neither of which ensure continuity of the heat flux.

B. Transient Temperatures - 1-D

Consider the one-dimensional slab with an insulated back surface and a front surface whose temperature is suddenly reduced to zero. Figure 8 illustrates the results obtained using several FD methods and several FEM elements[32]. In the FEM we write,

$$T(x,y,t) = < N(x,y) > \{\hat{T}(t)\} \qquad (15)$$

$$T^{n+1} = T^n + [\alpha \dot{T}^{n+1} + (1-\alpha)\dot{T}^n]\Delta t$$

$$\left[\alpha K + \frac{C}{\Delta t}\right]\{\hat{T}\}^{n+1} = (1-\alpha)[C]\{\hat{\dot{T}}\}^n + \left[\frac{C}{\Delta t}\right]\{\hat{T}\}^n + \alpha\{R\}^{n+1} \qquad (16)$$

or

$$\left[\alpha K + \frac{C}{\Delta t}\right]\{\hat{T}\}^{n+1} = \left[-(1-\alpha)K + \frac{C}{\Delta t}\right]\{\hat{T}\}^n + \alpha\{R\}^{n+1} + (1-\alpha)\{R\}^n \qquad (17)$$

One may also use a combined space-time interpolant [34]

$$T(x,y,t) = \sum N_i(x,y,t)\hat{T}_i \qquad (18)$$

but this method does not yet appear to be in regular use for producton codes. In figure 8, the use of equation 16 with $T(x,0)=0$ is referred to as 'without $\dot{T}$'. If $T(x,0) \neq 0$ or if equation 17 is used, the solution is referred to as 'with $\dot{T}$'. The relatively poor FEM results are due to the use of the consistent capacitance matrix and the impulsive start. Figure 9 compares the FEM with consistent matrices and the FDM with the lumped capacitance[35]. Because the FEM overpredicts the eigenvalues (i.e., yields an overly stiff system) while the FDM underpredicts, the FEM results show an early time overshoot. This can be corrected by lumping the capacitance terms [36,37] or by expressing $\dot{T}$ by using a different basis which is orthogonal to the shape functions for T[38,39]. The effect of lumping is shown most clearly by examining the element eigenvalues for the 8 node quadrilateral as illustrated in figure 10. Lumping may also be effected by using a reduced order of integration, which softens the transient response, but care must be taken to avoid creating singular capacitance matrices[40]. The best results have been found when using the same order of integration as for the conductance. For non-linear problems, a one point integration has been found to be sucessful[41]. Lumping has the further advantage of simplifying the algorithm and reducing the execution time and has been extensively used for structural problems[42]. Figure 11 shows the error in the effective thermal diffusivity for several different methods. Cubic Hermite elements[43] are seen to be the best, in agreement with the results of figure 8.

Another solution is to permit the discontinuous change to take place over several time increments, a technique commonly used in hydrodynamic shock analyses. Figure 12 shows that while this smooths the results, it does not reduce the lag in the response which is due to the excessive dissipation of the FDM and lumped FEM solutions. A better way is to evaluate the right hand side at the half time intervals or by a consistent time interpolation[44]. As illustrated, this gives the best performance. Table 2 lists the execution times needed for the different methods for an equal number of time steps. For non-linear problems in which the matrices must be reformed and re-solved, the FEM is substantially slower than the FDM, particularly when the SSOR method is used.

C. Singular Problems - Standard Methods [32]

Consider a two-dimensional plate as shown in figure 13. Because of the sharp edge of the insulated splitter plate, the heat flux at the tip is infinite and of the form

$$q = K_1 f_1(\theta)/\sqrt{h} + K_2 f_2(\theta)\sqrt{h} \qquad (19)$$

This problem was treated by using the MFDM and the FEM.

1. FDM

Figure 13 illustrates the results obtained with different mesh sizes using a 5 point stencil. Substantial errors in the temperature and in the heat fluxes, figure 14, were found and even a reduction from $\Delta x=\Delta y=1/4$ to $\Delta x=1/20$ did not yield convergence, even for mesh points further than $2\Delta x$ from the singularity.

2. FEM

Figures 13 and 14 also show the comparable results found by using different elements and sizes. Of the three elements used, the quadratic is seen to give the best results. Table 4 lists the execution times. Although the quadratic element times are long, the times necessary to produce an accuracy in the temperature comparable to that of the linear element or the MFDM are approximately equal. On the other hand, the accuracy in evaluating the heat flux is much better than that of the FDM, with the result that the comparable execution times are much less.

TABLE 4
RELATIVE EXECUTION TIMES FOR THE SINGULAR PROBLEM
2:1 RECTANGULAR NODAL GRID, STANDARD PROGRAMMING

number of nodes	$\Delta x(=\Delta y)$	FEM linear	FEM quadratic	FEM cubic	FDM SOR	FDM SSOR
15	1/4	1	1		0.8	1.1
45	1/8	3.6	10.4	110	4.5	3
66	1/10	6.0			8.1	4.3
153	1/16	20.5	121	3097	21	20
231	1/20	39.1	274		53	31
361	1/32	172	1531		207	183

D. Singular Problems - Lagrangian Variables

As described above, problems with concentrated heat sources or with singular heat fluxes can rarely be treated with standard FDM or FEM programs since the interpolating functions are incapable of adequately representing the singular temperature field. In both methods, the interpolating functions can be expanded to include the singular behavior. If the strength of the singularity is known, the extra term simply serves as an additional heat source. If the strength is unknown, the FDM and FEM must be modified. For the FDM, let $\diamond T$ be the FD approximation to the field equation and BT be the FD approximation to the boundary condition. Then by expressing the temperature as

$$T = \bar{T}(\text{smooth}) + K_1 S \tag{20a}$$

we find

$$\diamond T \doteq \nabla^2 T + K_1 (\diamond S - \nabla^2 S) \tag{20b}$$

$$BT \doteq \beta T + K_1 (BS - \beta S) \tag{20c}$$

It is thus apparent that the effect of the singularities can be considered by adding the pseudo heat source and boundary heat fluxes.

For the FEM, the appropriate formulation is[45]:

$$T = <n> \{\hat{T}\} + K_1(S - <N> \{\hat{S}\}) \tag{21}$$

In this solution the value of K_1 is considered to be a Lagrangian variable which is determined by differentiating the integral to yield extra equations of the form:

$$\frac{\partial I}{\partial \{\hat{T}\}} = [K]\{\hat{T}\} + K_1(S_x - <N_x> \{\hat{S}\})\{N_x\} \tag{22a}$$

$$\frac{\partial I}{\partial K_1} = (S_x - <N_x> \{\hat{S}\}) <N_x> \{\hat{T}\} + (S_x - <N_x> \{\hat{S}\})^2 K_1 \tag{22b}$$

Figure 13 illustrates a problem in which the magnitude of the singularity is unknown. In the FDM, the value of K_1 is found by applying the field equation to the closest boundary points, in addition to using the expanded boundary conditions. When using the FEM, it is not necessary that all elements contain the singular terms, only those near the singularity. However it has been found important[46] that a smooth transition between the singular elements and the regular elements be provided by establishing transition elements in which the temperature is of the form

$$T(x,y) = <N> \{\hat{T}\} + f(x,y) K_1 (S - <N> \{\hat{S}\}) \tag{23}$$

in which the function $f(x,y)$ has the value of 1 along edges common with the singular elements and 0 along edges common with the regular mesh. As seen from figure 15, relatively coarse meshes for both the FDM and the FEM are sufficient to determine q_x with good accuracy if the singular term is included, but even increasing the number of elements by a factor of 16 is insufficient if it is omitted.

Table 5 gives a comparison of the values of the temperature at point P as determined by several different FDM and FEM solutions and the equivalence of the two singular methods is apparent. Table 6 lists the values of the singular strengths, K_1 and K_2. Although K_2 varies considerably, K_1 (which is the dominant singularity) is quite constant. If K_2 were to be computed more exactly, then it would be necessary to include the next term in the singular series[47]. The superiority of the singular approach is evident.

TABLE 5

COMPARISON OF T_p BY DIFFERENT METHODS (Figure 13)

Δx	FEM 1st order triangle	FEM 2nd order triangle	FEM 1st order quadrilateral	FEM Singular	FDM 1st order	FDM 2nd order	FDM Singular
1/2	0.1429		0.1176	0.170			
1/4	0.1591	0.1737	0.1579		0.083	0.133	
1/6	0.1659	0.1771	0.1674				
1/8	0.1697	0.1784	0.1717	0.181	0.151	0.174	0.1815
1/10	0.1721		0.1741				
1/12	0.1738		0.1756		0.166	0.178	0.1824
1/16	0.1759		0.1775		0.173	0.180	0.1827
1/20	0.1772				0.176	0.181	0.1827
1/24	0.1781						
1/32	0.1793						

TABLE 6

SINGULARITY STRENGTHS AND T_p

Δx	number of singular elements	FEM K_1	FEM K_2	FEM T_p	FDM K_1	FDM K_2	FDM T_p
1/2	2(all)	1.285	0.091	0.170			
1/4	2	1.276	0.067	0.179			
	6	1.234	-0.032	0.179			
	8(all)	1.224	-0.045	0.178			
1/8	2	1.275	-0.056	0.182	1.1325	-0.492	0.1815
	6	1.232	-0.048	0.182			
	12	1.212	-0.077	0.181			
	18	1.203	-0.086	0.181			
	32(all)	1.195	-0.094	0.181			

Under some conditions, the analytical form of the singular terms may be so complex that the integration required for the FEM may be difficult to perform with acceptable accuracy. In this case, the singular FDM, which satisfies only at the nodal points, is a more useful method.

One of the interesting features of the FEM is that it is often possible to distort an element to produce the desired singularity[48]. Referring the the splitter plate problem, if an 8 node quadrilateral is used and the mid side nodes are shifted to the quarter points as indicated in figure 16, the element automatically includes a square root singularity. The figure compares these quarter point element results with those of the singular FDM and the comparison is excellent.

E. Radiation Problems

Regions with internal voids which have strongly radiating boundaries pose a problem, particularly for the FEM, even if the boundary conditions are linearized, since all of the nodes interact simultaneously and the resultant band width is very large and portions of the matrix are dense. One approach is to consider the problem as two problems, the radiation problem, and the conduction problem. The radiation problem is solved separately ensuring radiative equilibrium among the nodes. The radiative heat flux is then considered as a known heat flux and applied to the right hand side of equation 5. Because the boundary condition is so non-linear, it often proves necessary to use a strong under-relaxation factor to limit the flux change to assure convergence and avoid the overshoot observed with impulsively changed boundary conditions as indicated by the transient problem. If the rest of the problem is linear, this approach is satisfactory. If the rest of the problem is highly non-linear, an iterative FDM appears to work more efficiently.

F. Phase Changes

Probably the most difficult problem to treat effectively by either method is that of a transient phase change. Three methods, figure 17, appear to be the most common. In the first, the interface motion is computed on the basis of the conservation equations, just as is done for fluid shock calculations[49]. If an implicit time solution is used, this method requires the solution of non-linear algebraic equatons[50,51]. The second method is to assume that the phase change occurs over a small temperature range, T. The latent heat is then approximated by a large specific heat value, or the enthalpy may be used as the primary dependent variable. An iterative solution is often needed to ensure that the interface is maintained at the correct fusion temperature and special care must be taken in evaluating the capacity[51,52]. In a third method, based upon the use of the enthalpy and the temperature, the interface position is not explicitly determined. In this method an artificial specific heat is not used. This method has not been fully developed and some problems have been noted if the enthalpy interpolation is other than a step function[53]. In its present form, this method may be better suited for use with the BFDM.

Figure 18 compares some typical results of the first two methods using FDM and FEM. Comini's solution[51] used Lee's time integration and gave slightly greater solidification depths than the other methods. The FDM interface was slightly in error at the earliest time because of the lumped capacitance, although it quickly gave correct values. The FDM enthalpy solution gave essentially identical results.

G. Conclusions

Having used both FDM and FEM for more than 2 decades to solve a variety of thermal problems, ranging from simple 1-D transient cases to 3-D singular biological problems, we have drawn the following general conclusions.

1. The mathematical FDM appears to be best suited to research problems, especially ones for which the analyst wishes to ensure that the boundary conditions are treated with high order accurate schemes or ones in which special algorithms are used at specified nodal points.

2. The heat balance FDM appears to be best for:
   1. Highly non-linear problems, for which iterative solutions are efficient.
   2. Problems in which the continuity of the heat flux is important.
   3. Multi-dimensional problems involving change of phase.

3. The FEM is best suited to:
   1. Irregular regions for which the automatic mesh generation and a library of highly accurate elements permit good modelling of the region and the consequent temperature profile.
   2. Mildly non-linear problems for which the iterations are few.
   3. Problems for which graphical output is important.
   4. Problems in which special temperature profiles are desired, since these may be easily obtained with special elements.
   5. Problems involving singular temperature fields or concentrated heat sources.
   6. Problems in which different approximations are to be used in different regions or problems which involve the joining of several parts.

Although each method may appear to be best for a particular class of problems, we have also reached the rather general conclusions that:

1. The analyst should be knowledgeable about both methods, at least to the extent that their general characteristics are understood.
2. Because thermal FEM elements are still being developed, their characteristics are not generally known. The analyst should experiment with such elements until their behavior, singly or in concert with other elements, is clearly understood[54]. In particular, the performance of any one element is not intuitively obvious and may not be representative of a 'similar element' for another problem[55].
3. Except for very special problems, either approach is satisfactory and which method used depends more upon the availability and familiarity of codes (and pre- and post-processors) than upon any intrinsic differences between the two methods.

We recognize that this last conclusion is rather fuzzy, but since the results obtained with either method will vary as the stencil or the element is changed, we have found that nothing will compensate for the analyst's insight into the detailed characteristics of the method used. Since such insight is normally developed only by exercising a program on a variety of problems and since it is rare that any one person is comfortably familiar with more than one method, it is not surprising that there is a considerable tendency to continue using a familiar code, even if some shortcomings exist.

A recent round robin test of FDM and FEM[56] applied to a typical mixed boundary condition, transient problem (fig. 19) showed that either method was satisfactory and that the apparent value of either depended more upon the pre- and post-processors available than upon the intrinsic characteristics or accuracies of the methods. On the other hand, for some conditions, particularly anisotropic problems, the results of reference 57 indicate that the FDM may be more accurate, but when there was no cross coupling, the FEM was superior.

Readers should consult references 58 and 59 for more information about the FDM and references 60 and 61 for detailed insight into the FEM. Current work in the FDM is concentrated on improving the accuracy and execution times, with special emphasis on thermal network correction methods[62]. Extensions of the FEM to treat combined convection-conduction problems and to merge the thermal FEM with the structural FEM, taking the different mesh requirements and time responses into consideration, are also being studied[63]. In addition, some recent work[64,65] has shown that elliptical PDE's and conduction-convection problems can be accurately solved by combining collocation at the Gaussian points with the FEM elements. Finally, we should note that the development of small-scale personal computers, which are slow but may be dedicated to specific tasks, can be expected to have profound influences upon the structure of the computer codes, the use of interactive execution, and the specific algorithms used[66]. In this context, it should be recognized that the substructuring method, which is commonly used on these small computers, cannot be used for transient problems.

NOMENCLATURE

A	Area
B	FDM boundary condition operator
c	Specific heat capacity
C	Capacitance matrix
f	Transition function
I	Integral
J	Jacobian of transformation
k	Thermal conductivity
K	Thermal conductance matrix
l	Distance between nodes
n	Interpolating function
N	Shape function
q'	Generated heat density
q_n	Boundary heat flux
q_x	Directional heat flux
Q_i	Net heat input to a node
R	Right hand vector
S	Singular function
t	Time
T	Temperature
$\dot{T}$	Time derivative of temperature
$\bar{T}$	Smooth temperature distribution
α	Weighting parameter
β	Boundary condition operator
ε	Error
ξ, η	Coordinates of unit square
ρ	Density
Δ	Change
$\wedge$	Nodal values
$\diamondsuit$	Approximation to the Laplacian

REFERENCES

1. Courant,R., Friedricks,K. and Lewy,H., "On the Partial Difference Equations of Mathematical Physics", Math. Annalen., pp32-74, 1928
2. Visser,W., "A Finite Element Method for the Determination of Non-Stationary Temperature Distributions and Thermal Deformations", Proc. Conf. Matrix Meth. Struct. Mech., USAF Inst. of Tech., Wright-Patterson AFB, pp925-943, 1965
3. Fox,L., Numerical Solution of Ordinary and Partial Differential Equations, Pergamon Press, Oxford, 1962
4. Forsythe,G.E., and Wasow, N.R., Finite Difference Methods for Partial Differential Equations, J. Wiley, N.Y., 1960
5. Babuska,I., Prager,M. and Vitasek,E., Numerical Processes in Differential Equations, J. Wiley, N.Y., 1966
6. Leipholz,H., Direct Variational Methods and Eigenvalue Problems in Engineering, Noordhoff Publ., Netherlands, 1977
7. Turner,M.J., Clough,R.W., Martin,H.C., and Topp,L.J., "Stiffness and Deflection Analysis of Complex Structures", J.Aero. Sci., pp805-823, 1956
8. Smith,R.E., Numerical Grid Generation Techniques, NASA Conf. Publ. 2166, 1980
9. Dusinberre,G.M., Numerical Analysis of Heat Flow, McGraw-Hill, N.Y., 1949
10. Bickley, W.G., "Finite Difference Formulas for the Square Lattice", Q. J. Mech. Appl. Math., pp35-42, 1948
11. Strang,G. and Fix,G.J., An Analysis of the Finite Element Method, Prentice Hall, N.J., 1973
12. Bushnell,D., "Finite Difference Energy Models versus Finite Element Models: Two Variational Approaches in One Computer Program", Numr. and Comp. Meth. Struct. Mech., (Fenves,S.J.,et al, eds) Academic Press, N.Y., pp291-336, 1973
13. Hildebrand,F.B., Finite Difference Equations and Simulation, Prentice Hall, N.J., 1968
14. Lau,P.C.M., "Finite Difference Approximations for Ordinary Derivatives", Intl. J. Num. Meth. Engng., pp663-668, 1981
15. Jensen,P.S., "Solution of Two-Dimensional Boundary Value Problems by Arbitrary Grid Finite Difference Methods", Adv. Comp. Meth. PDE., (Vichnevetsky,R., ed), AICA, pp80-85, 1975
16. Greenstadt,J., "Cell Discretization", Conf. Appl Num. Anal., (Morris,J.L., ed) Springer-Verlag, Berlin, 1971
17. Payne,N.A.and Irons,B., ref in Zienkiewics,O.C., The Finite Element Method, 1st ed., McGraw Hill, N.Y., 1967
18. Eisenstat,S.C., Schultz,M.H. and Sherman,A.H., "Efficient Implementation of Sparse Symmetric Gaussian Elimination", Adv. Comp. Meth. PDE., (Vichnevetsky,R., ed) AICA, pp33-45, 1975
19. Irons,B.M. and Kan, D.K.Y., "Equation Solving Algorithms for the Finite Element Method", Num. Comp. Meth. Struct. Mech., (Fenves,S.J., et al, eds.) Academic Press, N.Y., pp497-511, 1973
20. Young,D.M., Iterative Solutions of Large Linear Systems, Academic Press, N.Y., 1971

21. Sheldon,J.W., "On the Numerical Solution of Elliptical Difference Equations", Math. Tables, pp101-111, 1955
22. Zedan,M. and Schneider,G.E., "3-D Modified Strongly Implicit Procedure for Finite Difference Heat Conduction Modelling", paper AIAA-81-1136, AIAA 16th Thermophysics Conf., Palo Alto, Calif, 1981
23. Schauer,D.A., "A Finite Element Bandwidth Minimizer'" Lawrence Livermore Lab., Livermore, Calif., 1973
24. Hogge,M.A., "Secant versus Tangent Methods in Non-Linear Heat Transfer Analysis", Int. J. Num. Meth. Engng., pp51-64, 1980
25. Suendermann,J., "The Application of Finite Elements and Finite Difference Techniques in Hydrodynamical Numerical Models", Formulation and Computational Algorithms in FEM, US-German Symp., (Bathe,K-J., et al, eds) MIT Press, Ma, 1977
26. Chung,B.T.F. and Chang,T.Y., "Heat Transfer in Solids with Variable Thermal Properties and Orthotropic Conductivity", paper AIAA-81-1137, AIAA 16th Thermophysics Conf., Palo Alto, Calif., 1981
27. Aguirre-Ramirez,G. and Oden,J.T., "Finite Element Technique Applied to Heat Conduction in Solids with Temperature Dependent Thermal Conductivity", Int. J. Num. Meth. Engng., pp345-355, 1973
28. Franke,R. and Salinas,D., "An Efficient Method for Solving Stiff Transient Field Problems Arising From FEM Formulations", Comp. Math. with Appl., pp15-21, 1980
29. Young,R.C., "Efficient Nonlinear Analysis by Factored Matrix Modification", paper M4/3, Fourth SMiRT Conf., San Francisco, Calif, 1977
30. Patterson,M.R., "CONTUR- A Subroutine to Draw Contour Lines for Randomly Located Data", ORNL/CSD/TM-59, 1978
31. Emery.A.F., "PLOT- A General Purpose Plotting Program with Smoothing", Mech. Engng. Dept, Univ. of Washington, Seattle, Wash., 1980
32. Emery,A.F. and Carson,W.W., "An Evaluation of the Use of the Finite Element Method in The Computation of Temperature", ASME J. Heat Trans., pp136-145, 1971
33. Tocher,J.L. and Hartz,B.J., "Higher Order Finite Element for Plane Stress", ASCE J. Engng. Mech., pp149-172, 1967
34. Chung,K.S., "The Fourth-Dimension Concept in the Finite Element Analysis of Transient Heat Transfer Problems", Int. J. Num. Meth. Engng., pp315-325, 1981
35. Emery,A.F., Sugihara,K. and Jones,A.T., "A Comparison of Some of the Thermal Characteristics of Finite Element and Finite Difference Calculations of Transient Problems", Num. Heat Trans., pp97-113, 1979
36. Schreyer,H.L. and Fedock,J.J., "Orthogonal Base Functions and Consistent Diagonal Mass Matrices for Two Dimensional Elements", Int. J. Num. Meth. Engng., pp1379-1398, 1979
37. Tong.P, Pian,T.H.H. and Bucciarelli,L.L., "Mode Shapes and Frequencies by Finite Element Methods Using Consistent and Lumped Masses", Comp. and Struct., pp623-638, 1971

38. Hinton,E., Rock,T. and Zienkiewicz,O.C., "A Note on Mass Lumping and Related Processes in the Finite Element Method", Earthquake Engng. and Struct. Dynamics, pp245-249, 1976
39. Fried,I. and Malkus,D.S., "Finite Element Mass Matrix Lumping by Numerical Integration with no Convergence Rate Loss", Int. J. Solids Struct., pp461-466, 1975
40. Jackson,C.P., "Singular Capacity Matrices Produced by Low Order Gaussian Integration in the Finite Element Method", Int. J. Num. Meth. Engng., p871-877, 1981
41. Huebner,K.H., The Finite Element Method for Engineers, J. Wiley, N.Y., 1975
42. Key,S.W., Beisinger,Z.E., and Krieg,R.D., "HONDO-II, A Finite Element Computer Program for the Large Deformation Dynamic Response of Axisymmetric Solids", Sand 78-0422, Sandia Laboratories, Albuquerque, N.M, 1978
43. Vichnevetsky,R. and De Schutter,F., "A Frequency Analysis of Finite Difference and Finite Element Methods for Initial Value Problems", Adv. Comp. Meth. for PDE, (Vichnevetsky,R., ed) AICA, pp46-52, 1975
44. Bettencourt,J.M., Zienkiewicz,O.C. and Cantin,G., "Consistent Use of Finite Elements in Time and the Performance of Various Recurrence Schemes for the Heat Diffusion Equation", Int. J. Nun. Meth. Engng., pp931-938, 1981
45. Emery,A.F., "The Use of Singularity Programming in Finite Difference and Finite Element Computation of Temperature", ASME J. Heat Trans., pp344-351, 1973
46. Benzley,S.E. and Beisinger,Z.E., "CHILES- A Finite Element Computer Program that Calculates the Intensities of Linear Elastic Singularities", SLA-73-0894, Sandia Laboratories, Albuquerque, N.M., 1973
47. Emery,A.F. and Segedin,C.S., "Singularity Programming- A Numerical Technique for Determining the Effect of Singularities in Finite Difference Solutions- Illustrated by Application to Plane Elastic Problems", Int. J. Num. Meth. Engng., pp367-380, 1972
48. Emery,A.F., Neighbors,P.K., Kobayashi,A.S. and Love,W.J., "Stress Intensity Factors in Edge-Cracked Plates Subjected to Transient Thermal Singularities", ASME J. Press. Vess. Tech., pp100-105, 1977
49. Morretti,G., "Floating Shock Fitting Techniques for Embedded Shocks in Unsteady Multi Dimensional Flows", Proc. 1974 HTFMI, Stanford Univ. Press, pp184-201, 1974
50. Fisher,I. and Medland,I.C., "The Multi Dimensional Stefan Problem: A Finite Element Approach", Finite Element Meth. in Engng., Univ. N.S. Wales, Australia, pp767-783, 1974
51. Comini,G. Del Guidice,S., Lewis,R.W. and Zienkiewicz,O.C., "Finite Element Solution of Non-Linear Heat Conduction Problems with Special Reference to Phase Change", Int. J. Num. Meth. Engng., pp613-624, 1974

52. Labdon,M.B. and Guceri,S.I., "Heat Transfer of Phase Change Materials in Two Dimensional Cylindrical Coordinates", paper AIAA-81-1046, AIAA 16th Thermophysics Conf., Palo Alto, Calif., 1981
53. Ronel,J. and Baliga,B.R., "A Finite Element Method for Unsteady Heat Conduction with or without Phase Change" ASME paper 79-WA/HT-54, ASME Winter Annual Mtg., N.Y., 1979
54. Robinson,J., "Element Evaluation- A Set of Assessment Points and Standard Tests", FEM in the Commercial Environment, (Robinson,J., ed) Robinson and Assoc., England, pp218-241, 1978
55. Haggenmacher,G.W. and Lahey,R.S., "Practical Aspects of the Finite Element Method", FEM in the Commercial Environment, (Robinson,R., ed). Robinson and Assoc., England, pp70-102,1978
56. Emery.A.F., Workshop on FE and FD Methods, Sandia Laboratories, Livermore, Calif., 1979
57. Sen Gupta,S.K. and Akin,J.E., "A Numerical Study of Coefficient Modelling", Adv. Comp. Meth. for PDE, (Vichnevetsky,R., ed) AICA, pp285-291, 1975
58. Mitchell,A.R. and Griffiths,D.F., The Finite Difference Method in Partial Differential Equations, J. Wiley, N.Y., 1980
59. Smith,G.D., Numerical Solution of Partial Differential Equations: Finite Difference Method, Oxford Univ. Press, Oxford, 1978
60. Strang,G., "Variational Crimes in the Finite Element Method", Math. Found. of the FEM with Appl. to PDE, (Aziz,A.K., ed) Academic Press, N.Y., pp689-710, 1972
61. Mitchell,A.R. and Wait,R., The Finite Element Method in Partial Differential Equations, J. Wiley, N.Y., 1977
62. Shimoji,S., "A Comparison of Thermal Network Correction Methods", paper AIAA-81-1139, AIAA 16th Thermophysics Conf., Palo Alto, Calif., 1981
63. Thornton,E.A. and Wieting,A.R., "Integrated Transient Thermal-Structural Finite Element Analysis" AIAA/ASEM/ASCE/AHS 22nd Struct. Structural Dynamics and Materials Conf., Atlanta, Ga., 1981
64. Houstis,E.N., Lynch,R.E., Papatheodorou,T.S. and Rice,J.R., "Development Evaluation and Selection of Methods for Elliptical Partial Differential Equations", Adv. Comp. Meth. PDE., (Vichnevetsky,R., ed) AICA, pp1-6, 1975
65. Herbst,B.M., "Collocation Methods and the Solution of Conduction-Convection Problems", Int. J. Num. Meth. Engng., pp1093-1101, 1981
66. Sorensen,M., "A Case Study Based on TOPAS",FEM in the Commercial Environment, (Robinson,J.,ed) Robinson and Assoc, England, pp1-5, 1978

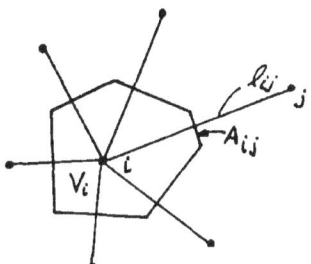

Figure 1.- Finite difference mesh for the heat balance method.

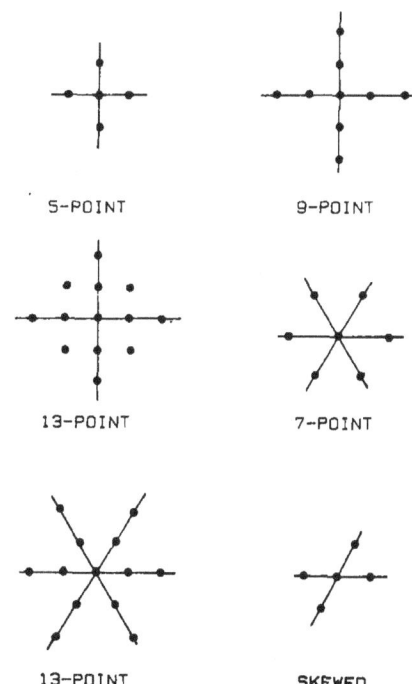

Figure 2.- Finite difference nodal point arrangements.

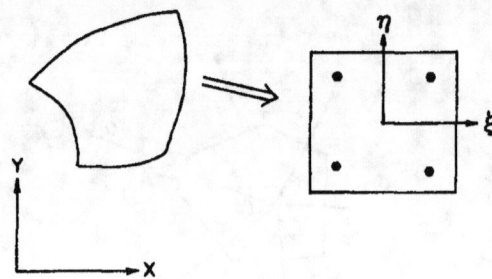

Figure 3.- Mapping an element on a unit square (showing the Gaussian integration points).

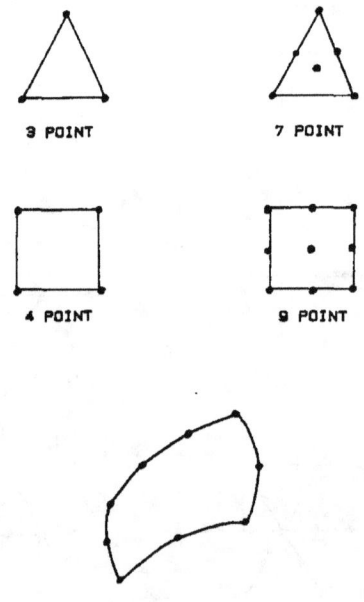

Figure 4.- Common two dimensional finite elements.

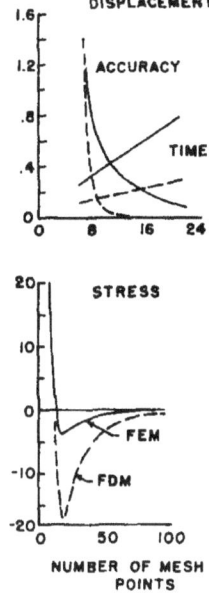

Figure 5.- Error (%) and computer time (sec) for a clamped shell using an FEM and an FDM [12].

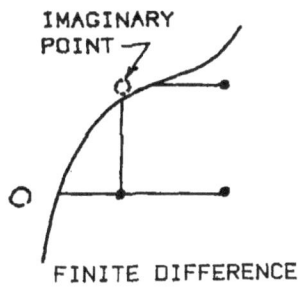

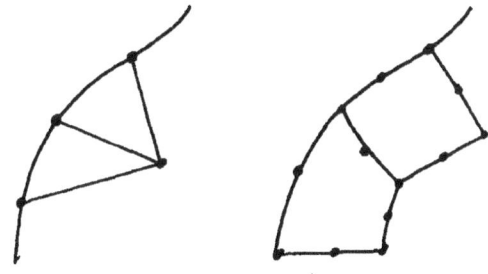

Figure 6.- Treatment of irregular boundaries.

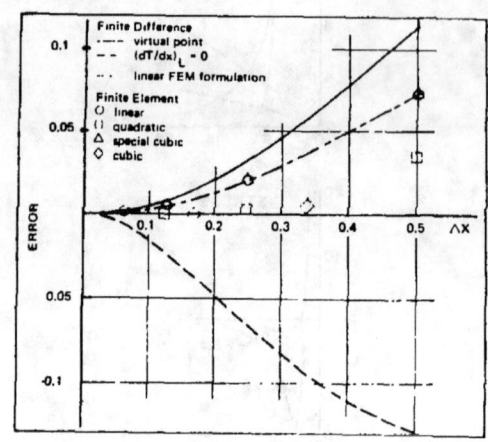

(T(exact) = 0.1429)

Figure 7.- Temperature error at the insulated surface for a source strength = X^m.

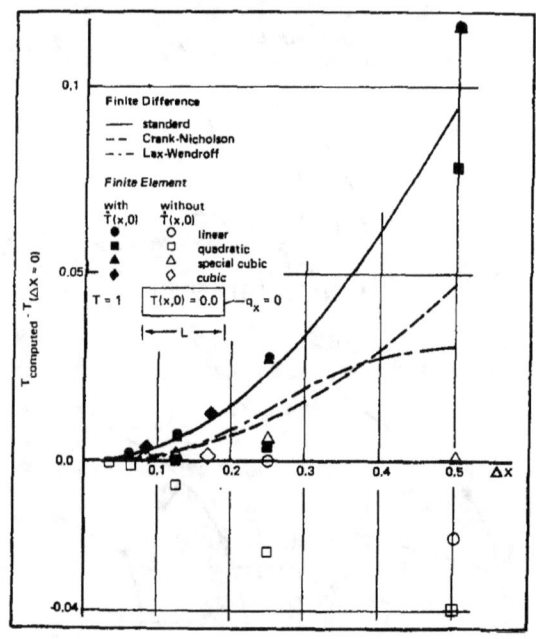

(T(exact) = 0.1573)

Figure 8.- Temperature error for $\alpha T/L^2 = 0.0625$ and $X/L = 0.5$.

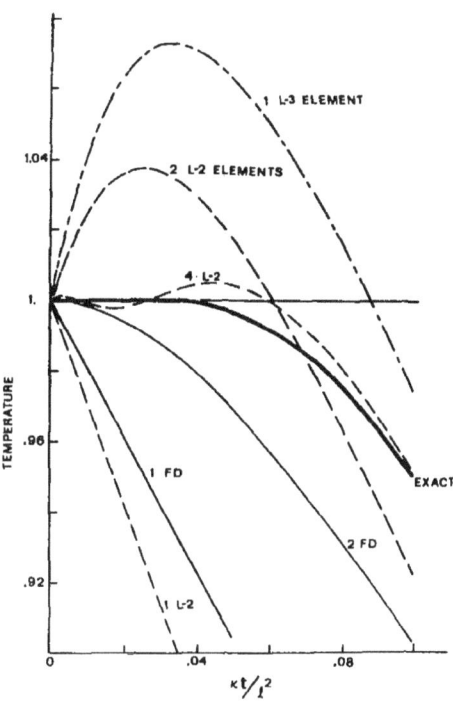

Figure 9.- Temperature at the insulated surface using an FDM and an FEM with 2 node linear (L-2) and 3 node linear (L-3) elements.

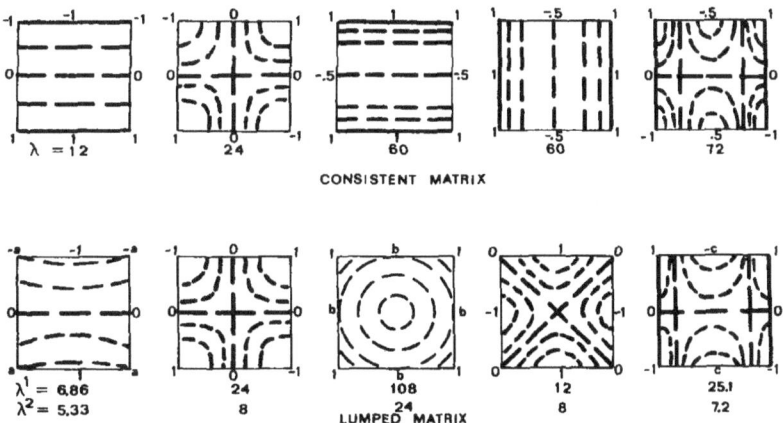

Figure 10.- Eigenvectors for the eight node element showing the effect of lumping.

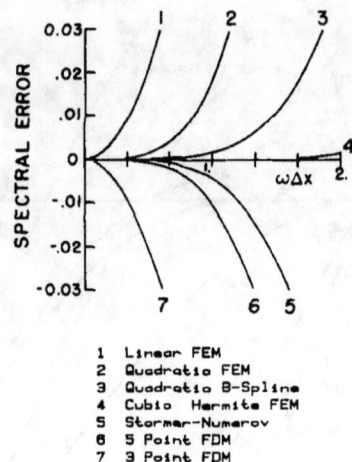

Figure 11.- Error in the numerical diffusivity for different methods [43].

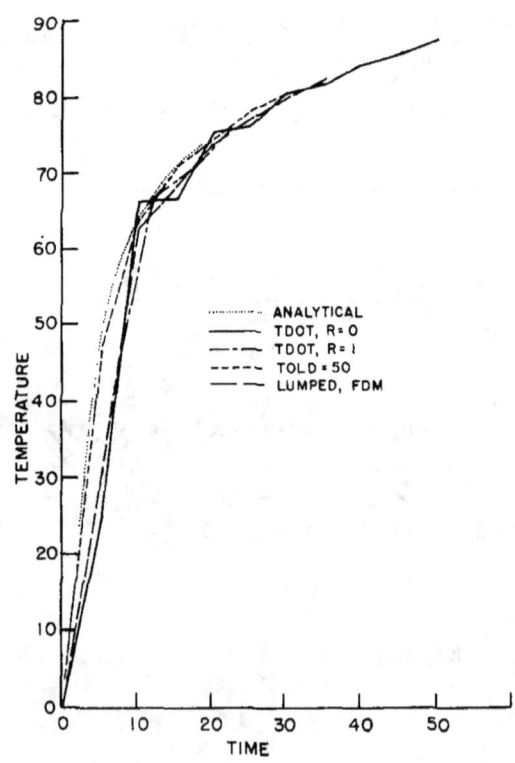

Figure 12.- Transient temperature response for a slab using $\dot{T}(X,0) = 0$ with ramps of 0 and 1 Δt and a consistent method using a modified initial surface temperature (TOLD = 50).

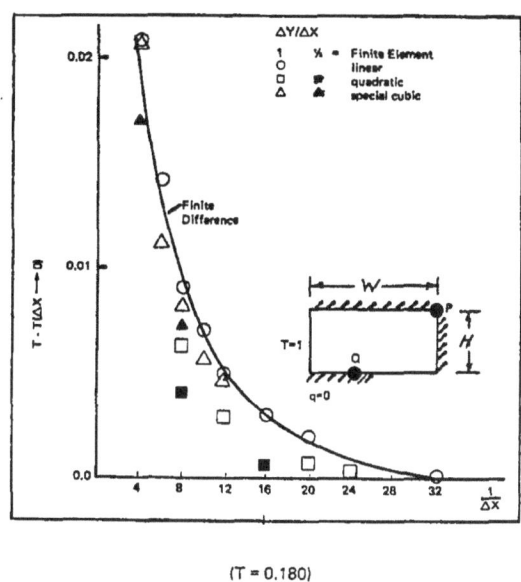

(T = 0.180)

Figure 13.- Temperature at point P for the singular problem.

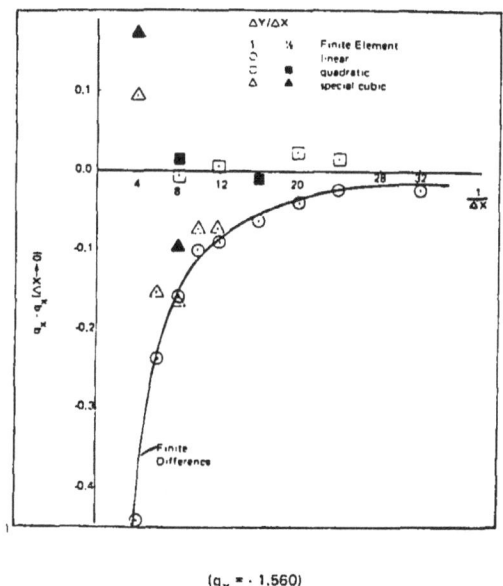

($q_x = -1.560$)

Figure 14.- Heat flux at point Q for the singular problem.

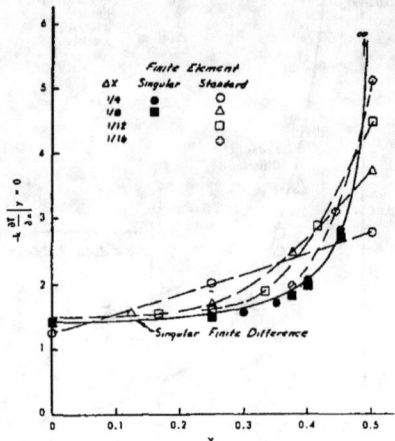

Figure 15.- Values of the heat flux along the insulating splitter plate.

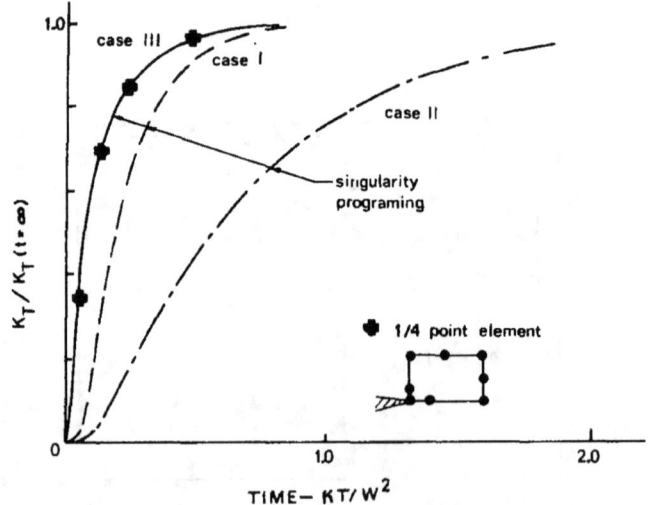

Figure 16.- Calculation of the time dependent singularity strength for the problem of figure 13.

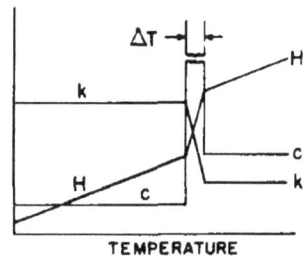

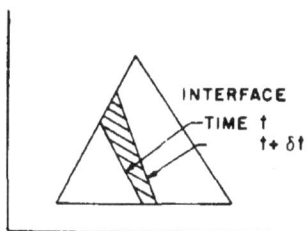

Figure 17.- Enthalpy and interface models for phase change calculations.

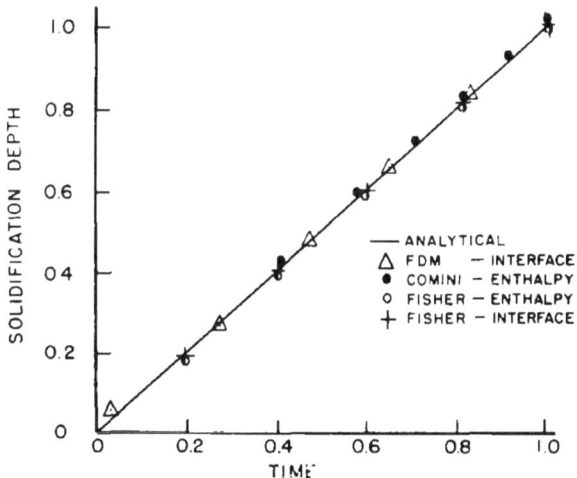

Figure 18.- Solidification front position for a plane problem.

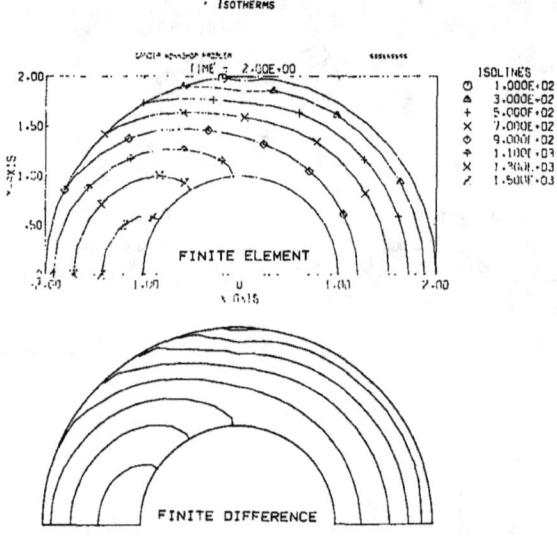

Figure 19.- Comparison of FDM and FEM results for a mixed boundary condition problem.

A METHOD TO MODEL LATENT HEAT FOR TRANSIENT ANALYSIS USING NASTRAN

Robert L. Harder
The MacNeal-Schwendler Corporation
Los Angeles, California

SUMMARY

A sample heat transfer analysis is shown which includes the heat of fusion. The method can be used to analyze a system with nonconstant specific heat. The enthalpy is introduced as an independent degree of freedom at each node. The user input consists of a curve of temperature as a function of enthalpy, which may include a constant temperature phase change. The basic NASTRAN heat transfer capability is used to model the effects of latent heat with existing direct matrix output (DMI) and nonlinear load (NØLIN) data cards. Although some user care is required, the numerical stability of the integration is quite good when the given recommendations are followed. The theoretical equations used and the NASTRAN techniques are shown in the paper.

INTRODUCTION

The problem of heat transfer with latent heat or nonconstant specific heat is one of current interest. Methods based upon introduction of enthalpy[1] and upon a very high specific heat in a small temperature range[2] have been published. These methods can be implemented in either finite difference or finite element formulations. The numerical stability of the transient integration must be analyzed, since failure in this area makes the method very expensive, due to excessive time steps required.

SYMBOLS

Values are given in SI Units.

A	area, m^2
B	capacity matrix, $J/°C$
c	specific heat, $J/kg°C$
h	enthalpy, J/kg
K	conduction matrix, $J/s°C$
L	latent heat, $h_\ell - h_s$, J/kg
M	mass, kg
N	nonlinear function
N_e	number of finite elements
P	thermal load, J/s
T	temperature, $°C$
t	time, s
x	coordinate, m
β	Newmark beta parameter
γ	stability parameter
κ	thermal conductivity, $J/s\ m\ °C$
λ	eigenvalue, growth factor
ρ	mass density, k/m^3

Subscripts:

ℓ	liquidus
s	solidus
w	wall
n	time step

MATHEMATICAL ANALYSIS

The derivation will be made for one-dimensional heat conduction, however the method is general. The temperature distribution is found from solution of the diffusion equation

$$\kappa \frac{\partial^2 T}{\partial x^2} = \rho c \frac{\partial T}{\partial t} \qquad (1)$$

When this equation is analyzed by finite element techniques, it is written

$$[B]\{\dot{T}\} + [K]\{T\} = \{P\} \qquad (2)$$

where the terms of the B (capacitances) matrix are $\rho c A \Delta x$ and the terms of the K (finite element conductivities) matrix are $\Delta \kappa / \Delta x$.

In the general case the conductivity and the heat capacity are not constant. In the case of phase change at constant temperature (see fig. 1) the method fails since the specific heat is effectively infinite. Introduction of the enthalpy gives two simultaneous equations

$$[M]\{\dot{h}\} + [K]\{T\} = \{P\} \qquad (3)$$

$$T = N(h) \qquad (4)$$

A Newmark beta numerical integration scheme is used, where the velocity terms are replaced by

$$\dot{h} = (h_{n+1} - h_n)/\Delta t \qquad (5)$$

and the constant terms by

$$T = \beta T_{n+1} + (1-\beta) T_n \qquad (6)$$

This parameter β is a stability parameter. If $\beta = 0$, the method is called foreward differencing, and $\beta = 1$ corresponds to backward differencing. For β greater than 0.5, integration of equation (2) is numerically stable for any mesh size and time step. Integration of (3) and (4) shows that there is a tendency to instability at large time steps.

STABILITY ANALYSIS

The basic method of stabilizing the equations is to evaluate the terms at the advanced time (t_{n+1}) rather than the time (t_n). We replace equation (4) by

$$T - h/c_{min} = N(h) - h/c_{min} \qquad (7)$$

where c_{min} is the minimum specific heat. The terms on the left side of (7) are evaluated as shown in (6), while the terms on the right are evaluated at t_n. An entirely equivalent method to (7) is

$$T - \left(\gamma \Delta t/c_{min}\right) \dot{h} = N(h) \qquad (8)$$

If (8) is derived from (7), the parameter γ would be β; however, we shall treat γ as an independent stability parameter.

Stability analysis requires a long derivation and will not be done here. The two basic steps are linearization and modal analysis. The equations are linearized by replacing the nonlinear curve by a linear fit. The mode analysis replaces a multi-degree-of-freedom problem with a series of two-degree-of-freedom problems. The short wave length modes are most unstable. Introduce a growth factor

$$T_{n+1} = \lambda\, T_n \qquad (9)$$

$$h_{n+1} = \lambda\, h_n \qquad (10)$$

There are two roots to the characteristic equation

$$\lambda = \begin{matrix} 1 - \beta^{-1} \quad \text{and} \\ 1 - \left(\dfrac{\rho c L^2}{4 N_e^2 \kappa \Delta t} + \gamma \dfrac{c}{c_{min}}\right)^{-1} \end{matrix} \qquad (11)$$

Since $-1 < \lambda < +1$ for stability,

$$\beta > 1/2 \qquad (12)$$

and

$$\frac{\rho c\ L^2}{4N_e^2 \kappa \Delta t} + \gamma \frac{c}{c_{min}} > 1/2 \qquad (13)$$

Thus $\beta = \gamma = 0.55$ gives good stability for any Δt. For $\Delta t < \rho c L^2/2N_e^2\kappa$, the solution is stable without the need of γ, but this Δt is usually much smaller than needed for accuracy.

NASTRAN RESULTS

The analytic solution given in reference (1) was chosen, since an exact solution allows analysis of accuracy. This is a one-dimensional model. The initial condition is ice at freezing temperature. Starting at time zero, the end is heated to a constant temperature. The end temperature is chosen so the enthalpy change after melting has the same value as the latent heat. The melting proceeds until such time that the water-ice interface has moved 1.24 meters, and then the temperature profile is examined. The model is 1.3 meters long and has either 26 or 130 elements. This arrangement was chosen to allow comparison of results (see Table I).

CONCLUSIONS

The NASTRAN results are of good accuracy. By using the stability factor, the accuracy has not been degraded. These results can easily be adapted to other geometries with the finite element method. No changes were required in NASTRAN to solve problems with latent heat.

REFERENCES

1. Wood, A.S., Ritchie, S.I.M., and Bell, G.E.; "An Efficient Implementation of the Enthalpy Method"; International Journal for Numerical Methods in Engineering, Vol. 17, pp. 301-305; 1981.

2. Morgan, K., Lewis, R.W., and Zienkiewicz, O.C.; "An Improved Algorithm for Heat Conduction Problems with Phase Change"; International Journal for Numerical Methods in Engineering, Vol. 12, pp. 1191-1195; 1978.

TABLE I. COMPARISON OF METHODS

Dimensionless Temperature at the Final Time

$(T - T_s) / (T_w - T_s)$

All results are for $\beta = \gamma = 0.55$

$\Delta x = 0.05$ m (26 elements)

Method Number of Steps		Exact	NASTRAN 800	NASTRAN 200	Ref 1 800	Ref 1 200
Distance	0.2	0.8184	0.8184	0.8181	0.8207	0.8242
from hot	0.6	.4695	.4694	.4686	.4758	.4871
end, m.	1.2	.0252	.0321	.0230	.0335	.0612

$\Delta x = 0.01$ m (130 elements)

Method Number of Steps		NASTRAN 1600	NASTRAN 800	NASTRAN 200	Ref 1 20000	Ref 1 5000
Distance	0.2	0.8184	0.8184	0.8181	0.8189	0.8195
from hot	0.6	.4694	.4693	.4686	.4707	.4726
end, m.	1.2	.0256	.0263	.0273	.0262	.0316

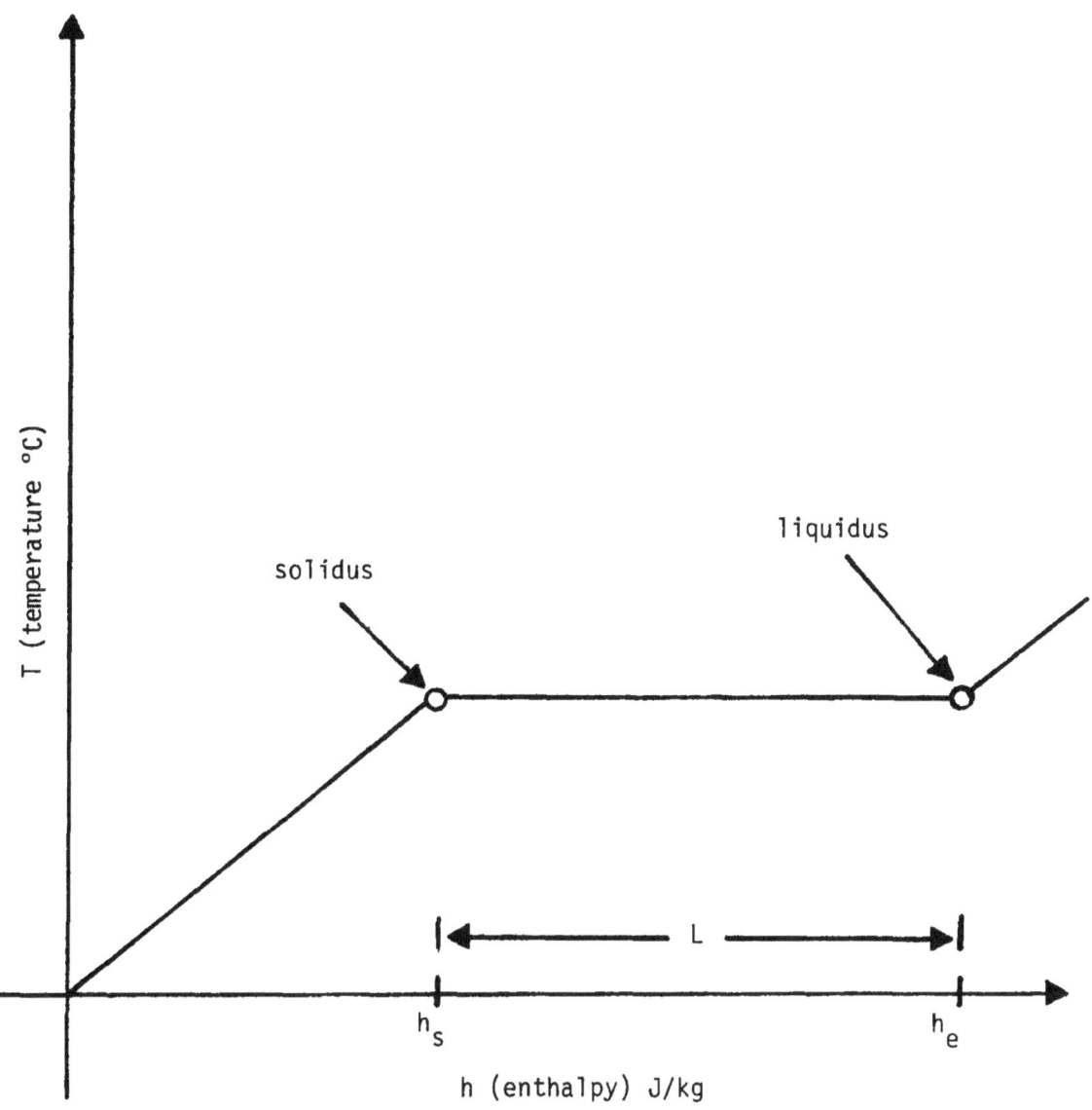

Figure 1.- Latent heat; typical curve of temperature as a function of enthalpy, showing change of phase. The specific heat, $c = dh/dT$, becomes infinite.

Howard M. Adelman
Research Engineer
NASA Langley Research Center
Hampton, Va.
Mem. ASME

Raphael T. Haftka
Professor
Department of Aerospace
and Ocean Engineering
Virginia Polytechnic Institute
and State University
Blacksburg, Va.

James C. Robinson
Research Engineer
NASA Langley Research Center
Hampton, Va.

SOME ASPECTS OF ALGORITHM PERFORMANCE AND MODELING
IN TRANSIENT THERMAL ANALYSIS OF STRUCTURES

The status of an effort to increase the efficiency of calculating transient temperature fields in complex aerospace vehicle structures is described. The advantages and disadvantages of explicit and implicit algorithms are discussed. A promising set of implicit algorithms with variable time steps, known as the GEAR package is described. Four test problems, used for evaluating and comparing various algorithms, have been selected and finite-element models of the configurations are described. These problems include a Space Shuttle frame component, an insulated cylinder, a metallic panel for a thermal protection system, and a model of the Space Shuttle Orbiter wing. Results generally indicate a preference for implicit over explicit algorithms for solution of transient structural heat transfer problems when the governing equations are "stiff" (typical of many practical problems such as insulated metal structures).

NOMENCLATURE

C   capacitance matrix

DT   time step size

h_n n-th time step

K   conductivity matrix

Q   thermal load vector

R   residual of the system of equations generated by the implicit method

t   time

t_n n-th time point

T   vector of temperatures

α   thermal diffusivity or coefficient in Adams-Moulton formula

β   coefficient in backward difference method

INTRODUCTION

An effort is in progress at the NASA Langley Research Center to improve capability to predict and optimize the thermal-structural behavior of aerospace vehicle structures. The focus of this activity is on space transportation vehicles such as the Space Shuttle Orbiter. A principal task is to reduce the computing effort for obtaining transient temperature fields. Current activity is focused on evaluation and comparison of explicit and implicit solution algorithms.

In reviewing current literature, a preference is evident among researchers for implicit algorithms for solution of stiff[1] sets of ordinary differential equations (2-7). Many engineering analysts, however, prefer to use the longer-established explicit algorithms. A partial explanation for this dichotomy is that the full power of the implicit approach has not been transferred from researchers to engineering analysts. In the explicit algorithms, the time step is limited (often severely) in order for the technique to be stable. In the implicit algorithms, there is no stability-imposed limitation on step size. The step size is limited by solution accuracy only, so that implicit algorithms can, in general, use much larger time steps than explicit algorithms. Because a single explicit time step is computationally faster than a single implicit time step, the key to the advantageous use of implicit algorithms is to use the largest possible time step size. As presently implemented in production thermal analysis computer programs, implicit algorithms generally require a user-specified fixed time step (8-11). The step size must be determined by trial and error.

The strategy being advocated in the solution of large problems by implicit methods is to use algorithms with variable step size and order and to automatically select both throughout the solution process (12-15). A promising set of algorithms, developed for the purpose of implementing the aforementioned

---

[1] Stiff sets or ordinary differential equations are characterized by solutions with widely varying time-constants. The typical case is when the solution to the homogeneous problem has very small time constants compared to those of the forcing function (1).

strategy, is denoted the GEAR algorithms (13-14). A version of the GEAR algorithms well-suited to heat transfer analysis denoted GEARIB has been recently installed in the SPAR finite-element thermal analyzer (8) for testing.

The purpose of the present paper is to describe some ongoing evaluations and demonstrations of the use of explicit and implicit algorithms for transient thermal analysis of structures using the finite-element method. A Shuttle frame test article, an insulated cylinder, a metallic multiwall thermal protection system panel, and a model of the Shuttle Orbiter wing are analyzed using SPAR. Comparisons between implicit and explicit algorithms are presented. The performance of the GEARIB algorithms and especially the value of variable step size and order is demonstrated. For benchmark checks, the cylinder is also analyzed with the MITAS lumped parameter program (16). It is a characteristic of thermal analysis by finite-element and lumped-parameter techniques that modeling affects the stiffness. Since stiffness is one of the key factors in the performance of implicit and explicit algorithms, the paper contains a study of the effects of modeling on the performance of the explicit and implicit algorithms. The present work focuses on the implicit and explicit algorithms implemented in production programs such as SPAR and MITAS. The authors do not evaluate but are aware of and hereby recognize recent developments underway which are still at the research stage. These include, for example, the mixed implicit-explicit techniques (17) and the use of quasi-Newton methods to solve the nonlinear algebraic equations associated with implicit algorithms (18).

NATURE OF ALGORITHMS USED IN TRANSIENT THERMAL ANALYSIS

A transient heat transfer problem when discretized by finite-element, finite-difference, or similar techniques, is governed by the following system of equations

$$C\dot{T} = Q(T,t) - K(T,t)T = F(T,t) \quad T(0) \text{ given} \quad (1)$$

where F is generally a nonlinear function. It is usually impractical to obtain an analytical solution to eq. (1) so that numerical integration methods are used. The simplest numerical integration technique is the Euler method which uses the first two terms in a Taylor series to predict T at time t_{n+1} as

$$T(t_{n+1}) = T(t_n) + h_n \dot{T}(t_n)$$
$$= T(t_n) + h_n C^{-1} F(T(t_n), t_n) \quad (2)$$

Euler's method is an example of an explicit integration technique, so-named because $T(t_{n+1})$ is given explicitly in terms of known quantities. Another approach is the backward-difference method which is an example of an implicit method. In this approach

$$T(t_{n+1}) = T(t_n) + h_n \dot{T}(t_{n+1})$$
$$= T(t_n) + h_n C^{-1} F(T(t_{n+1}), t_{n+1}) \quad (3)$$

Eq. (3) is a system of implicit equations for $T(t_{n+1})$, which is generally nonlinear. The explicit algorithm is therefore easier to implement but must be bounded to avoid numerical instability (unbounded propagation of numerical errors during the solution). Implicit techniques are generally stable and thus can take larger time steps which are determined from accuracy considerations.

Most practical transient thermal analysis problems in flight structures have the following characteristics which profoundly affect the choice of a solution method:

(1) The thermal response may be divided into regions of slowly and rapidly varying temperatures. Steep transients accompany initial conditions or sudden changes in the heat load.

(2) The rapidity of variation of the transient portion of the temperature history is proportional to the quantity L^2/α where L is a characteristic conduction length and α is thermal diffusivity. During the transient, time steps much smaller than L^2/α must be taken no matter what type of integration technique is used.

During a period of slowly-varying temperature large time steps may be taken by implicit integration techniques but explicit techniques must still use time steps which are less than L^2/α. When L^2/α values for some elements in the structure are small compared to the time scale of the slower temperature variation, the problem is stiff. It follows that stiff problems are usually best solved by implicit methods. The effort involved in solving a system such as eq. (3) is usually cost-effective if a small number of large time steps are used.

The Euler method and the backward-difference method are presented as representatives of a large class of explicit and implicit techniques, respectively. Higher-order methods (i.e., multistep) typically use more previous information to predict the temperature at the current time but the stability properties of explicit multistep methods are similar to those of the Euler method. Most explicit methods are unstable for time steps much larger than L^2/α. Accordingly, thermal analysis computer programs generally select the explicit time step automatically based on the stability requirement. For implicit methods, the analyst is left to select the implicit time step and order without a great deal of guideline information and usually several trial runs are needed. There is an emerging consensus that the approach to take for integrating stiff systems of ordinary differential equations would be to use implicit methods which automatically select the order and the step size based on desired accuracy. One package denoted the GEARIB algorithms has these features and is discussed next.

THE GEARIB ALGORITHMS

Several software packages based on the work of Gear have been developed for general use (13). The package most appropriate for application to finite-element thermal analysis is denoted GEARIB. This package is intended to solve systems of ordinary differential equations of the form

$$C(T,t) \dot{T} = F(T,t) \quad (4)$$

The package employs two classes of implicit multistep methods, Adams-Moulton and backward difference. For nonstiff equations, the Adams-Moulton method of order one through twelve is used. This method has the general form

$$T(t_{n+1}) = T(t_n) + h_n \sum_{i=0}^{q} \beta_i \dot{T}(t_{n+1-i}) \qquad (5)$$

where q is the order. For stiff equations, the backward difference algorithms of orders one through five are used. These algorithms have the general form

$$T(t_{n+1}) = h_n \beta_0 \dot{T}(t_{n+1}) + \sum_{i=1}^{q} \alpha_i T(t_{n+1-i}) \qquad (6)$$

The coefficients α_i and β_i are given in (15). The user selects the class of methods (Adams-Mouton or backward differences), and as described in (13) GEARIB automatically selects the appropriate time step and the order based on a user-specified error tolerance.

Use of the GEARIB algorithms is illustrated using the backward difference option. Applied to eq. (4), eq. (6) gives

$$R = C[T(t_{n+1}) - \sum_{i=1}^{q} \alpha_i T(t_{n+1-i})]$$
$$- h_n \beta_0 F(T(t_{n+1}), t_{n+1}) = 0 \qquad (7)$$

This system of nonlinear algebraic equations is solved by the modified Newton's method. That is

$$T^{i+1}(t_{n+1}) = T^i(t_{n+1}) - [\frac{\partial R}{\partial T}]^{-1} R \qquad (8)$$

where

$$[\frac{\partial R}{\partial T}] = C - \beta_0 h_n J$$

and $J = \partial F/\partial T$ is the Jacobian of the system at a previous time point. Methods used in GEARIB for computing J are described in (13) and (19).

DESCRIPTION OF TEST PROBLEMS AND RESULTS[2]

Insulated Shuttle test frame

A Shuttle Orbiter frame component analyzed and tested under transient heating as described in (20) is shown in figure 1 and consists of an aluminum frame surrounded by insulation. The principal purpose of the study of the configuration as discussed in (20) was to evaluate the thermal performance of the insulation during a simulated Shuttle flight. A secondary purpose was to evaluate the adequacy of thermal analysis techniques applicable to the Shuttle.

---
[2] Additional details of the test problems are given in (19). All calculations were performed on the Langley CDC Cyber 173 computer.

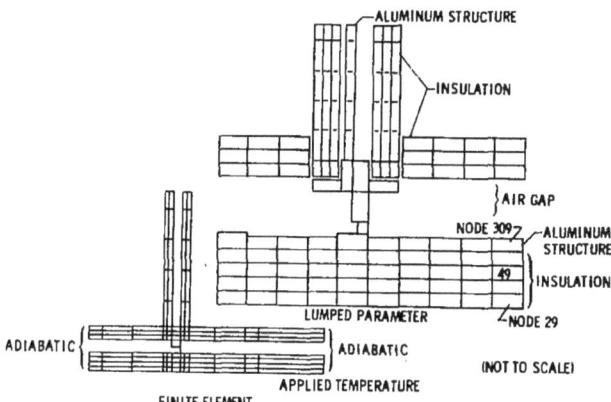

Fig.1 Finite-element and lumped parameter models of Shuttle frame

The lumped parameter model from (20) consists of a two-dimensional section of a symmetric half of the structure and contains 118 nodes (see figure 1). The unknown temperatures are located at the centroids of the lumps. The lumped parameter model was converted to a finite-element model for analysis using the SPAR program (8). The corresponding SPAR finite-element model contains 149 grid points located at the ends or corners of the elements. The model contains 148 elements including one-dimensional elements which account for conduction in the aluminum structure and radiation across the air gap and two-dimensional elements which model conduction in the insulation and across the gap. The difference in numbers of elements and grid points is due to the different modeling approaches of the two methods.

Minor modifications were made to the finite-element model following the conversion. These consisted of eliminating or consolidating some extremely thin or short finite elements in the aluminum structure in order to reduce the stiffness of the equations and to increase the allowable time step for the explicit solution algorithm. The properties of the aluminum structure are functions of temperature and the properties of the insulation are functions of temperature and pressure. The pressure dependence is treated in SPAR as time dependence using the pressure vs. time variation from the trajectory data for the simulated flight conditions. The applied heating is specified by tabulations of temperatures at the outer surface of the insulation.

The temperature history for the frame was computed using explicit (Euler) and implicit techniques (Crank-Nicholson and backward differences) and GEARIB. Comparisons of solution times are given in Table I. The explicit procedure using a time step of 0.16 s required 1723 s of CPU time. This time step was controlled by conduction through most of the aluminum elements along the center and front of the frame. Solution time using the Crank-Nicholson algorithm varied from 475 s to 65 s as the time step was varied between 1.0 and 50 s. The solution times for backward differences were close to those of Crank-Nicholson and are not shown. The GEARIB algorithm used time steps from 50 to 170 s and the solution time was 54 s. As indicated in Table I(b), there is very little loss of accuracy in either the structure or insulation temperatures with increased time step size.

TABLE I.- PERFORMANCE OF VARIOUS ALGORITHMS FOR TRANSIENT THERMAL ANALYSIS OF SHUTTLE FRAME

(a) Solution Time Comparison

Explicit		Implicit			
Euler		Crank-Nicholson		GEARIB	
Time Step (s)	Solution Time (s)	Time Step (s)	Solution Time (s)	Time Step (s)	Solution Time (s)
0.16	1723	1	475	50-170	54
		10	249		
		25	106		
		50	65		

(b) Effect of Time Step on Accuracy of Implicit Algorithms

Step Size (s)	Temp. of Node 309** at 1200 s		Temp. of Node 49** at 1200 s	
	K	°F	K	°F
1.0	442.1	335.7	477.0	398.6
10.0	442.0	335.6	476.9	398.5
25.0	439.4	331.6	475.6	396.0
50.0	437.9	328.3	474.8	394.7
0.16*	442.1	335.7	477.0	398.6
50-170***	443.1	337.5	477.9	400.3

*Explicit Algorithm
**See figure 1
***GEARIB

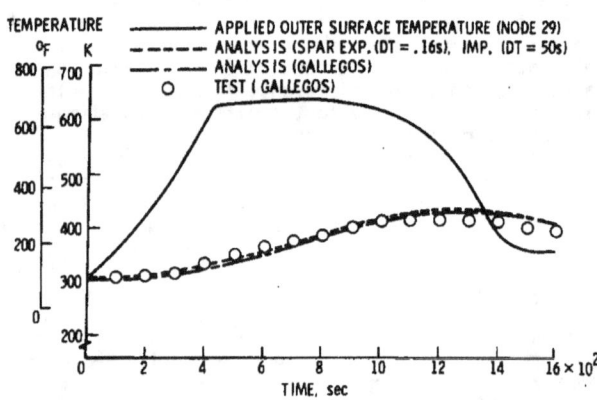

Fig.2 Temperature history in outer structural surface of Shuttle frame (Node 309)

The accuracy of the solutions by the various techniques is further assessed in figure 2 which displays temperature histories at a point in the outer layer of the aluminum structure corresponding to node 309 (see figure 1). The solid line in figure 2 represents the applied temperatures at the outer surface of the insulation (node 29). The dotted line shows temperatures obtained by the SPAR analysis. The SPAR temperatures are plotted as a single curve since there is little difference between the results. The dashed-dot line shows analytical results from the lumped parameter analysis of (20) which are also in close agreement with the SPAR temperatures. The circular symbols represent test data from (20). The closeness of all the results indicates that the models are adequate to simulate the temperature history in the test article.

Multiwall thermal protection system panel

The next example problem is one which grew out of a study of the thermal performance of a titanium multiwall thermal protection system (TPS) panel which is under study for future use on space transportation systems (21). The configuration as depicted in figure 3(a) consists of alternating layers of flat and dimpled sheets fused at the crests to form a sandwich. The representation of a typical dimpled sheet is shown in figure 3(b). For the purpose of this analysis, it is assumed that the heat load does not vary in directions parallel to the plane of the panel. This assumption in addition to the regular geometry of the structure leads to the modeling simplification wherein only a triangular prismatic section of the panel needs to be modeled; fig. 3(a). The intersection of this prism with a typical dimpled layer is indicated by the shaded triangle in fig. 3(b).

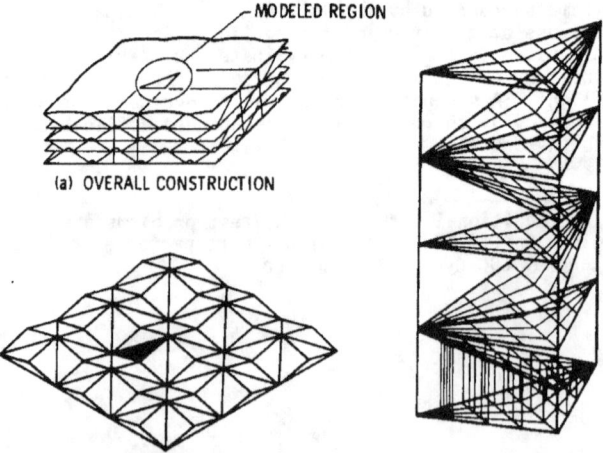

Fig.3 Multiwall thermal protection system panel

The finite-element model shown in fig. 3(c) contains 333 grid points located on nine titanium sheets (five horizontal and four inclined). The model contains 288 triangular and quadrilateral metal conduction elements, 264 solid air conduction elements which account for gas conduction between the layers and 544 triangular and quadrilateral radiation elements which account for radiation heat transfer between adjacent horizontal and inclined sheets. Thermal properties of titanium and air are functions of temperature. Radiation exchange (view) factors were computed and supplied to SPAR using the TRASYS II computer program (22).

The temperature history of the panel in response to an imposed transient temperature at the outer surface of the panel was computed for 3200 s. Results were obtained with SPAR using explicit, Crank-Nicholson, backward difference and GEARIB algorithms. Solution-time comparisons are presented in Table II. The explicit algorithm required a time step of .007 s. This time step was dictated by conduction of heat through the short heat paths between the vertices of adjacent triangular layers and indicates that this is an extremely stiff problem. Required solution time for the explicit algorithm was estimated to be 98368 s.

TABLE II.- COMPARISON OF ALGORITHMS FOR TRANSIENT THERMAL ANALYSIS OF TITANIUM MULTIWALL TPS
(3200 s temperature history)

Explicit		Implicit			
Euler		Crank-Nicholson		GEARIB	
Time Step (s)	Solution Time (s)	Time Step (s)	Solution Time (s)	Time Step (s)	Solution Time (s)
.007	98368*	1	28412**	1.0-113	2754
		5	6352		

*Extrapolated value based on 12296 s for 400 s of temperature history
**Extrapolated value based on 8879 s for 1000 s of temperature history

The Crank-Nicholson solution was carried out using time steps of 1 and 5 s which led to solution times of 28412 s and 6352 s, respectively. Backward difference was used with the same time steps and had the same solution times. GEARIB took time steps ranging between 1.0 and 113 seconds and required a solution time of 2754 seconds. This example shows again advantages of using implicit algorithms in general and the GEARIB algorithms in particular for thermal analysis of stiff problems. A plot of typical temperature histories for a point midway through the panel and the primary structure is shown in figure 4 along with the applied outer surface temperature. The results were obtained by the implicit algorithm with a time step of 5 s and are identical to results using a time step of 1 s and GEARIB.

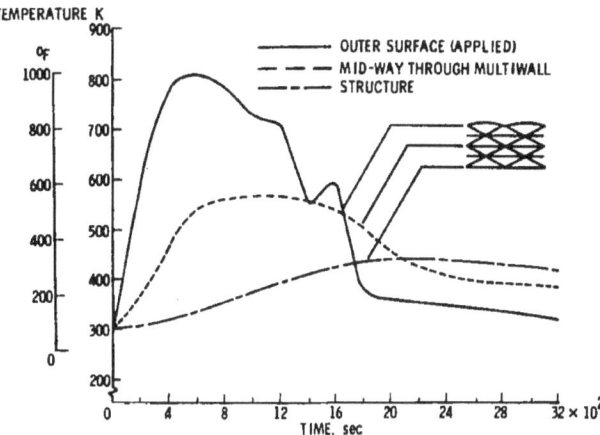

Fig.4 Transient temperatures in titanium multiwall TPS panel

Space Shuttle orbiter wing

The SPAR thermal model of the Shuttle orbiter wing (figure 5) consists of a relatively coarse model of the structure (327 grid points) augmented by layers of insulation attached to the upper and lower surfaces. The structure is modeled by rod, triangular and quadrilateral elements (K21, K31, K41 SPAR elements). The insulation on each surface is modeled by six layers of one-dimensional conduction elements (K21). Use of these elements neglects lateral heat transfer in the insulation--a reasonable assumption since the temperature gradients through the insulation are at least an order of magnitude greater than the lateral temperature gradients. The complete model contains 2289 grid points, 1400 one- and two-dimensional elements in the structure and 1962 one-dimensional elements in the insulation. Thermal properties of the aluminum structure are temperature-dependent; thermal properties of the insulation are temperature- and time-dependent.

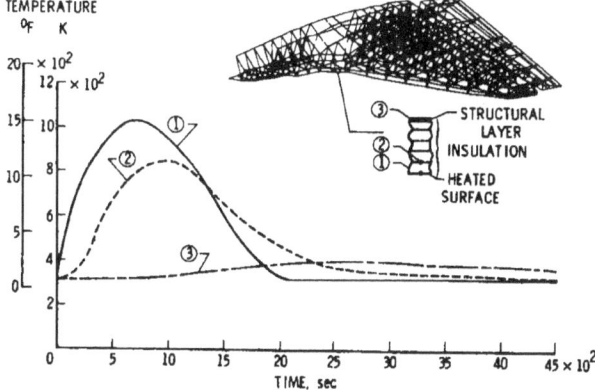

Fig.5 Transient temperatures in Shuttle orbiter wing

TABLE III.- COMPARISON OF ALGORITHMS FOR TRANSIENT
THERMAL ANALYSIS OF SPACE SHUTTLE ORBITER WING
(4500 s temperature history)

Explicit		Implicit			
Euler		Crank-Nicholson		GEARIB	
Time Step (s)	Solution Time (s)	Time Step (s)	Solution Time (s)	Time Step (s)	Solution Time (s)
10	2288	10	11730	1-528	557

For the purpose of this analysis, the applied heating on the wing is represented by a time-dependent temperature applied to the external surface of the insulation on the under side of the wing. The shape of this curve shown as the solid line in figure 5 is roughly indicative of atmospheric reentry heating. The temperature history of the wing for 4500 seconds was computed using the explicit, Crank-Nicholson, backward difference and GEARIB algorithms. Solution time comparisons are shown in Table III along with the time steps used to obtain comparable accuracy. The explicit algorithm used a time step of 10 seconds--in fact stability requirements actually permitted a time step of over 100 seconds but the step size was dictated by accuracy and the need to periodically update temperature-dependent material properties and not by stability requirements. The large permitted time step is due to the coarse modeling of the structure which did not include the thin, high-conducting or radiating elements present in the previous models. The implicit algorithms (Crank-Nicholson and backward difference produced the same results) were used with a time step of 10 s and required about five times as much computer time as the explicit algorithm. The GEARIB algorithms performed very well for this problem. By adaptively varying the time step from 1.0 second early in the temperature history to as large as 528 seconds toward the end, GEARIB required only 557 seconds to complete the solution. Figure 5 shows the temperature histories of a point on the structure and a point in the insulation 1/5 of the distance through the insulation of a typical cross section through the wing. The explicit, implicit, and GEARIB algorithms produced essentially the same results.

EFFECT OF MODELING ON ALGORITHM PERFORMANCE

This section of the paper describes a study of how modeling details can affect the performance of transient solution algorithms--especially explicit algorithms. Also, the influence of alternate ways of including the nonlinear effects of temperature-dependent material properties is studied. The structure chosen for the study is an insulated cylindrical shell shown in figure 6. The cylinder is 18 m (720 in.) in length and 4.5 m (180 in.) in diameter. The aluminum is 0.25 cm (0.1 in.) thick and the insulation is 5.0 cm (2.0 in.) thick. The outer surface of the insulation is heated over a region which consists of one-third the length and half the circumference.

Three finite-element models are used in the study. Due to symmetry, only half the cylinder is modeled in each case. In model I, solid (K81) elements are used exclusively--39 along the cylinder length, 4 around the circumference, and 3 through the depth (2 elements in the insulation and 1 in the structure). The outer surface has quadrilateral elements (K41) which receive the heat load and radiation elements (R41) which radiate to space. Model I contains 800 grid points and 650 elements. In model II, the solid elements in the structural layer are replaced by quadrilateral elements (K41) in which temperatures do not vary through the thickness. This is generally a good assumption for thin metal structures. Model II has an extra layer of solid elements in the insulation in order to preserve the number of grid points in the model at 800. In model III, the insulation is modeled with one-dimensional conductors (K21). This model neglects lateral heat conduction but as mentioned previously in connection with the Shuttle wing model, this effect is small for the class of insulated flight structures of interest in the present work.

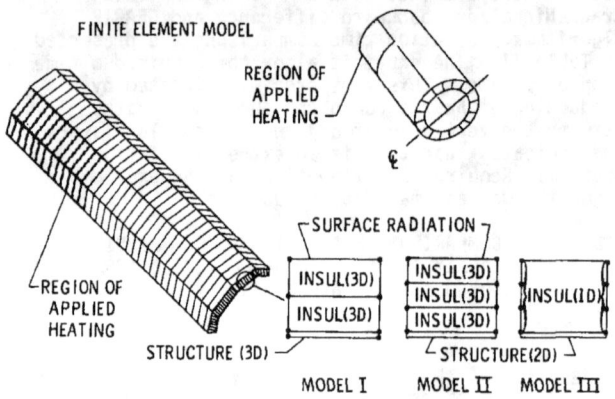

Fig.6 Finite-element models of insulated cylinder

Another aspect of the effect of modeling is comparison of results from finite-element and lumped-parameter models. To investigate this, the MITAS lumped parameter computer program (16) was applied to the analysis of the cylinder. The finite-element model I was converted to a lumped-parameter model by use of the CINGEN program (23). The resulting lumped-parameter model contained 625 nodes as compared to 800 grid points in the finite-element model. Recall the unknown MITAS temperatures are located only at the centroids of each lump.

TABLE IV.- EFFECT OF MODELING ON SOLUTION TIMES FOR
INSULATED CYLINDER PROBLEM

Problem and Model Algorithm	SPAR (Ref. 8)			MITAS (Ref. 16)
	Model I	Model II	Model III	lumped parameter model
Explicit (time step)	10107 (.06)	1518 (2.4-10)	279 (3.3-10)	226 (10)
Implicit* (DT=10 s)	1880	1920	536	320
GEARIB (time step)	1779 (1.0-83)	1707 (5-106)	266 (2-133)	

*Backward differences and Crank-Nicholson

The first 2000 seconds of the temperature history in the cylinder in response to a time-dependent heat load were computed in each model. The explicit

(Euler) and implicit (backward difference) algorithms were used for all models and in addition GEARIB was used for the three SPAR models. Solution times are shown in Table IV. Model I is extremely stiff as evidenced by the small time step of 0.06 seconds required for stability of the explicit algorithm. The high stiffness is due to the use of K81 elements to model the metal layer. In model II, the stiffness has been essentially eliminated by replacing the 3-D elements modeling the metal by 2-D elements. In this model, the explicit technique is faster than backward difference and GEARIB. In model III, due to low stiffness again, the explicit algorithm is faster than the implicit but GEARIB is slightly faster than the explicit technique. It is observed that in models I and II, GEARIB despite using much larger time steps was only marginally faster than the implicit method. This is due to the different ways of handling the temperature-dependent material properties. In the explicit and implicit methods, the properties are represented as being piecewise constant within time intervals specified by the user (by the input quantity TI) in SPAR. Material properties are evaluated at the beginning of each interval and the conductivity and capacitance matrices are regenerated at those times. Results for models I, II, and III in Table IV were obtained using TI = 20 s. In GEARIB, the material properties vary continuously and the residual R must be evaluated each time an iteration in solving eq. (8) is taken. The residual evaluation is much more costly in computer time than the regeneration of the conductance and capacitance matrices. This extra effort is the price paid for higher accuracy. However, this burden only shows up in problems which utilize solid (K81) elements due to the extreme cost of regenerating the matrices for those elements (note model III does not contain K81 elements). A way to eliminate the burden (for thermally isotropic elements) has been identified and is easily implemented. The method is to generate the matrices only once for unit values of the appropriate property and simply scale the matrices by the property whenever it is updated.

MITAS computation times are shown in the last column of Table IV. Because none of the SPAR models is equivalent to the MITAS model in terms of the number of unknown temperature or nodal connections, no direct comparison of MITAS and SPAR solution times is appropriate. However, some trends evident in Table IV are noted. The MITAS model is not particularly stiff as evidenced by the large time step used in the explicit solution technique. SPAR models II and III which begin to resemble the MITAS model in certain respects are also less stiff and favor explicit algorithms. It is noted that the way MITAS treats temperature-dependent material properties is by the scaling method cited above.

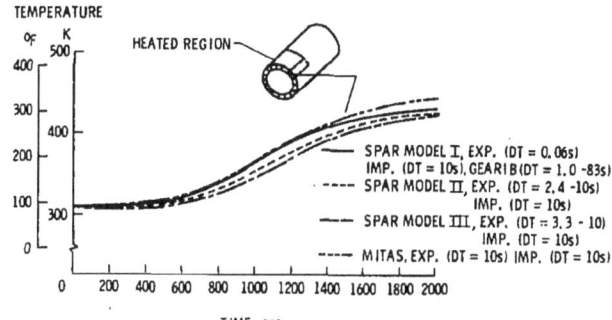

Fig.7 Effects of choice of algorithm and model changes on temperature history of insulated cylinder. Model I: all 3-D elements. Model II: insulation - 3-D, metal - 2-D. Model III: insulation - 1-D, metal - 2-D.

Figure 7 contains comparisons of temperature histories of a point in the cylinder. Model II is considered to be the best of the models (recall the additional insulation elements used) and thus the temperatures represented by the dotted line are thought to be the most accurate. These results are bracketed by results from model I and MITAS (from above) and by model III (from below). There are negligible differences between temperatures from the implicit and explicit solutions for any given model. Results from models II and III are different from that of model I because of the extra layer of elements through the insulation. The MITAS temperature history agrees well with that of model I (on which the MITAS model is based) except for some differences beginning at 1400 s.

CONCLUDING REMARKS

This paper discusses the status of an effort to obtain increased efficiency in calculating transient temperature fields in complex aerospace vehicle structures. Explicit solution techniques which require minimal computation per time step and implicit techniques which permit larger time steps because of better stability are reviewed. A promising set of implicit solution algorithms having variable time steps and order, known as the GEARIB package, is described. Four test problems for evaluating the algorithms have been selected and finite-element models of each one are described. The problems include a Shuttle frame component, an insulated cylinder, a metallic panel for a thermal protection system, and a model of the Space Shuttle Orbiter wing. Calculations were carried out using the SPAR finite-element program and the MITAS-lumped parameter program. Results generally indicate that implicit algorithms are more efficient than explicit algorithms for solution of transient structural heat transfer problems when the governing equations are stiff. Stiff equations are typical of many practical problems such as insulated metal structures and are characterized by widely differing time constants and cause explicit methods to take small time steps. As evidenced by their excellent performance in solving the test problems, the GEARIB algorithms offer high potential for providing increased computational efficiency in the solution of stiff problems. Studies were also made of the effect on algorithm performance of different models of the same cylinder test problem. These studies revealed that the stiffness of the problem is highly sensitive to modeling details and that careful modeling can reduce the stiffness of the resulting equations to the extent that explicit methods may become advantageous.

REFERENCES

1 Willoughby, R. A., *Stiff Differential Systems*, Plenum Press, New York, 1974.

2 Wilson, E. L., and Nickell, R. E., "Application of the Finite Element Method to Heat Conduction Analysis," *J. Nuclear Engineering and Design*, Vol. 4, 1966, pp. 276-286.

3 Wilson, E. L., Bathe, K. J., and Peterson, F. E., "Finite Element Analysis of Linear and Nonlinear Heat Transfer," *J. Nuclear Engineering and Design*, Vol. 29, 1974, pp. 240-253.

4 Hughes, T., "Unconditionally Stable Algorithms for Nonlinear Heat Conduction," *Comp. Meth. Appl. Mech. Eng.*, Vol. 10, 1977, pp. 135-139.

5 Bathe, K. J. and Khoshogoftaar, M. R., "Finite Element Formulation and Solution of Non-Linear Heat Transfer," *J. Nuclear Engineering and Design*, Vol. 51, 1979, pp. 389-401.

6 Hogge, M. A., "Integration Operators for First-Order Linear Matrix Differential Equations," *J. Comp. Meth. in Appl. Mech.*, Vol. 11, 1976, pp. 281-294.

7 Wood, W. L. and Lewis, R. W., "A Comparison of Time Marching Schemes for the Transient Heat Conduction Equation," *Int. J. Num. Meth. in Eng.*, Vol. 9, 1975, pp. 679-689.

8 Marlowe, M. B., Moore, R. A., and Whetstone, W. D., "SPAR Thermal Analysis Processors Reference Manual, System Level 16," NASA CR-159162, 1979.

9 Trent, D. S. and Welty, J. R., "A Summary of Numerical Methods for Solving Transient Heat Conduction Problems," Bulletin No. 49, Oct. 1974, Oregon State Univ., Engineering Experiment Station.

10 Krinke, D. C. and Huston, R. L., "A Critical Evaluation of Computer Subroutines for Solving Stiff Differential Equations," Report No. UC-ES-101578-8, Oct. 15, 1978, Office of Naval Research.

11 Enright, W. H., Hull, T. E., and Lindberg, B., "Comparing Numerical Methods for Stiff Systems of O.D.E.'s," *BIT*, Vol. 15, 1975, pp. 10-48.

12 Shampine, L. F and Gear, C. W., "A User's View of Solving Stiff Ordinary Differential Equations," *SIAM Review*, Vo. 21, No. 1, Jan. 1979, pp. 1-17.

13 Hindmarsh, A. C., "A Collection of Software for Ordinary Differential Equations," Rept. No. UCRL-82091, Jan. 1979, Lawrence Livermore Laboratory.

14 Gear, C. W., *Numerical Initial Value Problems in Ordinary Differential Equations*. Prentice-Hall, Englewood Cliffs, New Jersey, 1971.

15 Byrne, G. D. and Hindmarsh, A. C., "A Polyalgorithm for the Numerical Solution of Ordinary Differential Equations," *ACM Trans. on Num. Software*, Vol. 1, No. 1, March 1975, pp. 71-96.

16 Anonymous, "Martin Interactive Thermal Analysis System, Version 1.0," MDS-SPLPD-71-FD238 (Rev. 3), March 1972, Martin-Marietta Corp., Denver Colorado.

17 Liu, W. K. and Hughes, T., "Implicit-Explicit Finite Elements in Transient Analysis: Implementation with Numerical Examples," *J. Appl. Mech.*, Vol. 45, 1978, pp. 375-378.

18 Bathe, K. J. and Cimento, A. P., "Some Practical Procedures for the Solution of Nonlinear Finite Element Equations," *J. Comp. Meth. in Appl. Mech. and Eng.*, Vol. 22, 1980, pp. 59-85

19 Adelman, H. M. and Haftka, R. T., "On the Performance of Explicit and Implicit Algorithms for Transient Thermal Analysis of Structures," NASA TM 81880, September 1980.

20 Gallegos, J. J., "Thermal Math Model of FRSI Test Article Subjected to Cold Soak and Entry Environments," AIAA Paper 78-1627, 1978.

21 Jackson, L. R. and Dixon, S. C., "A Design Assessment of Multiwall, Metallic Stand-Off, and RSI Reusable Thermal Protection Systems Including Space Shuttle Applications," NASA TM 81780, 1980.

22 Jensen, C. L. and Goble, R. G., "Thermal Radiation Analysis System (TRASYS II) Users Manual," NASA CR-159273-1, 1979.

23 Anonymous, "Creation of Lumped Parameter Thermal Models by the Use of Finite Elements," NASA CR-158944, 1978.

ALGORITHMIC ASPECTS OF TRANSIENT HEAT TRANSFER PROBLEMS IN STRUCTURES

Raphael T. Haftka
Aerospace and Ocean Engineering
Virginia Polytechnic Institute and State University
Blacksburg, Virginia

M. Hassan Kadivar
Illinois Institute of Technology
Chicago, Illinois

ABSTRACT

The application of finite element or finite difference techniques to the solution of transient heat transfer problems in structures often results in a stiff system of ordinary differential equations. Such systems are usually handled most efficiently by implicit integration techniques which require the solution of large and sparse systems of algebraic equations. Most of the computation time required for the solution is spent in assembling and solving these algebraic equations. The present paper is mainly concerned with efficient assembly and solution of these systems using the incomplete Cholesky conjugate gradient algorithm. Several examples are used to demonstrate the advantage of the algorithm over other techniques.

INTRODUCTION

The analysis and design of high speed reentry vehicles such as the space shuttle require the prediction and optimization of the thermal-structural behavior. This means that the analyst needs to solve the heat transfer equation in a structure with complex boundary conditions, irregular geometries and variable thermal properties. The finite element method is one of the more effective approaches available for numerical solution of the transient heat transfer in complex structures. It is therefore expected that finite element systems such as SPAR (Ref. 1) will play a growing role in the analysis and design of such vehicles.

The application of the finite element method to transient heat transfer problems often results in a system of stiff ordinary differential equations (ODE's). Stiff ODE's are characterized by solutions with widely varying time constants. The typical case is when the solution to the homogeneous problem has very small time constants compared to those of the forcing function.

A great deal of effort was devoted in recent years to the development of integration techniques that are suitable for the solution of stiff systems of ODE's. In general, these are variable step size (and, sometimes, variable order) implicit techniques such as the Gear algorithms (Refs. 2,3). The application of such techniques to structural heat transfer problems has been recently shown to be very efficient (Ref. 4) compared to explicit algorithms.

The use of implicit integration techniques requires the repeated solution of large systems of algebraic equations. Because of radiation effects and temperature dependent material properties, these equations are nonlinear. These nonlinear equations are solved typically by the Newton Raphson method or its modified variant which replaces them by systems of linear equations. It is the assembly and solution of these large systems of equations which consumes the bulk of the computation time in the solution of a transient heat transfer problem. This topic is also the focus of this paper.

The solution of systems of linear algebraic equations can be handled by direct methods such as Gaussian elimination or by iterative methods such as successive over-relaxation (SOR). Problems in solid mechanics and structures are usually discretized by a finite element method and the associated systems of linear equations solved by elimination techniques. On the other hand, in fluid mechanics problems, finite difference methods are more common and the associated linear equations are solved by iterative techniques. Two reasons for the preference of structural analysts for elimination techniques are worth noting. The first is the ill conditioning which is typical of the systems of linear equations generated by a structural finite element model. This ill conditioning results in very slow convergence rates of iterative solution methods. The second reason is the typically good band structure of the system matrices which results from the use of one and two dimensional finite element models. This property allows efficient solution by elimination using band or skyline solvers.

In applications to transient heat transfer in structures, the finite element codes such as SPAR (Ref. 1) tend to use elimination techniques while finite difference (or lumped parameter) codes such as MITAS (Ref. 5) lean toward the use of iterative techniques. However, neither approach is entirely satisfactory, the iterative methods because of poor convergence and elimination techniques because of poor performance for wide band systems associated with radiation interconnectivities and three dimensional elements.

A promising new technique which is a cross between elimination techniques and iterative techniques is the incomplete Cholesky conjugate gradient (ICCG) method developed by Meijerink and Van der Vorst (Ref. 6), and extended by Kershaw (Ref. 7) to asymmetric matrices. The method has been successfully applied to finite difference modeled transport problems in plasma physics (Refs. 8-10), to finite difference and finite element modeled boundary value problems (Ref. 11-13), and to finite element modeled groundwater flow problems (Ref. 14).

The present paper is concerned with the implementation of the ICCG method to transient heat transfer problems in structures modeled by finite elements. Because of the repeated need to assemble and solve a similar system of equations. it is possible to reduce the computational effort by preliminary calculations. The ICCG method is compared to a conventional band-matrix elimination technique as well as to iterative techniques. A two dimensional space shuttle frame model and an insulated cylinder are used to demonstrate the efficiency of the method.

ANALYSIS

Numerical Integration Technique

A discretized transient heat transfer problem is governed by the following system of ordinary differential equations (ODE)

$$C(T,t)\dot{T} = Q(T,t) - K(T,t)T \qquad T(0) \text{ given} \qquad (1)$$

where T is the vector of nodal temperatures, C is the capacitance matrix, K is the total conduction matrix (including radiation and convection effects), Q is the thermal load vector and a dot indicates derivatives with respect to time, t. The dependence of the matrices C and K on time and temperature is due to time and temperature dependent material properties and to radiation. However, the cost of recalculating these matrices whenever T or t is changed is exorbitant, especially when three dimensional finite elements are involved. To alleviate this problem, the integration time is divided into time intervals and the material properties assumed to be constant in each time interval. As a result, in each time interval the matrix C is constant and the only variable part of the matrix K is due to radiation effects.

The ODE system (1) is most efficiently solved by a variable order, variable time step algorithm such as employed in the GEAR package (Ref. 3). However, in the present work, a simple fixed-step mid-difference (Crank Nicholson) algorithm is used. It was shown in Ref. 15 that the performance of the algorithm is quite satisfactory for the problems solved here.

Using a numerical integration algorithm, we evaluate the temperature T at a sequence of time points $t_1, t_2, \ldots$. Denoting as T_n the approximate solution for $T(t_n)$, the mid-difference algorithm replaces Eq. (1) by

$$\psi(T_n) = 2C \frac{T_n - T_{n-1}}{h_n} + K(T_n)T_n + K(T_{n-1})T_{n-1}$$

$$- Q(T_n, t_n) - Q(T_{n-1}, t_n) = 0 \qquad (2)$$

We assume that T_{n-1} has already been calculated so that Eq. (2) is a system of nonlinear algebraic equations for T_n. This system of equations is solved using the Newton Raphson method

$$J[T_n^{(m+1)} - T_n^{(m)}] = -\psi(T_n^{(m)}) \tag{3}$$

where $T_n^{(m)}$ is the m-th iterate and J is the Jacobian

$$J = \frac{\partial \psi}{\partial T_n} \tag{4}$$

If J is not recalculated as a function of $T_n^{(m)}$ but is kept constant, we have the modified Newton method. In the present work J was calculated at the beginning of each time interval (when material properties are updated) and was not updated inside a time interval unless the number of iterations in Eq. (3) exceeded three. As was noted before, the only nonlinearity in Eq. (2) is due to radiation. In the problems considered herein, only radiation to space was considered so that the Jacobian was symmetric and positive definite.

Solution of Linear Equations

Eqs. (3) constitutes a system of linear equations of the form

$$Ax = b \tag{5}$$

where in our case, A is symmetric, positive definite and sparse. Eq. (5) may be solved by elimination techniques or by iterative methods. Herein, several methods of solution were compared. The first is an elimination technique, the Gauss-Doolittle factorization, whereby A is factored as

$$A = LDL^T \tag{6}$$

where L is a lower triangular matrix with diagonal elements equal to one and D is a diagonal matrix. Once A has been factored the solution process proceeds easily.

Most iterative methods proceed by splitting the matrix A into two parts

$$A = M-N \tag{7}$$

and rewriting Eq. (5) as

$$(I - M^{-1}N)x = M^{-1}b \tag{8}$$

From Eq. (8), the following simple fixed point iterative process can be defined

$$x^{m+1} = M^{-1}(b + Nx^m) \qquad (9)$$

It is desirable to choose M so that it is close to A, so that we get fast convergence (note that if M=A, N=0, a single iteration is enough). On the other hand, we want to choose M so that M^{-1} or its equivalent can be formed cheaply. One well known choice is to take M equal to the diagonal of A (the Jacobi method) and another is to take M equal to the lower triangular part of A (the Gauss Seidel method). Asymptotically, the error in the solution is reduced by a factor equal to the largest eigenvalue of $M^{-1}N$. For most finite element generated matrices, this eigenvalue is very close to one so that convergence is very slow.

A method which is a cross between elimination techniques and iterative techniques is based on obtaining M from an incomplete elimination process. For a positive definite matrix A, the incomplete Cholesky decomposition M is defined as

$$M = LL^T \qquad (10)$$

where L is a lower triangular matrix with the same sparsity structure as A (that is, no fill-up permitted). The matrix M is a good approximation to the matrix A so that most of the eigenvalues of $M^{-1}N$ are very small. Also, for a very sparse matrix A, the cost in computation time and storage for obtaining L is much smaller than that required for a complete decomposition. However, even though most of the eigenvalues of $M^{-1}N$ are small, it is possible for a few to be close to one so that the convergence of the fixed point iteration, Eq. (9), is slow.

Another method which is sometimes used for the solution of linear equations is the conjugate gradient method.

The conjugate gradient method can be used to solve the system (5) by applying it to minimize the following error measure

$$e = (x-x_e)^T A(x-x_e) \qquad (11)$$

where x_e is the exact solution and x is the current approximation. Theoretically, the conjugate gradient method should reduce e to zero in no more than n iterations so that it is a deterministic method like Gaussian elimination. Because of round-off error, however, it does not terminate in exactly n iterations and may be regarded as an iterative method. Its convergence, while dependent on the ratio of the maximum to the minimum eigenvalues of A, is more favorable than those of iterative methods like Gauss-Seidel. This is because it tends to eliminate the error components corresponding to extreme eigenvalues in the first few iterations and attain a high convergence rate later. It has a decided advantage over the standard iterative methods for matrices with very few extreme eigenvalues and a large number of eigenvalues which are bunched together.

Meijerink and Van der Vorst (Ref. 6) put together a very clever combination of all the above techniques which can be very efficient for sparse poorly banded matrices. The idea is to apply the conjugate gradient method to Eq. (8) rather than Eq. (5) where the matrix M is obtained from an incomplete Cholesky decomposition of A, Eq. (10).

Because the matrix M is a good approximation to A, the matrix $I-M^{-1}N$ is close to the unit matrix and most of its eigenvalues are close to one. This provides a good setting for a very fast convergence of the conjugate gradient method.

To take advantage of the incomplete Cholesky conjugate gradient (ICCG) algorithm, sparse matrix storage techniques should be used for the matrix A. The method selected here and its implementation are discussed next.

Matrix Storage, Retrieval and Assembly Technique

The storage and retrieval technique used herein is due to Gustavson (Ref. 16) and Tewarson (Ref. 17). One array EJ is used to store by row all the nonzero elements of the lower triangular part of the Jacobian. Another array IC of the same size contains the column numbers of the entries of EJ. Finally, a third array IA stores the position in EJ of the last nonzero entry in each row so that IA (i)- IA (i-1) is the total number of nonzero elements in the i-th row. A method for generating IC and IA from element data was developed in Ref. 15. The use of the IC and IA arrays is compatible with an efficient implementation of the ICCG algorithm and avoids any operations on zero elements of A or L. It is not convenient, however, for assembling the Jacobian from the individual element matrices.

To expedite the generation and assembly of the Jacobian an additional storage system is used. The element conductivity matrices are calculated and stored for a unit value of the conduction coefficient. For each element matrix, another array IPLACE is generated which stores the destination of each entry of the conductivity matrices in the matrix EJ. During assembly the actual conduction coefficient is computed based on the average temperature of the element and this value is used to multiply the unit matrices. The use of the IPLACE array together with the unit matrices reduces the assembly time for the Jacobian considerably.

Computer Implementation

A computer program that implements the analysis methods denoted SMITT (sparse Matrix Iterative Techniques for Thermal Analysis) was written for the IIT Prime 400 minicomputer (Ref. 15). A parallel program using conventional band matrix storage and Gauss-Doolittle solution was also implemented. The Prime 400 is a virtual memory machine and theoretically each common block or subroutine can have up to 640,000 32 bit words. In practice it was found that

beyond 64,000 words the program did not work, so that this limit controlled the
maximum problem size. These programs were exercised on two example programs
described in the next section.

RESULTS AND DISCUSSION

Shuttle Orbiter Frame

The first test problem which was used for demonstrating the efficiency of
the ICCG algorithm is a shuttle orbiter frame, Fig. 1, tested under transient
heat loads by Gallegos (Ref. 18). The finite element model contains 190 grid
points and 199 elements, including two dimensional conduction elements and one
dimensional radiation elements.

The properties of the aluminum structure and the insulation are functions
of temperature (Ref. 18). The band width of the Jacobian is 39.

The problem has been solved for 500 seconds of response time using the ICCG
and Gauss Doolittle (GD) algorithms. Material properties were updated every 50
seconds and the integration time step for the Crank-Nicholson method was 25
seconds. The CPU times required for the solution are tabulated in Table 1. It
is seen that the ICCG algorithm is about twenty percent faster than Gauss Doolittle. This is remarkable because the problem is only mildly sparse with more
than twenty percent of the elements in the band being nonzero.

Insulated Cylinder

For the second test problem, a configuration was sought which was larger
(in terms of number of unknown temperatures) than the shuttle frame and exhibited some of the characteristics of an insulated air frame structure. These
considerations led to the insulated aluminum cylindrical shell depicted in Fig.
2. The cylinder is 18.3m (720 in.) in length and 2.3 m (90 in.) in diameter.
The aluminum is 2.5 cm (1 in.) thick and the insulation is 5.0 cm (2.0 in) thick.
The outer surface of the insulation is heated over a region which consists of
1/3 the length of half the circumference. The finite element model consists of
a symmetric half of the cylinder and is composed of solid brick elements (K81
elements in SPAR). Additionally, the outer surface of the insulation has quadrilateral radiation elements (R41) which radiate to free space. The time-dependence of the heat load on the cylinder is shown in Fig. 3. In all calculations, material properties of the metal and insulation are temperature dependent and are given in Table 2. The material properties were updated every
fifty seconds. The number of nodes, the bandwidth of the Jacobian and the
sparsity can be easily changed by varying the number of axial, radial and circumferential elements. Thus the problem is suited for checking the performance
of the ICCG algorithm vs. other algorithms for a wide range of these parameters.

The cylinder problem was used to compare the performance of the following
algorithms:

(i) Band Gauss-Doolittle elimination

(ii) The conventional static over-relaxation (SOR) method with an over-relaxation parameter of 1.4

(iii) The fixed point iteration given by Eq. (9) with the matrix M given by the incomplete Cholesky, Eq. (11)

(iv) The ICCG algorithm

Three sizes of problems are considered, with the number of nodes being 400, 720 and 1100 respectively. For each size results are obtained by keeping the total number of nodes approximately fixed and varying the band width. The total CPU time for calculating 200 seconds of the response is shown in Figure 4 for 400 nodes and in Figure 5 for 720 nodes (the SOR algorithm failed to converge for one point which is shown as a break in the curve).

For the largest problem the core storage requirements for the Gauss-Doolittle elimination could not be met. An explicit formula for predicting the computation time for the Gauss-Doolittle algorithm was devised on the basis of the available data (see Ref. 15). This formula was used to estimate the run time for that algorithm for 1100 nodes. The other three algorithms were actually run for 1100 nodes and the results are given in Figure 6.

From Figures 4, 5, 6, it is clear that the iterative algorithms are quite superior to Gaussian elimination for this problem. The advantage increases with increasing number of nodes and increasing band width. The ICCG algorithm is superior to the other two iterative algorithms. This was due to an average number of iterations of about 2.75 compared to 5 or more for the two other algorithms. An examination of the detailed run times for each subroutine (Ref. 15) revealed that most of the gains over the elimination algorithm were due to difference in the matrix decomposition.

An additional advantage of the iterative algorithms over elimination algorithms results from the insensitivity to band-width. The analyst does not have to worry about the best numbering of the nodes to reduce band width. The sensitivity to nodal numbering is one of the major disadvantages of implicit techniques (when used with Gaussian elimination) compared to explicit integration techniques.

CONCLUDING REMARKS

The system of ordinary differential equations generated by discretizing a transient heat transfer problem in structures is typically stiff. Such systems are most efficiently handled by implicit integration techniques which typically require the solution of large and sparse systems of linear algebraic equations. The generation and solution of these equations account for the major part of the computation in the solution of the differential equations. Traditionally, these algebraic equations are solved either by elimination techniques or by iterative techniques. Herein, it is suggested that a recently developed partial elimination algorithm can be more efficient for three dimensional problems where the algebraic equations are poorly banded.

The partial elimination algorithm - the incomplete Cholesky conjugate gradient (ICCG) algorithm is implemented in a finite element program for transient heat transfer in structures. It is coupled with a sparse matrix assembly and storage systems. The techniques are demonstrated for a two dimensional space shuttle frame and an insulated cylinder modeled by three dimensional finite elements. The ICCG algorithm was compared to the Gauss-Doolittle elimination algorithm as well as to two iterative algorithms. For the cylinder problem the ICCG algorithm was shown to be greatly superior to the elimination algorithm and significantly better than the iterative algorithms. For the well banded two dimensional problem, the ICCG algorithm is still marginally better. The results indicate that the ICCG algorithm has a potential for large saving in computational resources when implemented in computer programs for transient heat transfer in structures.

ACKNOWLEDGEMENT

This work was supported in part by NASA grant NSG-1266.

REFERENCES

1. Marlowe, M.B., Moore, R.A., and Whetstone, W.D., "SPAR Thermal Analysis Processors Reference Manual", NASA CR - 159162, October 1979.

2. Gear, C.W., Numerical Initial Value Problems in Ordinary Differential Equations, Prentice Hall, 1971.

3. Hindmarsh, A.C., "GEARB-Solution of Ordinary Differential Equations Having Banded Jacobian", Lawrence Livermore Laboratory Report UCID 30059-Rev. 1, 1975.

4. Adelman, H.M., and Haftka, R.T., "On the Performance of Explicit and Implicit Algorithms for Transient Thermal Analysis of Structures", NASA TM-81880, September, 1980.

5. Anonymous, "Martin Interactive Thermal Analysis System, Version, 1.0", Martin-Marietta Corp., Denver, Colo., MDS-SPLPD-71-FD238 (Rev 3), March 1972.

6. Meijerink, J.A., and Van der Vorst, H.A., "An Iterative Solution Method for Linear Systems of which the Coefficient Matrix is a Symmetric M-Matrix", Mathematics of Computation, Vol. 31, No. 137, pp. 148-162, 1977.

7. Kershaw, D.S., "The Incomplete-Choleski-Conjugate Gradient Method for the Iterative Solution of Systems of Linear Equations", Journal of Computational Physics, Vol. 26, pp. 43-65, 1978.

8. Petravic, M. and Kuo-Petravic G., "An ILUCG Algorithm Which Minimizes in the Euclidean Norm" Journal of Computational Physics, Vol. 32, pp. 263-269, 1979.

9. G. Kuo-Petravic and M. Petravic "A Program Generator for the Incomplete LU Decomposition-Conjugate Gradient (ILUCG) Method", Computer Physics Communications, Vol. 18, pp. 13-25, 1979.

10. Christiansen J.P., and Winsor N.K., "CASTOR 2: A Two Dimensional Laser Target Code", Computer Physics Communications, Vol. 17, pp. 397-412, 1979.

11. Axelson, O., and Munskgaard N., "A Class of Preconditioned Conjugate Gradient Methods for the Solution of a Mixed Finite Element Discretization of the Biharmonic Operator", Int. J. Num. Meth. Eng., Vol. 14, pp. 1001-1019, 1979.

12. Gustafson I., "On Modified Incomplete Cholesky Factorization Methods for the Solution of Problems with Mixed Boundary Conditions and Problems with Discontinuous Material Coefficients", Int. J. Num. Meth. in Eng. Vol. 14, pp. 1127-1140, 1979.

13. Manteuffel, T.A., "An Incomplete Factorization Technique for Positive Definite Linear Systems", Mathematics of Computation, Vol. 34, No. 150, pp. 473-497, 1980.

14. Gambolatti, G., "Fast Solution to Finite Element Flow Equations by Newton Iteration and Modified Conjugate Gradient Method," Int. J. Num. Meth., in Eng., Vol. 15, pp. 661-675, 1980.

15. Kadivar, H.M., "Algorithmic Aspects of the Solution of Transient Heat Transfer in Structures", PhD. Thesis, Illinois Institute of Technology, Chicago, Illinois, December 1980.

16. Gustavson, F.G., "Some Basic Techniques for Solving Sparse Systems of Linear Equations" in Sparse Matrices and Their Applications (Rose, P.J. and Williamsburg, R.A., eds), pp. 41-52, Plenum Press, N.Y. 1972.

17. Tewarson, R.P., Sparse Matrices, Academic Press, N.Y., 1973.

18. Gallegos, J.J., "Thermal Math Model of FRSI Test Article Subjected to Cold Soak and Entry Environments", AIAA Paper 78-1627, 1978.

Table 1. Solution Time for Shuttle Frame*

DURATION TIME SEC.	GAUSS-DOOLITTLE CPU SEC.	ICCG CPU SEC.
200.	52.54	42.95
500.	128.92	106.12

*Results obtained on a PRIME 400 minicomputer

Table 2. Material Properties for Insulated Cylinder

(a) Insulation: $\rho = 160$ kg/m^3 (0.00582 lbm/in^3)

T		C		K	
°K	°R	J/kg-°C	Btu/lbm-°R	W/m-°C	Btu/in-s-°R
200	360	523	0.125	0.0381	5.1×10^{-7}
367	660	↓	↓	.0546	7.3
478	860			.0711	9.5
589	1060			.0898	1.2×10^{-6}
700	1260	↓	↓	.112	1.5
811	1460			.142	1.9
922	1660			.180	2.4

(b) Aluminum $\rho = 2770$ kg/m^3 (0.0101 lbm/in^3)

T		C		K	
°K	°R	J/kg-°C	Btu/lbm-°R	W/m-°C	Btu/in-s-°R
200	360	769	0.184	99.5	0.00133
311	560	861	.206	125.0	.00167
367	660	903	.216	138.0	.00185
422	760	937	.224	154.8	.00207
478	860	974	.233	171.3	.00229
533	960	1012	.242	178.8	.00239
589	1060	1045	.250	181.1	.00242

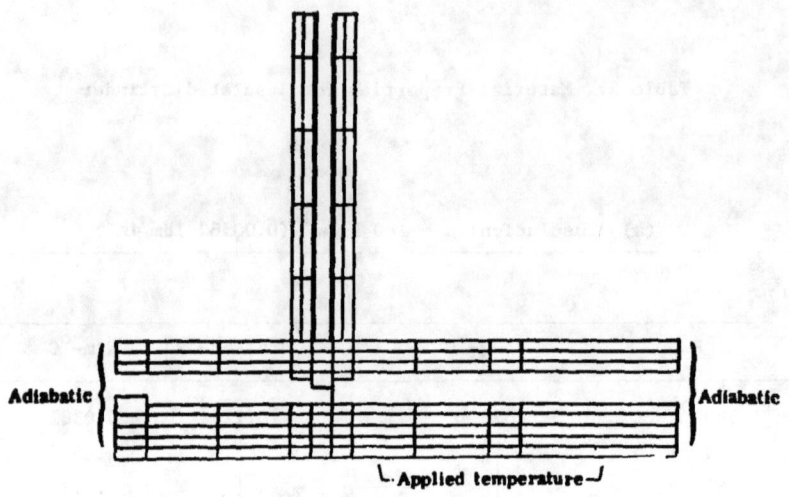

Figure 1.- Finite element model of shuttle frame.

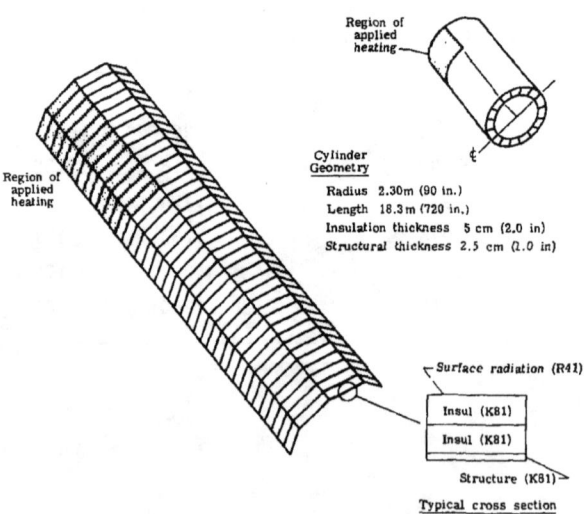

Figure 2.- Finite element model of insulated cylinder.

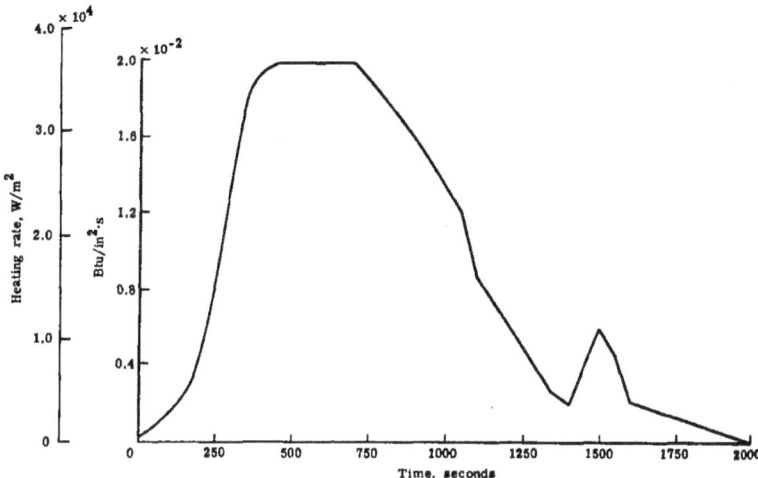

Figure 3.- Heating load at outer surface of insulated cylinder.

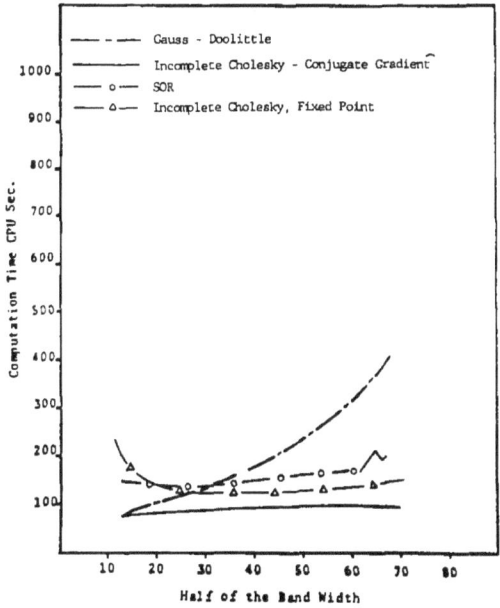

Figure 4.- Effect of Jacobian band width on total computation time for 400 node cylinder models.

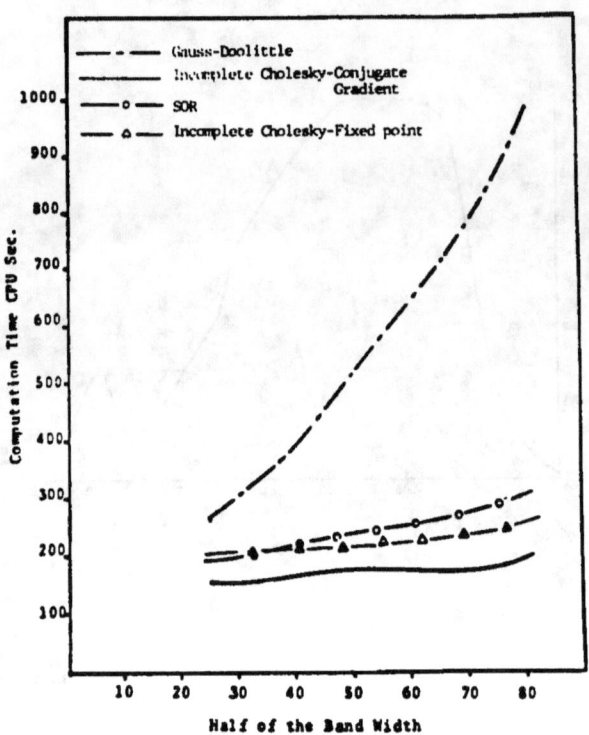

Figure 5.- Effect of Jacobian band width on total computation time for 720 node cylinder models.

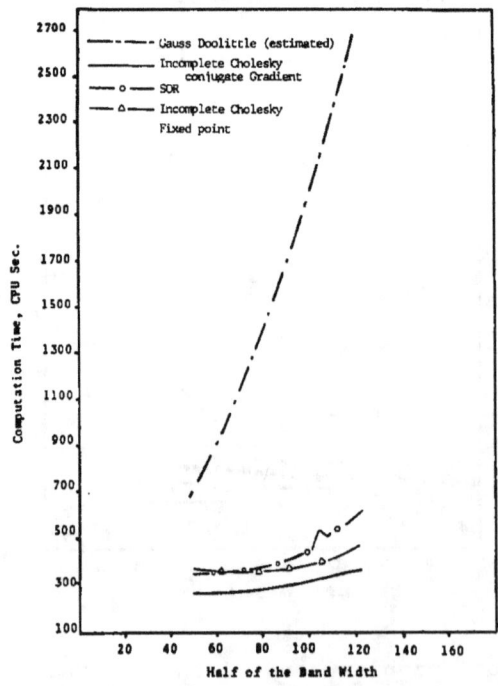

Figure 6.- Effect of Jacobian band width on total computation time for 1100 node cylinder models.

EVALUATION OF AN IMPROVED FINITE-ELEMENT THERMAL STRESS
CALCULATION TECHNIQUE

Charles J. Camarda
Aerospace Engineer, Langley Research Center
Hampton, Virginia

INTRODUCTION

The accurate calculation of thermal stresses in complicated airframe structures often requires a refined finite-element grid and a corresponding large computational time. In looking toward combined thermal-structural design and optimization calculations, it is essential to avoid having to solve an excessively large system of equations since such calculations are performed many times during optimization procedures.

A procedure for generating accurate thermal stresses with coarse finite-element grids was developed and described by Ojalvo (ref. 1). The procedure is based on the observation that for linear thermoelastic problems, the thermal stresses may be envisioned as being composed of two contributions--the first due to the strains in the structure which depend on the integral of the temperature distribution over the finite element and the second due to the local variation of the temperature in the element. Ojalvo's key idea was that the first contribution could be accurately predicted with a coarse finite-element mesh. The resulting strain distribution could then be combined via the constitutive relations with detailed temperatures from a separate thermal analysis. The result would be accurate thermal stresses from coarse finite-element structural models even where the temperature distributions have sharp variations.

Although this intriguing idea was proposed in 1974, its use has not been documented in the open literature except for the original AIAA Technical Note. It has recently received attention by this author because of the current interest at Langley in rapid analysis and design-oriented analysis techniques. The range of applicability of the method for various classes of thermostructural problems such as in-plane or bending type problems and the effect of the nature of the temperature distribution and edge constraints was not fully documented. These questions are addressed in this paper. Ojalvo's method is used in conjunction with the SPAR finite-element program (ref. 2) and extensive calculations are carried out and are described. Results are obtained for rods, membranes, a box beam and a stiffened panel.

SUMMARY OF OJALVO'S METHOD

The following is a brief summary of "Ojalvo's" method (ref. 1) for thermal stress calculations using a finite-element (F.E.) analysis procedure. For a detailed explanation of the following subject topics see references 1 and 3. The equations shown in figure 1 represent the system of matrix equations involved in a typical structural F.E. analysis. The system stiffness matrix [K] relates the vector of element node point displacements $\{\delta\}$ and mechanical and thermal load vectors ($\{P_{mech}\}$ and $\{P_T\}$ respectively) by the first equation. The stiffness matrix can be calculated from the strain shape function matrix [B] and the constitutive relations [D] as shown. If a structure, idealized by finite elements, has sufficient nodal constraints so the stiffness matrix [K] is nonsingular, the node-point displacements $\{\delta\}$ can be obtained from the first equation. These displacements are then used to calculate corresponding strains $\{\epsilon\}$ and stresses $\{\sigma\}$.

As explained in reference 1, the dependence of $\{\epsilon\}$ upon the temperature T for a linear elastic analysis is through $\{P_T\}$ and $\{P_T\}$ is a function of the spatial integral of temperature throughout the element. Hence it appears $\{\epsilon\}$ is related to integrals of T. The stresses $\{\sigma\}$, however, are not only related to integrals of T through $\{\epsilon\}$ but are also directly related to the local temperature. Thus, one can expect a greater accuracy in the numerical calculation of strain than stress in thermomechanical problems since some of the errors of approximations in T are self-cancelling in the integrals which determine strain. In essence one should calculate accurate strains with as coarse a F.E. grid as possible to approximate the thermal load vector. Theoretically one could then use a coarser structural F.E. grid than a corresponding thermal F.E. grid to obtain the same desired accuracies in each analysis. One could improve the coarse-grid F.E. stress results by using the coarse-grid structural results for $\{\epsilon\}$ and coupling them with the fine-grid thermal results for T. This last statement and the last equation of the figure summarize Ojalvo's method. By comparison, in the conventional F.E. procedure there is no separation in the strain and stress calculation; the same F.E. grid is used to calculate consistent nodal thermal load, and nodal strains and stresses. Also, in most finite elements the temperature field in the element is similar to the displacement field; hence, sharp variations in temperature are not accounted for.

FOR FINITE ELEMENT ANALYSIS:

$$[K]\{\delta\} = \{P_{mech}\} + \{P_T\} \quad \text{WHERE} \quad \{P_T\} = \int T(x,y,z)[B]^T[D]\{\alpha\}\,dV$$

$$[K] = \int [B]^T[D][B]\,dV$$

$$\{\epsilon\} = [B]\{\delta\} = [B]\left\{[K]^{-1}\left(\{P_{mech}\} + \int T(x,y,z)[B]^T[D]\{\alpha\}\,dV\right)\right\}$$

$$\{\sigma\} = [D]\left(\{\epsilon\} - T\{\alpha\}\right)$$

- $\{\epsilon\}$ IS RELATED TO INTEGRAL OF T
- $\{\sigma\}$ IS RELATED TO THE LOCAL TEMPERATURE
- THEREFORE $\{\epsilon\}$ SHOULD BE NUMERICALLY MORE ACCURATE THAN $\{\sigma\}$

HENCE CALCULATE STRESS AS: $\{\sigma\} = [D]\left(\underbrace{\{\epsilon\}}_{\text{COARSE-GRID STRUCT. ANALYSIS}} - \underbrace{T\{\alpha\}}_{\text{FINE-GRID THERMAL ANALYSIS}}\right)$

Figure 1

SHUTTLE ORBITER VERTICAL FIN STRUCTURAL IDEALIZATION ANALYZED BY OJALVO

To demonstrate the usefulness of Ojalvo's method, reference 1 presents results for a hot structure concept for the vertical fin of the Space Shuttle Orbiter (ref. 4). Temperatures were calculated by a detailed lumped parameter analysis and used as input to a coarse-grid F.E. structural analysis. As shown in figure 2, only a section of the fin main structural box between ribs 9 and 10 was idealized by finite elements using rod, membrane and shear elements. Rene' 41 was as the structural material at locations where the temperature exceeded 923K (1200°F) and Inconel 718 was used for locations below that temperature. The spars were corrugated to partially alleviate excessive thermal stresses.

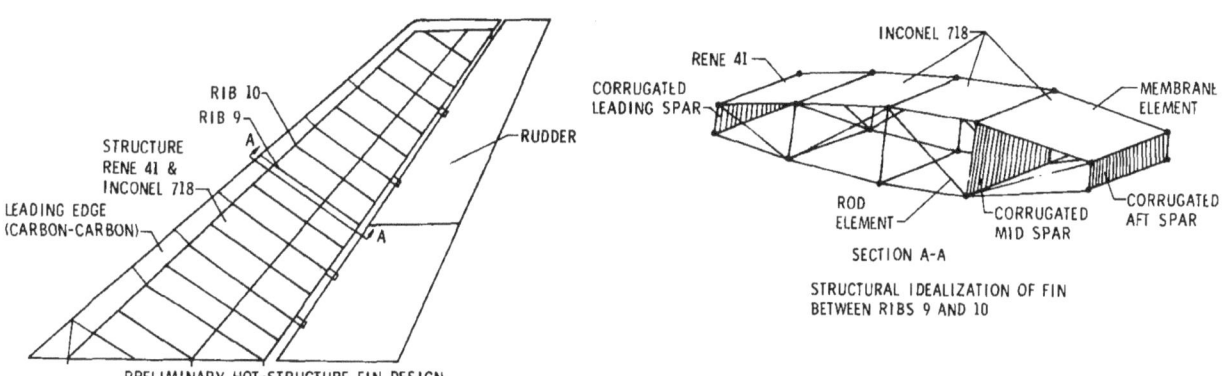

Figure 2

RESULTS OF OJALVO'S METHOD FOR THE SHUTTLE ORBITER VERTICAL FIN

A chordwise temperature distribution from a detailed lumped-parameter transient thermal analysis is shown as a solid curve in the upper left-hand part of figure 3. The massive spars and spar caps act as heat sinks causing sharp temperature drops in their vicinity during transient heating. The dashed lines are the integrated average values of temperature which were used as input for the cover panels in the structural F.E. model. Since the fin is long and slender and since the section analyzed is far enough away from the ends beam theory may be used. If z is the coordinate direction through the depth of the beam (along the chord in this example), and assuming $T = T(z)$ and beam symmetry in two directions, the following equation can be used to predict strains (ref. 1):

$$\varepsilon = \frac{\int_A E\alpha T dA}{\int_A E dA} + \frac{z \int_A E\alpha T z dA}{\int_A E z^2 dA}$$

Because of the above relationship, the sharp drops in temperature near the spars have little effect on the strains. Hence, as shown in figure 3, the strain distribution in the covers predicted by coarse-grid F.E. is linear and very close to the predicted beam theory solution regardless of the nature of the temperature distribution. Thus a coarse grid is sufficient to obtain accurate strains for a thermal stress problem which would require a fine grid to predict the temperature distribution. Also, mesh refinement of the structural grid may not be necessary for performing quasi-static thermal/structural analysis.

A comparison of stress distributions using beam theory, conventional F.E., and Ojalvo's method is shown in figure 3 at the right. The conventional F.E. procedure, using constant-strain constant-temperature quadrilateral membrane elements and integrated average temperature values miss large peak stresses in the covers. However, stress results using the strains from the coarse-grid structural F.E. analysis and the existing detailed temperature distribution (Ojalvo's method) agree closely with beam theory results. Ojalvo's method works well for this problem because the strain distribution is linear and the thermal load vector can be closely approximated by a coarse grid (4 elements).

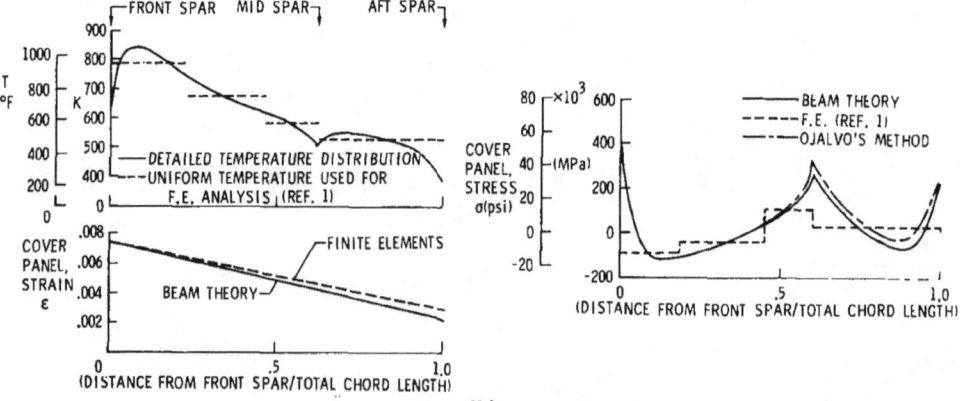

Figure 3

QUALIFICATION OF OJALVO'S METHOD

With regard to earlier statements in reference 1, it is true that the node point displacements $\{\delta\}$ are directly related to the spatial integral of the temperature distribution as shown by the first equation in figure 4. However, the strains are directly related to the strain shape function matrix [B] which relates strains to node point displacements $\{\delta\}$. As shown in the figure [B] is related to the strain operator matrix [L] which performs the operation of differentiation and the shape function matrix [N] which relates the displacement vector $\{u\}$ to the vector of node point displacements $\{\delta\}$. The [B] matrix operates on the $\{\delta\}$ and transforms it into $\{\varepsilon\}$; embedded in this operation is the differentiation of the shape function matrix [N]. The direction of differentiation (implied in [L]) is dependent upon the type of element. Hence, depending on the arguments of integration in $\{P_T\}$ and the direction of differentiation in [L] the strains may be directly related to the local temperatures and not their integrals as stated in reference 1.

As stated in reference 5, deflections caused by heating were "acceptable" but "special attention" was necessary for the "interpolation of the stresses which could deviate considerably from the true stress state." Here, reference 5 indicates that deflections were less sensitive to temperature distributions than their corresponding stresses, as stated in reference 1; however, no mention is made of the sensitivity of strains to temperature distributions. It appears that for bending problems, beams, or plates, there is a direct relationship between ε and integrals of T. The reason for this relationship is that for beam-type problems the argument of integration for $\{P_T\}$ is through the depth of the beam while the desired stresses are in the longitudinal direction. This means the direction of differentiation implied by [B] is along the length and hence does not affect the integration. The direction of differentiation for membrane-type problems, however, is the same as the arguments of integration in $\{P_T\}$ and hence the strain variation should be similar to the temperature variation. This reasoning suggests that while there appears to be an advantage in using Ojalvo's method for bending problems, it does not appear to be appropriate for membrane or plane-stress problems.

$$\{\delta\} = [K]^{-1} \left(\{P_{mech}\} + \int_V T(x,y,z)[B]^T[D]\{\alpha\}dv \right)$$

$$\{\varepsilon\} = [B]\{\delta\} = [L][N]\{\delta\}$$

WHERE FOR EXAMPLE

$$[L]\{u\} = \begin{Bmatrix} \varepsilon_x \\ \varepsilon_y \\ \gamma_{xy} \end{Bmatrix} = \begin{bmatrix} \partial/\partial x & 0 \\ 0 & \partial/\partial y \\ \partial/\partial y & \partial/\partial x \end{bmatrix} \begin{Bmatrix} u \\ v \end{Bmatrix} ; \quad [L] = \text{STRAIN OPERATOR (DIFFERENTIATION) MATRIX}$$

$[N]\{\delta\} = \{u\}$; N = SHAPE FUNCTION MATRIX

HENCE [B] PERFORMS DIFFERENTIATION

- BENDING PROBLEM $\quad \{\varepsilon\} \sim \int T dv$
- MEMBRANE OR PLANE STRESS PROBLEM $\quad \{\varepsilon\} \sim T$

Figure 4

COMPARISON OF STRESS CALCULATION METHODS FOR CLAMPED ROD PROBLEM

To qualify the usefulness of Ojalvo's method, analytical and numerical results for several classes of thermostructural problems are presented beginning with the simplest one-dimensional example, a rod. As shown in figure 5, an aluminum rod of length ℓ is clamped at both ends and subjected to a sinusoidal temperature variation along its length; the governing equations are as follows:

(1) $\quad \dfrac{d^2 u}{dx^2} - \alpha \dfrac{dT}{dx} = 0 \qquad u(0) = u(\ell) = 0$

Integrating and substituting $T = T_0 \sin \dfrac{\pi x}{\ell}$

(2) $\quad u(x) = -\dfrac{\alpha T_0 \ell}{\pi} \cos \dfrac{\pi x}{\ell} - \dfrac{2\alpha T_0 x}{\pi} + \dfrac{\alpha T_0 \ell}{\pi}$

and

(3) $\quad \varepsilon_x(x) = \alpha T_0 \sin \dfrac{\pi x}{\ell} - \dfrac{2\alpha T_0}{\pi} = \alpha T - \dfrac{2\alpha T_0}{\pi}$

$\sigma = -2 E \alpha T_0 / \pi$

From eq. (3) and figure 5 the strain distribution is directly related to the local temperature T rather than the integral of T. Thus, to obtain accurate results for strains and stresses one would need as fine a structural F.E. grid as the thermal F.E. grid. Analytical results for stress from eq. 4 give $\sigma = -179.8$ MPa (-26.07 ksi); conventional F.E. results using SPAR for 2, 4, 8, and 16 elements per length converge from -141.18 MPa (-20.475 ksi) for the 2-element grid to -184.38 MPa (-26.741 ksi) for the 16-element grid. A comparison of analytical, conventional F.E., and Ojalvo's method for stresses is shown in figure 5 at the right. Ojalvo's method was employed using strain results from two different F.E. model, the 4-element and the 8-element models. From the figure note that conventional F.E. results (dashed line) are comparable to results using Ojalvo's method (shown by symbols). In some instances Ojalvo's method is more accurate than conventional results and in some instances (where the coarse strains are not close to exact values) Ojalvo's method is less accurate. For other boundary conditions similar results were obtained; the strains varied as the local temperature and no benefit was realized using Ojalvo's method.

COMPARISON OF STRESS CALCULATION METHODS FOR CLAMPED ROD PROBLEM

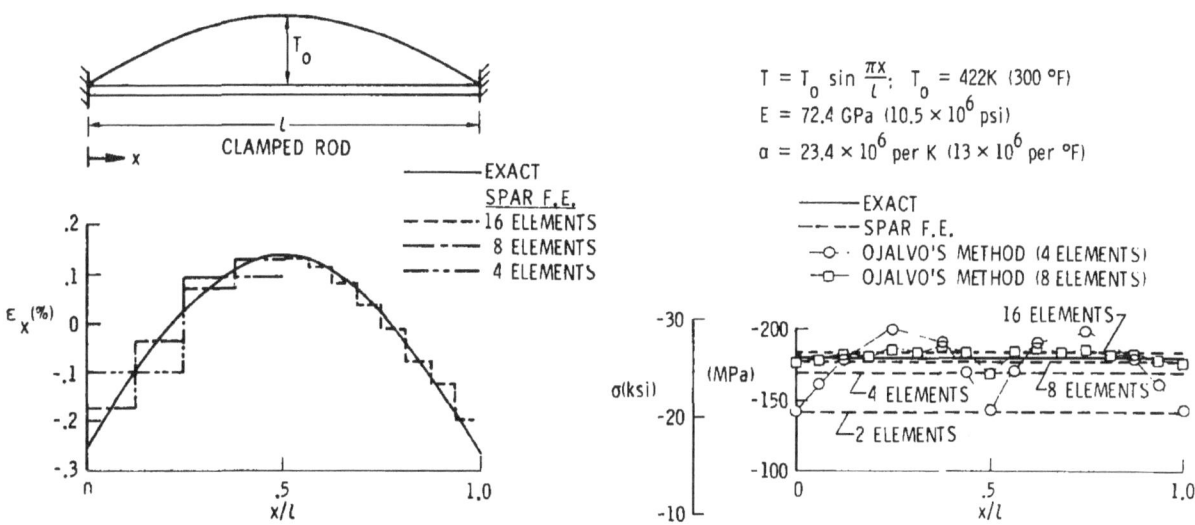

Figure 5

THIN FREE MEMBRANE SUBJECT TO A PARABOLIC
TEMPERATURE DISTRIBUTION

The problem shown in figure 6 is that of a thin rectangular titanium plate of uniform thickness, free to expand, in which the temperaure is an even function of y:

$$T = T_0(1 - y^2/c^2)$$

From reference 6

$$\sigma_x = \frac{1}{2c} \int_{-c}^{c} \alpha T E \, dy - \alpha T E \quad \text{and} \quad \sigma = 0$$

From the thermoelastic constitutive relations and the above equations

$$\sigma_x = \frac{E}{1-\nu^2} [\varepsilon_x + \nu \varepsilon_y - (1 + \nu)\alpha T] = \frac{1}{2c} \int_{-c}^{c} \alpha T E \, dy - \alpha T E$$

$$\sigma_y = \frac{E}{1-\nu^2} [\varepsilon_y + \nu \varepsilon_x - (1 + \nu)\alpha T] = 0$$

solving for ε_x and ε_y gives

$$\varepsilon_x = \frac{2\alpha T_0}{3} = \text{Constant}$$

$$\varepsilon_y = (1+\nu)\alpha T - \frac{2}{3}\nu\alpha T_0$$

From the above equations the strain in the x-direction is a constant, but the strain in the y-direction is directly related to the local temperature T. Hence for accurate results for σ_x accurate values of ε_x, ε_y, and T are needed; since the ε_y varies as the local temperature in the y-direction as fine a discretization as that used in the thermal analysis is needed. If accurate or converged values of ε_y are not used, Ojalvo's method will produce less accurate results for stresses as seen in the previous problem. Hence if the linear strain distribution results of the 8 x 1 grid (triangular symbol) are used with the fine or exact temperature distribution (Ojalvo's method) the results for stresses are not as accurate as the conventional F.E. method in regions where strains are not converged as shown in the right side of the figure. Also, conventional F.E. results for the 8 x 2 grid are close to the exact solution; therefore if Ojalvo's method is used with this grid, it will give little improvement.

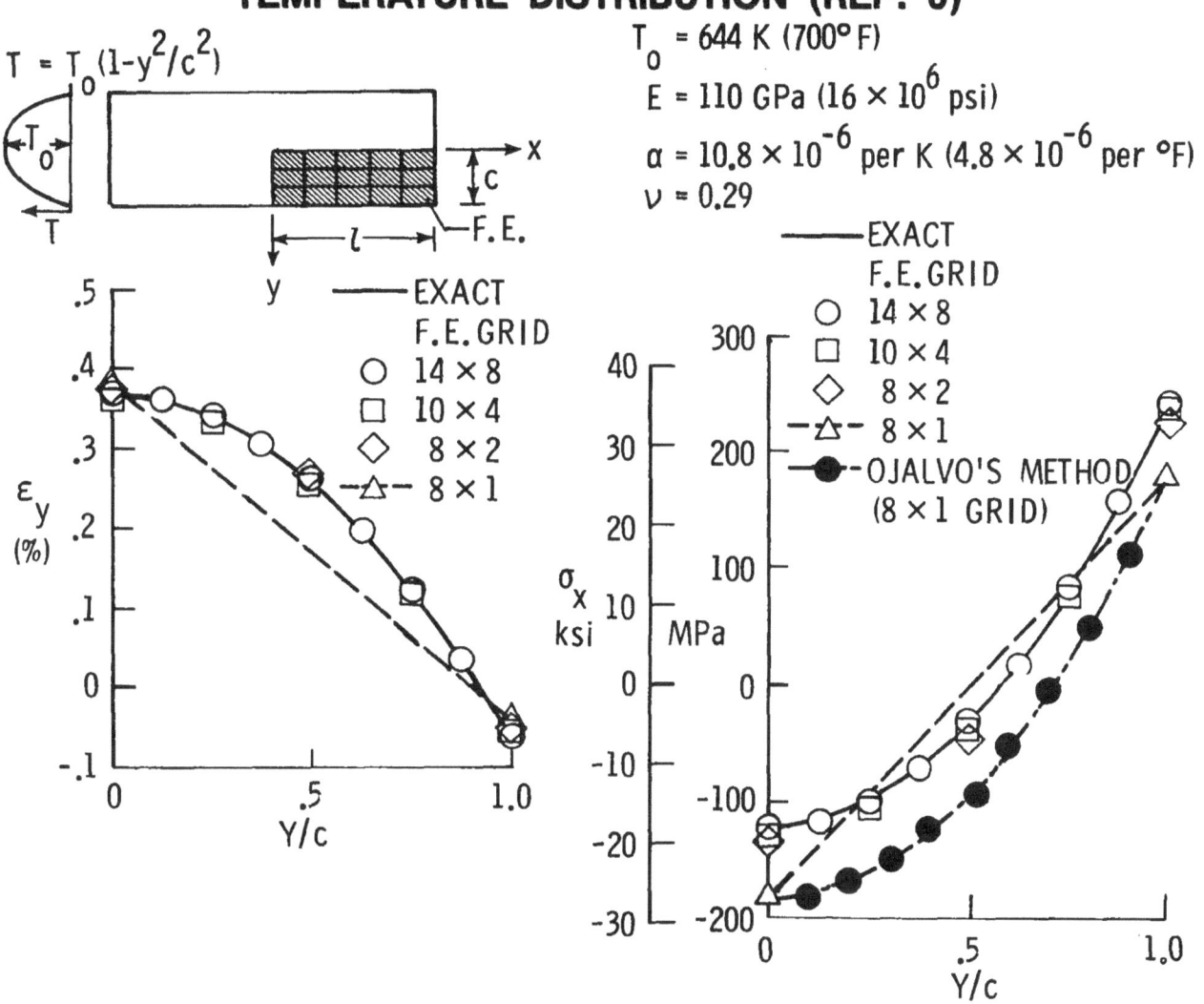

Figure 6

CANTILEVERED TITANIUM BOX BEAM WITH
TEMPERATURE VARYING THROUGH THE DEPTH

A titanium box-beam was idealized as a built-up structure consisting of rod and membrane elements. The fine-grid F.E. model consists of 1555 nodes, 1512 membrane elements and 216 rod elements representing one-half of the entire box beam. The load applied to the structure is the temperature which varies in the y-direction only as shown in figure 7. This temperature distribution (solid curve) is a least-squares fit of a fifth-degree polynomial curve to the temperature distribution of reference 1. The dashed lines represent integrated average values of temperature which were used in an approximate beam theory solution to compare with F.E. results from several different models.

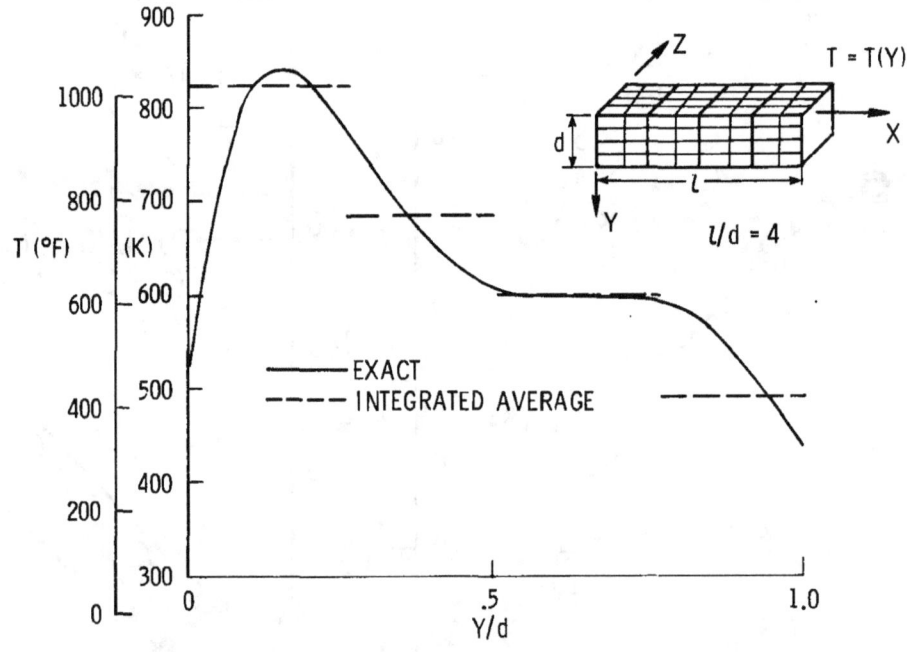

Figure 7

COMPARISON OF OJALVO'S METHOD WITH CONVENTIONAL F.E. FOR A CANTILEVERED BOX BEAM

The box beam is cantilevered as shown in figure 8. Finite-element results for strain in the membranes at $Z = 0$ at various stations along the X-axis using the fine grid (48 x 12 grid) are shown as the dashed curves. Beam theory results are shown as a solid line, and results of the coarser grids are shown by the symbols. Far from the ends of the beam, plane sections remain plane; the strains are very nearly linear and results are close to predicted beam theory results. At the free end ($x/\ell = 1.0$), however, the strain distribution is similar to the temperature distribution; this is necessary to insure the stress-free end condition.

Nondimensionalized stress results at the center of the beam ($x/\ell = 0.5$) are shown in figure 8 at the right. Notice that results from the coarse structural grid (diamond symbol and dashed line) miss critical (peak) stresses located within the element predicted by beam theory (solid curve) and the fine-grid F.E. results (circular symbol). The use of Ojalvo's method, shown by the single-dashed curve, gives a good representation of the actual stress distribution consistent with results of reference 1.

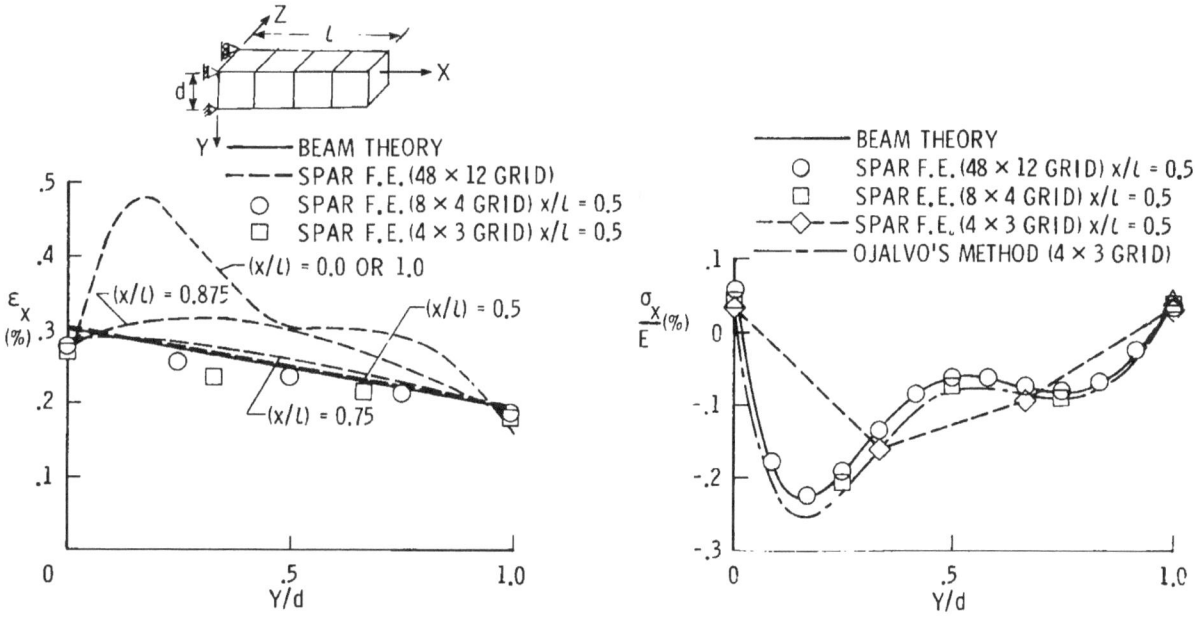

Figure 8

COMPARISON OF OJALVO'S METHOD WITH CONVENTIONAL F.E.
FOR A CANTILEVERED BOX-BEAM (FREE END)

A comparison of the exact solution ($\sigma_x=0$) with conventional F.E. and Ojalvo's method for the free edge of the beam ($x/\ell=1.0$) is shown in figure 9. Since the strain distribution at $x/\ell=1.0$ is similar to the temperature distribution, it is not surprising that the coarse-grid F.E. results for ε are inaccurate. For this particular case the results from Ojalvo's method are worse than the results using the Conventional F.E. with a coarse grid.

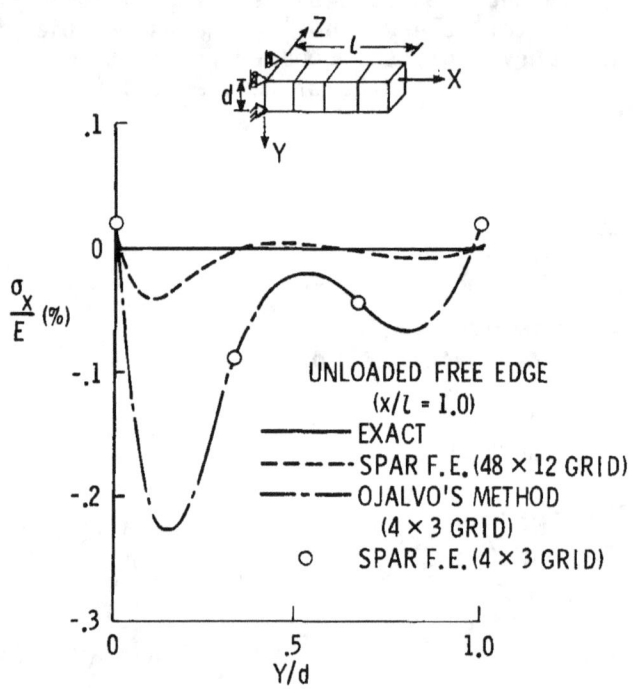

Figure 9

TEMPERATURE DISTRIBUTION IN SKIN AND STIFFENER OF A HEATED STIFFENED PANEL

Another application for which Ojalvo's method might be useful, other than for wing and fin-type structures, would be in the stiffeners of a heated stiffened panel (ref. 7). As indicated in reference 8, stresses in stiffeners can be critical during transient heating. Shown in figure 10 are temperature distributions in the skin and stiffener of a stiffened panel subjected to a hypersonic heating simulation (ref. 7). The specimen was cooled prior to heating to simulate a cold soak condition which explains the sub-ambient temperatures over a portion of the stinger. These temperatures were used as input for a F.E. structural analysis.

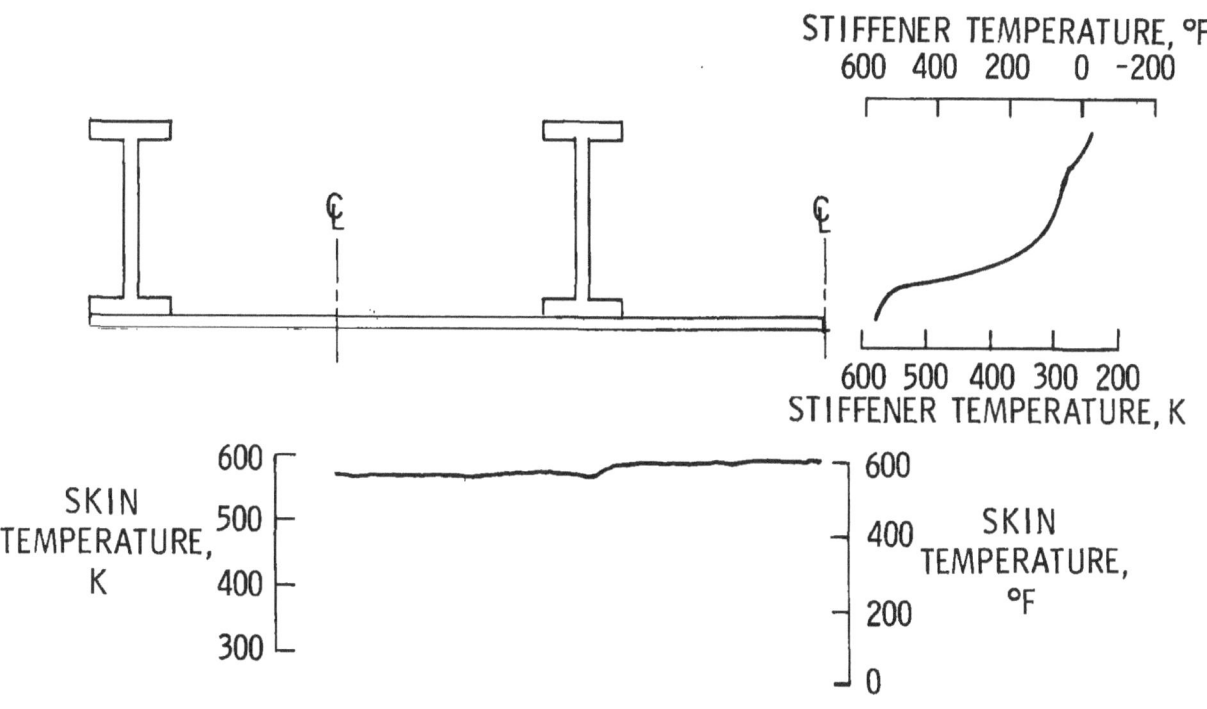

Figure 10

FINITE-ELEMENT IDEALIZATIONS OF A SQUARE STIFFENED TITANIUM PANEL

One-quarter of a stringer-stiffened panel was idealized by two different F.E. grids (fig. 11). The fine-grid model has ten membrane elements through the stringer depth and the coarse-grid model has one membrane element through the depth. Rod elements were used to represent the upper and lower flanges of the stiffener. Symmetry boundary conditions were imposed along the x- and y- axes and the point x=y=z=0 was fully constrained.

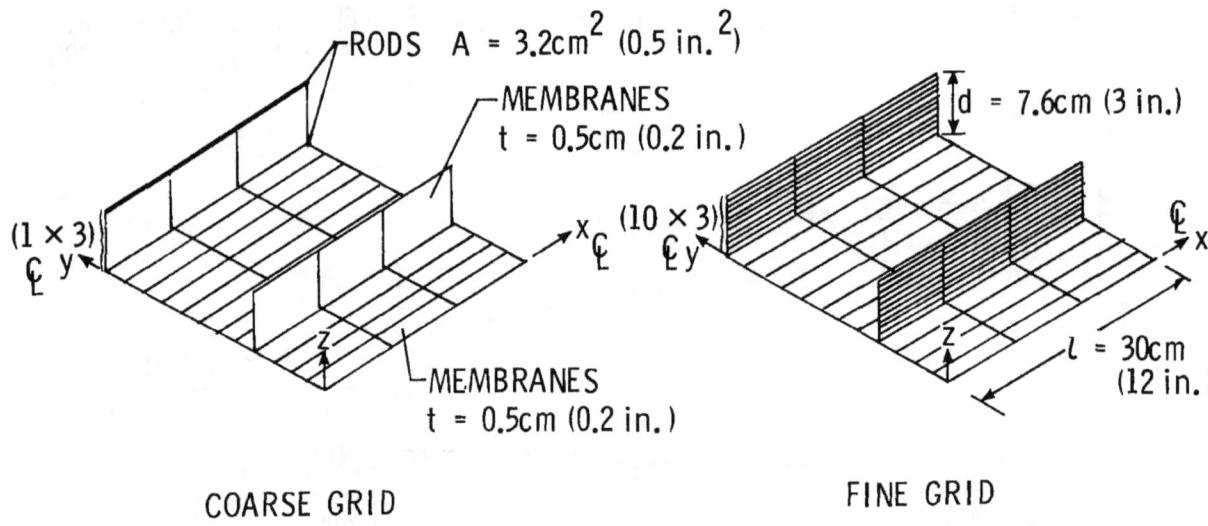

Figure 11

COMPARISON OF OJALVO'S METHOD WITH CONVENTIONAL F.E. FOR A STIFFENED PANEL

A comparison of strains of the fine- and coarse-grid models indicates that the coarse grid is sufficient in obtaining a good approximation to the linear strain distribution. The coarse-grid results for stresses (solid symbol and dashed curve) badly miss the peak stresses which occur in the center of the element as shown in figure 12 by the solid line. However, when the coarse grid strains are coupled with the detailed temperatures (Ojalvo's method) a good representation of the actual stress distribution (the square symbols) is obtained. This indicates that stiffeners in a heated panel are appropriate applications of Ojalvo's method in thermal/structural analysis and design.

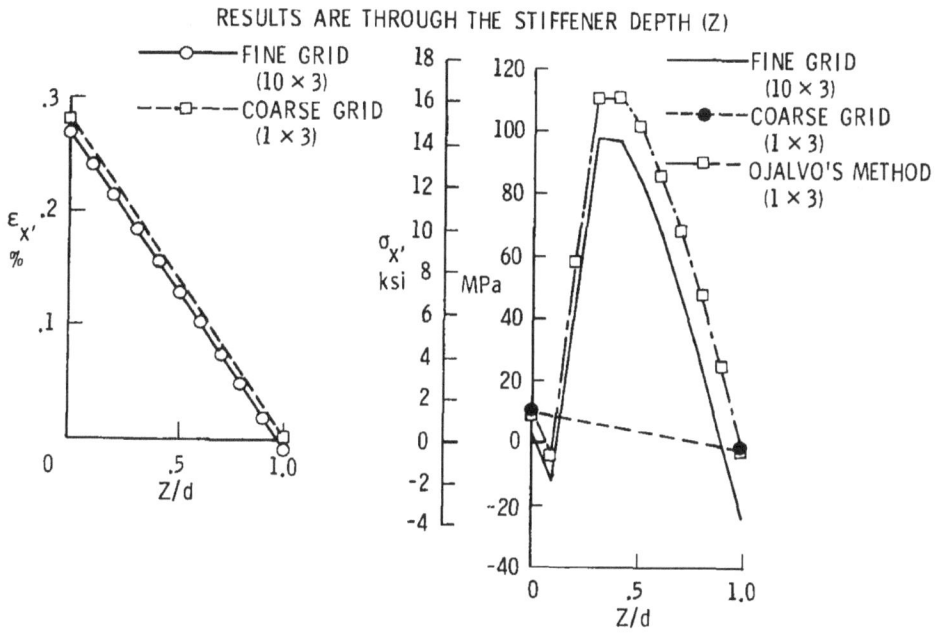

Figure 12

CONCLUDING REMARKS

The usefulness of a thermal/structural analysis technique for improving thermal stress calculations termed Ojalvo's method was investigated by numerical examples of several classes of thermostructural problems (fig. 13). The problems investigated include a rod, a thin membrane, a box beam, and a stiffened panel. The basis of Ojalvo's method is an observation that "strains in heated structures idealized by conventional components are generally less sensitive to spatially distributed temperature variations than are their corresponding stresses." Results of most bending-type problems indicate that Ojalvo's method is useful since the strains are related to the integrals of temperature and hence are less sensitive to local temperature variations. This means that for those problems where Ojalvo's method is appropriate the structural F.E. idealization may be coarser than the thermal idealization and also that the same coarse structural representation can be used for many different time slices in a quasi-static thermostructural analysis. For plane stress or membrane type problems the strain distributions are similar to the temperature distribution and a finite-element grid fine enough to calculate accurate temperatures would be necessary to calculate accurate stresses. This negates the usefulness of Ojalvo's method for this class of problems. Several useful areas for application of Ojalvo's method include built-up structures which can be idealized as bending elements (beams or plates) and stiffeners in a stiffened panel.

- EVALUATE OJALVO'S METHOD FOR THERMAL STRESS ANALYSIS
- KEY FEATURE - STRAINS ARE LESS SENSITIVE TO TEMPERATURE VARIATIONS THAN STRESSES
- PROBLEMS ANALYZED
    - ROD
    - MEMBRANE
    - BOX BEAM
    - STIFFENED PANEL
- FOR MEMBRANE AND PLANE STRESS PROBLEMS STRAINS ARE RELATED TO LOCAL TEMPERATURES AND OJALVO'S METHOD IS GENERALLY NOT USEFUL
- FOR BENDING PROBLEMS STRAINS ARE RELATED TO THE INTEGRALS OF TEMPERATURE AND OJALVO'S METHOD IS BENEFICIAL

Figure 13

REFERENCES

1. Ojalvo, I. U.: Improved Thermal Stress Determination by Finite Element Methods. Technical Note, AIAA Journal, Vol. 12, No. 8, August 19, 1974.

2. Whetstone, W. D.: SPAR Structural Analysis System Reference Manual-System Level II. Volume I-Program Execution. NASA CR-145096-1, 1977.

3. Zienkiewicz, O. C.: The Finite Element Method. Third ed., McGraw-Hill, New York, 1977.

4. Ojalvo, I. U.: Thermal Dynamic Modeling Study. NASA CR-2125, Oct., 1972.

5. Webber, J. P. H.: Thermo-Elastic Analysis of Rectangular Plates In Plane Stresses by the Finite Displacement Method. Journal of Strain Analysis, Vol. 2, No. 1, 1967, pp. 43-51.

6. Timoshenko, S. P.; and Goodier, J. N.: Theory of Elasticity. Third ed., McGraw-Hill, New York, 1970.

7. Jenkins, Jerald M.: Correlation of Predicted and Measured Thermal Stresses on an Advanced Aircraft Structure with Similar Materials. NASA TM-72862, April 1979.

STATUS REPORT ON DEVELOPMENT OF A REDUCED BASIS TECHNIQUE FOR
TRANSIENT THERMAL ANALYSIS

Charles P. Shore
Langley Research Center
Hampton, Virginia

INTRODUCTION

For some time researchers in structural analysis have recognized that the large number of degrees of freedom required in the solution of structural problems has often been the result of geometry and structural arrangement rather than complexity of the response behavior. This fact has led to considerable research into methods to reduce the degrees of freedom in structural problems and hence computer resources and costs. These methods have become known as reduction methods and are thoroughly reviewed in reference 1. One technique to reduce the degrees of freedom in static and dynamic problems is the reduced basis method which combines the classical Rayleigh-Ritz approximation with contemporary finite-element methods to retain modeling versatility as the degrees of freedom are reduced. The present paper reviews the reduced basis method and its applications to a nonlinear dynamic response problem presented in reference 1 and then summarizes the status of a research effort to apply the method to nonlinear transient thermal response problems.

SUMMARY OF METHOD FOR
NONLINEAR DYNAMIC RESPONSE PROBLEMS

The equation of motion for a nonlinear dynamic response problem neglecting damping is shown at the top of figure 1. In the equation $[M]$ represents the mass matrix, $\{\ddot{X}\}$ is a vector of nodal accelerations and $\{Q\}$ and $\{F\}$ are the applied loads and internal nodal forces, respectively. The total number of degrees of freedom in the problem is denoted by m. The internal nodal forces are comprised of a linear portion and a vector of nonlinear displacement dependent terms as indicated by the expression for $\{F\}$. The essence of the reduction method is to use a few known modes or global basis vectors to represent the displacements in the structure. Thus, $\{X\}$ is replaced by the expression $[\Gamma]\{\Psi\}$ where $[\Gamma]$ is a matrix whose columns are the known structural mode shapes and $\{\Psi\}$ is a vector of modal participation coefficients which become the new unknowns in the problem. For practical application to dynamic response problems, $[\Gamma]$ is composed of only the first few vibration modes; thus, n is much smaller than m. To reduce the equations, the expression for $\{X\}$ is substituted into the equation of motion and both sides of the equation premultiplied by the transpose of $[\Gamma]$.

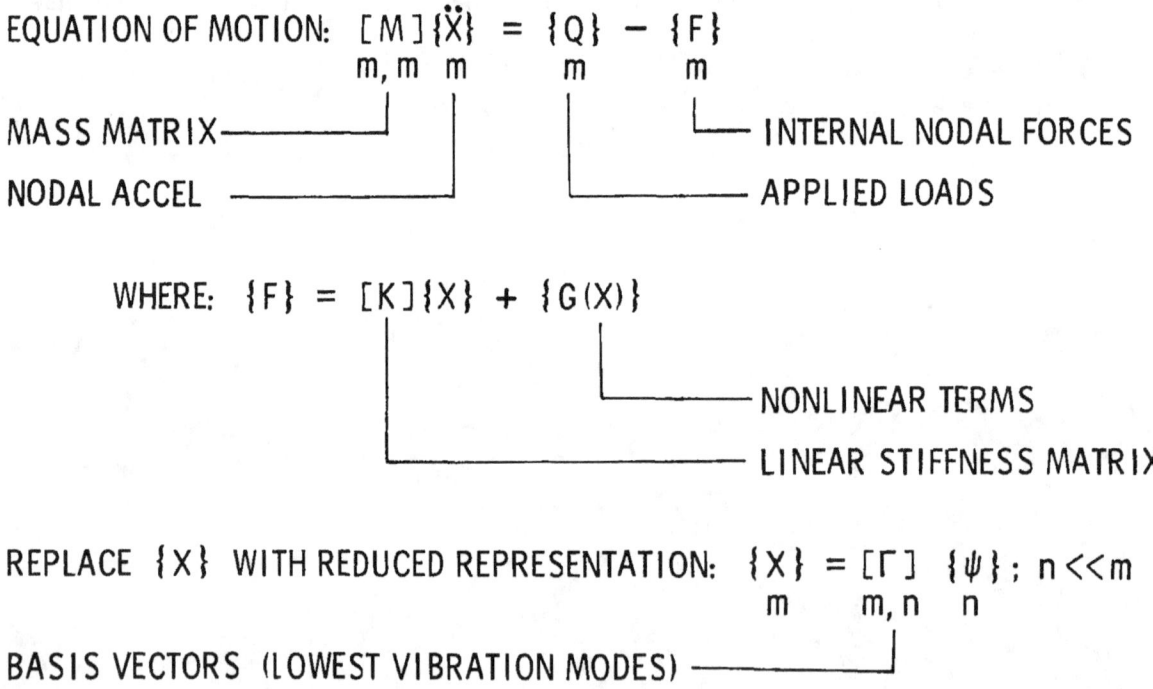

Figure 1

REDUCED EQUATIONS FOR NONLINEAR DYNAMIC RESPONSE PROBLEMS

The reduced equation of motion, expressed in terms of the unknown modal participation coefficients, is shown at the top of figure 2. In this equation n represents the number of basis vectors in $[\Gamma]$ and hence the number of unknowns in the reduced problem. The barred quantitites represent the reduced matrices or vectors and are obtained by the indicated matrix multiplications. As shown at the bottom of the figure, the solution process consists of solving eigenvalue problems to obtain the basis vectors, using the basis vectors to reduce the equations and then integrating the reduced equations to obtain the modal participation coefficients and thus, the dynamic response of the structure. This technique was applied to a shallow spherical cap subjected to a step load in reference 1 as described in the next two figures.

REDUCED EQUATION OF MOTION: $[\bar{M}]_{n,n} \{\ddot{\psi}\}_n = \{\bar{Q}\}_n - \{\bar{F}\}_n$

WHERE: $[\bar{M}]_{n,n} = [\Gamma]^T_{n,m} [M]_{m,m} [\Gamma]_{m,n}$

$\{\bar{Q}\}_n = [\Gamma]^T_{n,m} \{Q\}_m$

$\{\bar{F}\}_n = [\bar{K}]_{n,n} \{\psi\}_n + [\Gamma]^T_{n,m} \{G(\psi)\}_m$

$[\bar{K}]_{n,n} = [\Gamma]^T_{n,m} [K]_{m,m} [\Gamma]_{m,n}$

SOLUTION PROCESS}
- SOLVE EIGENVALUE PROBLEM FOR BASIS VECTORS
- REDUCE EQUATIONS
- INTEGRATE REDUCED EQUATIONS TO OBTAIN DYNAMIC RESPONSE

Figure 2

SELECTION OF BASIS VECTORS

As indicated in figure 3, a combination of two sets of basis vectors were considered for step loaded dynamic response problems in reference 1. The first consisted of a few eigenvectors from the solution of a linear eigenvalue problem based on initial conditions. The second set was comprised of a few vectors from the linear problem and a few from the solution of a steady-state (static) nonlinear eigenvalue problem where the structural stiffness matrix has been modified to contain the nonlinear stiffness terms associated with the steady-state nonlinear deflections.

CASE OF STEP LOADING

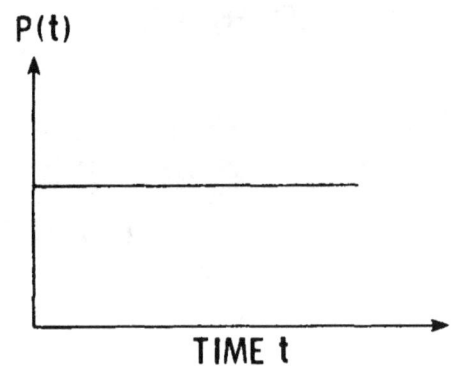

BASIS VECTORS CONSIST OF:

- FEW EIGENVECTORS OF LINEAR PROBLEM

$$[K]\{X\} = \lambda[M]\{X\}$$

- FEW EIGENVECTORS OF STEADY-STATE (STATIC) NONLINEAR PROBLEM

$$\left[[K] + \left[\frac{\partial G_i}{\partial X_j}\right]\right]\{X\} = \lambda[M]\{X\}$$

Figure 3

CLAMPED SHALLOW SPHERICAL CAP

The problem shown in figure 4 consists of a clamped spherical cap subjected to a point load of 177.93 Newtons at the apex applied as a step function in time. The shell is axially symmetric and the meridian was modeled by 10 shear-flexible curved elements with quintic interpolation functions for each of the displacement and rotation components (for a total of 148 nonzero displacement degrees of freedom). Nondimensional motion histories for the shell apex from the full system equations (148 degrees of freedom) and two sets of reduced equations (10 initial modes and 5 initial + 5 steady-state modes) are shown on the right of figure 4. The 10 initial or linear modes track the full system solution for a short time but fail to duplicate the full response of the shell. The combined linear and steady-state nonlinear modes, however, do a very good job of duplicating the response except for a slight shift in phase after about 200 microseconds. This good agreement has led to consideration of the modal reduced basis technique for nonlinear transient thermal analysis as outlined in figure 5.

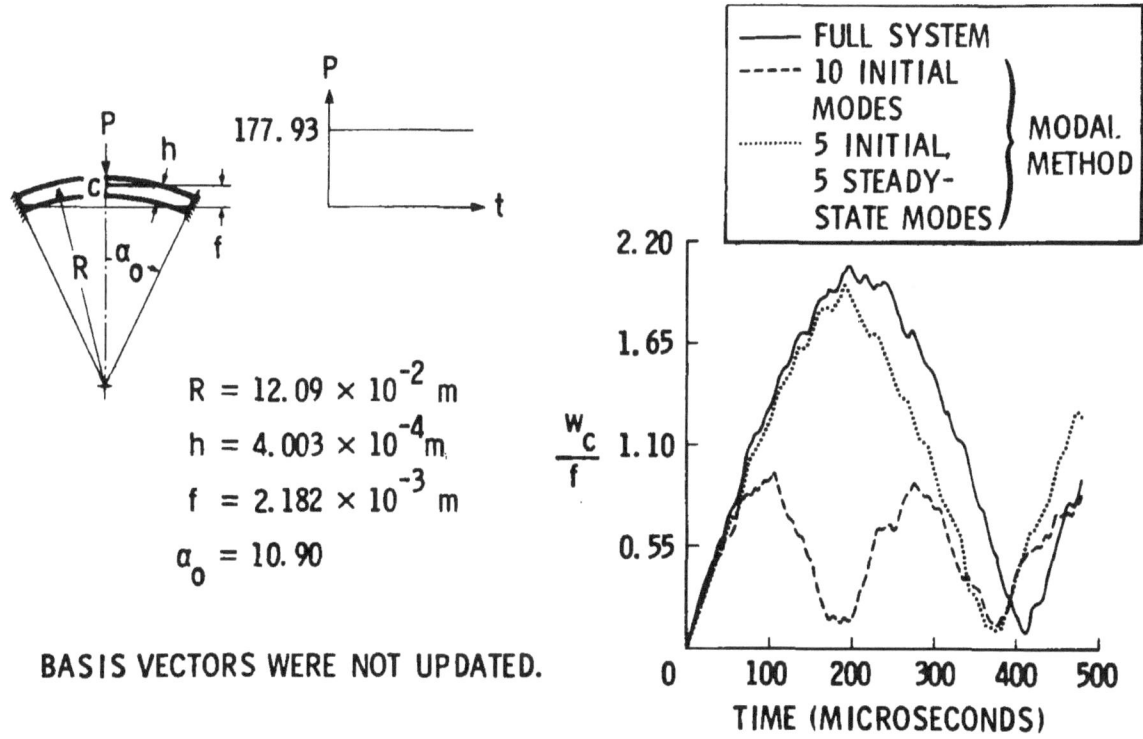

Figure 4

APPLICATION TO NONLINEAR TRANSIENT THERMAL ANALYSIS

Matrix equations describing heat transfer in a heated structure are shown at the top of figure 5. In the equations $[K]$ is the conductance matrix, $\{T\}$ the nodal temperatures, $[C]$ the capacitance matrix, $\{\dot{T}\}$ the time rate of change in the nodal temperatures and $\{Q\}$ the applied heat load. The total number of degrees of freedom is denoted by m. To reduce the equations, $\{T\}$ is replaced by a modal representation where $[\Gamma]$ contains vectors of thermal mode shapes and $\{\Psi\}$ is a vector of unknown modal participation coefficients. The vectors in $[\Gamma]$ may be obtained from solution of two thermal eigenvalue problems associated with the full system of equations. When $\{T\}$ is replaced with the modal representation in the heat transfer equation and both sides of the equation multiplied by the transpose of $[\Gamma]$, a set of reduced equations in terms of the unknown modal participation coefficients is obtained. The barred quantities represent the reduced matrices and vectors obtained by the indicated matrix multiplications. Similar to the dynamic response problem, it is assumed that local temperatures can be represented by a few global modes or basis vectors so that n will be much smaller than m.

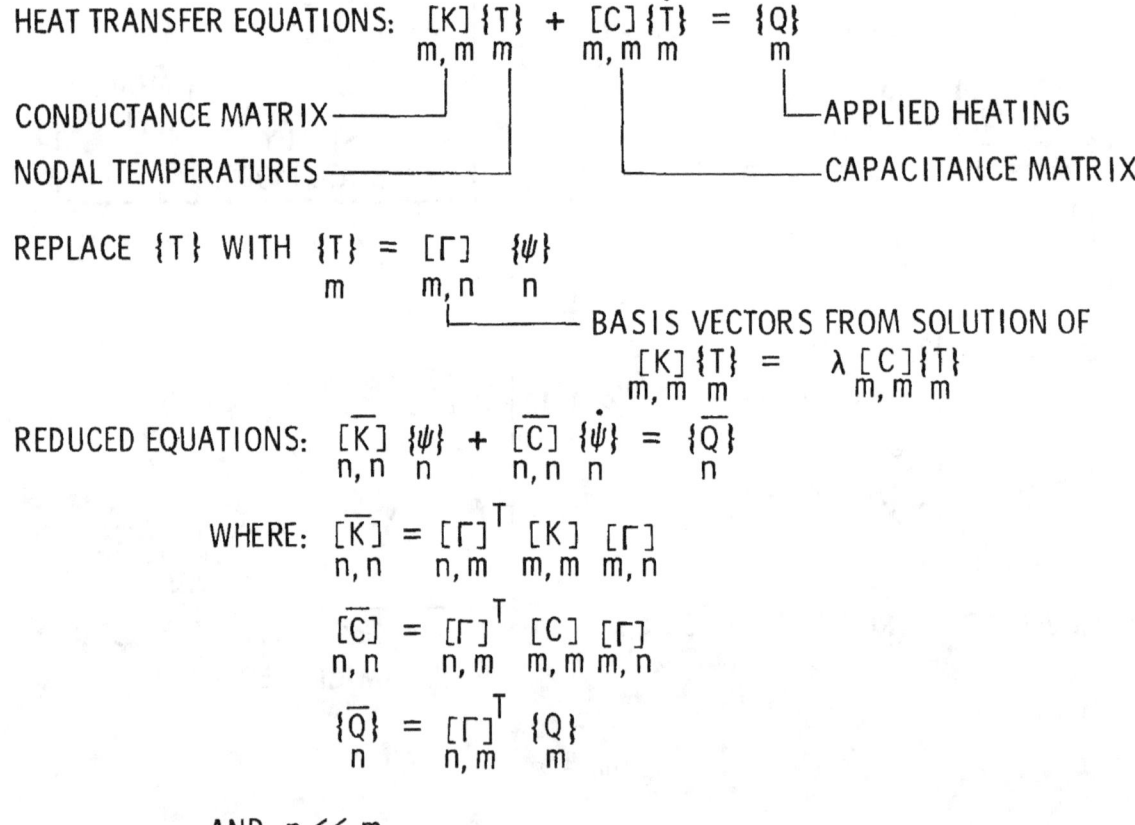

Figure 5

IMPLEMENTATION OF REDUCED BASIS TECHNIQUE

Implementation of the reduced basis technique for thermal problems is outlined on figure 6. The SPAR Finite Element Thermal Analyzer (ref. 2) was used to generate full system conductance and capacitance matrices and heat load vectors and save them for use in auxiliary computer programs. An existing eigenvalue extraction routine was used to solve the thermal eigenvalue problems to obtain thermal mode shapes used as basis vectors. These basis vectors were then used in a pilot computer program to reduce the full system equations and integrate them using the Crank-Nicholson algorithm to obtain the unknown modal participation coefficient $\{\psi\}$ and thus the thermal response. This process was evaluated by applying it to the sample problem described in figure 7.

- OBTAIN FULL SYSTEM MATRICES WITH SPAR THERMAL ANALYZER

- SOLVE EIGENVALUE PROBLEMS TO OBTAIN BASIS VECTORS

- USE TEST CODE TO REDUCE EQUATIONS

- USE A CRANK-NICHOLSON ALGORITHM TO INTEGRATE REDUCED EQUATIONS

Figure 6

SAMPLE THERMAL PROBLEM

The problem shown in figure 7 represents a 147.32 cm segment of the lower surface of the Space Shuttle wing and consists of a 0.39 cm thick aluminum skin covered by a 3.81 cm thick layer of Reusable Surface Insulation (RSI). The combined structure was modeled with two-dimensional finite elements as shown on the left of figure 7. The RTV adhesive-Strain Isolator Pad (SIP)-RTV adhesive bonding mechanism used to attach the RSI to the aluminum was also included in the model. The grid shown has 84 node points and hence 84 degrees of freedom since the elements used to model the structure have only temperature as the nodal degrees of freedom. The edges and aluminum structure were assumed to be adiabatic and the surface was heated by the heat pulse shown on the right of figure 7. The heat pulse is reasonably representative of Shuttle reentry and is sufficient to produce surface temperatures where radiation becomes appreciable and, thus, causes the heat transfer equations to become highly nonlinear. Thermal properties of the RSI are also nonlinear as indicated in figure 8.

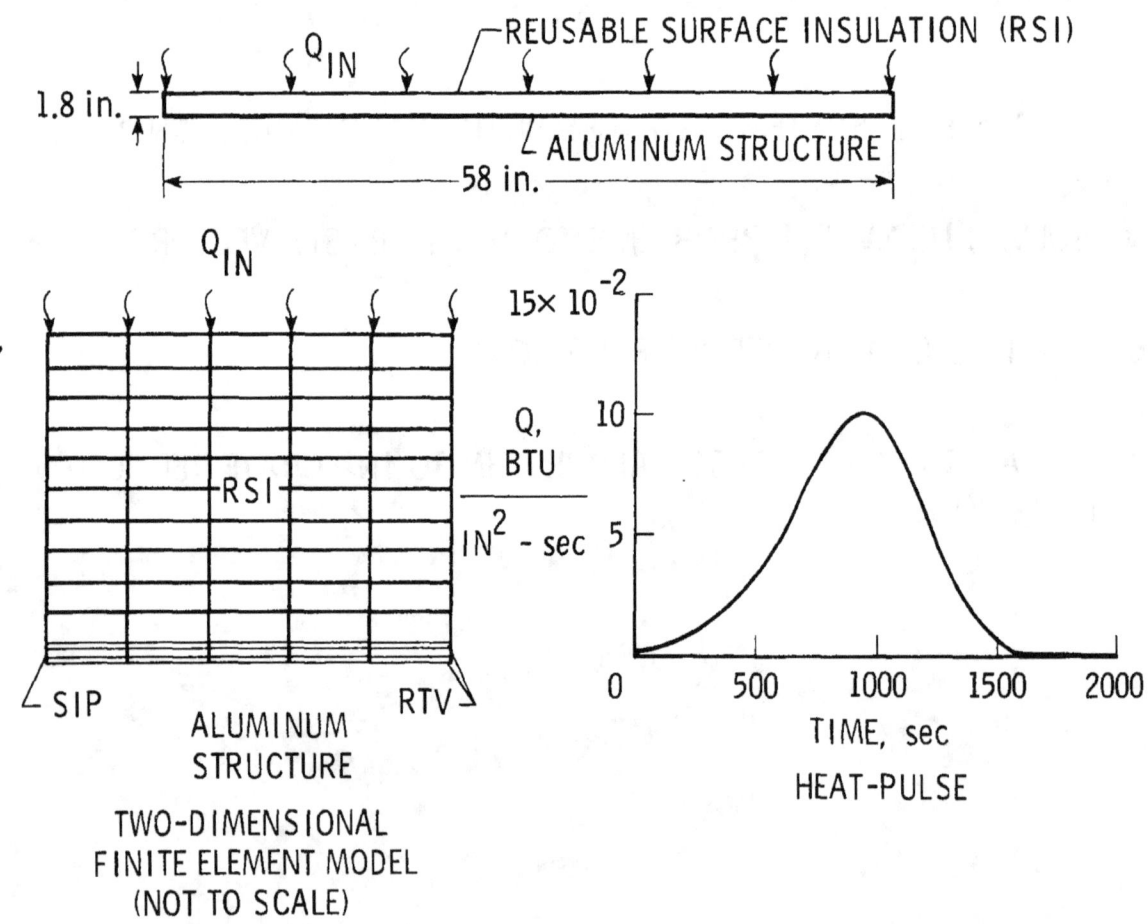

Figure 7

RSI THERMAL PROPERTIES

Specific heat and conductivity for the RSI are shown as functions of temperature in figure 8. The specific heat varies with temperature and because the RSI is very porous, the conductivity varies with pressure as well as temperature. The version of the SPAR Thermal Analyzer used in this investigation accomodates only temperature and time dependent properties. Consequently, the pressure dependency was converted to a time dependency by utilizing the known pressure history for a typical Shuttle reentry trajectory. Thus, the nonlinear material properties of the RSI also contribute to the overall nonlinearity of the problem.

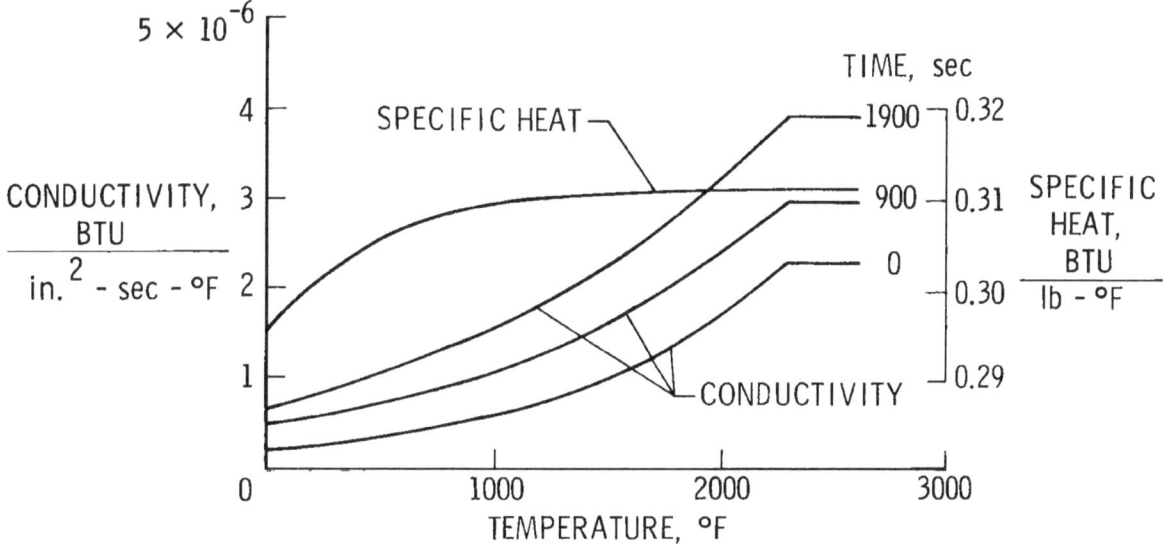

Figure 8

TEMPERATURE PROFILES FOR SAMPLE PROBLEM

A series of temperature distributions through the depth of the sample problem from a full SPAR analysis are shown in figure 9 for several discrete times during the heat pulse. These distributions indicate the type of behavior the basis vectors must approximate to be useful. Initially the entire structure is at a constant temperature of 311 K. As heating is applied, the RSI surface experiences a rapid temperature rise which gradually diffuses through the RSI and SIP to the aluminum skin. After peak heating occurs, the surface begins to cool while the interior of the RSI and the aluminum skin continue to experience a temperature increase. To be useful, the basis vectors used to reduce the degrees of freedom must characterize this nonlinear response, give accurate solutions and be easily and inexpensively generated. The approach used to generate basis vectors for this problem is shown in figure 10.

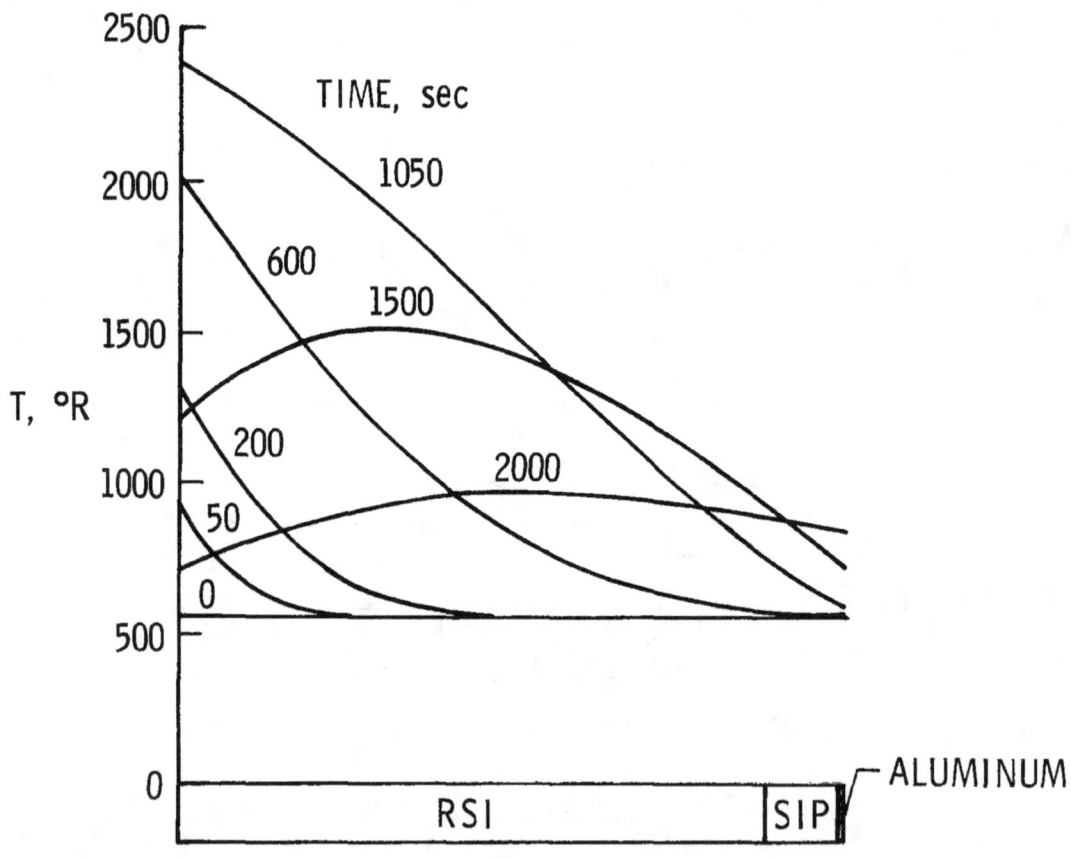

Figure 9

GENERATION OF BASIS VECTORS

Since the use of eigenvectors from the structural eigenvalue problem proved to be a useful set of basis vectors in the dynamic response problem, a similar approach was taken to generate basis vectors for the thermal response problem. In general, the thermal eigenvalue problem indicated in figure 10 would be solved for two temperature states of the system. The first state corresponds to the initial temperature condition and the second state corresponds to a temperature distribution from a "pseudo" steady-state problem for time averaged thermal properties and heating where the aluminum temperature was held constant at some selected value. A few thermal mode shapes from the first eigenvalue problem and a few from the second eigenvalue problem (which include the nonlinear temperature effects) would be combined to form a set of basis vectors. Additionally, for reasons which are explained subsequently, the reciprocal of the first vector from the two eigenvalue problems and a constant vector might also be included as basis vectors. Thermal mode shapes from the eigenvalue problem based on initial conditions are shown in figure 11.

- SOLVE THE EIGENVALUE PROBLEM: $[K]_{m,m} \{T\}_m = \lambda [C]_{m,m} \{T\}_m$

- USE THE THERMAL MODE SHAPES AS BASIS VECTORS

    (1) FIVE VECTORS FROM PROBLEM INITIAL CONDITIONS

    (2) FIVE VECTORS FROM PSEUDO STEADY STATE SOLUTION

    (3) RECIPROCAL OF FIRST VECTOR FROM (1) AND (2)

Figure 10

THERMAL MODE SHAPES (BASIS VECTORS)

Normalized thermal mode shapes from the linear eigenvalue problem (in which matrices were evaluated at an initial temperature of 311 K) are shown in figure 11. Although numbered sequentially, these modes do not, in fact, represent the first five modes from the two-dimensional eigenvalue problem associated with the finite element model shown in figure 7. Because of the two-dimensional nature of the eigenvalue problem, most of the lower modes involve multiple waves in the lateral direction. A total of 84 eigenvalues were extracted and the five modes shown have only a single wave in the lateral direction with multiple waves through the depth of the structure. As a first attempt to approximate the temperature distributions shown in figure 9, twelve modes from the eigenvalue problem for the initial temperature condition were selected as basis vectors. Additionally, to enhance the representation of the diffusion character of the temperature distributions up to 600 sec (see fig. 9), the reciprocal of the first mode shape was also used as a basis vector. Finally, to accommodate a uniform temperature change, a constant vector was included for a total of 14 basis vectors. Temperatures from the reduced basis method are compared with temperatures from full SPAR analysis in figure 12.

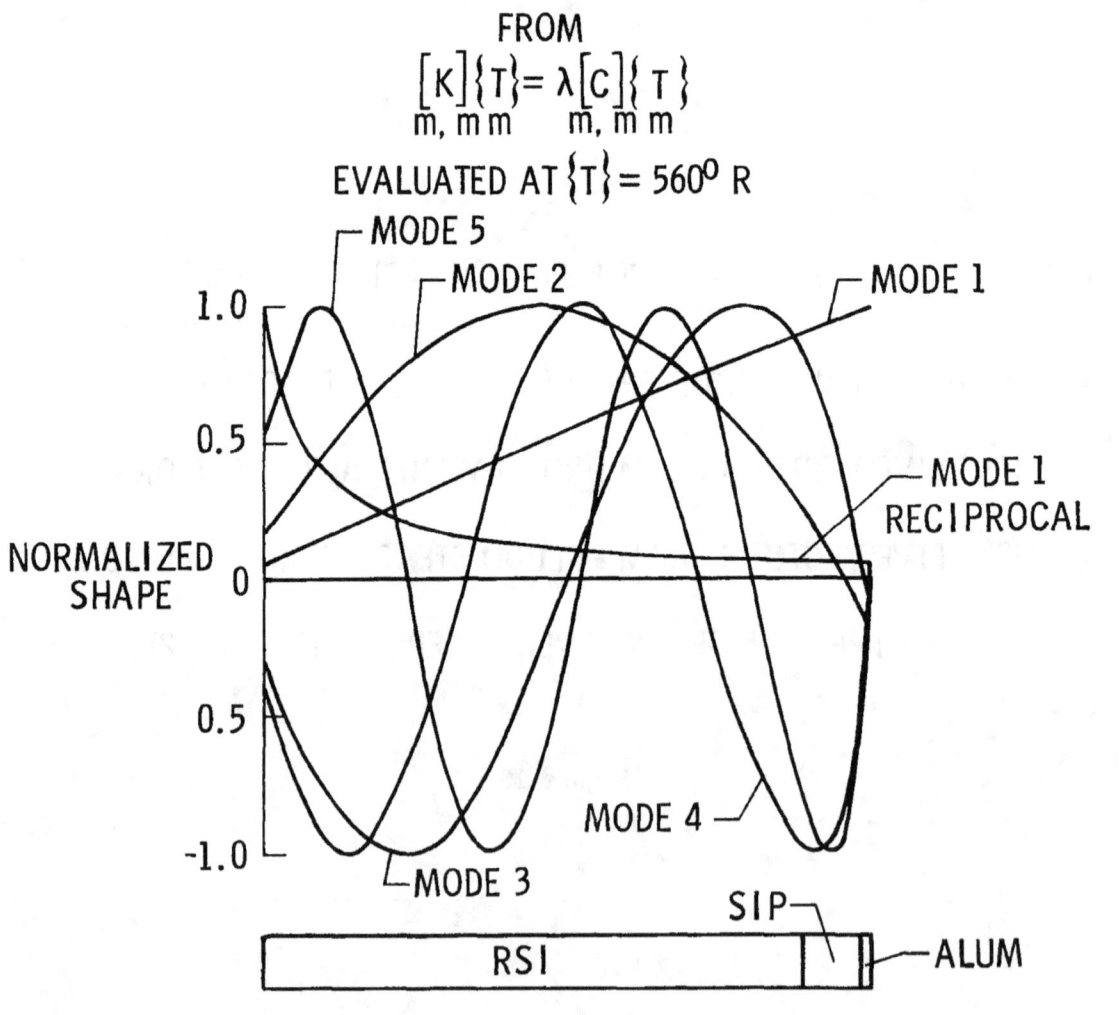

Figure 11

COMPARISON OF REDUCED BASIS AND FULL SYSTEM RESULTS

Temperature histories for the sample problem are shown for the RSI surface, RSI mid-point, and the aluminum structure in figure 12. The solid curves represent results from the full system of equations obtained with the SPAR Thermal Analyzer and the solid symbols are the reduced basis results based on the 14 modes discussed in figure 11. The results from the reduced basis method agree very well with those from the full SPAR analysis. However, it should be noted that the uniform heating and symmetry of the sample problem result in a one-dimensional problem in which a 14 degree of freedom model (i.e., a single vertical slice through the model in figure 7) would be sufficient for the problem. Thus, use of 14 basis vectors would be expected to give excellent results.

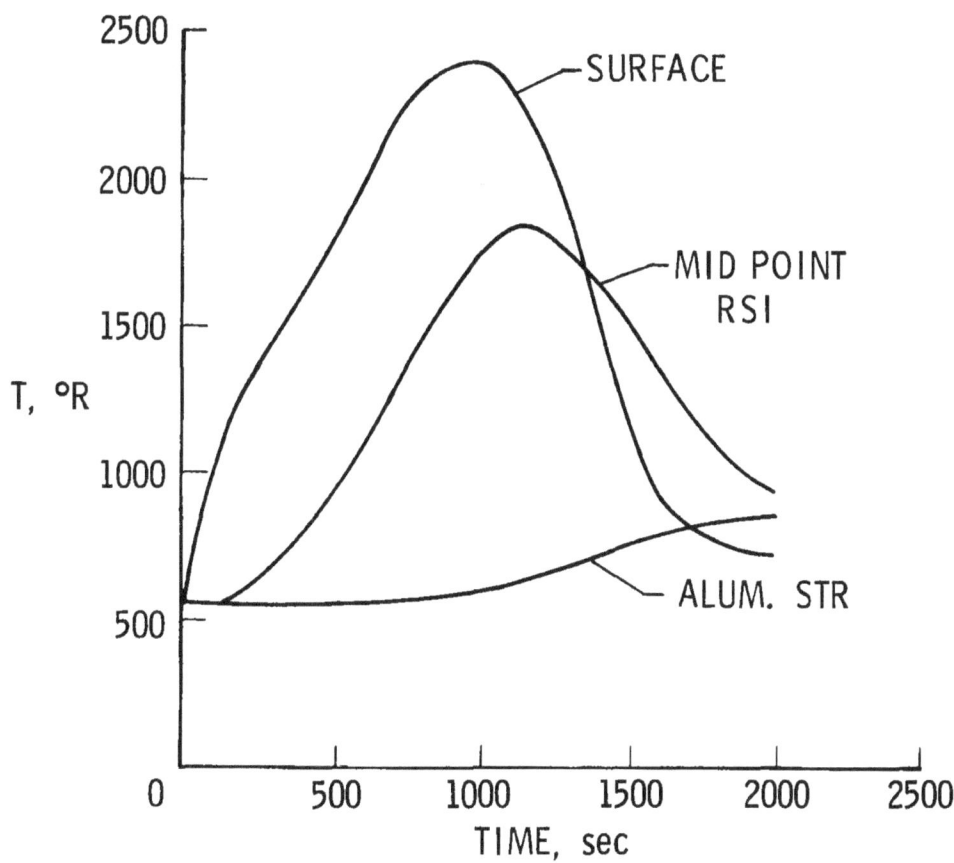

Figure 12

SUMMARY

The effort described in the current paper is directed toward applying the reduced basis method to nonlinear transient thermal analysis. Quite obviously the success of the method depends on the choice of basis vectors used to reduce the system of equations. Initial efforts used a set of 14 basis vectors consisting of modes from a thermal eigenvalue problem where the matrices were evaluated at the initial temperatures. This set of basis vectors gave excellent results for a one-dimensional 14 degrees of freedom thermal problem. Future work will focus on use of additional or alternate basis vectors including modes from the previously described eigenvalue problems, time derivatives of such eigenvectors, and possibly one-dimensional eigenvectors (analogous to the use of beam vibration modes in plate vibration problems). The type and number of basis vectors needed for approximate solutions to more complex problems beginning with two-dimensional nonsymmetric transient thermal problems will be studied.

REFERENCES

1. Noor, Ahmed K.: Recent Advances In Reduction Methods For Nonlinear Problems. Computers & Structures, Vol. 13, pp 31-44, 1981.

2. Marlowe, M. B.; Moore, R. A.; and Whetstone, W. D.: SPAR Thermal Analysis Processors Reference Manual, System Level 16, NASA CR-159162, 1979.

APPLICATIONS OF PERTURBATION TECHNIQUES TO HEAT-TRANSFER PROBLEMS

Osama A. Kandil
Department of Mechanical Engineering and Mechanics
Old Dominion University, Norfolk, Va. 23508

Summary

Two perturbation techniques are applied to two singular perturbation problems in heat transfer to obtain uniformly valid solutions which can serve as benchmarks for the numerical techniques: finite-difference and finite-element techniques. In the first problem, the method of strained parameters coupled with the application of a solvability condition is used to obtain a uniform solution for the problem of unsteady heat conduction in a long nearly circular cylinder. In the second problem, the method of matched asymptotic expansion coupled with Van Dyke's matching principle is used to obtain a uniform solution for the problem of one dimensional conduction-convection heat transfer of a uniform fluid flow.

I. Introduction

The main purpose of this paper is to demonstrate the capabilities of the perturbation techniques in developing approximate closed-form solutions for heat transfer problems involving difficulties which preclude their solutions exactly or require resorting to computational techniques such as finite-difference, finite-element, and panel techniques. These difficulties may be due to nonlinear governing equations, equations with variable coefficients, nonlinear boundaries, and existance of boundary layers near portions of the boundaries.

Although computational techniques in the areas of fluid dynamics, heat transfer, and structures are rapidly advancing and are capable of developing excellent solutions for realistic problems, one always needs bench-mark solutions, if experimental data are not available, to check the developed computer code or to check the accuracy of the computed results. In this regard, a closed form perturbation solution for a simplified problem which retains the same difficulties (weakly nonlinear equations and boundary conditions and weakly irregular boundaries) can best serve this purpose.

Among the perturbation techniques, the straightforward expansion in terms of a parameter in the problem leads to satisfactory results if one is dealing with a regular perturbation problem or if its region of nonuniformity is avoided (ref. 1 and 2). However, for singular perturbation problems, the straightforward expansions yield nonuniform solutions and one has to use other perturbation techniques to obtain uniform solutions (ref. 3-6). Infinite domains in a problem, a small parameter multiplying the highest derivative of the governing equation, type change of a partial differential equation, and existence of singularities are some of the sources of nonuniformities of straightforward solutions.

In this paper, two applications in the area of heat transfer are considered and closed-form uniform perturbation solutions are developed. In the first application, the problem of unsteady heat conduction in a long nearly circular cylinder is considered and a straightforward solution is shown to breakdown. The method of strained parameters coupled with the application of a solvability condition is used to develop a uniform solution.

Similar problems in the areas of duct acoustics (ref. 7) and vibrations (ref. 3) were considered where the methods of multiple scales and the method of strained parameters were used, respectively.

In the second application, the problem of one dimensional conduction-convection heat transfer of a uniform fluid flow in a single channel is considered. For small ratios of conduction to convection heat transfer, the problem is shown to possess a thermal boundary layer where large temperature gradients exist. The method of matched asymptotic expansion coupled with Van Dyke's matching principle is used to develop a uniform solution. This problem was considered in reference 8 for single and merging flows by using the finite-element technique. Steady two-dimensional problems with different locations of the boundary layer can be found in reference 6.

II. Unsteady Heat Conduction in a Long Nearly Circular Cylinder

We consider the two-dimensional unsteady heat conduction in a long cylinder whose cross sectional area is nearly circular. Initially, the cylinder is at temperature $g(r^*,\phi)$ and at any later time the surface is kept at zero temperature. The radius of the cylinder is expressed as

$$r_o^* = R + a\,f(\phi) \quad \text{where} \quad \int_0^{2\pi} f(\phi)\,d\phi = 0 \quad \text{and} \quad a \ll R \tag{1}$$

Dimensionless quantities are introduced by using the mean radius of the cylinder R, the characteristic temperature T_c, and the time R^2/α (α is the thermal diffusivity) as reference quantities. The dimensionless form of the problem is given by

$$\tilde{\theta}_{rr} + \frac{1}{r}\tilde{\theta}_r + \frac{1}{r^2}\tilde{\theta}_{\phi\phi} = \tilde{\theta}_t \tag{2}$$

$$\tilde{\theta}(r_o,\phi,t) = 0 \quad \text{on } r_o = 1 + \varepsilon f(\phi) \text{ and } \int_0^{2\pi} f(\phi)\,d\phi = 0 \tag{3}$$

$$\tilde{\theta}(r,\phi,0) = g(r,\phi) \tag{4}$$

The parameter ε is a small quantity characterizing the small deviation of the cross sectional area from the circular shape. The temporal variation is separated by assuming a solution of the form

$$\tilde{\theta}(r,\phi,t) = \theta(r,\phi)\,e^{-\beta^2 t} \tag{5}$$

Substituting equation (5) into equations (2) and (3), we obtain

$$\theta_{rr} + \frac{1}{r}\theta_r + \frac{1}{r^2}\theta_{\phi\phi} + \beta^2\theta = 0 \tag{6}$$

$$\theta(r_o,\phi) = 0 \quad \text{on } r_o = 1 + \varepsilon f(\phi) \tag{7}$$

Equation (6) is the Helmholtz equation. Although equation (6) is linear, the problem is not separable because the boundary condition, in the present form, is not separable. Since ε is a small parameter, one can expand $\theta(r,\phi;\varepsilon)$ in the form of a power series in terms of ε as follows

$$\theta(r,\phi;\varepsilon) = \theta_0(r,\phi) + \varepsilon\,\theta_1(r,\phi) + \text{-----} \tag{8}$$

In equation (8), only two terms are considered and hence a first-order solution is intended. Since ε appears in the argument of θ, equation (7), one needs to extract ε from this argument so that the process of equating coefficients of like powers of ε can correctly be accomplished. Therefore, the boundary condition of equation (11) is expanded around $r = 1$ using a Taylor-series expansion. This process is well known as the "transfer of the boundary condition." Thus, we get

$$\theta(r_o,\phi) = \theta(1,\phi) + \varepsilon f(\phi) \theta_r(1,\phi) + \cdots = 0 \tag{9}$$

Substituting equation (8) into equations (6) and (9) and equating coefficients of like powers of ε, we obtain the following two sets of problems:

$O(\varepsilon^0)$ - Problem

$$\theta_{orr} + \frac{1}{r}\theta_{or} + \frac{1}{r^2}\theta_{o\phi\phi} + \beta^2\theta_o = 0 \tag{10}$$

$$\theta_o(1,\phi) = 0 \tag{11}$$

$O(\varepsilon)$ - Problem

$$\theta_{1rr} + \frac{1}{r}\theta_{1r} + \frac{1}{r^2}\theta_{1\phi\phi} + \beta^2\theta_1 = 0 \tag{12}$$

$$\theta_1(1,\phi) = -f(\phi)\theta_{or}(1,\phi) \tag{13}$$

In the perturbation expansion used above, we note that only the dependent variable θ is expanded in terms of the small parameter ε. Such a perturbation method is called a "straightforward-perturbation method." Straightforward expansions break down when we deal with singular perturbation problems. Next, it is shown that the straightforward expansion breaks down for this problem.

The solution of the $O(\varepsilon^0)$ problem is obtained by using the method of separation of variables. The solution is found as

$$\theta_o = J_n(k_{nm}r)(A_{nm}e^{in\phi} + \bar{A}_{nm}e^{-in\phi}) \tag{14}$$

where $k_{nm}(=\beta)$ are the zeros of the $J_n(\beta) = 0$, and A_{nm} are complex constants. Setting $\beta^2 = k^2_{nm}$ in equation (12) and substituting equation (14) into equation (13), we get

$$\theta_{1rr} + \frac{1}{r}\theta_{1r} + \frac{1}{r^2}\theta_{1\phi\phi} + k_{nm}^2 \theta_1 = 0 \tag{15}$$

$$\theta_1(1,\phi) = -k_{nm} J_n'(k_{nm}) f(\phi) (A_{nm} e^{in\phi} + \bar{A}_{nm} e^{-in\phi}) \tag{16}$$

The solution of the $O(\varepsilon)$ problem is obtained by expanding each of θ_1 and $f(\phi)$ in a Fourier series as

$$\theta_1(r,\phi) = \sum_{-\infty}^{\infty} G_t(r) e^{it\phi} \tag{17}$$

$$f(\phi) = \sum_{-\infty}^{\infty} f_p e^{ip\phi} \tag{18}$$

where $f_p = \frac{1}{2\pi} \int_0^{2\pi} f(\phi) e^{-ip\phi}$ and $f_0 = 0$ according to the condition in equation (3). Substituting equation (17) and (18) into equations (15) and (16), multiplying the results by $\exp(-is\phi)$ and integrating from 0 to 2π, we get

$$G_s'' + \frac{1}{r} G_s' + (k_{nm} - \frac{s^2}{r^2}) G_s = 0 \tag{19}$$

$$G_s(1) = -k_{nm} J_n'(k_{nm}) [A_{nm} f_{s-n} + \bar{A}_{nm} f_{s+n}] \tag{20}$$

If $s \neq n$, equations (19) and (20) have a unique solution since the only solution for the corresponding homogeneous problem is the trivial solution. If $s = n$, we obtain

$$G_n'' + \frac{1}{r} G_n' + (k_{nm} - \frac{n^2}{r^2}) G_n = 0 \tag{21}$$

$$G_n(1) = -k_{nm} f_{2n} \bar{A}_{nm} J_n'(k_{nm}) \tag{22}$$

we note that the corresponding homogeneous problem has a nontrivial solution and hence the inhomogeneous problem will not have a solution unless a solvability condition is satisfied. However, a solvability condition does not exist and hence there is no solution to equations (21) and (22). Therefore, the solution of the $O(\varepsilon)$ problem breaks down.

The reason behind this trouble is due to the straightforward method used here. To obtain a uniform solution, we use the method of strained parameters. In this method, we expand β, in addition to the expansion given by equation (8), as follows

$$\beta = \beta_0 + \varepsilon \beta_1 + \cdots \tag{23}$$

where β_1 is to be determined in the course of the solution. Substituting equations (8) and (23) into equation (6), substituting equation (8) into equation (9), and equating coefficients of like powers of ε, we again obtain two sets of problems. The $O(\varepsilon^0)$ problem is the same as that given by equations (10) and (11) with the exception of replacing β^2 by β_0^2. The $O(\varepsilon)$ problem is given by

$$\theta_{1rr} + \frac{1}{r}\theta_{1r} + \frac{1}{r^2}\theta_{1\phi\phi} + \beta_0^2 \theta_1 = -2\beta_0\beta_1\theta_0 \tag{24}$$

and by equation (13).

Substituting equation (14) into equation (24) and replacing β_0 by k_{nm}, we get

$$\theta_{1rr} + \frac{1}{r}\theta_{1r} + \frac{1}{r^2}\theta_{1\phi\phi} + k_{nm}^2 \theta_1 = -2 k_{nm} \beta_1 J_n(k_{nm}r)(A_{nm}e^{in\phi} + \overline{A}_{nm}e^{-in\phi}) \tag{25}$$

The boundary condition is still given by equation (16).

The solution of equations (25) and (16) is obtained by substituting equations (17) and (18) into equations (25) and (16), multiplying the result by $\exp(-is\phi)$ and integrating from 0 to 2π. Again, we obtain two cases corresponding to $s \neq n$ and $s = n$. The former case is the same as that given by equations (19) and (20) in which a unique solution exists. In the latter case, we have

$$G_n'' + \frac{1}{r} G_n' + \left(k_{nm}^2 - \frac{n^2}{r^2}\right) G_n = -2 k_{nm} \beta_1 A_{nm} J_n(k_{nm}r) \tag{26}$$

$$G_n(1) = -k_{nm} f_{2n} \overline{A}_{nm} J_n'(k_{nm}) \tag{22}$$

As we mentioned before, the problem given by equations (26) and (22) has a solution if and only if a solvability condition exists. To obtain the solvability condition, we write equation (26) in the self-adjoint form by multiplying both sides of the equation by r. Next, we multiply both sides by the adjoint u(r) and integrate the result by parts over the range of r. Thus, we get

$$\int_0^1 G_n [(ru')' + (k_{nm}^2 r - \frac{n^2}{r})u] \, dr + [r u G_n' - r u' G_n]_0^1$$

$$= - 2 k_{nm} \beta_1 A_{nm} \int_0^1 r u J_n(k_{nm}r) \, dr \qquad (27)$$

The adjoint equation is obtained by setting the coefficient of G_n to zero. Thus, we obtain

$$(r u')' + (k_{nm}^2 r - \frac{n^2}{r}) u = 0 \qquad (28)$$

The adjoint boundary conditions are obtained by choosing

$$u(1) = 0 \quad \text{and} \quad u(0) < \infty \qquad (29)$$

in equation (27). Equation (27) reduces to

$$u'(1) G_n(1) = 2 k_{nm} \beta_1 A_{nm} \int_0^1 r u J_n(k_{nm}r) \, dr \qquad (30)$$

The solution of the adjoint problem, equations (28) and (29), is given by

$$u(r) = J_n(k_{nm}r) \qquad (31)$$

Substituting equations (22) and (31) into equation (30), and performing the integration on the right hand side, we obtain the equation defining β_1 as

$$\beta_1 = - k_{nm} f_{2n} \overline{A}_{nm}/A_{nm} \qquad (32)$$

since f_{2n}, $\overline{A}_{nm}$ and A_{nm} are complex constants, we assume

$$f_{2n} = b_{2n} e^{i\gamma_{2n}}$$

$$A_{nm} = \frac{1}{2} a_{nm} e^{i\lambda_{nm}}, \quad \overline{A}_{nm} = \frac{1}{2} a_{nm} e^{-i\lambda_{nm}} \qquad (33)$$

Substituting equation (33) into equation (32), and equating the real and imaginary parts, we get

$$\lambda_{nm} = \frac{1}{2}\gamma_{2n} \quad \text{or} \quad \frac{1}{2}(\gamma_{2n} - \pi) \tag{34}$$

$$\beta_1 = -k_{nm} b_{2n} \quad \text{or} \quad k_{nm} b_{2n} \tag{35}$$

Substituting equation (34) into equation (33) and substituting the result into equation (4), we get

$$\theta_o^{(1)} = a_{nm} J_n(k_{nm}r) \cos(n\phi + \gamma_{2n}/2) \tag{36}$$

and

$$\theta_o^{(2)} = a_{nm} J_n(k_{nm}r) \sin(n\phi + \gamma_{2n}/2) \tag{37}$$

Substituting equation (35) into equation (23), we get

$$\beta^{(1)} = k_{nm} - \varepsilon k_{nm} b_{2n} \tag{38}$$

$$\beta^{(2)} = k_{nm} + \varepsilon k_{nm} b_{2n} \tag{39}$$

Substituting equation (36) into equation (8), substituting this result and equation (38) into equation (5), repeating the same process with equations (37) and (39), and forming a linear combination of the two, we obtain

$$\theta(r,\phi,t) = \sum_{n,m} a_{nm} J_n(k_{nm}r) \{\cos(n\phi + \gamma_{2n}/2) \mathrm{Exp}[-k_{nm}(k_{nm} - 2\varepsilon b_{2n})t]$$

$$+ c_{nm} \sin(n\phi + \gamma_{2n}/2) \mathrm{Exp}[-k_{nm}(k_{nm} + 2\varepsilon b_{2n})t]\} + \text{----} \tag{40}$$

The constants a_{nm} and c_{nm} are found from the initial condition.

III. One Dimensional Conduction-Convection Heat-Transfer in a Uniform Flow

We consider the one dimensional convective diffusion equation modified by a convective surface loss term. This equation has been used to model the farfield behavior of thermal regime for single and merging fluid flows. Here, only the problem of single channel flows with specified temperatures at the upstream and downstream boundaries is considered. The governing equation of the average temperature $T(x^*)$ is given by

$$-k A T'' + \rho C A u T' + h p(T - T_e) = 0 \tag{1}$$

The boundary conditions are

$$T(0) = T_1 \tag{2}$$

$$T(L) = T_2 \tag{3}$$

In equations (1) and (2), k is the coefficient of thermal conductivity in the flow direction, A is the flow cross-sectional area, ρ is the fluid density, C is the fluid specific heat, u is the flow average velocity, h is the convection heat exchange coefficient, p is the convection perimeter, T_e is the convection exchange temperature, and L is the channel length.

Dimensionless parameters are introduced by using the pipe length L, and the temperature difference $T_1 - T_e$ as reference quantities. The dimensionless form of the problem is given by

$$-(1/P_e) y'' + y' + (N_u/P_e) y = 0 \tag{4}$$

$$y(0) = 1, \quad y(1) = \theta_2 \tag{5}$$

where $P_e = \rho C u L/k$ is the Peclet number, $N_u = hd/k$ is the Nusselt number, $d = 4A/b$ is the hydraulic diameter, and $y(x) = [T(x) - T_e]/(T_1 - T_e)$. For large Peclet numbers (small ratio of conduction to convection heat transfer) we let $1/P_e = \varepsilon$, where ε is now a small parameter. Moreover, we assume $N_u/P_e = b = O(1)$.

Thus, we obtain the following problem describing the spatial variation of the temperature:

$$\varepsilon y'' - y' - b y = 0 \tag{6}$$

$$y(0) = 1, \quad y(1) = \theta_2 \tag{7}$$

Equations (6) and (7) describe a typical boundary-layer problem where a small parameter multiplies the highest derivative. This problem can successfully be treated by using any of the several suitable perturbation techniques, namely, the method of matched asymptotic expansion, the method of multiple scales, and the method of composite expansions, among others. In this paper, we develop a uniformly valid solution by using the method of matched asymptotic expansion.

Since the coefficient of y' is negative in the interval $0 \leq x \leq 1$, the boundary layer exists at the boundary $x = 1$. Next, we develop outer and inner solutions and match them by using Van Dyke's matching principle to obtain a composite solution which is uniformly valid everywhere.

Outer Solution:

The outer solution $y^o(x;\varepsilon)$ is expressed in the form

$$y^o(x;\varepsilon) = y_0(x) + \varepsilon\, y_1(x) + \text{----} \tag{8}$$

Dropping the second boundary condition of equation (7), substituting equation (8) into equation (6) and into the first boundary condition of equation (7), and equating coefficients of like powers of ε, we obtain the following problems:

$O(\varepsilon^o)$ Problem

$$y_0' + b\, y_0 = 0 \tag{9}$$

$$y_0(0) = 1 \tag{10}$$

$O(\varepsilon^1)$ Problem

$$y_1' + b\, y_1 = y_0'' \tag{11}$$

$$y_1(0) = 0 \tag{12}$$

The outer solution of equations (9)-(12) is given by

$$y^o = e^{-bx}(1 + \varepsilon\, b^2\, x) + O(\varepsilon^2) \tag{13}$$

Inner Solution:

To develop an inner solution valid near the boundary layer, we drop the boundary condition at $x = 0$ and introduce the following stretching transformation for the inner variable

$$\xi = \frac{1-x}{\varepsilon} \quad (14)$$

In terms of the inner variable, the inner solution $y^i(\xi;\varepsilon)$ is governed by

$$y^{i''} + y^{i'} - \varepsilon b y^i = 0 \quad (15)$$

$$y^i(0) = \theta_2 \quad (16)$$

Next, we expand the inner solution in the form

$$y^i(\xi;\varepsilon) = Y_0(\xi) + \varepsilon Y_1(\xi) + \cdots \quad (17)$$

Substituting equation (17) into equations (15) and (16) and equating coefficients of like powers of ε, we obtain the following problems:

$O(\varepsilon^0)$ Problem

$$Y_0'' + Y_0' = 0 \quad (18)$$

$$Y_0(0) = \theta_2 \quad (19)$$

$O(\varepsilon^1)$ Problem

$$Y_1'' + Y_1' = b Y_0 \quad (20)$$

$$Y_1(0) = 0 \quad (21)$$

The inner solution of equations (18)-(21) is given by

$$y^i = \theta_2 + A_0(1 - e^{-\xi}) + \varepsilon\{b \xi[\theta_2 + A_0(1 + e^{-\xi})] + A_1(1 - e^{-\xi})\} + O(\varepsilon^2) \quad (22)$$

We note that the outer solution is completely known while the inner solution contains the two unknown coefficients A_0 and A_1. They are determined by applying Van Dyke's matching principle to the inner and outer solutions. The principle states that:

the m-term inner expansion of (the n-term outer expansion) = the n-term outer expansion of (the m-term inner expansion)

where m and n are any two integers. To find the left hand side, we write the n-term outer expansion in terms of the inner variable, equation (14), expand the functions of ε keeping the inner variable fixed, and keep m-terms of the resulting expansions. An opposite procedure is applied to the right hand side. For m = n = 2, we obtain

$$A_o = e^{-b} - \theta_2, \quad A_1 = b^2 e^{-b} \tag{23}$$

Substituting equation (23) into equation (22), we obtain the inner solution as

$$y^i = \theta_2 + (e^{-b} - \theta_2)(1 - e^{-\xi}) + \varepsilon\{b\,\xi[\theta_2 + (e^{-b} - \theta_2)(1 + e^{-\xi})] + b^2 e^{-b}(1-e^{-\xi})\} + O(\varepsilon^2) \tag{24}$$

Next, we express the composite expansion in the form

$$y^c = y^o + y^i - (y^o)^i \tag{25}$$

Substituting equations (13) and (24) and the result of the left hand side of Van-Dyke's matching principle in equation (25), we obtain the composite solution as

$$y^c = (1 + \varepsilon b^2 x)e^{-bx} + [(1-b+bx)(\theta_2 - e^{-b}) - \varepsilon b^2 e^{-b}]e^{-(1-x)/\varepsilon} + O(\varepsilon^2) \tag{26}$$

It should be noted that an exact solution exists for equations (6) and (7) which is given by

$$y = [(e^{m_2} - \theta_2)e^{m_1 x} - (e^{m_1} - \theta_2)e^{m_2 x}]/(e^{m_2} - e^{m_1}) \tag{27}$$

where $m_{1,2} = [1 \pm \varepsilon(1 = 4\,\varepsilon\,b)^{1/2}]/2\,\varepsilon$ \hfill (28)

The solution of the unsteady flow problem described by

$$-k\,A\,T_{x^*x^*} + \rho\,C\,A\,U\,T_{x^*} + h\,p(T-T_e) + \rho\,C\,A\,T_{t^*} = 0 \tag{29}$$

$$T(0, t^*) = T_1 \quad, \quad T(L, t^*) = T_2 \quad, \quad T(x^*, t^*) = T_0 \tag{30}$$

is an easy extension to the steady solution given by equation (26). The problem is divided into steady and transient problems. The steady problem with the inhomogeneous boundary conditions is already considered. Upon assuming an exponential time-decay solution, the transient problem with homogeneous boundary conditions reduces to an eigenvalue problem coupled with a boundary layer. A uniform solution of the problem is obtained by using the method of matched asymptotic expansions (as shown before) coupled with strained eigenvalues which are expressed as a power series of ε.

IV. Concluding Remarks

The methods of strained parameters and matched asymptotic expansions are successfully used to obtain closed-form perturbation solutions for heat-transfer problems with irregular boundaries and with boundary layers, respectively. The techniques given here are applicable to a large class of similar problems where various geometrical shapes and three-dimensional dependence are considered. Nevertheless, such solutions serve as bench marks for the computational techniques. Moreover, useful solutions can be obtained by combining a computational technique with a perturbation technique in a certain problem. The computational technique is used to solve the zeroth-order problem while the perturbation technique is used to solve the higher-order problems.

References

1. Ash, R. L.; Crossman, G. R.: Influence of Temperature Dependent Properties on a Step-Heated Semi-Infinite Solid. Trans. of the ASME, May 1971, pp. 250-253.

2. Cooper, L. Y.: Constant Heating of a Variable Conductivity Half Space. Quart. of Applied Math., Vol. 27, 1969, pp. 173-183.

3. Nayfeh, A. H.: Introduction to Perturbation Techniques. John Wiley & Sons, Inc., N.Y., 1981.

4. Van Dyke, M.: Perturbation Methods in Fluid Mechanics. The Parabolic Press, Stanford, Calif., 1975.

5. Nayfeh, A. H.: Perturbation Methods. John Wiley & Sons, Inc., N.Y., 1973.

6. Cole, J. D.: Perturbation Methods in Applied Mathematics. Blaisdell Publishing Company, Waltham, Mass., 1968.

7. Neyfeh, A. H.; Kandil, O. A.: Propagation of Waves in Cylindrical Hard-Walled Ducts with General Weak Undulations. AIAA Journal, Vol. 16, No. 10, October 1978, pp. 1041-1044.

8. Thornton, E. A.; Dechaumphai, P.: Convective Heat Transport in Merging Flows. Third Int. Conference on Finite Elements in Water Resources, Univ. of Miss., Oxford Campus, May 19-23, 1980.

DEVELOPMENT OF MIXED TIME PARTITION PROCEDURES FOR THERMAL ANALYSIS OF STRUCTURES[*]

Wing Kam Liu
Assistant Professor
Department of Mechanical and Nuclear Engineering
Technlogical Institute
Northwestern University
Evanston, Illinois 60201

ABSTRACT

The computational methods used to predict and optimize the thermal-structural behavior of aerospace vehicle structures are reviewed. In general, two classes of algorithms, implicit and explicit, are used in transient thermal analysis of structures. Each of these two methods has its own merits. Due to the different time scales of the mechanical and thermal responses, the selection of a time integration method can be a difficult yet critical factor in the efficient solution of such problems.

Therefore mixed time integration methods for transient thermal analysis of structures are being developed. This proposed methodology would be readily adaptable to existing computer programs for structural thermal analysis.

1. INTRODUCTION

Over the last two decades, significant attention has been devoted to the development of lightweight, durable thermal protection systems (TPS) for future space transportation systems. Research programs are currently under way at the Langley Research Center to investigate various metallic TPS concepts [1]. One of the proposed candidates is the titanium multiwall tile (see [2] and references therein for a discussion). Early design procedures of the TPS concept involved both analytical and experimental studies. In particular, a degree of confidence has been established in the TPS concept due to the design studies by Jackson and Dixon [3] and Blair et. al. [4].

A titanium multiwall tile consists of alternating layers of superplastically formed dimpled sheets and flat septum sheets of titanium foil. As described in reference [3], this multiwall concept impedes all three modes of

---

[*] The partial suppport of NASA under Grant No. NAG-1-210 to this research is gratefully acknowledged.

heat transfer----conduction, radiation and convection. The superplastically formed dimpled sheets and the long thin conduction path tend to minimize heat conduction. The flat septum sheets of titanium foil impede radiation. The small individual volumes created by the dimpled layers virtually eliminate air convection. The optimal design of such thermal protection systems requires effective techiques in coupled thermal and stress analyses. Finite element methods offer the greatest potential in modeling such complicated problems. However, the resulting semi-discrete equations may involve many thousand degrees of freedom. Since the problem to be solved is transient and non-linear, the selection of an appropriate time integration method is an essential step in the solution of such a complicated problem. Adelman and Haftka [5] recently conducted a survey study on the performance of explicit and implicit algorithms for transient thermal analysis of structures. Calculations were carried out using the SPAR finite element computer program [6] and a special purpose finite element program incorporating the GEARB and GEARIB algorithms. Based upon their studies, they concluded that, generally, implicit algorithms are preferable to explicit algorithms for "stiff" problems, though non-convergence and/or wide-banding of the resulting matrix equations may decrease the advantage of the implicit methods.

These difficulties are similar to those found in fluid-structure problems. Over the past few years, several remedies have been proposed for these difficulties. Belytschko and Mullen [7] have proposed an explicit-implicit method where the mesh is partitioned into domains by nodes and the partitions are simultaneously integrated by explicit and implicit methods. Hughes and Liu [8] have proposed an alternate implicit-explicit finite element method where the mesh is partitioned into domains by elements and this element partition concept simplifies the computer-implementation and enhances its compatibility with the general purpose finite element software.

Although the implicit-explicit method has been proven to be very successful in some fluid-structure interaction problems (see e.g., [8-10]), the size and complexity of the program are increased because of the addition of the implicit method. To overcome these difficulties, Belytschko and Mullen [11] have proposed an E^m-E partition, in which explicit time integration is used throughout. However, different time steps within different parts of the mesh can be employed simultaneously. Partitioned and adaptive algorithms for explicit time integration have also been proposed by Belytschko [12].

Recently, Liu and Belytschko [13] put forward a general mixed time implicit-explicit partition procedure within a linear context. It incorporates the mentioned algorithms as special cases and is shown to have better stability properties than those in E^m-E partition [11]. Similar concepts can also be used in transient conduction forced-convection analysis (see Liu and Lin [14]).

In the present paper, we extend these implicit-explicit concepts (nodes and elements) to transient thermal analysis of structures where different time integration methods with different time steps can be used in each element group. The aim of this approach is to achieve the attributes of the various time integration methods.

For example, in transient structural analysis, explicit methods require the size of the time step to be proportional to the length of the shortest element, while in transient thermal analysis, explicit methods require the step size to be proportional to the the square of the length of the shortest element. So it is more advantageous to employ this mixed time implicit-explicit technique for transient thermal analysis of structures since the E^m-E partition proposed in [11,12] is often inefficient for this kind of problem though it is very efficient in structural analysis.

In section 2 we review the finite element formulation for transient heat conduction. In section 3 we describe the mixed time integration procedures viz two element groups "A" and "B". A family of integration partitions can then be deduced by selecting the appropriate definitions for the quantities of "A" and "B". Five useful partitions which are of practical importance are presented. In section 4 the stability characteristic of the algorithm is discussed. In section 5 we generalize the mixed time methods described in section 3 to NUMEG element groups. A computational algorithm for this mixed time implicit-explicit integration is also presented. Numerical results are presented in section 6 and conclusions and suggestions for further research are presented in section 7.

2. FINITE ELEMENT FORMULATION FOR TRANSIENT HEAT CONDUCTION

We consider a body Ω enclosed by surface Γ which consists of two parts: Γ_g and Γ_q. The Cartesian coordinates of the body will be denoted by x_i.

The governing equations for transient heat conduction are:

$$\theta,_{ii} = \frac{1}{c^2} \dot{\theta} \qquad \text{in } \Omega \qquad (2.1)$$

$$\theta = g \qquad \text{for } x_i \text{ in } \Gamma_g \qquad (2.2)$$

$$\theta,_i n_i + h\theta = q \qquad \text{for } x_i \text{ in } \Gamma_q \qquad (2.3)$$

and

$$\theta = \theta_o \qquad \text{for } x_i \text{ in } \Omega \text{ and } t = 0. \qquad (2.4)$$

Here a comma designates a partial derivative with respect to x_i; a superscript dot designates time (t) derivative; n_i is the component of the outward unit normal vector; c^2 is the thermal diffusivity (the ratio of thermal conductivity to specific heat times density); θ is the temperature; h is the convective heat transfer coefficient; and g, q and θ_o are given functions. Repeated indices denote summations over the appropriate range.

The variational or weak form of equations (2.1)-(2.4) is:

$$(\dot{\theta},v) + a(\theta,v) = (q,v) \qquad \text{in } \Gamma_q \qquad (2.5)$$

where v is the test function; and

$$(\dot{\theta},v) = \int_\Omega \frac{1}{c^2} \dot{\theta} \, v d\Omega \qquad (2.6)$$

$$a(\theta,v) = \int_\Omega \theta_{,i} v_{,i} d\Omega + \int_{\Gamma_q} h\theta v \, d\Gamma \qquad (2.7)$$

and

$$(q,v)_{\Gamma_q} = \int_{\Gamma_q} qv \, d\Gamma \qquad (2.8)$$

The finite element equations are obtained by approximating the trial functions by shape functions (N_i) so that

$$v = \sum_{i=1}^{NEQ} N_i(x_j) d_i(t) \qquad (2.9)$$

$$g = \sum_{i=NEQ+1}^{NUMNP} N_i(x_j) g_i(x_j,t) \qquad (2.10)$$

and

$$\theta = v + g \qquad (2.11)$$

Here NUMNP is the total number of nodal points used in the finite element mesh and NEQ is the number of trial functions used (for this particular case it is equal to the number of equations to be solved).

The resulting semidiscrete equation for transient heat conduction is then:

$$\underline{M}\dot{\underline{\theta}} + \underline{K}\underline{\theta} = \underline{F} \qquad (2.12)$$

with initial condition

$$\dot{\underline{\theta}}(0) = \underline{\theta}_o \qquad (2.13)$$

where

$$\underline{M} = [M_{ij}] = (N_i, N_j) = \int_\Omega \frac{1}{C^2} N_i N_j d\Omega \qquad (2.14)$$

$$\underline{K} = [K_{ij}] = a(N_i, N_j) = \int_\Omega N_{i,k} N_{j,k} d\Omega + \int_{\Gamma_q} h N_i N_j \, d\Gamma \qquad (2.15)$$

and

$$\underline{F} = [F_i] = (q, N_i)_{\Gamma_q} - (N_k, N_i)\dot{g}_k - a(N_k, N_i) g_k \qquad (2.16)$$

$\underline{M}$ and $\underline{K}$ are assumed to be symmetric and positive definite. The thermal parameters, C^2 and h, are in general temperature dependent. However for discussion purposes herein, they are assumed constant throughout.

3. MIXED TIME PARTITION PROCEDURES

In this section, mixed time integration methods are employed to solve equations (2.12) and (2.13). For the purpose of describing these mixed time integration techniques we subdivide the mesh into element groups A and B, each of which is to be integrated by a different method. Let n be the time step number, $\underline{\theta}_n$, $\underline{v}_n$ and $\underline{F}_n$ be approximations to $\underline{\theta}(t_n)$, $\dot{\underline{\theta}}(t_n)$ and $\underline{F}(t_n)$ respectively. Let $m\Delta t$ and Δt be the time steps used for element group A and element group B respectively, where m is an integer and is greater or equal to 1. A time step cycle (mΔt) can then be defined by an increment of m substeps with a time step of Δt each, so that one time step cycle is defined by step n to step n+m. The portions of the matrices obtained by assembling element group A and element group B are denoted by superscripts "A" and "B", respectively. Hence it follows that any global matrix is the sum of the two matrices, cf. $\underset{\sim}{M} = \underset{\sim}{M}^A + \underset{\sim}{M}^B$ and $\underset{\sim}{K} = \underset{\sim}{K}^A + \underset{\sim}{K}^B$. Nodes associated with <u>only</u> element group B are denoted by superscript "B", whereas those which are in contact with at least one element of group A are denoted by superscript "A"; nodes which are connected to both group A and group B are designated by "C", so "C" is a subset of "A". To simplify the presentation, we further denote those element matrices associated with at least one node C by superscript "C", so $\underset{\sim}{M}^C$ and $\underset{\sim}{K}^C$ are subsets of $\underset{\sim}{M}^B$ and $\underset{\sim}{K}^B$ respectively. However, in actual computer implementation this element group is not necessary. We further denote $\underset{\sim}{M}^R = \underset{\sim}{M}^A + \underset{\sim}{M}^C$ and $\underset{\sim}{K}^R = \underset{\sim}{K}^A + \underset{\sim}{K}^C$.

Similarly, all vectors are then partitioned accordingly into "A" and "B" parts, cf. $\underset{\sim}{\theta} = (\theta^A \theta^B)^T$, $\underset{\sim}{V} = (V^A V^B)^T$ and $\underset{\sim}{F} = (F^A F^B)^T$. The superscript "T" denotes the transpose. The vector $\underset{\sim}{\theta}$ is sometimes redefined by augmented matrices, $\underset{\sim}{\theta} = \underset{\sim}{\theta}^{*A} + \underset{\sim}{\theta}^{*B}$ where $\underset{\sim}{\theta}^{*A} = (\theta^A 0)^T$ and $\underset{\sim}{\theta}^{*B} = (0\ \theta^B)^T$. Similar definitions are used for $\underset{\sim}{V}$ and $\underset{\sim}{F}$. Any nonzero terms in $\underset{\sim}{F}^A$ obtained in a computation of $\underset{\sim}{F}^{*B}$ are neglected; they are assumed to be zero.

As an example, consider a one dimensional mesh depicted in figure 1, it consists of 8 nodes and 7 elements. Then the set of nodes "A" will be 1,2,3,4,5; the set of nodes "B" will be 6,7,8; and the set of nodes "C" will be 5.

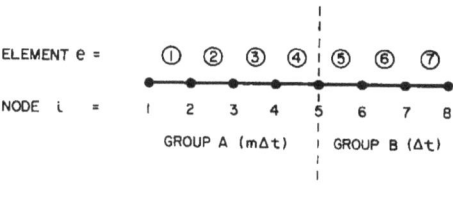

Figure 1.

Let $\underset{\sim}{M}^e$, $\underset{\sim}{K}^e$ and $\underset{\sim}{F}^e$ be the eth element mass, stiffness and force contributions to the global arrays respectively; then

$$\underset{\sim}{M}^A = \sum_{e=1}^{4} \underset{\sim}{M}^e \qquad \underset{\sim}{K}^A = \sum_{e=1}^{4} \underset{\sim}{K}^e \qquad \underset{\sim}{F}^A = \sum_{e=1}^{4} \underset{\sim}{F}^e$$

$$\underset{\sim}{M}^B = \sum_{e=5}^{7} \underset{\sim}{M}^e \qquad \underset{\sim}{K}^B = \sum_{e=5}^{7} \underset{\sim}{K}^e \qquad \underset{\sim}{F}^B = \sum_{e=5}^{7} \underset{\sim}{F}^e$$

$$\underset{\sim}{M}^C = \underset{\sim}{M}^5 \qquad \underset{\sim}{K}^C = \underset{\sim}{K}^5 \qquad \underset{\sim}{F}^C = \underset{\sim}{F}^5$$

and

$$\underset{\sim}{M}^R = \sum_{e=1}^{5} \underset{\sim}{M}^e \qquad \underset{\sim}{K}^R = \sum_{e=1}^{5} \underset{\sim}{K}^e \qquad \underset{\sim}{F}^R = \sum_{e=1}^{5} \underset{\sim}{F}^e$$

If we let P_i^{*B} be the ith component of the global assembled vector $\underset{\sim}{P}^{*B}$ then
$$\underset{\sim}{P}^{*B} = (0,0,0,0,P_5 \equiv 0, P_6, P_7, P_8)$$

With these definitions, the mixed time partition is given as follows.
• Governing equation
for j=0,m;

$$\underset{\sim}{M} \underset{\sim}{V}_{n+j} + \underset{\sim}{K}^R \hat{\underset{\sim}{\theta}}^{*A}_{n+j} + \underset{\sim}{K}^B \hat{\underset{\sim}{\theta}}^{*B}_{n+j} = \underset{\sim}{F}_{n+j} \qquad (3.1)$$

and

for j=1,...,m-1;

$$\underset{\sim}{M}^B \underset{\sim}{V}^{*B}_{n+j} + \underset{\sim}{K}^B \hat{\underset{\sim}{\theta}}^{*B}_{n+j} + \underset{\sim}{K}^C \hat{\underset{\sim}{\theta}}^{*A}_{n+j} = \underset{\sim}{F}^{*B}_{n+j} \qquad (3.2)$$

where $\hat{\underset{\sim}{\theta}}^{*x}_{n+j}$ is a suitable extrapolator (and/or interpolator) of $\underset{\sim}{\theta}^{*x}$ (and/or $\underset{\sim}{\theta}^{*x}_{n+m}$) for x=A and B. In actual computation, equation (3.2) is implicitly included in equation (3.1), and for j=1,...,m-1 no quantities of A are being solved. A family of integration partitions can then be deduced from equations (3.1) and (3.2) if $\underset{\sim}{M}$ is assumed to be lumped. Some members which are of practical importances are shown in table 1.

Table 1

Designation	Time Integration In Element Group A	Time Integration In Element Group B	Extrapolator/Interpolator Node A	Extrapolator/Interpolator Node B
E-E	explicit with Δt	explicit with Δt	$\underset{\sim}{\theta}^{*A}_n$	$\underset{\sim}{\theta}^{*B}_n$
mE-E	explicit with $m\Delta t$	explicit with Δt	$\underset{\sim}{\theta}^{*A}_n$	$\underset{\sim}{\theta}^{*B}_n$
mE-I	implicit with $m\Delta t$	explicit with Δt	$\underset{\sim}{\theta}^{*A}_{n+m}$	$\underset{\sim}{\theta}^{*B}_n$
E-I	implicit with Δt	explicit with Δt	$\underset{\sim}{\theta}^{*A}_{n+1}$	$\underset{\sim}{\theta}^{*B}_n$
I-I	implicit with Δt	implicit with Δt	$\underset{\sim}{\theta}^{*A}_{n+1}$	$\underset{\sim}{\theta}^{*B}_{n+1}$

For purposes of describing the computer implementation and stability analysis, the modified generalized trapezoidal rule will be used to carry out the time temporary discretization of equations (3.1) and (3.2) though other implicit integration methods can also be used; they are:

• Modified generalized trapezoidal rule
for j=1,...,m;

$$\tilde{\underset{\sim}{\theta}}^A_{n+j} = \underset{\sim}{\theta}^A_n + (1-\alpha)j\Delta t \underset{\sim}{V}^A_n \qquad (3.3)$$

for 1 < j < m define the set "C" only,

$$\tilde{\theta}^B_{n+j} = \theta^B_{n+j-1} + (1-\alpha)\Delta t\, V^B_{n+j-1} \tag{3.4}$$

$$\theta^A_{n+m} = \tilde{\theta}^A_{n+m} + \alpha m\, \Delta t\, V^A_{n+m} \tag{3.5}$$

and

$$\theta^B_{n+j} = \tilde{\theta}^B_{n+j} + \alpha \Delta t\, V^B_{n+j} \tag{3.6}$$

In the above equations, α is a free parameter which governs the stability and accuracy of the method. We now illustrate some useful partitions which have been depicted in table 1.

Example 1: E-E partition

In this case, $m=1$, $\hat{\theta}^x_{n+1} \equiv \tilde{\theta}^x_{n+1}$ for $x = A$ and B. Equations (3.1) and (3.2) reduce to:

$$M V_{n+1} + K \tilde{\theta}_{n+1} = F_{n+1} \tag{3.7}$$

$$\tilde{\theta}_{n+1} = \theta_n + (1-\alpha)\Delta t\, V_n \tag{3.8}$$

and

$$\theta_{n+1} = \tilde{\theta}_{n+1} + \alpha \Delta t\, V_{n+1} \tag{3.9}$$

Equations (3.7) to (3.9) represent the predictor-corrector explicit algorithms with equation (3.8) as the predictor and equation (3.9) as the corrector.

Example 2: mE-E partition

In this case, $m > 1$, $\hat{\theta}^A_{n+j} \equiv \tilde{\theta}^A_{n+j}$ and $\hat{\theta}^B_{n+j} \equiv \tilde{\theta}^B_{n+j}$. Equations (3.1) to (3.6) reduce to:

PREDICTOR PHASE:

    equation (3.3)   (3.10)

and

    equation (3.4)   (3.11)

GOVERNING EQUATIONS:

    equation (3.1)   (3.12)

and

    equation (3.2)   (3.13)

CORRECTOR PHASE:

$\qquad$ equation (3.5) $\hfill$ (3.14)

and

$\qquad$ equation (3.6) $\hfill$ (3.15)

<u>Example 3</u> mE-I partition

In this case, m > 1 ; in equation (3.1), $\hat{\theta}^A_{n+1} \equiv \tilde{\theta}^A_{n+m}$ for element group A only, $\hat{\theta}^A_{n+1} \equiv \tilde{\theta}^A_{n+m}$ for the portion which is related to K^C for j=0 and m. This is done automatically if element group A is defined to be the implicit element group and element group B is defined to be the explicit element group. $\hat{\theta}^A_{n+1} \equiv \tilde{\theta}^A_{n+1}$ for 1 ≤ j ≤ m, and $\hat{\theta}^B_{n+j} \equiv \tilde{\theta}^B_{n+j}$ for 1 ≤ j ≤ m . Equations (3.1) to (3.6) reduce to:

PREDICTOR PHASE:

$\qquad$ equation (3.3) $\hfill$ (3.16)

and

$\qquad$ equation (3.4) $\hfill$ (3.17)

GOVERNING EQUATIONS:

for j=0,m;

$$M\underset{\sim}{V}_{n+j} + \underset{\sim}{K}^A \underset{\sim}{\theta}^{*A}_{n+j} + \underset{\sim}{K}^B \tilde{\underset{\sim}{\theta}}_{n+j} = \underset{\sim}{F}_{n+j} \qquad (3.18)$$

and

for j=1,...,m-1;

$$\underset{\sim}{M}^B \underset{\sim}{V}^{*B}_{n+j} + \underset{\sim}{K}^B \underset{\sim}{\theta}^{*B}_{n+j} + \underset{\sim}{K}^C \underset{\sim}{\theta}^{*A}_{n+j} = \underset{\sim}{F}^{*B}_{n+j} \qquad (3.19)$$

CORRECTOR PHASE:

$\qquad$ equation (3.5) $\hfill$ (3.20)

$\qquad$ equation (3.6) $\hfill$ (3.21)

<u>Example 4</u>: E-I partition

This is a special case of example 3. Equations (3.1) to (3.6) reduce to:

$$M\underset{\sim}{V}_{n+1} + \underset{\sim}{K}^A \underset{\sim}{\theta}_{n+1} + \underset{\sim}{K}^B \tilde{\underset{\sim}{\theta}}_{n+1} = \underset{\sim}{F}_{n+1} \qquad (3.22)$$

$$\tilde{\underset{\sim}{\theta}}_{n+1} = \underset{\sim}{\theta}_n + (1-\alpha)\Delta t \, \underset{\sim}{V}_n \qquad (3.23)$$

$$\underset{\sim}{\theta}_{n+1} = \tilde{\underset{\sim}{\theta}}_{n+1} + \alpha \Delta t \, \underset{\sim}{V}_{n+1} \qquad (3.24)$$

Equations (3.22) to (3.24) represent the implicit-explicit algorithms developed by Hughes and Liu (see e.g., [8,10]) in which equation (3.23) is the predictor and equation (3.24) is the corrector.

Example 5: I-I partition
In this case, m=1, $\theta_{n+1}^A \equiv \hat{\theta}_{n+1}^A$ and $\hat{\theta}_{n+1}^B \equiv \theta_{n+1}^B$. Equations (3.1) to (3.6) reduce to the usual implicit formulation and it is:

$$M V_{n+1} + K \theta_{n+1} = F_{n+1} \tag{3.25}$$

$$\tilde{\theta}_{n+1} = \theta_n + (1-\alpha)\Delta t \, V_n \tag{3.26}$$

$$\theta_{n+1} = \tilde{\theta}_{n+1} + \alpha \Delta t \, V_{n+1} \tag{3.27}$$

4. STABILITY CRITERION

Our aim in this section is to deduce the stability characteristic of these mixed time partition algorithms. It suffices to restrict ourselves to the case in which $F=0$ and all mass matrices are lumped for purposes of stability analysis. An energy balance technique (see [8] for a discussion) is employed to carry out the stability analysis. To simplify the subsequent writing, the following notations will be used.

$$[x_{n+m}] = x_{n+m} - x_n \tag{4.1}$$

$$\langle x_{n+m} \rangle = (x_{n+m} + x_n)/2 \tag{4.2}$$

$$[x_{n+j}] = x_{n+j+1} - x_{n+j} \tag{4.3}$$

and   for $j=0,1,\ldots,m-1$

$$\langle x_{n+j} \rangle = (x_{n+j+1} + x_{n+j})/2 \tag{4.4}$$

We have not made a complete satisfactory stability analysis of these mixed time partition procedures. However, if we assume:

1. $\langle V_{n+m}^{*A} \rangle^T K^C \langle V_{n+m}^{*A} \rangle / z^{*B^T} K^C z^{*B} \geq 1/m^2$ where

$$z^{*B} = V_n^{*B} + \sum_{j=1}^{m-2} \langle V_{n+j}^{*B} \rangle \tag{4.5}$$

2. $\langle V_n^{*B} \rangle^T K^C \langle V_n^{*B} \rangle / V_n^{*A^T} K^C V_n^{*A} \geq (1+(m-1)\alpha)^2 \tag{4.6}$

3. $\langle V_{n+j}^{*B} \rangle^T K^C \langle V_{n+j}^{*B} \rangle / V_n^{*A^T} K^C V_n^{*A} \geq (1-\alpha)^2$ for

   $j=1,\ldots,m-1$ \hfill (4.7)

and let

4. $S_n = V_n^{*A^T} M^{*A} V_n^{*A} + V_n^{*B^T} M^{*B} V_n^{*B} \geq 0 \tag{4.8}$

5. $P_{n+m}^A = \langle V_{n+m}^{*A} \rangle^T K^A \langle V_{n+m}^{*A} \rangle \geq 0 \tag{4.9}$

6.
$$P^B_{n+j} = \langle V^{*B}_{n+j}\rangle^T \bar{K}^B \langle V^{*B}_{n+j}\rangle > 0 \qquad (4.10)$$

the energy expression of these mixed time partition procedures can be shown to be:

$$S_{n+m} \leq S_n - 2m\Delta t\, P^A_{n+m} - 2\Delta t \sum_{j=0}^{m-1} P^B_{n+j} \qquad (4.11)$$

Here, $\underset{\sim}{K}^{\bar{B}} = \underset{\sim}{K}^B - \underset{\sim}{K}^C$ and the stability is governed by $\underset{\sim}{M}^{*R}$ and $\underset{\sim}{M}^{*B}$ provided $\alpha \geq 1/2$. Let

$$\underset{\sim}{Q}^X_j = \underset{\sim}{M}^X - 1/2\, j\Delta t\, \underset{\sim}{K}^X \qquad (4.12)$$

and

$$\underset{\sim}{W}^X_j = \underset{\sim}{M}^X + (\alpha - 1/2)j\Delta t\, \underset{\sim}{K}^X \qquad (4.13)$$

the definitions of $\underset{\sim}{M}^{*R}$ and $\underset{\sim}{M}^{*B}$ for the five cases discussed in section 3 are:

Example 1: E-E partition
$$\underset{\sim}{M}^{*R} = \underset{\sim}{Q}^R \quad \text{and} \quad \underset{\sim}{M}^{*B} = \underset{\sim}{Q}^B_1 \qquad (4.14)$$

Example 2: mE-E partition
$$\underset{\sim}{M}^{*R} = \underset{\sim}{Q}^R_m \quad \text{and} \quad \underset{\sim}{M}^{*B} = \underset{\sim}{Q}^B_1 \qquad (4.15)$$

Example 3: mE-I partition
$$\underset{\sim}{M}^{*R} = \underset{\sim}{W}^A_m + \underset{\sim}{Q}^C_m \quad \text{and} \quad \underset{\sim}{M}^{*B} = \underset{\sim}{Q}^B_1 \qquad (4.16)$$

Example 4: E-I partition
$$\underset{\sim}{M}^{*R} = \underset{\sim}{W}^A_1 + \underset{\sim}{Q}^C_1 \quad \text{and} \quad \underset{\sim}{M}^{*B} = \underset{\sim}{Q}^B_1 \qquad (4.17)$$

Example 5: I-I partition
$$\underset{\sim}{M}^{*R} = \underset{\sim}{W}^R_1 \quad \text{and} \quad \underset{\sim}{M}^{*R} = \underset{\sim}{W}^B_1 \qquad (4.18)$$

These mixed time partition procedures are stable if $\alpha \geq 1/2$ and $\underset{\sim}{M}^{*R}$ and $\underset{\sim}{M}^{*B}$ are both positive definite. A summary of the results is as follows:

Example 1: E-E partition
$$\Omega^A_{crit} = \Omega^B_{crit} \leq 2 \qquad (4.19)$$

Example 2: mE-E partition
$$\Omega^{mA}_{crit} = \Omega^{mC}_{crit} \leq 2 \quad \text{and} \quad \Omega^B_{crit} \leq 2 \qquad (4.20)$$

Example 3: mE-I partition

$$\Omega^{mC}_{crit} \leq 2 \quad \text{and} \quad \Omega^{B}_{crit} \leq 2 \tag{4.21}$$

Example 4: E-I partition

$$\Omega^{C}_{crit} \leq 2 \quad \text{and} \quad \Omega^{B}_{crit} \leq 2 \tag{4.22}$$

Example 5: I-I partition

$$\text{unconditionally stable} \tag{4.23}$$

In equations (4.19) to (4.22), Ω^{jx} is defined to be $j\Delta t \lambda^{x}_{crit}$ where λ^{x}_{crit} denotes a typical eigenvalue of the eigenproblem

$$\underset{\sim}{M}^{x} \underset{\sim}{\dot{\theta}} + \underset{\sim}{K}^{x} \underset{\sim}{\theta} = \underset{\sim}{0} \tag{4.24}$$

5. IMPLEMENTATION ASPECTS

In this section, we generalize the mixed time integration methods described in section 3 to NUMEG element groups. Different time integration methods (implicit/explicit) with different time steps can be used in each element group. Let Δt_{NEG} and T_{NEG} be the element group time step and element group time respectively for NEG = 1,...,NUMEG. There are NUMEL elements in each element group. We denote Δt as the minimum time step amount for all these element groups. In this formulation all element group time steps are required to be integer multiples of Δt and the time steps for adjacent groups are integer multiples of each other. Furthermore, no two implicit groups with different time steps can be adjacent to each other. In addition, for each implicit group that element group time step must be greater than those of the adjacent explicit groups. The main advantage of this m_1 implicit - m_2 explicit - m_3 implicit -... etc. technique is to minimize the semi-bandwidth of complicated problems especially in the three-dimensional case. To illustrate the idea, consider the one dimensional mesh shown in figure 2. It consists of NUMEG element groups and NUMNP nodes. In this case NUMEG is equal to 4 and NUMNP is equal to 12. We assumed that node 1 is an essential boundary condition node and hence the number of equations, NEQ, is equal to 11. The essence of the present development can be deduced graphically by considering

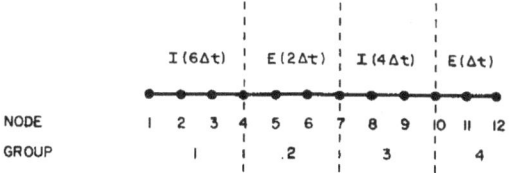

Figure 2.

the solution procedures of the matrix equations. The "active column equation solver" is the key to the success of this technique (see[8,13] for a description of this equation solver). The profile of the effective stiffness matrix $\tilde{K}^*$ of this one dimensional mesh is shown in figure 3. We can observe from figure 3 the following:

Group 1: implicit with $\Delta t_1 = 6\Delta t$, five words of storage (1-5), 3 elements and 3 equations.

Group 2: explicit with $\Delta t_2 = 2\Delta t$, two words of storage (6-7), 3 elements and 2 equations.

Group 3: implicit with $\Delta t_3 = 4\Delta t$, seven words of storage (8-14), 3 elements and 4 equations.

Group 4: explicit with $\Delta t_4 = \Delta t$, two words of storage (15-16), 2 elements and 2 equations.

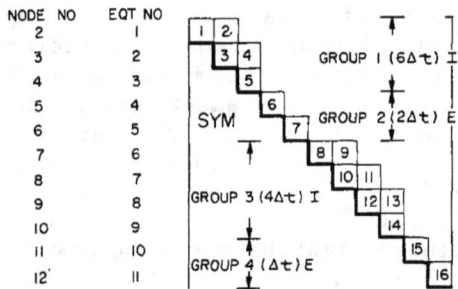

Figure 3.

The equation systems of each element group are uncoupled and hence each group can be integrated at its own group time step. For example, we assume the effective stiffness matrix $\tilde{K}^*$ is formed and factorized once. In a time interval of $6\Delta t$, group 1 will be integrated implicitly once, group 2 will be integrated explicitly three times, group 3 will be integrated implicitly once and group 4 will be integrated explicitly six times. In order to handle the forward reduction and backsubstitution and update procedures automatically, we required two arrays Δt_{NODE} and T_{NODE}; each has a dimension of NUMNP. Δt_{NODE} array contains the nodal time steps of each node. Nodes associated with only one element group NEG are assigned a time step of Δt_{NEG}, whereas those which are in common to other element groups are assigned to have the maximum time step from the adjacent groups. T_{NODE} array contains the nodal time of each node. From these two arrays (Δt_{NODE} and T_{NODE}) and the boundary condition codes, another time step array Δt_{NEQ} and an equation time array T_{NEQ} of the equation systems can then be generated. Both Δt_{NEQ} and T_{NEQ} have dimensions of NEQ. We require further a master time T_M which is incremented by the smallest time step Δt. For this particular example the Δt_{NODE} and Δt_{NEQ} arrays are:

$$\Delta t_{NODE} = (6\Delta t, 6\Delta t, 6\Delta t, 6\Delta t, 2\Delta t, 2\Delta t, 4\Delta t, 4\Delta t, 4\Delta t, 4\Delta t, \Delta t, \Delta t)$$

and

$$\Delta t_{NEQ} = (6\Delta t, 6\Delta t, 6\Delta t, 2\Delta t, 2\Delta t, 4\Delta t, 4\Delta t, 4\Delta t, 4\Delta t, \Delta t, \Delta t)$$

The T_{NODE} and T_{NEQ} arrays are incremented by time steps of Δt_{NODE} and Δt_{NEQ} respectively. With these definitions, the generalized mixed time integration is to proceed over the time interval $[0, T_{max}]$. The procedures are as follows:

1. Initialization

    Set T_M, T_{NEG}, T_{NODE} and $T_{NEQ} = 0$

2. Determine $\underset{\sim}{V}_o$

$$\underset{\sim}{V}_o = \underset{\sim}{M}^{-1} (\underset{\sim}{F}_o - \underset{\sim}{K}\underset{\sim}{\theta}_o) \tag{5.1}$$

3. Form and factorize $\underset{\sim}{K}^*$

$$\underset{\sim}{K}^* = \sum_{NEG=1}^{NUMEG} \underset{\sim}{K}^{*NEG} \tag{5.2}$$

where

$$\underset{\sim}{K}^{*NEG} = \underset{\sim}{M}^{NEG} + \alpha \Delta t_{NEG} \underset{\sim}{K}^{NEG} \text{ if implicit} \tag{5.3}$$

and

$$\underset{\sim}{K}^{*NEG} = \underset{\sim}{M}^{NEG} \text{ which is a diagonal matrix if explicit} \tag{5.4}$$

4. $T_M \leftarrow T_M + \Delta t$; set effective force $\underset{\sim}{F}^*$ equal to zero

5. Loop on element groups NEG=1,...,NUMEG
    If $T_{NEG} + \Delta t_{NEG} > T_M$ go to 5a

6. Loop on elements e=1,...,NUMEL

6a. Define predictor values $\underset{\sim}{\widetilde{\theta}}^e$

    If $T^e_{NODE} + \Delta t^e_{NODE} < T_M$

    then $\underset{\sim}{\widetilde{\theta}}^e_{NODE} = \underset{\sim}{\theta}^e_{NODE} + (1-\alpha)\Delta t_{NODE} \underset{\sim}{V}^e_{NODE} \tag{5.5}$

    If $T^e_{NODE} + \Delta t^e_{NODE} > T_M$

then $\tilde{\underline{\theta}}^e_{NODE} = \underline{\theta}^e_{NODE} + (1-\alpha)(T_M - T^e_{NODE})\underline{V}^e_{NODE}$ (5.6)

6b. Form element effective force $\underline{f}^{*e}$

$\underline{f}^{*e} = \underline{M}^e \tilde{\underline{\theta}}^e + \alpha \Delta t_{NEQ} \underline{F}^e$ if implicit (5.7)

and

$\underline{f}^{*e} = \underline{M}^e \tilde{\underline{\theta}}^e + \alpha \underline{W}(\underline{F}^e - \underline{K}^e \tilde{\underline{\theta}}^e)$ if explicit (5.8)

where

$\underline{W}$ = diagonal matrix with Δt^e_{NODE} along the diagonals (5.9)

6c. Sum up effective force from element contributions

$\underline{F}^* \leftarrow \underline{F}^* + \underline{f}^e$ (5.10)

6d. End of element loop

5a. End of element group loop

7. Solve for $\underline{\theta}$, i.e., forward reduction and backsubstitution

$\underline{\theta} = \underline{K}^{*-1} \underline{F}^*$ (5.11)

7a. Loop on equation number N=1,...,NEQ

If $T^N_{NEQ} + \Delta t^N_{NEQ} > T_M$ go to 7b

Forward reduction and backsubstitution for equation N

7b. End of equation number loop

8. Update $\underline{V}$ and $\underline{\theta}$

8a. Loop on N=1,...,NUMNP

If $T^N_{NODE} + \Delta t^N_{NODE} > T_M$ go to 8b

$\underline{\theta}^N \leftarrow$ solution from step 7
$\tilde{\underline{V}}^N \leftarrow (\underline{\theta}^N - \tilde{\underline{\theta}}^N)/\alpha \Delta t^N_{NODE}$

8b. End of nodal number loop

9. If $T_M > T_{MAX}$ stop, otherwise go to 4

----
‡ "←" means "is replaced by"

6. NUMERICAL EXAMPLES

The stability and accuracy of these mixed time partition procedures are confirmed by the following one dimensional heat conduction problem. The finite element mesh consists of a rod kept at a temperature of 0.0 for all time at the left end and insulated at the right end, and subjected to a constant initial temperature of 0.1. The finite element model consists of (from left to right) 10 elements, each with a length $\ell_1 = 10.0$, 10 elements, each with a length of $\ell_2 = 100.0$, and 10 elements, each with a length of $\ell_1 = 10.0$. The thermal diffusivity, C^2, is set to 200.0. A total of five computer runs are being made. They are (1) explicit, (2) explicit-implicit, (3) 10 explicit-explicit, (4) 10 explicit-implicit, and (5) implicit. All analyses are run with $\Delta t = 0.25$ which is the critical time step based on $\ell_1 = 10.0$. For the m explicit-implicit cases, the implicit elements are the middle 10 elements ($\ell_2 = 100.0$). The results obtained from these five analyses are compared to the analytical solution. They are depicted in figure 4. The temperature reported is at $x = 100.0$. The solution time ratios (normalized by the implicit time) for the above five cases are 0.892, 0.971, 0.578, 0.638 and 1.000 respectively.

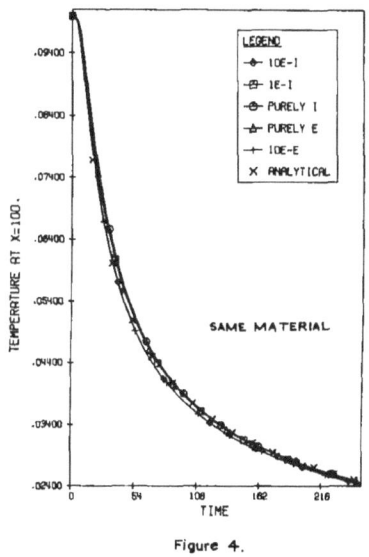

Figure 4.

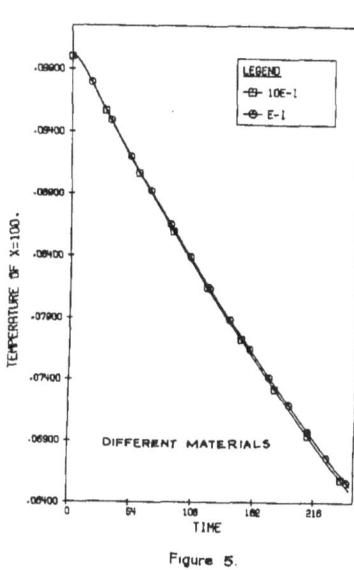

Figure 5.

In order to demonstrate the advantages of this mixed time technique C^2 is raised to 2.0×10^5 for the implicit elements. A time step of $\Delta t = 0.25$ is used for the explicit-implicit and 10 explicit-implicit runs. This problem would therefore not be stable with a 10E-E partition. The results are presented in figure 5.

All the above calculations performed with $\alpha = 0.5$ and lumped mass matrices are assumed throughout.

7. CONCLUSION

In this paper, we have developed a family of mixed time partition procedures for transient thermal analysis of structures. Both the stability criterion and the implementaton aspects of these methods are described. Numerical corroboration of the stability and accuracy of these techniques is also presented. The implementation procedures of these new algorithms are straightforward and are recommended for inclusion in current thermal analysis computer programs.

REFERENCES

1. H. N. Kelly; D.R. Rummler; and R. L. Jackson: Research in Structures and Materials for Future Space Transportation Systems - An Overview. Presented at the AIAA Conference on Advanced Technology for Future Space Systems, May 8-11, 1979, Langley Research Center, Hampton, Virginia, AIAA Paper No. 79-0859.

2. J. L. Shideler; H. N. Kelly; M. L. Blosser; and H. M. Adelman: Multiwall TPS - An Emerging Concept. AIAA/ASME/ASCE/AHS 22nd Structures, Structural Dynamics and Materials Conference, April 6-8, 1981, Atlanta, Georgia.

3. R. J. Jackson and S. C. Dixon: A Design Assessment of Multiwall, Metallic Stand-Off and RSI Reusable Thermal Protection System Including Space Shuttle Applicatons. NASA TM 81780, April 1980.

4. W. Blair; J. E. Meany; and H. A. Rosenthal: Design and Fabrication of Titanium Multiwall Thermal Protection System (TPS) Test Panels, NASA CR 159241, February 1980.

5. H. Adelman and R. Haftka: On the Performance of Explicit and Implicit Algorithms for Transient Thermal Analysis of Structures. NASA TM 81880, September 1980.

6. SPAR Structural Analysis System Reference Manual, Vol. 1, NASA CR 158970-1, December 1978.

7. T. Belytschko; and R. Mullen: Stability of Explicit-Implicit Mesh Partition in Time Integration. *International Journal Numerical Method in Eng.* 12, 1575-1586, 1978.

8. T. Hughes; and W. Liu: Implicit-Explicit Finite Elements in Transient Analysis. *Journal Applied Mechanics* 45, 371-378, 1978.

9. W. Liu: Development of Finite Element Procedures for Fluid-Structure Interactions. EERL 80-06, California Institute of Technology, Pasadena, California, August 1980.

10. T. Hughes; W. Liu; and A. Brooks: Review of Finite Element Analysis of Incompressible Viscous Flows by the Penalty Function Formulation. *Journal Computational Physics* 30, 1-60, 1979.

11. T. Belytschko and R. Mullen: Explicit Integration of Structural Problems. In P. Bergan et. al., (eds.) *Finite Elements in Nonlinear Mechanic*, 2, 697-720, 1977.

12. T. Belytschko: Partitioned and Adaptive Algorithms for Explicit Time Integration. Nonlinear Finite Element Analysis in Structural Mechanics, edited by W. Wunderlich et al., Springer-Verlag, Berlin, July 1980.

13. W. Liu; and T. Belytschko: Mixed Time Implicit-Explicit Finite Element for Transient Analysis. Accepted for publication in Computer and Structure, 1982.

14. W. Liu; and J. Lin: Mixed Time Integration Schemes for Transient Conduction Forced-Convection Analysis. Presented in 2nd Natonal Symposium on Numerical Methods in Heat Transfer. College Park, Maryland, September 28-30, 1981.

INTEGRATED THERMAL-STRUCTURAL ANALYSIS OF LARGE SPACE STRUCTURES

Jack Mahaney, Earl A. Thornton, and Pramote Dechaumphai
Old Dominion University
Norfolk, Virginia 23508

INTRODUCTION

The flight of Columbia marks the advent of large space structures. Soon orbiting structures as much as 1000 meters across may be deployed or constructed in orbit. Many of these structures may be antennas built of a lattice-work of graphite/epoxy truss members.

Optimum antenna performance requires very fine control of the shape of the antenna surface since the shape affects both frequency control and pointing accuracy. A significant factor affecting the antenna shape is the temperature of the structure and the resulting deformation. To accurately predict the temperature of the structure, it is necessary first to accurately predict thermal loads. As the structure orbits the Earth, the thermal loads change constantly so that the thermal-structural response varies continuously throughout the orbit.

The purpose of this paper is to present the results from recent applications of integrated finite-element methodology to heat-load determination and thermal-structural analysis of large space structures (fig. 1).

The paper will concentrate on four areas: (1) the characteristics of the integrated finite element methodology, (2) fundamentals of orbital heat-load calculation, (3) description and comparison of some radiation finite elements, and (4) application of the integrated finite-element approach to the thermal-structural analysis of an orbiting truss structure.

- MOTIVATION
  - SIZE OF STRUCTURES
  - ACCURATE PREDICTION OF HEAT LOADS AND TEMPERATURE
  - CONTROL OF DEFORMATION

- PURPOSE
  - DESCRIBE AN INTEGRATED FINITE-ELEMENT (FE) APPROACH FOR THERMAL-STRUCTURAL ANALYSIS OF LARGE-SPACE STRUCTURES

- SCOPE
  - CHARACTERISTICS OF INTEGRATED APPROACH
  - FUNDAMENTALS OF IN-ORBIT HEATING
  - DESCRIPTION AND COMPARISON OF FINITE ELEMENTS
  - ANALYSIS OF STRUCTURE

FIGURE 1

CHARACTERISTICS OF INTEGRATED FINITE ELEMENT ANALYSIS

One approach toward integrated thermal-structural analysis capability is a common methodology in a single program capable of both thermal and structural analysis. This approach, herein called the customary approach, has the disadvantage of inefficient data transfer between thermal and structural analyses because of inherent differences between the thermal and structural models. Another disadvantage of the customary approach is that basic differences between the thermal and structural analysis requirements are not exploited.

To exploit the capabilities of the finite element (F.E.) method the concept of integrated thermal-structural analysis was proposed in references 1 and 2. An integrated thermal-structural analysis is characterized by: (1) thermal and structural finite elements formulated with a common geometric discretization with elements formulated to suit the needs of their respective analyses, (2) thermal and structural finite elements which are fully compatible, and (3) equivalent thermal forces which are based upon a consistent finite element force vector (fig. 2).

Some of the benefits of the integrated approach are: (1) improved temperature distributions based upon new thermal elements, (2) more efficient analyses because of the elimination of data processing between dissimilar thermal and structural models, and (3) improved accuracy in the structural analysis through consistent incorporation of the improved temperature distributions.

- COMMON FE METHODOLOGY
- GEOMETRIC MODEL WITH COMMON DISCRETIZATION
- IMPROVED THERMAL ELEMENTS
- MINIMIZE DATA PROCESSING
- TEMPERATURES INTEGRATED INTO STRUCTURAL ANALYSIS

FIGURE 2

INTEGRATED THERMAL-STRUCTURAL FINITE ELEMENT ANALYSIS

Figure 3 shows schematically the concepts of the customary and integrated approaches applied to an orbiting structure. The sequence followed in the customary approach is shown on the left. First, the heating loads are calculated for an appropriate mathematical model. Next, a thermal analysis is performed based upon a thermal model selected to best represent the heat transfer problem. The thermal model may be based on the lumped parameter or F.E. method. Next the nodal temperature data are transferred to the structural analysis. In most cases these data must be processed to conform the input temperature vector to that required by the structural model. Often the structural and thermal models use different nodes and elements, and approximate thermal forces are computed from average element temperatures.

An integrated analysis is shown on the right side of the figure. The heating loads, thermal, and structural analyses are performed on a model based on a common discretization. Although the discretization is common, the thermal and structural elements are formulated to best suit their respective analyses. The transfer of data is compatible, with no data processing required. Consistent thermal forces are computed from thermal element and nodal input data supplied directly from the thermal analysis.

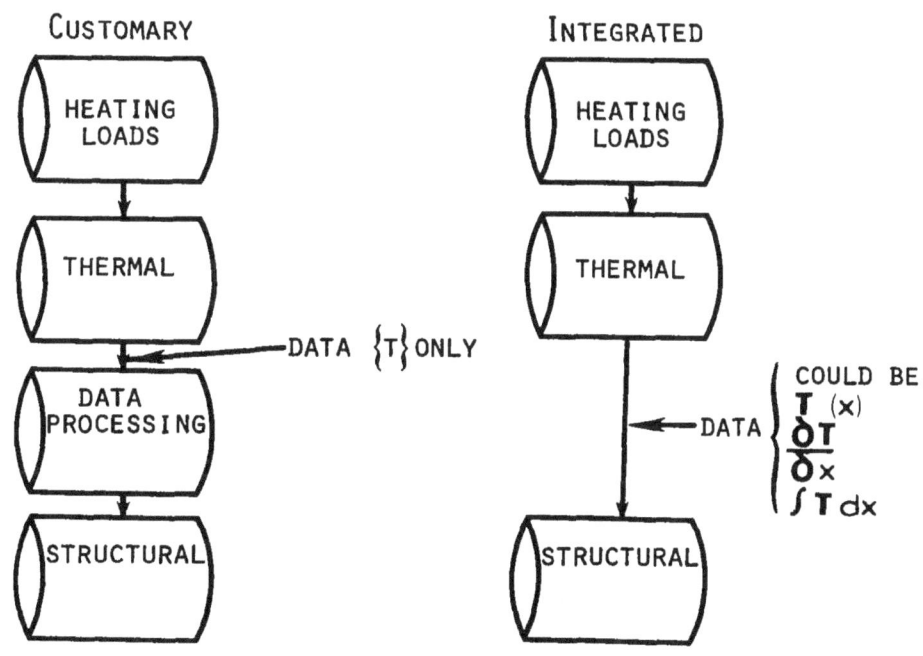

FIGURE 3

ORBITAL HEATING RATE GEOMETRY AND SOURCES

A surface in Earth orbit is heated by both the Sun and the Earth. The solar flux is approximately 1390 W/m², but the amount of heat absorbed is a function of the surface absorptivity, a_s, and the projected surface area normal to the flux. Thus, the total solar heating is

$$\dot{q}_s = 1390(W/m^2) \, a_s \, \cos\psi \qquad (1)$$

The Earth provides two sources of heat, emission and albedo (fig. 4). The emitted heating is computed by assuming the Earth to be a black body radiating at T = 289 K. The heat absorbed by a surface is a function of the surface absorptivity, a_e, and the view factor F. The view factor between an orbiting flat plate and a sphere was developed by Cunningham (ref. 3) and takes into account the altitude of the surface, size of the sphere, attitude ϕ, and other basic geometric quantities. The amount of heat absorbed by the surface from Earth emission is given by

$$\dot{q}_e = \sigma T^4 \, a_e \, F \qquad (2)$$

where σ is the Stefan-Boltzmann constant.

The Earth acts as a reflector of solar radiation. The albedo factor, AF, is a measure of the fraction of solar energy reflected. It has been shown by Modest (ref. 4) that while the heat is a complicated function of altitude and attitude (θ,ϕ) a good approximation can be obtained by using Cunningham's view factor F in the equation

$$\dot{q}_a = 1390(W/m^2) \, AF \, \cos\theta \, a_s \, F \qquad (3)$$

At any point in the orbit, then, the total heat absorbed by the surface is the sum of the three heating rates above, or

$$\dot{q} = \dot{q}_s + \dot{q}_e + \dot{q}_a \qquad (4)$$

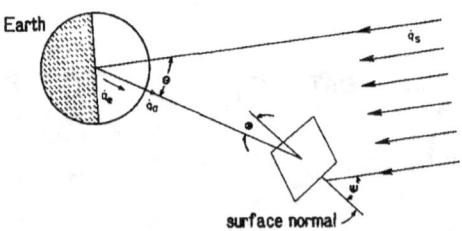

Heat sources to orbiting structure
  solar, $\dot{q}_s$
  Earth emission, $\dot{q}_e$
  Earth albedo (reflection of solar), $\dot{q}_a$
Heating rate depends on
  orbit and orientation (altitude, Θ,Φ,Ψ)
  surface absorptivity

FIGURE 4

DEMONSTRATION OF EARTH SHADOW EFFECTS

This picture of Saturn taken by Voyager I shows dramatically the effect of a planetary shadow (fig. 5). Since the thermal load on an Earth-orbiting structure is reduced as much as 95% when in the Earth's shadow, shadow dwell time is critical. The entry and exit times can be calculated from the period of the orbit and the dwell time, which can be calculated from the altitude of the structure and the angle between the structure's orbit and the plane of the ecliptic (ref. 5).

The umbra-penumbra effects are of little consequence for orbiting structures and are ignored. For example, in geosynchronous orbit a 1000 meter platform travels fast enough to transit the penumbra in less than 4 milliseconds. In lower orbits the time is even shorter.

FIGURE 5

ORBITAL HEAT LOADS OF EARTH-FACING SATELLITES

The heating of Earth-facing satellites in geosynchronous and low Earth orbits is considered (fig. 6). A satellite in a geosynchronous Earch orbit (GEO) has a period of 24 hours and an altitude of 35,876 km. In this orbit the effects of solar flux predominate since the structure is so far from the Earth, and they vary as the structure's orientation changes with respect to the Sun. Since the structure is Earth-facing, that is, maintaining a constant orientation with respect to the Earth, the amount of Earth-emitted heating is constant, although very low, less than 1% of the maximum solar heat. Albedo heating varies over the daylight side of the planet (over the night side it is zero), but due to the high altitude is also less than 1% of the maximum solar heating.

The low Earth orbit (LEO) considered has a period of 90 minutes at an altitude of 279 km, similar to the shuttle orbit. In this orbit all incident heating rates are important. As before, solar and albedo heating vary throughout the orbit while emitted heating is constant. In this LEO, the two Earth heating rates can be as much as 37% of the maximum solar heating.

The effects of transit through the Earth's shadow are significant in any orbit. Within the shadow, the only heating is from Earth-emitted heat. In a GEO, however, this is negligible (less than 15 W/m^2) while in a LEO it is significant. We will see later that the absence of heating during shadow transit can greatly affect the temperature history.

- GEOSYNCHRONOUS (GEO) (PERIOD=24 HOURS, ALTITUDE=35,875 KM)
  - SOLAR FLUX PREDOMINATES AND VARIES WITH ORIENTATION
  - EARTH-EMITTED HEATING RATE CONSTANT, LESS THAN 1% OF MAXIMUM SOLAR FLUX
  - ALBEDO HEATING RATE VARIES, LESS THAN 1% OF MAXIMUM SOLAR FLUX

- LOW EARTH ORBIT (LEO) (PERIOD=90 MINUTES, ALTITUDE=279 KM)
  - ALL INCIDENT HEATING RATES CONSIDERED AND VARY WITH ORIENTATION
  - EARTH HEATING RATE AS MUCH AS 37% OF TOTAL

- EARTH SHADOW SIGNIFICANT
  - EARTH EMITTED HEATING ONLY
  - IN GEO, EARTH EMITTED HEATING ALMOST NEGLIGIBLE
  - IN LEO, EARTH EMITTED HEATING CONSIDERABLE

FIGURE 6

ORBITING STRUCTURES THERMAL MODELING

A truss is one of the fundamental structural concepts under consideration for orbiting structures. A typical truss member experiences conduction heat transfer combined with emitted radiation and radiation heating from both nearby truss members and other satellite components (fig. 7). Radiation exchange between members is neglected because computational experience (ref. 6) has shown that the member-to-member radiation heat exchanges in a truss are negligible in comparison with the incident heating and emitted radiation. Although member-to-surface radiation exchanges may be important, they are not considered. In general, both material and surface properties are temperature dependent and vary throughout an orbit. Thus the basic heat transfer problem is inherently nonlinear because of the emitted radiation combined with temperature-dependent properties and transient because of the strong time-dependence of the heat loads. Three alternative F.E. thermal models of a truss member are presented in figure 8.

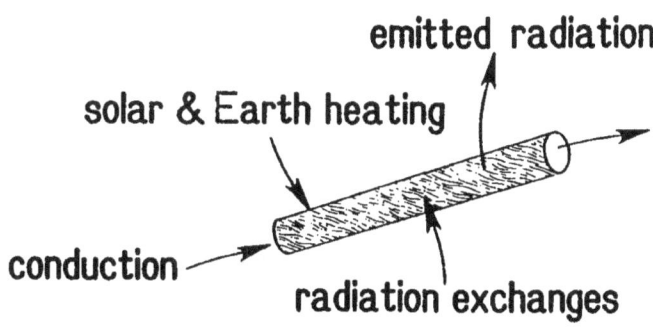

- **Considerations**
  - conduction combined with radiation
  - nonlinear, transient
  - member–to–member radiation exchanges
  - member–to–surface radiation exchanges
  - temperature dependence of material and surface properties

FIGURE 7

ROD ELEMENT THERMAL MODELS

Three thermal models of a truss member are considered: (1) a conventional two node element with a linear temperature distribution, (2) a nodeless variable higher order element with a quadratic temperature distribution, and (3) an isothermal element. The first two elements are useful in modeling members with significant member temperature gradients due to conduction, and the last element is useful for modeling members with negligible conduction. The isothermal element is similar to traditional lumped heat transfer models and does not transfer heat via conduction between adjacent members as with the first two elements. Thus with isothermal elements the solution of simultaneous equations is avoided, and the transient response of each member is computed separately (fig. 8).

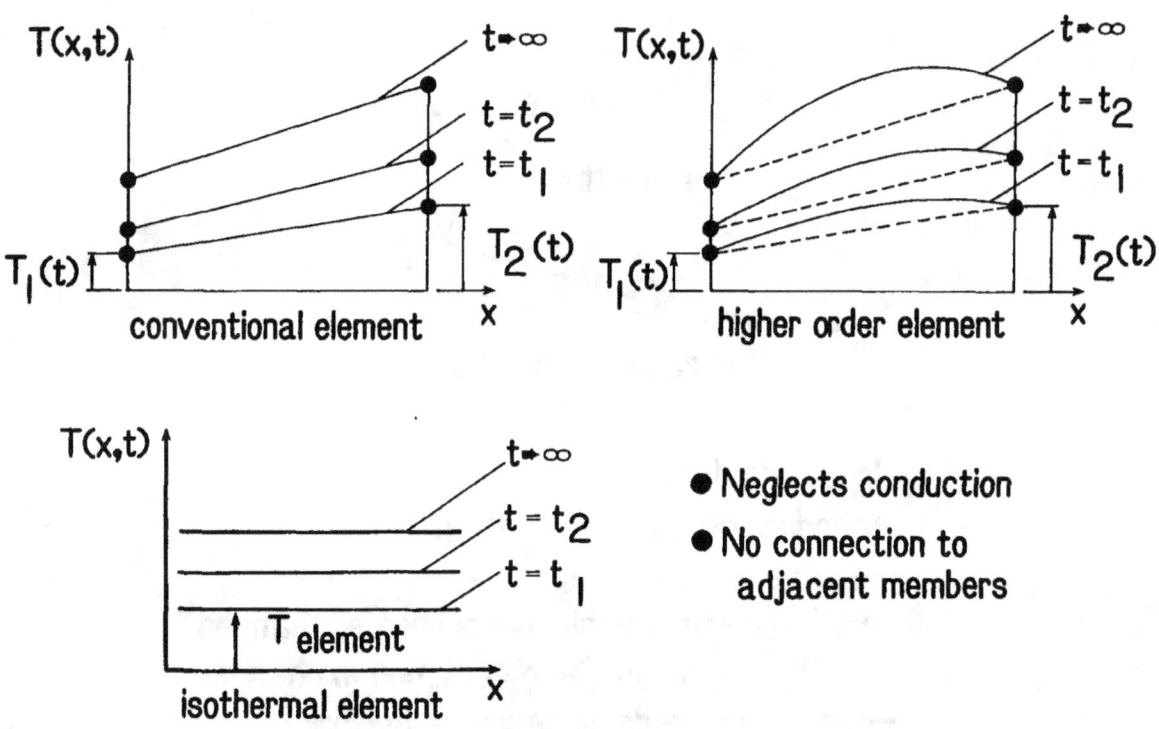

FIGURE 8

TRUSS MEMBER TEMPERATURE DISTRIBUTION

To determine the element to best model a typical space structure truss member, a repeating module of a space truss was analyzed at the noon orbit position. Member properties were representative of a tube fabricated from graphite-epoxy. Radiation equilibrium temperature distributions were computed for all members of the repeating module, but because of geometric and heat load symmetry typical results are represented by the three highlighted members (fig. 9). By symmetry, all joint temperatures are equal.

The repeating module was analyzed with: (1) one conventional element per member, (2) ten conventional elements per member, (3) one higher order element per member, and (4) one isothermal element per member.

Using the ten conventional element solution as the "exact" solution shows that a single conventional element predicts correct nodal temperatures but incorrectly predicts the temperature distribution within an element. The isothermal element, however, does an excellent job of predicting the nearly uniform member temperatures but does not predict nodal temperatures. The higher-order element (results not shown) did better than the single conventional element but tended to overestimate member interior temperatures.

For computation of the structural response, the results from the isothermal elements are superior for this low conductivity material since the average member temperature is predicted quite well. Use of these elements gives improved structural accuracy and also allows smaller, uncoupled thermal models with significant computational advantages.

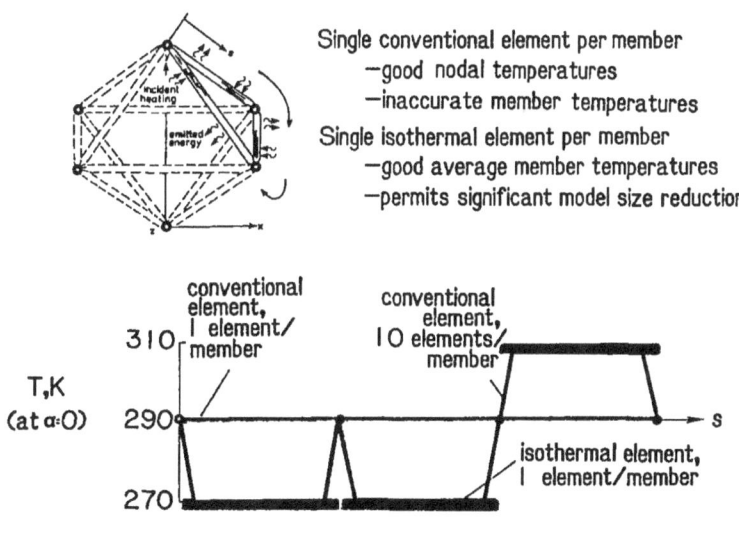

FIGURE 9

STRUCTURAL MODEL

The structural element employed is the standard two-node linear rod with a linear displacement distribution. Structural mass is neglected, so that at each point of interest in an orbit a linear quasi-static displacement analysis is performed. The average member temperatures computed by isothermal elements are used to compute equivalent thermal forces.

Note that for the heat load, thermal, and structural analyses one common geometric model has been employed (fig. 10). In each of these analyses, the analytical models have employed the geometry of the common discretization but each analysis has been "tailored" to best suit the problem. These features are fundamental characteristics of an integrated analysis.

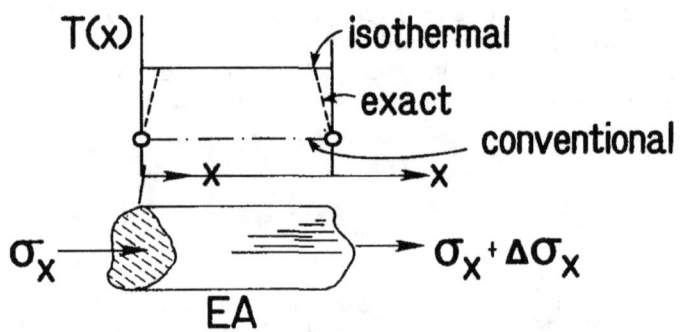

Thermal-stress model

- quasi-static, no structural mass
- use isothermal member temperature for equivalent thermal forces

FIGURE 10

36-MEMBER OCTETRUSS

To illustrate the integrated thermal-structural analysis approach, the octetruss designed and fabricated at LaRC is analyzed. The LaRC octetruss is a 36-member space truss designed and constructed for buckling tests (ref. 7). Each member consists of two truncated cones made of graphite-epoxy members connected with aluminum joints. Each member is 5.42 meters (213.4 inches) long from end to end.

The following assumptions were made in modeling and analyzing the octetruss: (1) each member was considered to be a uniform graphite-epoxy tube, (2) the aluminum joints were disregarded, (3) each member was considered to be isothermal (conduction between members was disregarded), and (4) material properties were considered constant (fig. 11).

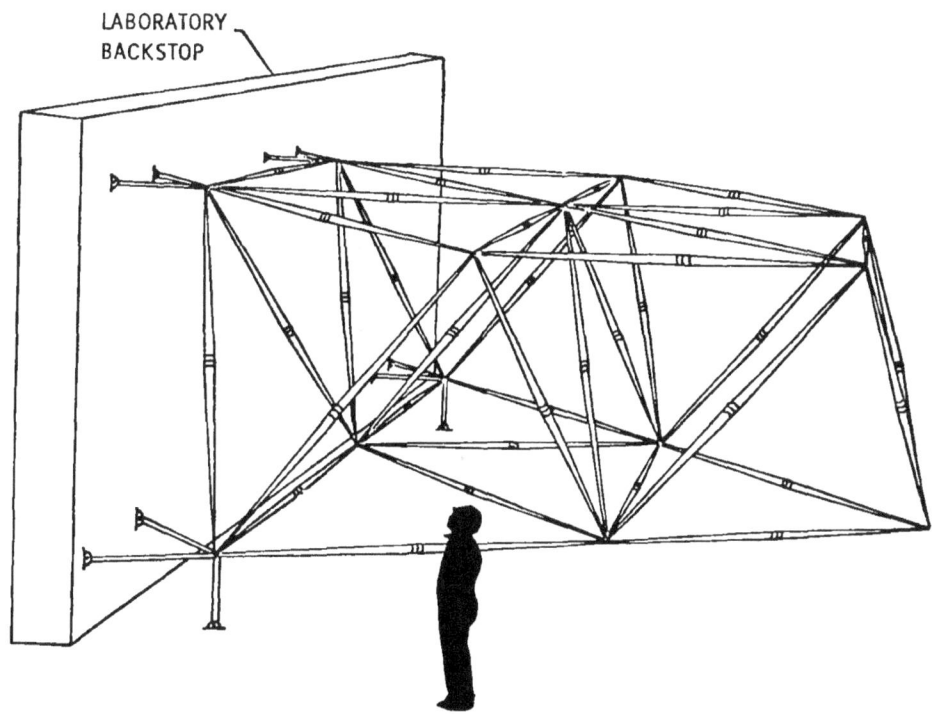

TYPICAL TRUSS MEMBER
- COMPOSITE MEMBER
- ALUMINUM JOINTS

FINITE ELEMENT MODEL
- ISOTHERMAL MEMBER
- ALUMINUM JOINTS NEGLECTED

FIGURE 11

TRUSS IN ORBIT

The Earth-facing octetruss is in a geosynchronous orbit in the ecliptic plane. In this orbit, the structure is in the Earth shadow for a longer time than for low Earth orbit, so that effects of the shadow on the thermal-structural response are most noticeable.

The heat load, thermal, and structural responses of the truss are computed in time increments starting from the satellite noon position ($\alpha = 0$ in fig. 12). The member temperature initial conditions are first computed from a steady-state radiation equilibrium analysis. These temperatures are used as reference temperatures for the structural analysis.

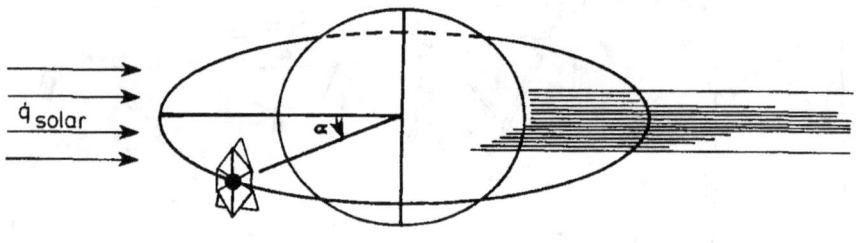

FIGURE 12

TYPICAL ELEMENT HEAT LOADS

Heating histories through one geosynchronous orbit are shown in figure 13 for three representative members of the octetruss. Note that the variation of heating rates is due solely to different member orientation with respect to incident solar heating. Because of geometric symmetry, there are six sets of members with each member of a set having the same heating history. The geometric symmetry and multiplicity of the same member heating histories occur because the octetruss is flat. In a curved space truss, such as a parabolic antenna structure, all member heating rates would differ.

The abrupt drop and rapid rise in heating occur during passage through the Earth shadow. The very low heating during shadow passage is due to Earth-emitted radiation.

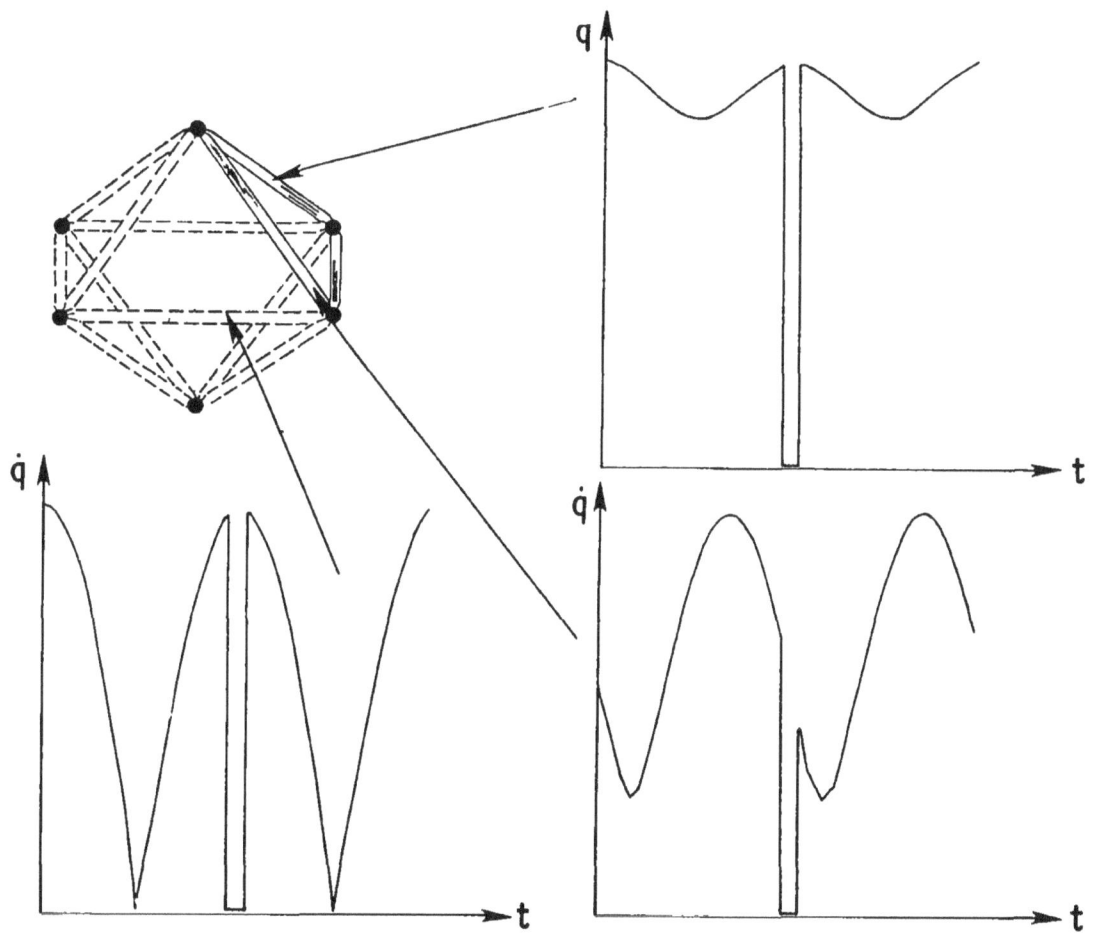

FIGURE 13

TYPICAL ELEMENT TEMPERATURE HISTORIES

Typical member temperature histories computed from the isothermal elements are shown in figure 14. The same member families are used in temperature histories as were used in the preceding heating histories. The member temperatures follow the heating rates very closely because the members have low thermal capacitance and so stay close to radiation equilibrium. The only exception to this is the period of shadow transit, where the heating falls to almost zero abruptly and the temperatures fall more slowly toward radiation equilibrium. Upon reentering the sunlight, each member experiences a high heating with a temperature rise in the neighborhood of 6 K per second.

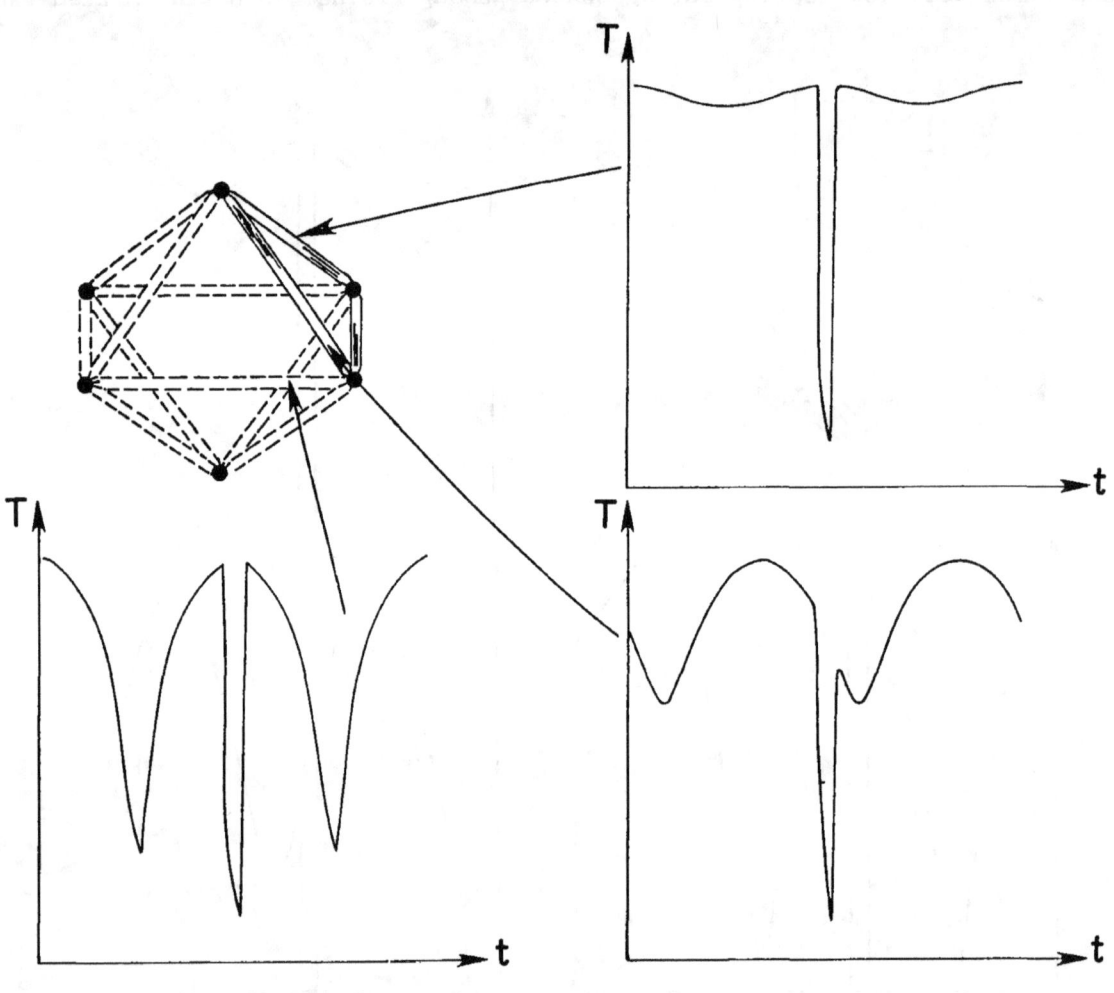

FIGURE 14

STRUCTURAL RESPONSE

Shown in figure 15 are outlines of the structure's shape at various points in the orbit. The structural response can be characterized by two features: (1) there is no out-of-plane bending due to the symmetry of the structure, and (2) the only distortion is a shearing of the top surface with respect to the bottom surface.

Notice that once the structure enters the Earth's shadow, the structure returns to its original shape. As it goes further into the shadow, however, it shrinks, getting smaller and smaller. Once the structure has reentered the sunlight, the shear deformation returns almost instantly.

Deformations shown here are greatly exaggerated; the actual deformations are just a few millimeters (the diameter of the truss is 10.8 m) due to the low thermal coefficient of expansion of the composite material. Recall, also, that the effects of aluminum joints have been neglected. Preliminary computations (not presented herein) suggest that aluminum joints with their relatively high coefficient of thermal expansion should be considered.

Member stresses exist because the truss is statically indeterminate, but they are quite small and can be neglected.

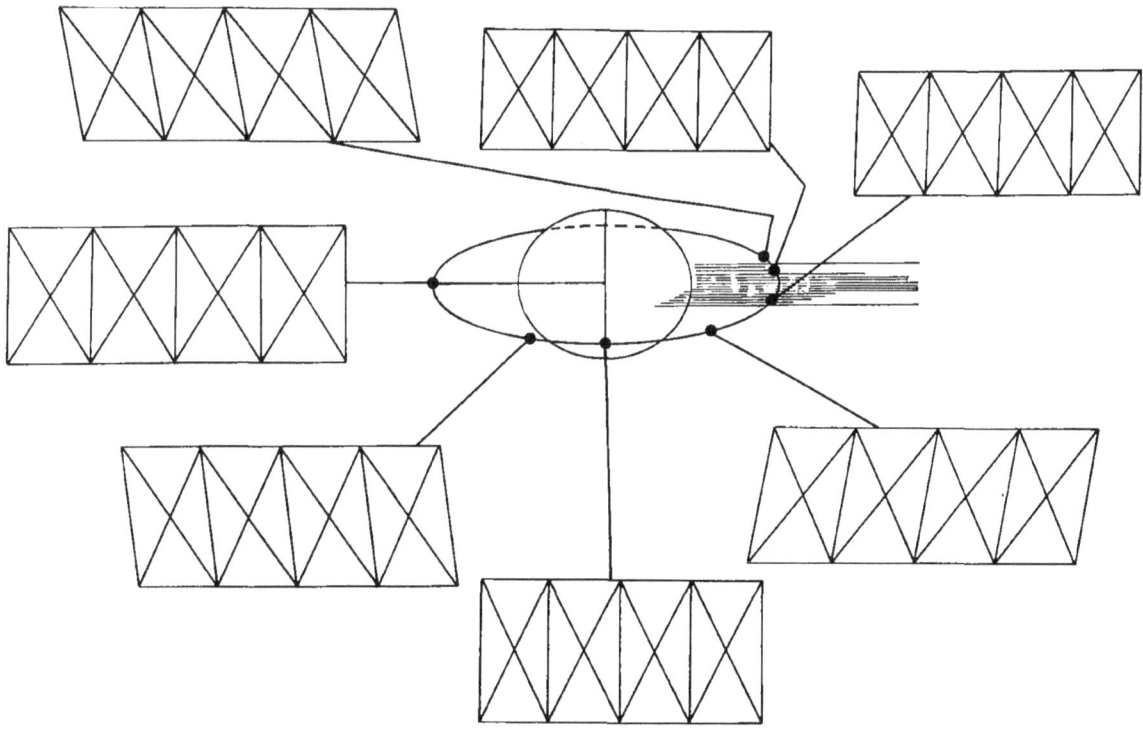

FIGURE 15

STRUCTURAL DEFORMATIONS

Two views of the octetruss structure are shown in figures 16 to 22 at the same points in the orbit that were highlighted in figure 15.

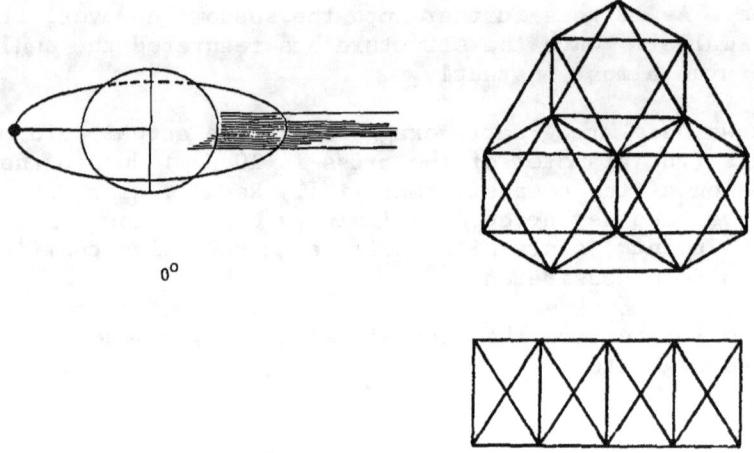

FIGURE 16

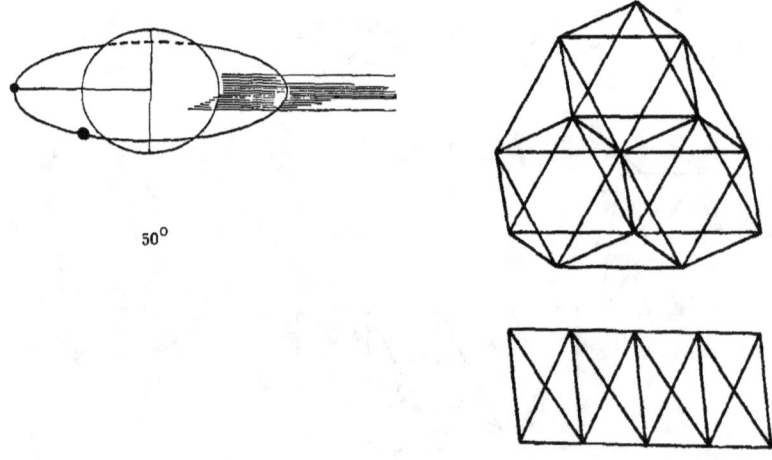

FIGURE 17

STRUCTURAL DEFORMATION

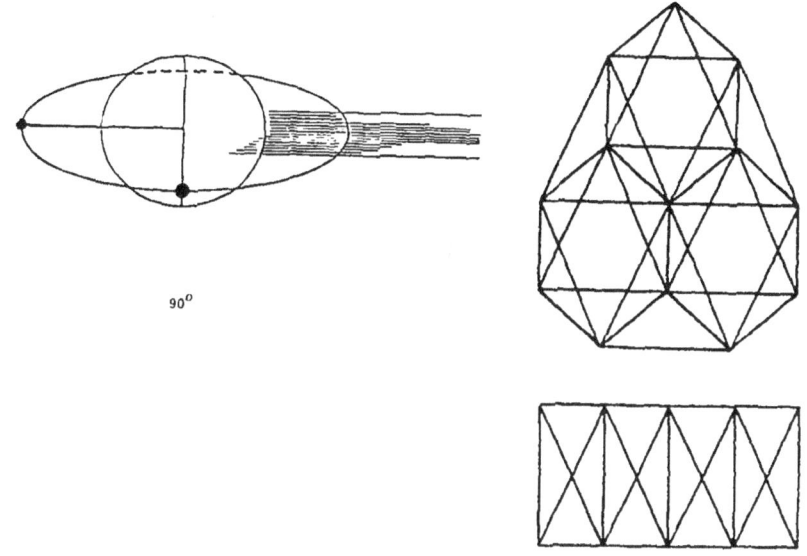

FIGURE 18

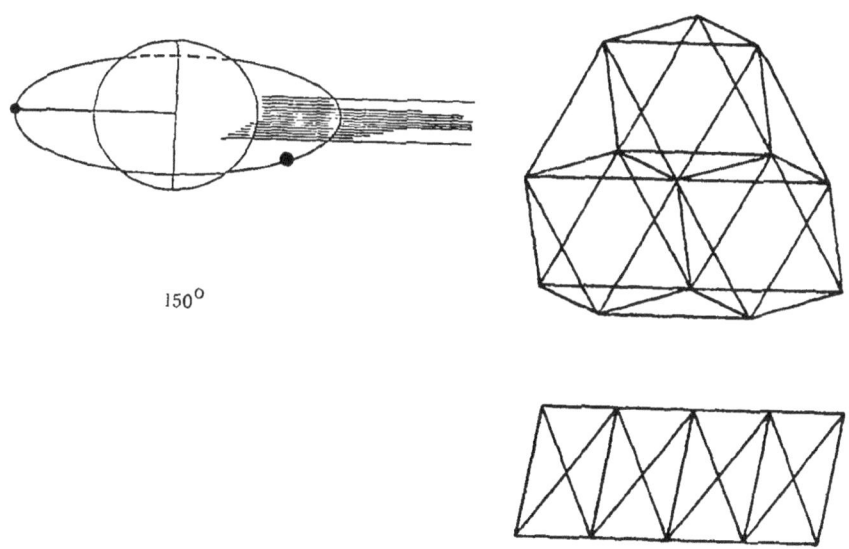

FIGURE 19

STRUCTURAL DEFORMATION

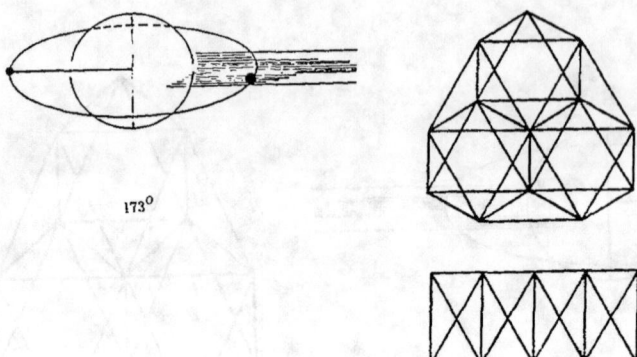

173°

FIGURE 20

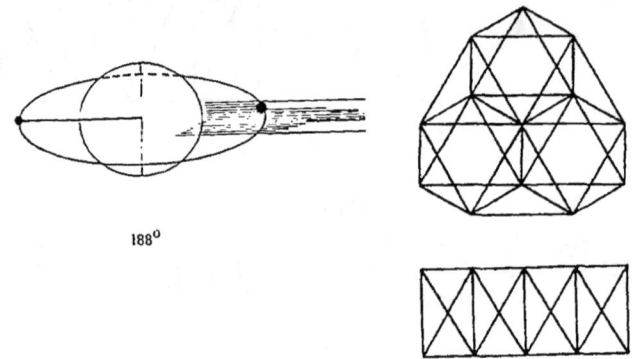

188°

FIGURE 21

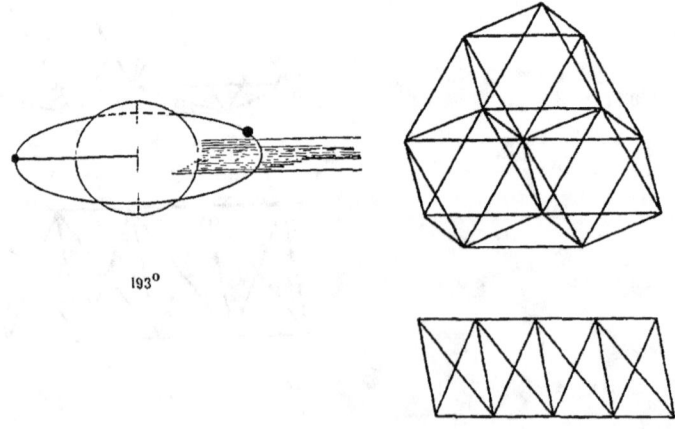

193°

FIGURE 22

CONCLUDING REMARKS

This paper demonstrates the characteristics of an integrated thermal-structural analysis approach which employs a geometric model with a common discretization for all analyses. It allows the use of improved thermal elements and uses the results from the thermal analysis directly in the structural analysis without any intervening processing of the data.

Comparative calculations for three thermal elements show that an isothermal element works best for low thermal conductivity materials. The isothermal element gives a good representation of the member temperatures and yields the best member forces.

An illustrative example with the LaRC octetruss gives typical thermal effects on an orbiting truss structure and shows that the integrated finite element approach is an attractive method for the thermal-structural analysis of large space structrues (see figure 23).

- CHARACTERISTICS OF INTEGRATED APPROACH
    - GEOMETRIC MODEL WITH COMMON DISCRETIZATION
    - IMPROVED THERMAL ELEMENTS
    - STRUCTURALLY INTEGRATED THERMAL RESULTS

- THREE TRUSS THERMAL ELEMENTS DESCRIBED FOR CONDUCTION COMBINED WITH RADIATION

- ISOTHERMAL ELEMENT BEST FOR LOW CONDUCTIVITY MATERIAL
    - GOOD REPRESENTATION OF MEMBER TEMPERATURES
    - BEST MEMBER FORCE

- OCTETRUSS EXAMPLE ILLUSTRATED TYPICAL ORBITING STRUCTURE RESULTS

- INTEGRATED FINITE ELEMENT APPROACH ATTRACTIVE FOR ORBITING STRUCTURES ANALYSIS

FIGURE 23

ACKNOWLEDGEMENT

This paper was based upon research supported by NASA Langley Research Center under grant NSG 1321.

REFERENCES

1. Thornton, Earl A.; Dechaumphai, Pramote; and Wieting, Allan R.: Integrated Thermal-Structural Finite Element Analysis. Proceedings of the AIAA/ASME/ASCE/AHS 21st Structures, Structural Dynamics, and Materials Conference, May 12-14, 1980, Seattle, Washington, AIAA Paper No. 80-0717, pp. 957-969.

2. Thornton, Earl A.; Dechaumphai, Pramote; Wieting, Allan R.; and Tamma, Kumar K.: Integrated Transient Thermal-Structural Finite Element Analysis. Proceedings of the AIAA/ASME/ASCE/AHS 22nd Structures, Structural Dynamics, and Materials Conference, April 6-8, 1981, Atlanta, Georgia, AIAA Paper No. 81-0480, pp. 16-32.

3. Cunningham, F. G.: Power Input to a Small Flat Plate from a Diffusely Radiating Sphere with Applications to Earth Satellites. NASA TN D-710, 1961.

4. Modest, M. F.: Solar Flux Incident on an Orbiting Surface after Reflection from a Planet. AIAA Journal, June 1980, pp. 727-730.

5. Kreith, Frank: Radiation Heat Transfer for Spacecraft and Solar Power Plant Design. International Textbook Co., pp. 95-96, 1962.

6. Chambers, B. C.; Jensen, C. L.; and Coyner, J. V.: An Accurate and Efficient Method for Thermal-Thermoelastic Performance Analysis of Large Space Structures. AIAA 16th Thermophysics Conference, June 23-25, 1980, Palo Alto, California, AIAA Paper No. 81-1178.

7. Card, M. F.; Bush, H. G.; Heard, W. L. Jr.; and Mikulas, M. M. Jr.: Efficient Concepts for Large Erectable Space Structures. Large Space Systems Technology, NASA CP-2035, 1978, pp. 627-656.

INTERACTIVE MODELING, DESIGN AND ANALYSIS OF LARGE SPACECRAFT

L. Bernard Garrett
NASA Langley Research Center
Hampton, Virginia

INTRODUCTION

Large space systems on the order of tens to hundreds of meters in size are projected to be operational in the future. The sizes will be driven by one or two principal considerations: economy of scale (e.g. antenna or sensor farms mounted on platforms with shared central utility supporting subsystems) or advanced systems which require large physical areas (e.g. high power solar arrays or remote sensing microwave radiometer antenna systems). These future spacecraft, unlike today's spacecraft which are generally enclosed monoque structrues with a few appendages, will have large expanses of lattice (truss-like) structures with hundreds or thousands of individual connecting members. The lightweight, flexible structures will be subjected to on-orbit environmental loads (gravity gradient, thermal, low-frequency transient vibrations, etc.) which usually were ignored in past spacecraft designs. Unless efficient design and analysis capabilities are developed for these advanced structures, the engineers will be severely taxed by the modeling, design, and analysis efforts. Further, computer resources will be rapidly consumed by the use of the prevalent large, single-discipline design and analysis codes.

The purpose of this presentation is to describe an efficient computer-aided design and analysis capability which has been developed to relieve the engineer of much of the effort required in the past. The automated capabilities can be used to rapidly synthesize, evaluate, and determine performance characteristics and costs for future large spacecraft concepts. The Interactive Design and Evaluation of Advanced Spacecraft Program (IDEAS) is used to illustrate the power, efficiency, and versatility of the approach. Although the IDEAS capabilities are by no means complete, the program has reached a certain level of maturity in the use of interactive data processing capabilities and spacecraft systems analysis oriented software to guide the design of future large space systems.

The coupling of space environment modeling algorithms with simplified analysis and design modules in the IDEAS program permits rapid evaluation of completing spacecraft and mission designs. The approach is particularly useful in the conceptual design phase of advanced space missions when a multiplicity of concepts must be considered before a limited set can be selected for more detailed analysis. Integrated spacecraft systems level data and data files are generated for subsystems and mission reexamination and/or refinement and for more rigorous analyses.

COMPUTER-AIDED DESIGN AND ANALYSIS CAPABILITIES - IDEAS PROGRAM OVERVIEW

The IDEAS program consists of about 30 interdisciplinary applications modules that include structural, thermal, and control system modeling; on-orbit static, dynamic, and thermal loading analysis; structural element design; surface accuracy analysis; antenna RF performance; and cost approximations. These modules are described in detail in references 1-3. They reside on both mainframe and super mini-computer systems. Data files are transferable between the two computer systems. These modules are executable from remote interactive graphics terminals. Processing and data control are accomplished via simple efficient executive and data base programs and file management routines. User prompts for file names and unformatted data inputs are provided. CRT graphic displays of finite element models and of summary information (temperature contours, element loading histograms, mode shapes, etc.) are presented to the user for immediate assessment and interactive modification of the spacecraft and/or mission as necessary.

The primary IDEAS modules and basic functions of each module are shown in Figure 1. The Aerospace Vehicle Interactive Design (AVID) program, developed by Wilhite (ref. 4) provides executive control and data base management capabilities for IDEAS. Additional procedure files and data file management routines reside in the individual IDEAS modules.

IDEAS was developed for multidiscipline spacecraft systems analysts as opposed to single discipline specialists or computer systems experts. The executive, data base/file management routines and applications modules were selected to provide a rapid, cost-effective computer-aided design and analysis capability for future large spacecraft systems concepts. The program is user friendly, prompting the analyst with queries or requests for unformatted input data, file names, processing paths, etc. The applications modules have been integrated to pass compatible, properly formatted files and data base information between single-discipline programs.

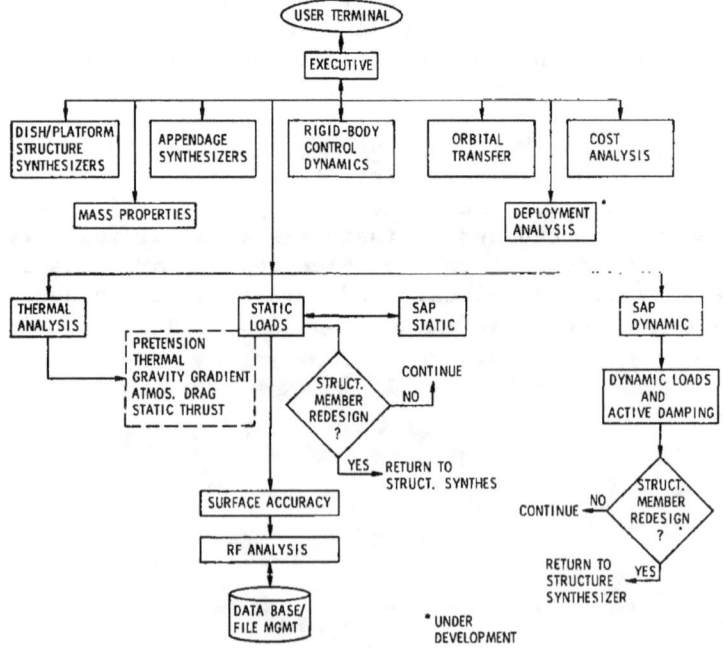

Figure 1

IDEAS MENU

The small executive program (50 to 100 lines of coded instructions) allows the analyst to run the applications modules individually or in any desired sequence by menu selection. The menu of spacecraft programs currently available is shown in Figure 2. All modules are set up to run in the interactive mode except for the ones prefaced with the word batch and the static and dynamic structural analysis modules. Note that all batch modules with the exception of the structure analysis modules have an interactive counterpart. Only the structural analysis programs have to be run in the batch mode because of the Langley Research Center 70,000 word memory interactive constraint for the mainframe computers. (Typical structural models analyzed to date require 150-200 K memory for these modules.) The minicomputer system does not have this constraint.

```
╔══════════════════════════════════════════════════════════════╗
║                      IDEAS PROGRAM                           ║
║   INTERACTIVE DESIGN AND EVALUATION OF ADVANCED SPACECRAFT   ║
╚══════════════════════════════════════════════════════════════╝
```

INPUT LETTER(S) OF PROGRAM(S) TO BE EXECUTED

A - AVID DATA MANAGEMENT PROGRAM - AVID DMP
B - LASS PREPROCESSOR - ANALOG
C - TETRAHEDRAL TRUSS STRUCTURE SYNTHESIZER - TTSS
D - GENERAL TRUSS SYNTHESIZER - GTS
E - RIGID-BODY CONTROL DYNAMICS - RCD
F - THERMAL ANALYSIS - TA
G - STATIC LOADS - SL
H - STRUCTURAL ANALYSIS PROGRAM (STATIC, DYNAMIC) - SAP
I - SURFACE ACCURACY - SA
J - ACTIVE DAMPING - ACTD
K - DYNAMIC LOADS - DL
L - LASS POST PROCESSOR - POST
M - LASS COST PROGRAM - COST
N - EXIT LASS PROGRAM

O - BATCH TTSS
P - BATCH RCD
Q - SYSTEM DESIGN AND COST MODULE - SDCM
R - BATCH GTS
S - GENERAL TRUSS SYNTHESIZER (NON-DISH)
T - BATCH TA
U - INTERACTIVE MODE PLOTTING MODULE
V - BOX RING
W - RADIAL RIB
X - HOOP AND COLUMN
Y - CONTIGUOUS BOX TRUSS
Z - MASS PROPERTIES
1 - ORBITAL TRANSFER
2 - RF ANALYSIS

Figure 2

STRUCTURE SYNTHESIZER MODULES

The Tetrahedral Truss Structure Synthesizer (TTSS) module, developed by W. D. Honeycutt of the General Dynamics Corporation, Convair Division, is used to rapidly model flat or curved tetrahedral truss structrues and to initially size the structural members. The module automatically generates the nodal geometry; the member connectivity, cross-sectional areas and masses; and the resultant finite-element model of the structure for a specified dish diameter, shape, number of bays, and a diagonal angle which defines the truss depth. The tetrahedral truss configuration and major hardware components are shown in Figure 3.

The truss structural members are assumed to be circular tubes, isogrids, or triangular truss struts. The surface and diagonal members are sized separately for Euler buckling from input material properties and initial loading conditions. Upper and lower surface members are pinned; diagonal elements may be pinned or clamped at the user's option. Structural members can be constrained to minimum material thicknesses and tube diameters so that the column buckling equations will not design members too small for practical use. An option also permits the sizing of the member (diameter and thickness) from user specified length over radius of gyration and a tube radius over thickness inputs.

The folding hinges, spiders, bearing, and end fitting masses are computed as functions of the structural member diameter. A mesh-reflective surface and the support system may be optionally included in the calculations and is automatically distributed at each nodal point on one of the surfaces. The mesh control system, used to maintain contour, is computed as a percentage of the mesh weight. A contingency mass is included which is defined as a percentage of the mass of all the structural components.

A total of 45 input variables is needed to run TTSS. The module calculates and outputs the mass of the structural components, mesh system, and total system; the center of gravity; and mass inertia properties. The displayed outputs also included structural member dimensions, hardware part counts, unit masses, total group masses, mesh area, and configuration packaging dimensions for inward and outward folded deployable surfaces.

Uniquely nameable data base and files, including a complete finite-element model of the tetrahedral truss, are created in TTSS for later use in other modules.

Structure synthesizer modules which have similar modeling characteristics as that of the tetrahedral are also incorporated in IDEAS. They include contiguous box trusses, box truss ring, radial rib, and hoop and column synthesizers. Some of these finite-element models are shown in Figures 4-6.

CONFIGURATION DEFINITIONS FOR THE TETRAHEDRAL TRUSS STRUCTURE SYNTHESIZER MODULE

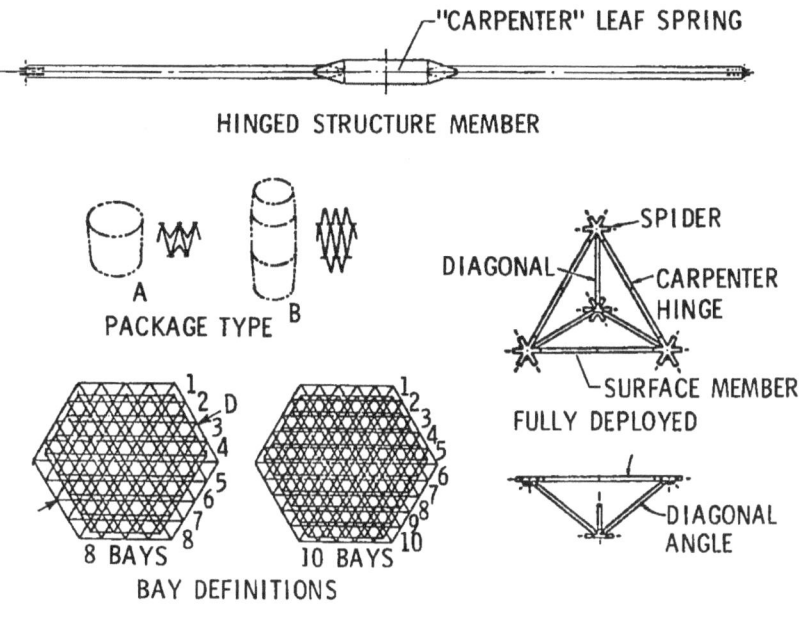

Figure 3

BOX TRUSS RING ANTENNA STRUCTURE
FOR A MICROWAVE RADIOMETER SATELLITE

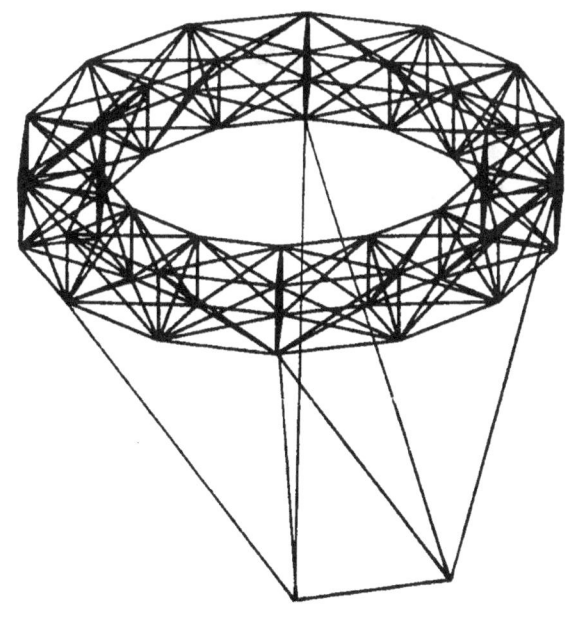

Figure 4.

RADIAL RIB ANTENNA STRUCTURE
FOR A LAND MOBILE COMMUNICATION SATELLITE

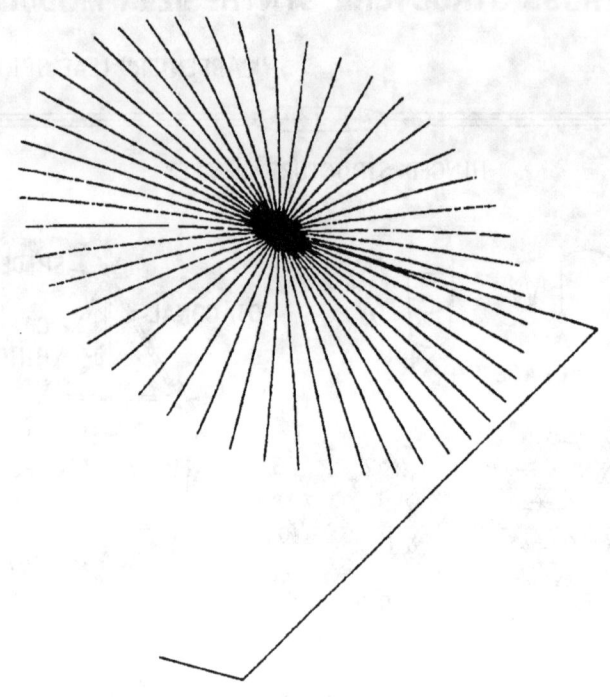

Figure 5

HOOP AND COLUMN ANTENNA STRUCTURE
FOR A LAND MOBILE COMMUNICATIONS SATELLITE

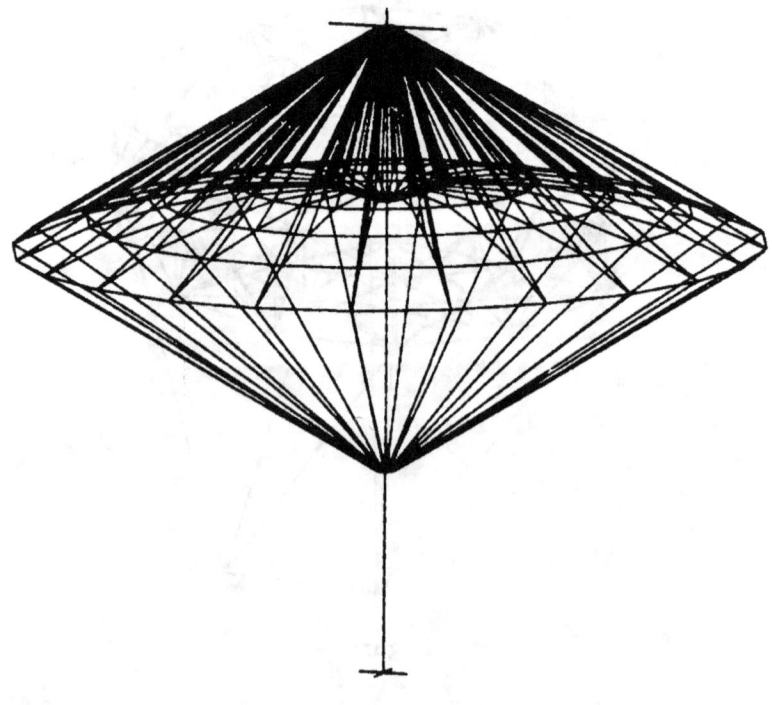

Figure 6

APPENDAGE SYNTHESIZER MODULE

The Appendage Synthesizer Module is used to complete the definition of the spacecraft structure, materials, and supporting subsystems. Required inputs include dish data files from the appropriate Structure Synthesizer Module; added structural members and connectivity, member design loads; subsystem locations, masses and areas; and mesh blockage factor. Hollow tubes, isogrids, triangular trusses (see Fig. 7), cables, and deployable astromast members are available structural appendage types. Outputs include the design of the added members; a finite-element model of the total spacecraft; an atmospheric drag approximation model; and updated mass, inertia, centers of gravity, and pressure properties. The mass-per-unit area of all elements needed for the thermal analysis is also generated in this module. The Appendage Synthesizer can be used to create spacecraft designs from keyboard input when no synthesized structure is desired. Alternatively, finite-element model data formatted by external structural analysis programs can be preprocessed and read directly into the module to save labor and time.

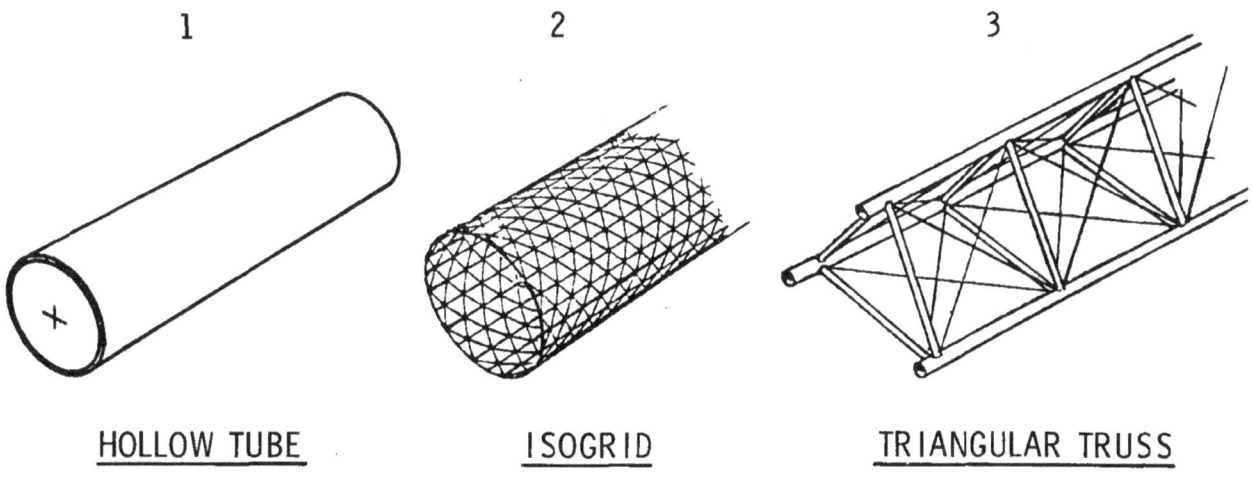

Figure 7

RIGID-BODY CONTROL DYNAMICS MODULE

The Rigid-body Control Dynamics (RCD) module calculates the on-orbit environment and maneuver forces and torques at user-specified circular orbital altitude and spacecraft orientation. The module then determines the momentum storage and desaturation requirements, and iterates the masses of the control systems, propellant, and tankage to meet the orbit-keeping, attitude control, and maneuver requirements of the spacecraft. Principal features of RCD are shown in Figure 8. The total torque and force time histories are analyzed to determine cyclic momentum for momentum exchange system sizing and accompanying momentum desaturation requirements. Momentum desaturation is accomplished by reaction control system (RCS) thrusters. RCS requirements for orbit keeping are also determined. Finally, RCS requirements are computed assuming RCS control in lieu of the momentum exchange plus desaturation systems. Technical capabilities for this module were provided by Chiarappa and Eggleston (ref. 5). References 6 and 7 also provide supporting information on satellite drag and referenced atmospheric data.

Spacecraft mass, inertia, areas, centers of gravity, and pressure are input from the synthesizer files. Those parameters are updated in RCD in accordance with the momentum exchange and propulsion systems sizing and mass computations. A total of 34 addition input variables plus a thruster force matrix are needed to run RCD. Input categories include the orbital parameters; the spacecraft orientation (inertial or Earth oriented), maneuver requirements, and pointing accuracy; the control system location and performance characteristics; and the propellant resupply periods. An arbitrary number of RCS thrusters may be located at multiple nodes. The thrust level and direction for each individual thruster is user specified. The program assumes that individual thrusters can fire in either a positive or negative direction along one of the principal axes.

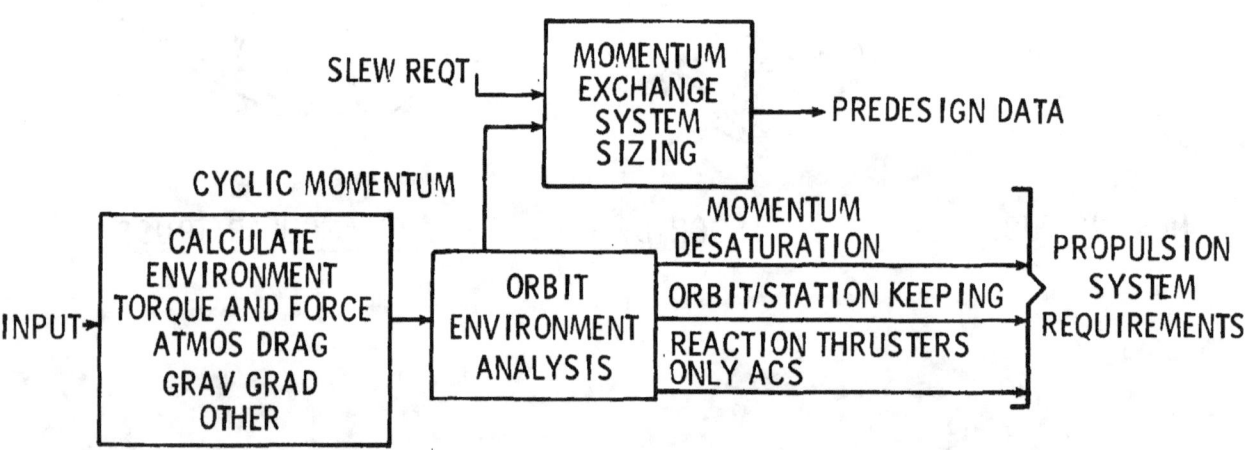

Figure 8

THERMAL ANALYSIS MODULE

The Thermal Analysis (TA) module is used to compute the radiation equilibrium temperature for each structural member at a given position in the spacecraft orbit. Technical capabilities for the module (fig. 9) were developed by G. A. Howell of the General Dynamics Corporation, Convair Division, from the original work of Ballinger and Christensen (ref. 8). Heat sources are solar radiation, Earth albedo, and Earth radiation. The thermal response of each is determined from the balance between absorption of energy from the three heat sources and reradiation of energy from the elements to deep space. The position of the members relative to the Sun and the Earth are varied at 36 intervals in the orbit. Earth shadowing is included. The members may be single or double shadowed by a translucent mesh. There is no radiation exchange or conduction between members and no shadowing of members by other members. Inputs to TA include the thermal properties of the members and mesh transmissivity constants; the finite element geometry and unit area data files from Appendage Synthesizer and/or RCD; and Sun-Earth-spacecraft geometry inputs from RCD. Outputs include the temperature of each member at a user-specified location in orbit, temperature contours of the dish upper and lower surface members, and a temperature file, TATMPS. Since one isothermal temperature is computed per member, the thermal model is completely compatible with the structural finite element model. Temperatures for each member are read into the static loads module from the TATMPS file for use in conjunction with the Structural Analysis Program for the generation of thermal loads and deflections. For 439 members, evaluated 12 minutes apart in the orbit, TA executes in 16 seconds CP time and in about 10 wall-clock minutes.

PURPOSE:

- TO COMPUTE EQUILIBRIUM HEATING RATES AND ISOTHERMAL TEMPERATURES OF EACH MEMBER IN ORBITAL ENVIRONMENT

HEAT BALANCE:

- SUN, EARTH ALBEDO, EARTH THERMAL RADIATION AND STRUCTURAL MEMBER ABSORPTION/RERADIATION
- VARIES POSITION OF MEMBERS RELATIVE TO SUN AND EARTH AT 36 INTERVALS WITHIN THE ORBIT (EARTH SHADOWING INCLUDED)
- MEMBERS ARE SHADOWED OR DOUBLE SHADOWED BY MESH

INPUTS:

- THERMAL PROPERTIES OF MEMBERS
- FINITE ELEMENT GEOMETRY, MASS/UNIT AREA DATA FILES
- SUN-EARTH-S/C GEOMETRY INPUTS FROM RCD

OUTPUTS:

- TEMPERATURE OF EACH MEMBER
- THERMAL CONTOURS
- TTAMPS FILE FOR SAP STATIC

MAINFRAME STATISTICS:

- 16 SECONDS CP TIME
- 10 MINUTES

439 ELEMENTS
12 ORBITAL MINUTES

Figure 9

RADIAL RIB ANTENNA TEMPERATURES

Temperature contour data are shown in figure 10 for a 55-meter diameter radial rib antenna. This is one of several concepts under study for a Land Mobile Communication Satellite mission. The satellite is in a circular equatorial orbit at geosynchronous Earth orbital altitude. The satellite has been in the sunlight about 1.5 minutes after a 1.3-hour cold soak in the Earth shadow.

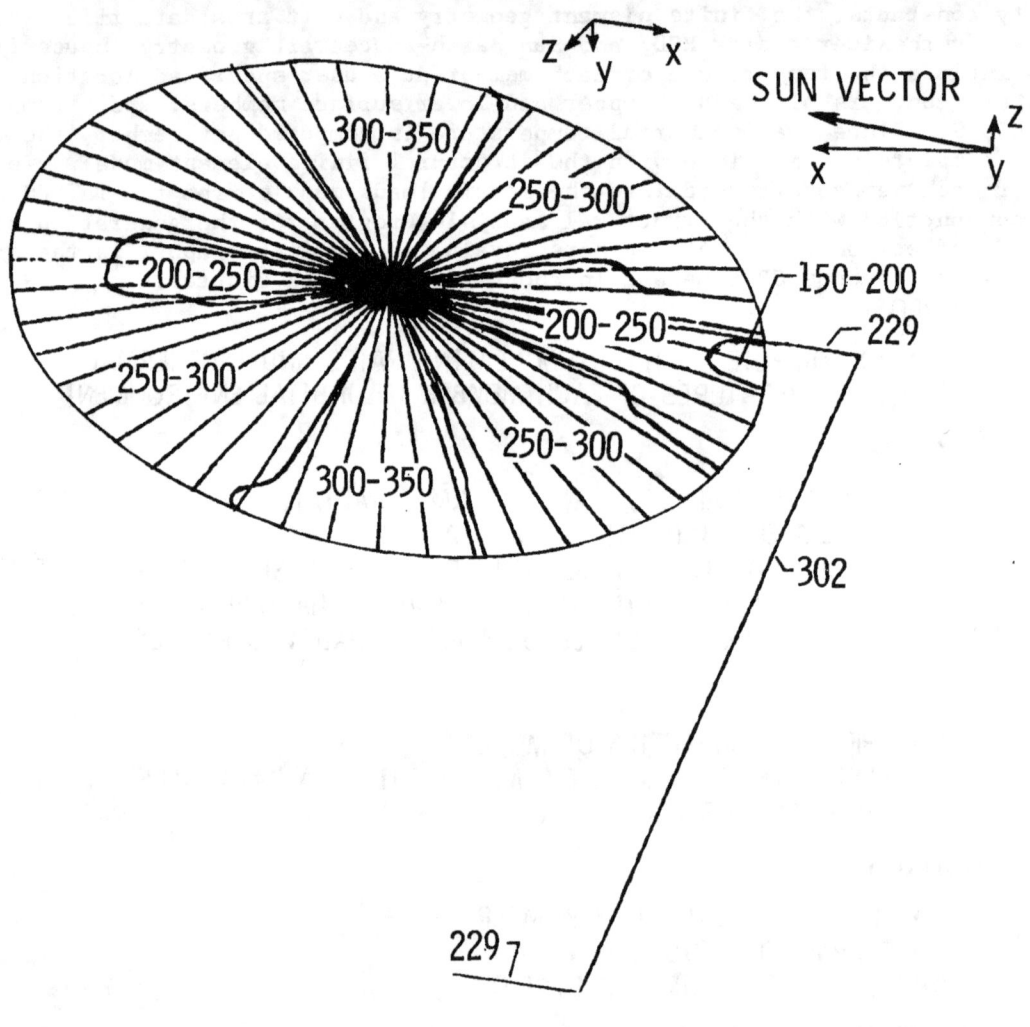

Figure 10

SELECTED MISSION AND SPACECRAFT DESIGN DETAILS

The capabilities of the IDEAS program to deal with the multidiscipline aspects of spacecraft preliminary design are illustrated by examples in the design, analysis, and parametric evaluation of an advanced spacecraft--the passive Microwave Radiometer Spacecraft (MRS) designed to perform soil moisture measurements from low-Earth orbit.

The MRS structure and supporting systems are shown schematically in Figure 11. The structure consists of a relatively stiff double-layered 750-m diameter tetrahedral truss dish (graphite epoxy composite structural members) with an RF reflective mesh (aluminized Kapton with a unit mass of 0.03 kg/m^2) attached to offsets on the concave surface. Support beams (graphite epoxy composite) and tension cables (Kevlar) provide stabilization and boresight control for the feed horns mounted on a curved beam located at the focal arc of the reflector. The spacecraft operates at a nominal altitude of 750 km. The dish points toward nadir with the feed beam oriented normal to the spacecraft velocity vector.

Attitude control is provided by a dual-ring annular momentum control device (AMCD) and eight one-newton liquid oxygen/liquid hydrogen thrusters. The AMCE rings are magnetically supported in races at the outer periphery of the convex surface of the dish to provide pitch, roll, and yaw control. Orbital velocity makeup is provided by four larger liquid oxygen/liquid hydrogen thrusters. Two are located on the dish structure providing 1,500-newtons thrust each and two are located at the extremities of the feed beam providing 500-newtons thrust each. Three propellant tanks are located on the convex side of the dish in a triangular arrangement at the three center-most nodes. Other subsystems included in the analysis are shown in Figure 11.

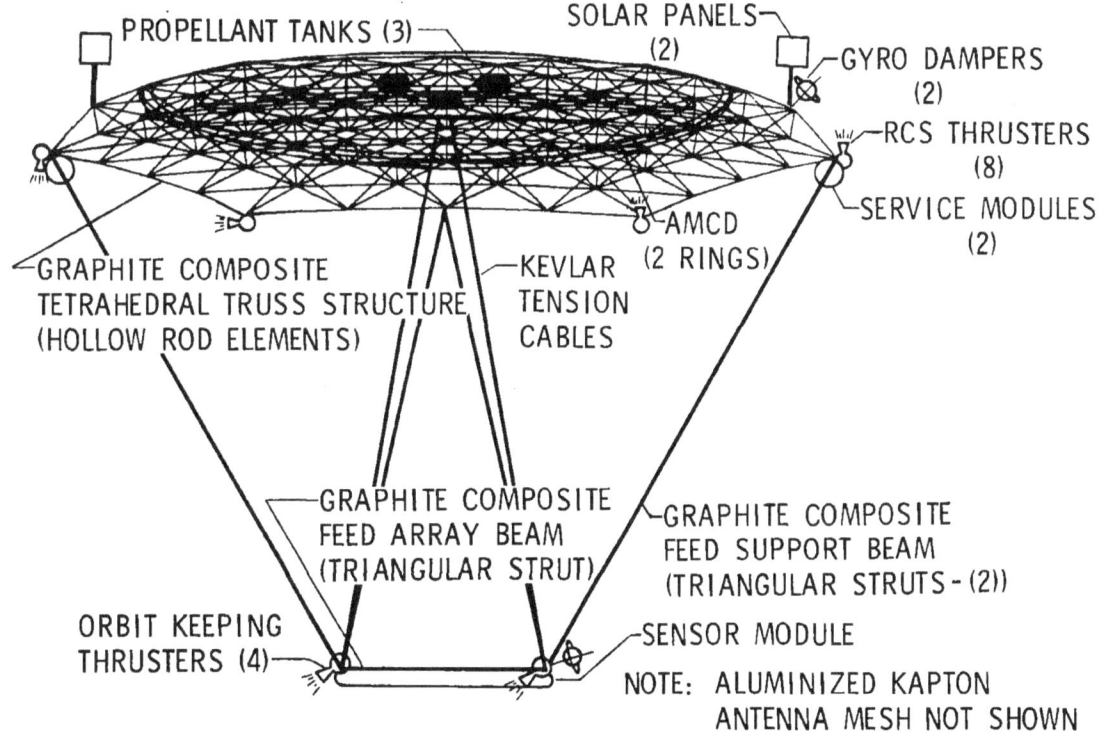

Figure 11

MICROWAVE RADIOMETER SPACECRAFT TEMPERATURE CONTOURS

Selected thermal contours for MRS dish members oriented in the same direction are shown in figure 12. The solid lines denote the structural members for which the temperature contours are applicable. Similar contours are also plotted for both concave surface members oriented in the other directions. The contours aid the analyst in rapidly visualizing approximate temperature ranges and distributions for the entire structure and are preferable in the interactive analysis to review of temperature printouts for hundreds of members.

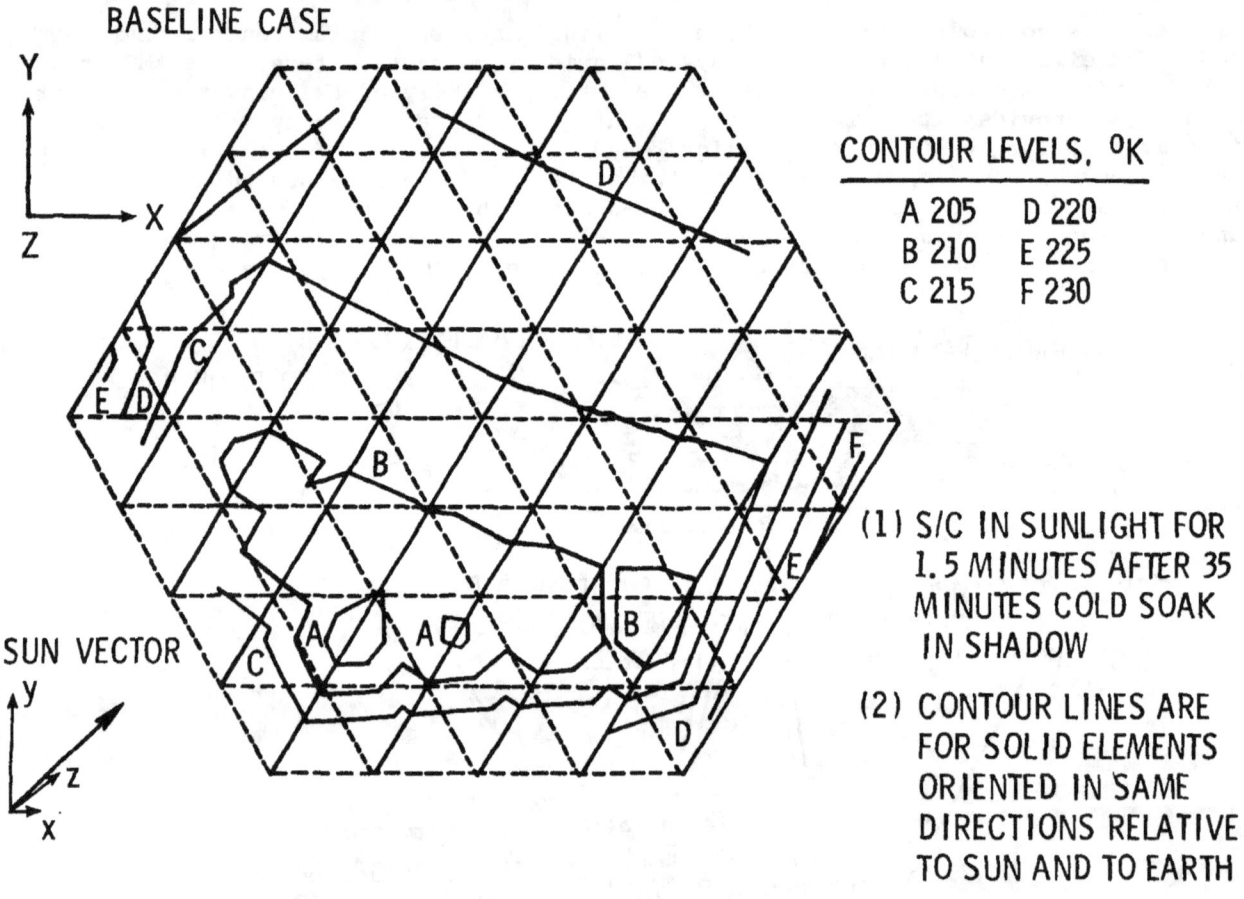

Figure 12

TEMPERATURE VARIATIONS OF THE
MICROWAVE RADIOMETER SPACECRAFT THROUGHOUT THE ORBIT

Thermal loads on the individual elements will vary throughout the orbit and it is not generally known a priori where in the orbit the loads reach the maximum. However, some insight on maximum thermal loading can be gained by calculating element temperatures and temperature differentials at selected orbital points. In this study, heating rates and temperatures of each MRS structural member were calculated in the IDEAS Thermal Analysis (TA) module at four points in the orbit:

Point a. Orbit Anomaly Angle = 1.5 radians, time = 0.40 hours
(just prior to S/C entry into Earth shadow)

Point b. Orbit Anomaly Angle = 3.7 radians, time = 0.98 hours
(just prior to S/C exit from Earth shadow)

Point c. Orbit Anomaly Angle = 3.9 radians, time = 1.03 hours
(just after S/C exit from Earth shadow)

Point d. Orbit Anomaly Angle = 5.8 radians, time = 1.53 hours
(midway in sunlight portion of orbit)

Start and end of Earth shadow were at anomaly angles of 1.6 and 3.8 radians (time 0.4239 hours and 1.005 hours), respectively. The orbit period was 1.667 hours at the 750-km altitude. Figure 13 gives summary results of maximum and minimum temperatures of the various members at each point in the orbit.

Element temperatures for Point c (just after exit of the spacecraft from Earth shadow) were selected for use in the static loads analysis. This point was selected on the combined basis of near-maximum temperature difference between elements and relatively low temperatures for all the elements.

MAXIMUM/MINIMUM ELEMENT TEMPERATURES, K

CASE	DISH MEMBERS			FEED BEAM	FEED SPT BEAM	TENSION CABLES
	CONCAVE SURF. (MESH SIDE)	CONVEX SURFACE	DIAGONAL			
A	312/232	312/237	312/236	313	312/308	310/286
B	191/183	184/183	179/176	191	179/179	180/180
C	251/201	264/196	255/214	267	261/256	262/229
D	332/324	328/324	325/300	321	298/271	301/275

Figure 13

MICROWAVE RADIOMETER SPACECRAFT ENVIRONMENTAL LOADS

<u>SAP Static</u>. The static-load-carrying capabilities and internal stresses in the individual structural members are evaluated in the Structural Analysis Program (SAP, ref. 9), which was originally the work of Professor Edward L. Wilson of the University of California at Berkeley, and the Static Loads (STLO) module. SAP is a general purpose structural analysis program for static and dynamic linear analyses of three dimensional structural systems. SAP static calculates nodal displacements and rotations, member forces and moments, and internal stresses for up to five separate load conditions and for the linear combination of all loads acting simultaneously. However, the program requires loading inputs to perform the analyses which are provided by the STLO module.

<u>STLO</u>. The STLO module operates in two parts. STLO, part 1 collects all the appropriate static loads data in a properly formatted file for SAP and generates environmental and spacecraft-induced loads for the following five loading conditions: (1) pretension; (2) thermal; (3) gravity gradient; (4) atmospheric drag; and (5) static thrust. The structural finite-element model is included in the STAMOD file. Inputs to STLO, part 1, include a full description of the mass points for the gravity gradient computations from the DYML file, the projected area approximations for the atmospheric drag loads from the Appendage Synthesizer files, the isothermal member temperatures from the Thermal Analysis module, and thrust and pretensioning forces. Following the SAP run, STLO, part 2 outputs summary data of the actual loads on the structural members, compares them to the design loads, and permits the user to redesign the elements if the actual loads differ considerably from the design loads. If many members are poorly designed, the user can instruct the program to recycle through the appropriate synthesizer modules with the updated design loads and revise the member sectional areas. If the spacecraft mass and inertia properties are significantly modified, the RCD module can redefine the control system requirements. Continuous iterations can be performed under user control until a satisfactory solution is obtained for the structural loads, member sizes, spacecraft mass, inertia and drag properties, and the control system requirements. At any step in the design and analysis process, the user may decide that he has a poor design and may revise the design or change subsystems (which may either be current space-qualified hardware or advanced technology subsystems) and continue with the design process.

The contributions of various environmental load conditions are illustrated in figures 14 and 15 for the three-load case (thermal, gravity gradient, and atmospheric drag) and for thermal load only. Note that there is little change in the loads levels or distribution from the thermal-only case to the three-load case. The environmental loads are extremely small. Minimum gage structural members designed on the basis of practical ground fabrication and handling loads are expected to have a much higher load-carrying capability.

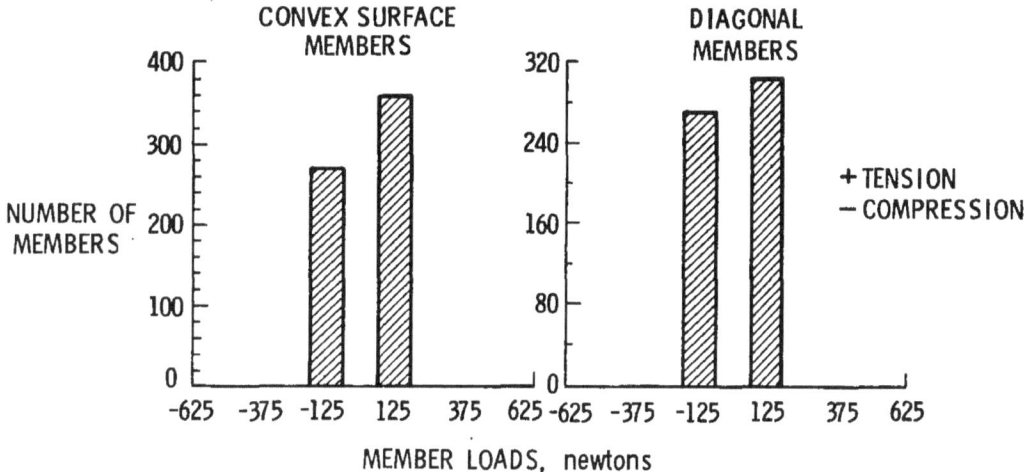

Figure 14

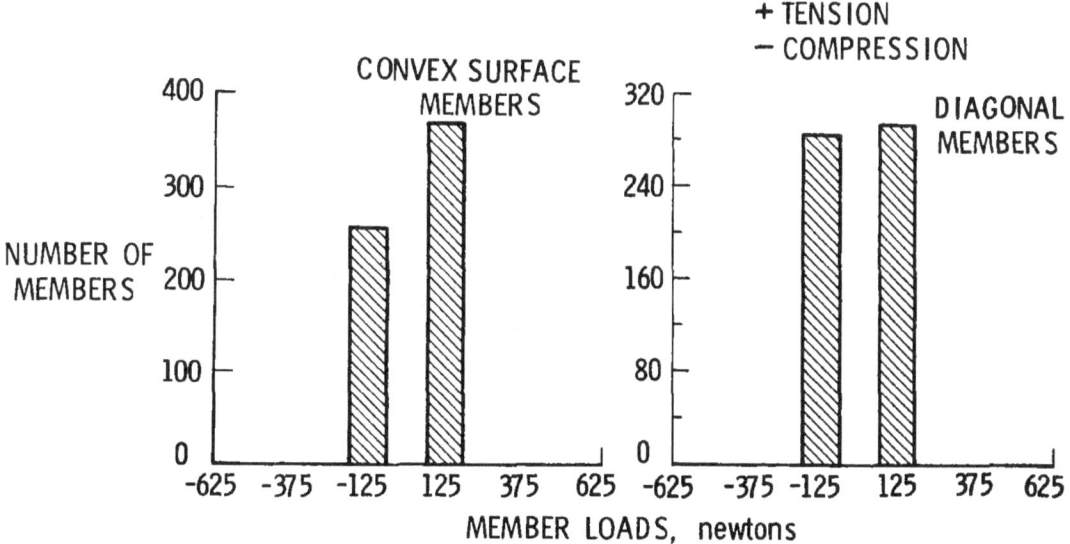

Figure 15

MICROWAVE RADIOMETER SPACECRAFT SURFACE ACCURACY

Performance must also be factored into the design evaluation process. For example, in the case of large aperture systems, surface distortions, boresight offset, and defocus are important parameters leading to the establishment of RF antenna or solar concentrator performance and figure control requirements. The IDEAS Surface Accuracy (SA) Model establishes these first-order effects on performance. SA computes the overall surface roughness (rms displacement), lines of constant derivation from an ideal surface (distortions), and changes in focal length, boresight direction, and boresight displacement for reflective surfaces. The SAP static module files supply SA with finite element model data for all original and statically displaced node point locations. SA plots the local normal displacement and distortion contours for the mesh surface nodes. The shapes of surfaces available are parabolic, spherical, or flat.

It should be noted that most spacecraft are free-free structures which require nonredundant translational constraints for purposes of static analysis. The STLO module has been coded to automatically provide nonredundant constraints at three node points on opposite corners of a tetrahedral truss dish structure (one node restrained in x, y, and z, one node restrained in x and z and one node restrained in z only) to arrive at static loads. The calculated loads and stresses are valid for the real free-free spacecraft; however, resulting static deflections at individual node points are sensitive to the method of restraint and the nodes which are restrained. The program is used to convert the artificially constrained nodal deflections into distortions that are independent of the constraints.

The MRS surface distortion contours due to the environmental loads are shown in figure 16. Defocus and boresight offset data are also shown. The resulting surface rms roughness is well within the 6-mm tolerance required for good microwave radiometry performance.

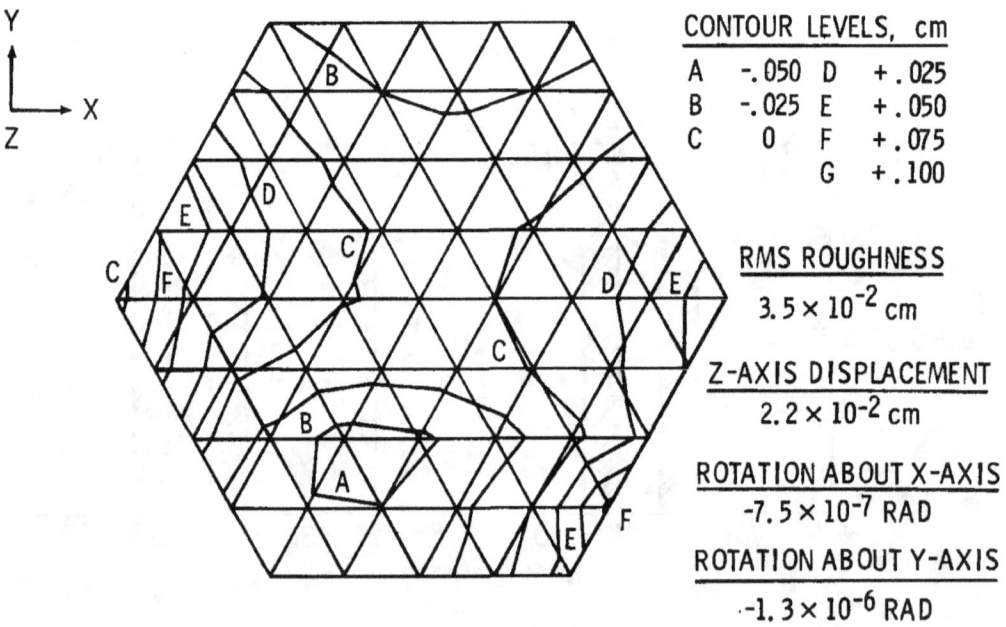

Figure 16

MICROWAVE RADIOMETER SPACECRAFT DYNAMIC ANALYSIS

In the IDEAS program the SAP Dynamic Analysis Module is used to generate modal frequencies, normalized forces, and stresses. The deflections in all cases are normalized to unit-generalized masses. All appropriate dynamic analysis data, accumulated from the structural synthesizer and other analysis modules, are combined in a properly formatted input file (DYML) for the SAP Dynamic Analysis module. An automated eigenvalue shift procedure has been employed in IDEAS to overcome most numerical instability or singularity problems associated with rigid body modes (normally there are six of these zero roots for the free-free structures). SAP outputs data on the number of modes requested by the user, including the first six rigid body modes for the linear system. A post-processor has been added to the Langley version of IDEAS to scale and plot flexible body mode shapes for the spacecraft structure.

Mode shapes for the first and fourth flexible body modes of the MRS are shown in figures 17 and 18. These figures were generated in the Interactive Plotting Module from SAP (Dynamic) solutions. The low frequencies noted in the figures are due to mast and feed beam action. The tetrahedral truss dish first flexible body frequency is an order-of-magnitude (~ 0.5 Hz) higher than the fundamental mast frequency. Although this study did not address minimum vibrational frequency or flexible-body-control-system requirements for the MRS, it should be noted that the triangular truss beams could be replaced with somewhat stiffer structural members to provide moderate increases (possibly a factor of 2 or 3) in the lower-order frequencies. However, even with technology advances in stiff, lightweight materials and structural design concepts, it is likely that these low frequencies and possibly high-amplitude vibrations will be typical of large future systems. Frequencies on the order of 0.01 to 0.1 are far below the design capabilities of space-qualified controllers/actuators and will require innovative control concepts and much more detailed analyses.

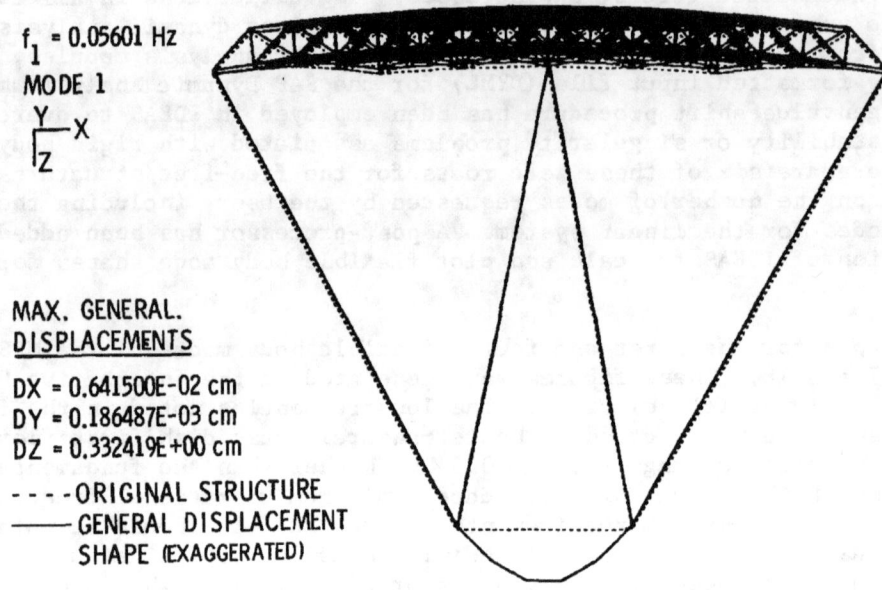

Figure 17

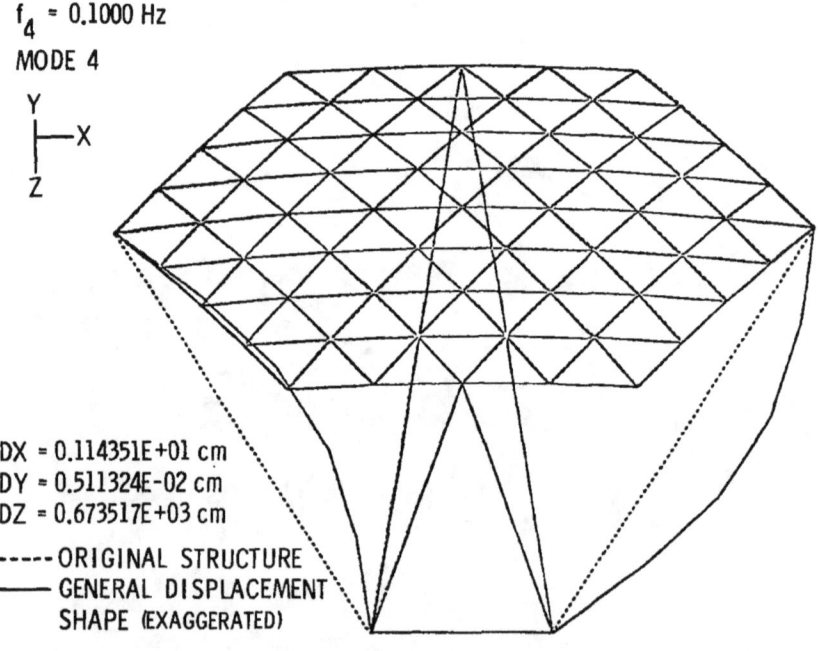

Figure 18

ADDITIONAL IDEAS CAPABILITIES

The IDEAS program has several additional capabilities for design, analysis, and evaluation of future large spacecraft concepts. These capabilities are summarized below.

Orbital Transfer. This module is used to compute the propellant required to perform an orbital plane change maneuver or to raise or lower a satellite orbit. Inputs include the satellite mass, thruster specific impulse, and initial and final orbit parameters.

RF Analysis. This module calculates the db loss for random distortions in an antenna surface. Inputs required include node point displacements or overall surface rms errors and the antenna operating frequency.

Dynamic Loads and Active Damping. Other capabilities include the Dynamic Loads (DYLO) module developed by Leondis (ref. 1) to provide dynamic deflection data at node points and dynamic loads on each member. Inputs include the flexible body modes from SAP dynamic, the finite-element files (DYML), and user-specified transient-force functions and structural damping characteristics. The use of active damping systems can be evaluated in this module.

Cost. Developmental and first unit costs are computed in the COST module principally from cost-estimating relationships. Subroutines calculated costs associated with large space structures comprised of many structural members of various types of materials and design complexity. Spacecraft subsystem costs are approximated from subsystem masses or performance data. Shuttle launch costs are based on both spacecraft/subsystem mass and on packaging volumes. On-orbit construction costs may be estimated from user inputs on construction time and crew size. All cost-estimating relationships are in 1976 dollars; however, totals costs are updatable to any subsequent year with appropriate inflation factors.

PROGRAM RUN TIMES AND THROUGHPUT STATISTICS

A 32-bay tetrahedral truss dish structure (6864 structural members) of any diameter, curvature, and truss depth can be mathematically synthesized in the Tetrahedral Truss Structure Synthesizer module in about 5 wall-clock minutes on the minicomputer system. The program incorporates all joints, pins, hinges, and mesh system and designs the structural members for a user-specified Euler buckling load. Then it generates and displays the finite-element model and summary data and writes the data and files to retrievable dish storage. The corresponding 8 Bay case (420 members) executes in 1 to 2 wall-clock minutes on the minicomputer (at 9600 Baud line rate) or about 6 wall-clock minutes and 27 seconds CP time on the mainframe (at a 1200 Baud line rate).

In general, for a typical case with about 500 structural members, the CP times for each applications module range from a few seconds to 2 minutes on the mainframe. Corresponding wall-clock times vary between 1 and 15 minutes. All modules can be run in series in 2 to 3 wall-clock hours on the mainframe. (See ref. 3 for more information on run time statistics for the individual modules.)

CONCLUDING REMARKS

Capabilities, performance, and advantages of a systems-oriented interactive computer-aided design and analysis system have been presented. A single user at an interactive terminal can create, design, analyze, and conduct parametric studies of Earth-orbiting spacecraft with relative ease. The approach is shown to be particularly useful in the conceptual design phase where various missions and spacecraft options are to be evaluated in a timely, cost-effective manner.

The Interactive Design and Evaluation of Advanced Spacecraft (IDEAS) program was developed specifically to provide spacecraft system analysts with the interactive capabilities to rapidly analyze and evaluate spacecraft performance across several disciplines. The primary emphases are on structures, thermal analyses, and controls. Simple and efficient executive data base and file management systems relieve the analyst of much of the tedium associated with computer system command protocol. Formatted data inputs reduce the possibility of input errors and greatly increase throughput capabilities. Extensive graphical displays let the analyst rapidly evaluate the results, make timely design changes, and continue in the interactive processing mode. Although the IDEAS capabilities are by no means complete, the program has reached a certain level of maturity in the use of interactive data processing capabilities and spacecraft systems analysis oriented software to guide the design of future large space systems. The coupling of space environmental modeling algorithms with first order design and analysis modules permits rapid evaluation of competing spacecraft and mission designs. Spacecraft redesign is easily accomplished and baseline designs can be altered in an orderly manner for subsystem and mission design trades. Integrated spacecraft systems level data and finite element model files are generated for more rigorous analysis.

Example problems of large flexible spacecraft in low-Earth and geosynchronous-Earth orbits have been used to illustrate some of the IDEAS capabilities. Study results lead this author to conclude that the space environmental loadings on most of the future spacecraft will be small. The structural strength requirements will be dictated by either ground-based manufacturing, assembly and testing, or launch or on-orbit deployment.

Optimally designed structures with physical sizes approaching 100 meters may have fundamental frequencies on the order of 0.01 to 0.1 Hz, which are below current control system capabilities. The desire on the part of the dynamicists to increase the structural stiffness will penalize the spacecraft design on both volume and mass basis. Further work is required in this area to achieve reasonable design compromise.

This author is also concerned with the application of some of the existing thermal analysis codes for these large structures which may have thousands of individual structural members. If some of the large conduction and radiation heat exchange computer programs are used the computer run time costs can become expensive. It is not clear that this concern has been adequately addressed by the thermal prohibitively analysis community.

REFERENCES

1. Leondis, A. F.: "Large Advanced Space Systems Computer-Aided Design and Analysis Program." NASA CR-159191, July 1980.

2. Leondis, A. F.: "Large Advanced Space Systems (LASS) Computer Program." Presented at the AIAA/NASA Conference on Advanced Technology for Future Space Systems, AIAA 79-0904, May 1979.

3. Garrett, L. B.: Interactive Design and Analysis of Future Large Spacecraft Concepts. Presented at the AIAA 16th Thermophysics Conference, AIAA 81-1177, June 1981.

4. Wilhite, A. W. and Rehder, J. J.: "AVID: A Design System for Technology Studies of Advanced Transportation Concepts." Presented at the AIAA/NASA Conference on Advanced Technology for Future Space Systems, AIAA 79-0872, May 1979.

5. Chiarappa, D. J. and Eggleston, D.: "Low Altitude Vertistat Equations of Motion." GDC Report No. DGS ERR-AN-977, December 1966.

6. Cook, G. E.: "Satellite Drag Coefficients." Royal Aircraft Establishment Technical Report No. 65005, January 1965.

7. COSPAR Working Group IV, CIRA 65, COSPAR International Reference Atmosphere 1965. North Holland Publishing Co., Amsterdam, 1965.

8. Ballinger, J. C. and Christensen, E. H.: "Environmental Control Study of Space Vehicles." GDC Report ERR-AN-017, January 1961.

9. Cronk, J. J.: "Solid SAP: User's Manual." GDC Document CASD-CIH-74-008, October 1977, Revised October 1978.

INTERACTIVE COMPUTATION OF RADIATION VIEW FACTORS*

A.F. EMERY
H.R. MORTAZAVI
C.J. KIPPENHAN

UNIVERSITY OF WASHINGTON
SEATTLE, WASHINGTON

ABSTRACT

The development of a pair of computer programs to calculate the radiation exchange view factors is described. The surface generation program is based upon current graphics capabilites and includes special provisions which are unique to the radiation problem. The calculational program uses a combination of contour and double area integration to permit consideration of radiation with obstructing surfaces. Examples of the surface generation and the calculation are given.

INTRODUCTION

The calculation of the radiation exchange between two surfaces by the usual engineering method

$$\frac{\sigma(T_1^4 - T_2^4)}{\frac{1-\epsilon_1}{\epsilon_1 A_1} + \frac{1}{A_1 F_{12}} + \frac{1-\epsilon_2}{\epsilon_2 A_2}} \qquad (1)$$

first requires the evaluation of the view factor, F_{12}, which represents the fraction of energy leaving one diffuse surface which is intercepted by the second surface. F_{12} is defined by[1]

$$A_1 F_{12} = \iint_{A_1 A_2} \frac{\cos\theta_1 \cos\theta_2}{\pi r_{12}^2} dA_1 dA_2 \qquad (2)$$

---

*This work was supported by NASA grant NAG-1-41.

Although the equation is simple in appearance, its evaluation is fraught with difficulties since it requires a full and precise description of both surfaces. In addition, if the view from surface 1 to surface 2 is obscured by an interposed surface or object, some provisions must be made to account for that porion of the view which is occluded. It is important to recognize that the obstructed view is not a constant, but varies, depending upon the position of the elemental area dA_1.

The calculation of the view factor is thus really two problems in one:

1. To devise an efficient way of describing the surfaces and their spatial relationship to each other.

2. To calculate the resulting view factors.

This paper describes the development of a pair of computer programs to do this. The calculational program was developed first and used with hand input in a batch mode. It later became apparent that efficient use of the program required some form of automatic surface generation and the appropriate subroutines were added. It also became painfully obvious that the average user made so many errors (both in key punching and in defining the surfaces) that an interactive program, with graphic capability, was necessary.

SURFACE GENERATION

The first impression was that a standard CAD/CAM program could be easily modified to provide the needed interactive capability. However, some of the unique requirements of the view factor calculation program led us to conclude that a special program would be more appropriate. This program was based upon the following assumptions:

1. Surfaces would be only 3 or 4 sided with straight sides
2. To simplify the obstruction calculations, all surfaces would be planar
3. A surface would radiate from one side only. Thus every plate, no matter how thin, would require 2 surfaces.
4. The direction of the surface normal must be uniquely and simply defined.
5. Surfaces could not penetrate one another.

Some of these requirements were needed to calculate the view factors. Others were imposed to simplify the surface generation. Several additional requirements were found to be useful, although not absolutely necessary, namely:

1. All surfaces are initially generated in the x,y plane, facing upwards. Three dimensional surfaces (cylinders etc) would be oriented along the z axis
2. Only a global coordinate system would be used. Individual surfaces would not carry along an embedded coordinate system.
3. Surface corner nodes would always be numbered in a counter clockwise direction to specify the surface normal.
4. Curved surfaces would be represented by a combination of triangles and quadrilaterals, the size of each adjusted by the user to give the best representation.

The different capabilities of the generation program are illustrated in figure 1. Because radiation is a surface phenomena, the surface orientation (i.e., the direction of the surface normal) must be uniquely defined. In testing the program, particularly with inexperienced users, we have found that it must not only be interactive but must permit the user to modify any command at any time and must have a very complete graphics capability. The following sections describe the different features of the program.

A. Surface Types

Three different surface types are used: minor, major and groups. A major surface is one that is generated by a single command. Typical major surfaces are listed in table 1. A major surface can be manipulated as a single entity. A group is a collection of major surfaces and is also a single entity. When the major surface is generated, it may be composed of many minor or sub-surfaces. Figure 2 illustrates the development of a group and shows the minor surfaces. Minor surfaces cannot be treated as independent entities since this would lead to possible erratic and unacceptable distortions of the major surfaces.

TABLE 1
Typical Major Surfaces

Two Dimensional
1. Triangle
2. Rectangle
3. Quadrilateral
4. Circle (annulus, sector)
5. Ellipse (annulus, sector, orientation)

Three Dimensional (inside or outside radiation, top and/or bottom)
1. Box
2. Cylinder (right, slant, annular, sector)
3. Cone (right, slant, frustrum, annular, sector)
4. Sphere (sector, cutting planes)

B. Command Data Base

Because even experienced users tend to make many mistakes or desire to make a substantial number of modifications, it is important that all commands be stored for future use and modification. Commands are either informative or functional. All functional commands are stored in a data base and the user can:

1. List any or all of the commands
2. Insert new commands between existing commands.
3. Delete commands
4. Temporarily suspend commands
5. Repeat the commands
6. Interrupt the repeat
7. Call for information or graphical display at any time during the repetition of the commands.

Since surface movements are non-commutative and because the current structure cannot be generated by only a portion of the commands, any changes in the data base must be accomplished by restarting the command sequence (i.e. repeating) and modifying during the subsequent generation process.

C. Surface Manipulations

The user must be able to manipulate each entity (major or group) by such movements as:

1. Translation
2. Rotation
3. Replication (i.e, duplicating an original set of surfaces and manipulating the new set)
4. Scaling in x,y or z coordinates

In addition, all surfaces (including minor surfaces) must be subject to:

1. Deleting (temporary or permanent)
2. Adding
3. Restoring (if previously deleted)
4. Joining to another surface to form a new entity
5. Separating from another
6. Reversing the normal direction
7. Combining with others to form a single radiating surface but still capable of independent manipulation.

D. Graphical Display

The key to effective surface generation is a high speed graphical display. Our experience has been that the typical user will call for a display after every one or two functional commands. Furthermore, the user will generally ask for more than one view. Thus the graphics must be fast and versatile. The minimum number of graphical views and commands appears to be:

1. Orthographic views (top,bottom,left,right,front,back) singly or in groups
2. Perspective or isometric views
3. Hidden line removal from any or all views
4. Variable field of view
5. Rotation and translation of the object.
6. Deletion of selected surfaces from the view
7. Numbering of selected surfaces or corner nodes
8. Display of surface normals
9. Views as seen from one surface to another

Although the isometric view gives the best overall picture, the orthographic views prove to be the most useful when precise movements are necessary. Because obstructions so completely change the radiation exchange, it is critical that the user be able to look from any one surface to any other to check for obstructions. This must be done with hidden line removal to yield the maximum information.

Figure 3 illustrates the incorrect movement of panels on a cylinder as shown by an isometric view and an orthographic view--the latter being necessary for the correct placement.

Although a wire frame display is useful, the usual structure has so many surfaces that a hidden line removal is necessary. Unfortunately hidden views are very time consuming and consequently the program incorporates a somewhat inexact, but very fast, routine. Part of its speed is based upon the use of planar, straight sided surfaces with counter clockwise numbering. Any other surface requires some type of raster scan hidden surface algorithm with an unacceptable increase in computing time[2]. It is interesting to note that a fast graphics display may not give the desired information. We have found that if the lines are drawn slowly enough so that the viewer can get a feeling for the order of generation, it is easier to visualize the structure. On the other hand, if the display is nearly instantaneous, many viewers cannot recognize the structure. This is particularly true if an old command data base is being reused. Finally the surface normals must be displayed since a surface has only one active side. Most user errors appear to be related to an incorrect orientation of a surface after a series of manipulations have been made. Figure 4 illustrates a set of double surfaces to represent panels and the erroneous orientation of the replicated set is clearly seen.

E. Surface Refinement

After the basic structure has been constructed, three additional functions are needed:

1. Refinement by subdividing surfaces
2. Assuring planarity of surfaces
3. Condensing nodes

Refinement is simply the subdivision of surfaces into triangular sub-surfaces. Since triangles are always plane, planarity is assured by dividing any non-planar quadrilateral into 2 triangles. Obviously, refinement also guarantees planarity. Finally, upon exit any two nodes which are within a prescribed distance ϵ are condensed into one node; all inactive surfaces and unused nodes are deleted. The data file is created for use by the view factor calculating progam and for subsequent plotting. The command data base is closed for future use.

CALCULATING THE VIEW FACTORS

Because analytical expressions for F_{ij} exist only for simple surfaces [3], the computation of the view factors for complex structures is practical only by:

1. Double area integration
2. Contour integration
3. Nusselt projection
4. Monte Carlo

The Monte Carlo method[4] tends to be too expensive for most surfaces even if one of the adaptive techniques [5] is used. Nusselt projection (i.e., the use of the unit sphere) generally produces a curved projected surface and consequently requires a complex integration to find the projected area. In addition, obstructions will distort the contour of the viewed surface, rendering the calculation of the projected edges very difficult. In this case, one simply projects the position of an elemental area, not the edges of the contour. Since the projection must be done from every point on the viewing surface, it essentially reduces to the double area integration method.

A. Double Area Integration

The double area integration is performed in two steps (figure 5).

1. Mapping the surface onto a unit square by

$$x' = \sum N_i(\xi,\eta) \hat{x}_i \qquad (3)$$

The functions N_i are the usual finite element isoparametric functions[6].

2. Evaluating the integral

$$A_i F_{ij} = \sum_{k=1}^{n_i} \sum_{\ell=1}^{n_j} h_k^i h_\ell^j I(P_k, P_\ell) J^i(P_k) J^j(P_\ell) \qquad (4)$$

where $P_k(\xi,\eta)$ are the Gaussian points in the unit square, $J_i(P_k)$ are the Jacobian of transformation evaluated at these points, h_k are the Gaussian weights and the superscripts i and j refer to the surfaces and

$$I(P_k, P_\ell) = \frac{\cos\Theta_k \cos\Theta_\ell}{\pi r_{ij}^2 (k,\ell)} \qquad (5)$$

B. Contour Integration

The double area method is the best of the three, but it is not sufficiently accurate when portions of the surfaces are close to each other since the denominator of equation 1 becomes very small and the integral nearly singular. For these reasons we adopted the contour integration method as described by Mitalas and Stephenson [7]. In this method, the double area integration can be replaced by

$$A_i F_{ij} = \frac{1}{2\pi} \oint_{A_i} \oint_{A_j} \ln r_{ij} (dx_i\, dx_j + dy_i\, dy_j + dz_i\, dz_j) \qquad (6)$$

If the edges of all surfaces are straight lines and divided into short segments, equation 6 can be expressed as

$$A_i F_{ij} = \frac{1}{2\pi} \sum_k \sum_\ell D(k,\ell) (L_\alpha \ln L_\alpha \cos\alpha + L_\beta \ln L_\beta \cos\beta + P\gamma - L) dt \qquad (7)$$

where the various terms are defined on figure 6 and

$$D(k,\ell) = \ell_s \ell_t + m_s m_t + n_s n_t \quad (\ell\ m\ n = \text{direction cosines})$$

This formulation is not sensitive to the separation distance of the edges since the term $L_\alpha \ell n L_\alpha$ is not singular as the distance r goes to zero. We note that planar or non-planar surfaces can be treated equally well.

C. Comparison

In general, when numerically implemented, either Contour integration or Double Area integration will produce acceptable results. However, when two surfaces have a common edge (the adjoint problem) the Double Area integration method may perform very poorly[8, 9]. Table 2 lists comparable values obtained by the two methods for the surface shown on figure 7. Because the greatest portion of the radiation occurs in the corner, where the surfaces are the most proximate, the Double Area method tends to be inaccurate. By contrast the Contour integration method is very accurate, even with very few edge elements. When the surfaces are separated slightly, there is some improvement in the Double Area results, but not sufficient to justify its choice over the contour method.

TABLE 2
Percentage Error in the Numerical Calculation of the View Factor
between Two Surfaces of Equal Breadth (L=H)
(see figure 7)

	Contour Integration		Double Area Integral			
	Infinite Strip	Finite Area	Infinite Strip		Finite Area	
d/L	S=0.0	S=0.0	S=0.0	S=0.1 L	S=0.0	S=0.1 L
1.0	0.2%	3.7	20.7	21.5	59.1	57.6
0.5	0.	0.8	12.9	11.2	32.0	24.1
0.3		0.4	9.1	6.6	21.8	12.1
0.2		0.1	5.7	2.8	13.2	4.9
0.1		0.	2.9	0.7	6.7	1.2
0.05			1.5	0.2	3.4	0.3

d = dx=dy=dz for the Double Area Integration
    edge segment length for the Contour Integration

S = Separation distance along the x coordinate

We note that the use of the Contour integral with as few as 5 segments on an edge gives acceptable results, while the use of the Double Integral with a corresponding spacing is quite unacceptable.

D. Program Structure

The basic program structure is illustrated in Table 3.

TABLE 3
PROGRAM FLOW

1. Geometrical Input (definition of surfaces)
2. Unobstructed View Factor Computations using contour integration
3. Obstructed View Factor Computations
    1. determining if two surfaces, i and j, can see each other
    2. determining if a 3rd surface is interposed between the pair
    3. subdividing surfaces i and j into elemental areas
    4. constructing a ray between points on surfaces i and j and determining if it is intercepted
    5. computing F_{ij} using the double area integral

Each of the three major functions of the program will be described in detail in the following sections.

D1. Input

Data input is from the output file of the Surface Generation program.

D2. Unobstructed View Factor Calculations

The values of F_{ij} are computed directly by using the Contour Integration equation. Calculational time is reduced by the use of the reciprocity relationship, at the cost of the loss of some additional information.

D3. Obstructed View Factor Calculations

(A) Elimination of non-obstructing Surfaces

The calculation of F_{ij} when one or more surfaces are interposed between surfaces i and j is the most difficult and time consuming part of the calculation. In order to accelerate this calculation, every possible effort must be made to eliminate all non-obstructing surfaces from consideration. Consider the situation schematically shown on figure 8. The best calculational procedure found to date is:

1. Using the original x,y,z coordinate system, define the smallest rectangular parallelepiped (the view prism) which contains surfaces i and j (indicated by the dashed lines). Eliminate all surfaces which do not penetrate this view prism. This process is usually referred to as "clipping". Check to see if any surface completely fills the view prism, since if the view is totally obscured no further computations are needed.
2. Define a new coordinate system, $\bar{x},\bar{y},\bar{z}$, with the $\bar{z}$ axis directed along the line connecting the centroids of surfaces i and j and eliminate all possible surfaces by another clipping pass. Again test for total blockage.
3. Define a third coordinate system, $\bar{\bar{x}},\bar{\bar{y}}$ by rotating the $\bar{x},\bar{y}$ coordinate system until the rectangular area which encloses both surfaces i and j is a minimum (figure 8b) and perform another clipping test and test for total blockage.

It cannot be emphasized too strongly that acceptance or rejection of a possible obstructing surface can only be done by the simple clipping operation associated with view prisms which are constructed from rectangular parallelepipeds. Any other procedure to test for penetration of the view prism calls for geometric calculations which are unacceptably expensive. Since the transformations involved in steps 2 and 3 require matrix operations on all of the corner coordinates of all candidate obstructions, it is imperative that each successive step eliminate as many obstructing surfaces as possible. We have considered using the graphic capability of the generating program to detect the obstructing surfaces, but it has proven to be too time consuming and inefficient.

Once the final set of possible obstructing surfaces has been determined, F_{ij} can be calculated in two ways. Consider the view of surface j as seen from a segment on the contour of surface i as shown in figure 9. The obstructing surfaces obscure part of surface j, leading to the formation of the two visible sub-areas indicated by the dotted line contours. We note that: a) the determination of the unobstructed part of the surface is the classical hidden surface problem of graphic display for which, currently, there are no efficient methods; b) the number of non-contiguous visible sub-areas and the number of straight line segments encompassing these sub-areas may vary considerably as seen from different points of the contour of surface i. For this reason, Contour Integration is not an acceptable method; hence, we must utilize Double Area Integration.

(B) Ray Interception

We express F_{ij} as

$$A_i F_{ij} = \sum_{k=1}^{n_i} \sum_{\ell=1}^{n_j} f_{k\ell} \, I(k,\ell) \, dA_k \, dA_\ell \qquad (8)$$

where $f_{k\ell} = 1$ if the ray between dA_k and dA_ℓ is unobstructed

$= 0$ if obstructed.

Since $f_{k\ell}$ is a discontinuous function of position, the higher accuracy of Gaussian quadrature is not always realized; and since Gaussian quadrature requires substantially more numerical operations, it has proven best to use Newton-Cotes integration.

Once each surface has been divided into elemental areas, the centroids of these areas determined, and the rays between each of the points on surface i to each of the points on surface j defined, the calculation of $f_{k\ell}$ proceeds as follows (figure 10):

1. Determine if the angle between the outward normals to surfaces i and j and the ray $P_{k\ell}$ is greater than 90 degrees. If so, set $f_{k\ell}=0$ since the surfaces cannot see each other. Note that this must be done for every ray, since the inability of one point on surface i to see any given point on surface j does not ensure that other parts of the two surfaces cannot see each other.
2. Determine the intersection of the ray $P_{k\ell}$ with each of the possible obstructing surfaces by examining each in turn. This is done by:

    1. Finding the intersection of the ray $P_{k\ell}$ with the infinite surface which contains the surface. Because the intersection requires an iteration for arbitrary surface, but not for planes, only plane surfaces are permitted. The determination of the intersection is best effected by pre-calculating the equation of the plane of each surface and the transformation R

    $$\begin{Bmatrix} x' \\ y' \\ z' \end{Bmatrix} = [R] \begin{Bmatrix} x \\ y \\ z \end{Bmatrix} \qquad (9)$$

    which produces the coordinate system x', y', z' for which x' and y' are in the plane of the obstructing surface and z' normal to it.

2. Determining if the point of intersection is within the surface or not. If it is, $f_{k\ell}=0$; if not, $f_{k\ell}=1$. Determining whether the intersection point is within the surface by mapping the surface to a square and checking the values of ξ and η is not efficient because the non-linear mapping requires an iterative solution for ξ and η. The most efficient way appears to be by computing the angle between the line drawn from the corner node to the intersection point and between the corner node and the next corner node. For convex surfaces, if this angle is negative for any corner, the point is not within the surface and the ray is not blocked.

The most efficient calculation is one in which the obstruction is detected early, thus eliminating the consideration of the remaining surfaces. This is only possible if the candidate obstructing surfaces can be sorted in the order of their included angle as seen from the elemental area, dA_k. Unfortunately, this determination of the included angle is equivalent to finding the view factor from the elemental area dA_k to that portion of the obstructing surface which is within the view prism between points dA_k and dA_ℓ. Thus far we have not found a sorting method which gives a net reduction in computational times. Our experience is that maintaining a fixed order of checking the rays for interception in a single view prism is an effective method. (It should be noted that this entire problem can be compared to the usual graphical display problem for perspective views of hidden surfaces, but since in essence the viewing point must range over the entire surface i, calculationally it is the square of this classical problem.)

That the Double Area Integration is inaccurate for surfaces with a common edge has already been discussed; yet only the Double Area Integration is useful for the obstructed view calculations. Therefore it is important to avoid this situation by appropriately defining the surfaces.

EXAMPLES

We present two examples of the use of the program. Figure 11 represents an enclosure with a dividing panel, with the surfaces as numbered. The dividing partition must be expressed as two surfaces, infinitesimally separated, not one. Because surface 1 has an obstructed view of surface 9 with which it has a common edge, it must be represented as two surfaces, 1A and 1B. In this way, surface 1A which adjoins surface 9 has an unobstructed view and is treated by Contour Integration. Surface 1B, which has an obstructed view, is not adjoining and can be successfully treated by Double Area Integration. Figure 12 represents an enclosed cylinder which obstructs the view of the other surfaces. Table 4 lists typical execution times. The obstruction calculations were carried out using the three different elimination methods and different numbers of rays per surface and the accelerating effect of the clipping routines is clearly shown. Because of the simple geometries and the orientation of the surfaces along coordinate axes, the second clipping was not effective since the slight reduction in computational time due to the elimination of a few additional surfaces was less than the time necessary to perform the coordinate transformations. When the x,y axes are rotated by about 22 degrees, the first clipping effectiveness is reduced by about 30% and the value of the second clipping is more pronounced. For general problems, all clipping procedures must be used.

TABLE 4
Calculation Times
for the Example Problems depicted in Figures 11 and 12

	Adjoint Surfaces figure 10 (12 surfaces)		Enclosed Cylinder figure 11 (48 surfaces)		
Number of rays per surface	9	16	9	16	16 (rotated coordinates)
Clipping routine	Calculational Times in Central Processor Seconds on CDC Cyber 175/750				
none	5.6	14.8	99.0	274.1	274.1
1st pass	2.9	6.3	35.7	82.8	110.8
2nd pass	3.0	6.3	38.9	82.5	82.5
2nd pass(opt=2)					76.3

CONCLUSIONS

Our experience in using the programs has emphasized that radiation view factors of complex structures can only be accomplished if a fast, interactive program with good graphics capability is used. Although it would be more satisfying if curved surfaces could be treated, high program efficiency requires that they be modelled by an assemblage of flat triangles or quadrilaterals.

The view-factor calculational algorithm described has proven to be very effective for surfaces which have an obstructed view of each other. Under some conditions, statistical methods may prove to be more efficient, but for general configurations the combination of the Contour Integration and Double Area Integration methods, in concert with ray intersection calculations, has proven to be about the fastest method currently available. Further increases in computing efficiencies are possible if hardware perspective view, hidden surface devices are used, but such devices are not currently available for digital computers used in thermal analyses. From another point of view, the use of the rays may be regarded as a highly adaptive form of the Monte Carlo method which bears the same relationship to the usual method that the quasi-deterministic Exodus [10] method bears to the usual Monte Carlo method for solving diffusion problems.

NOMENCLATURE

A_i	Area of surface i
dA_i	Elemental area of surface
F_{ij}	Viewfactor from surface i to surface j
h_k	Gaussian weight for point k in surface i
$J(P_k)$	Jacobian of transformation at point k
n_j	Number of elemental areas in surface j
N	Isoparametric shape functions
P_k	Gaussian point k on surface i
r_{ij}	Distance from surface i to surface j
R	Rotation matrix
x,y,z	Global coordinate system
$\bar{x},\bar{y},\bar{z}$	Rotated coordinate system
θ	Angle between r_{ij} and a surface normal
ξ,η	Surface coordinates of the unit square

REFERENCES

1. Sparrow, E.M. and Cess, R.D., Radiation Heat Transfer, Brooks/Cole Publishing Co., Belmont, Calif, 1966
2. Sutherland, I.E., Sproull, R.F. and Schumacker, R.A., 'A Characterization of Ten Hidden-Surface Algorithms' ONR Contract Report, Contract N00014-72-C-0346
3. Hamilton, D.C. and Morgan, W.R., 'Radiant Interchange Configuration Factors', NACA Technical Note 2836 (1952)
4. Siegel, R. and Howell, J.R., Thermal Radiation Heat Transfer, McGraw-Hill Book Company, New York, 1972
5. Hitzl, D.L. and Maltz, F.H.,'Adaptive Estimation Procedures for Multi-Parameter Monte Carlo Computations', Journal of Computational Physics, vol 21, 1980, pp 218-241
6. Zienkiewicz, O.C., The Finite Element Method in Engineering Science, McGraw-Hill Book Co., New York, 1971
7. Mitalas, G.P. and Stephenson, D.G.,'Fortran IV Programs to Calculate Radiant Energy Interchange Factors', National Research Council of Canada, Div. Bldg. Res., DBR Computer Program 25, CP25 Ottawa, 1966
8. Look, DC. and Love, T.J., Numerical Quadrature and Radiative Heat-Transfer Computations, AIAA, vol8, pp819-820, 1970
9. Rasmussen, M.L. and Jischke, M.C., Radiant Heat Transfer at the Vertex of Adjoint Plates, AIAA, vol 8, pp 1360-1361, 1970
10. Emery, A.F. and Carson, W.W. 'A Modification of the Monte Carlo Method--the Exodus Method' ASME J. of Heat Transfer, August, 1968, pp. 328-333

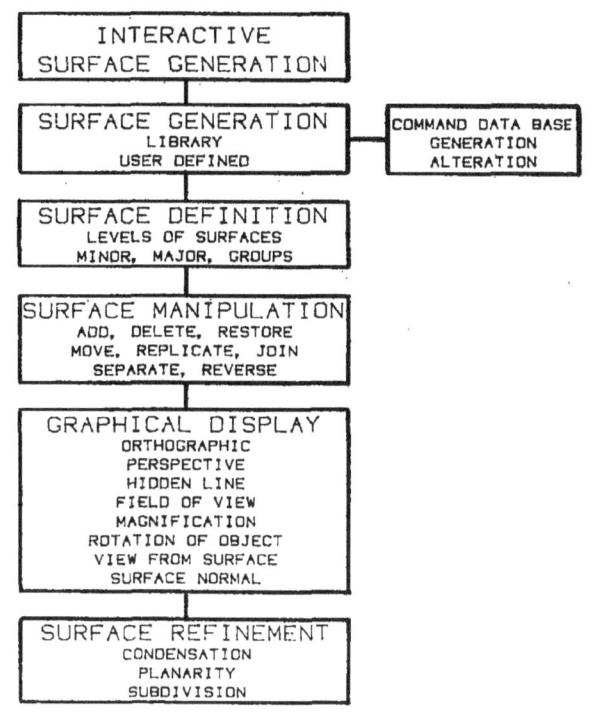

Figure 1.- Surface generation-program capabilities.

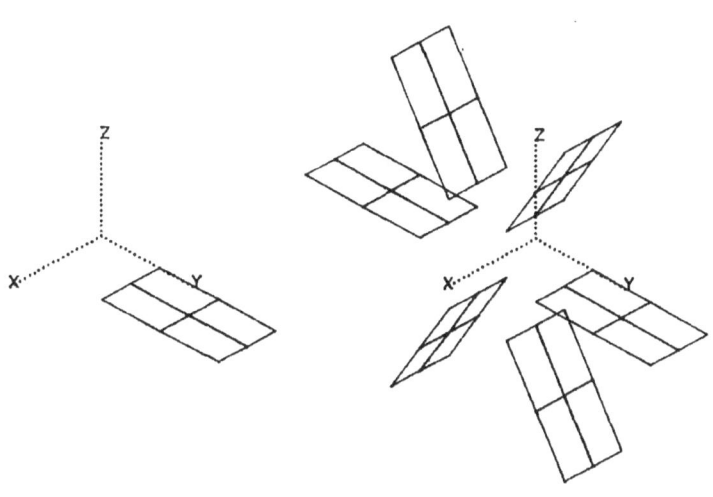

Figure 2.- Development of a group of surfaces.

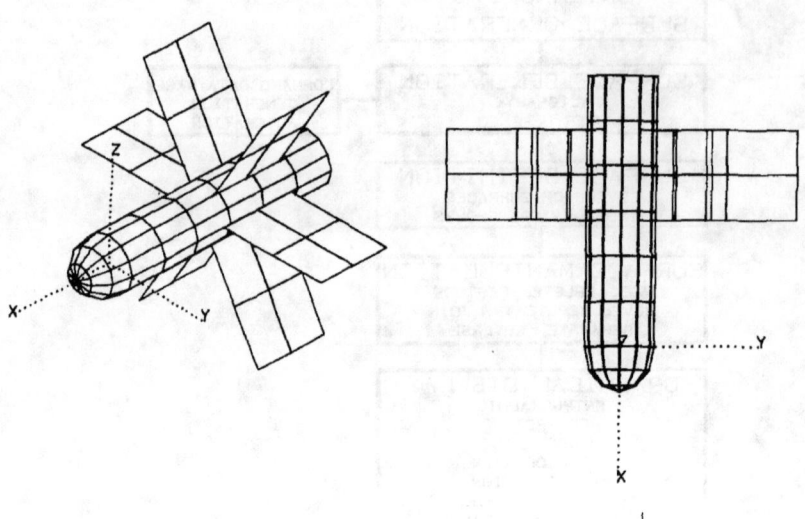

Figure 3.- Incorrect placement of panels, isometric and orthographic views.

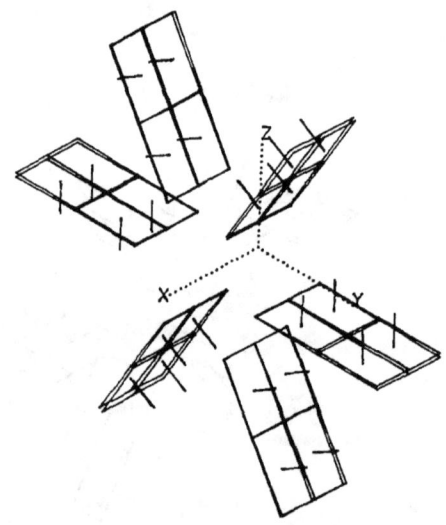

Figure 4.- Display of surface normals showing incorrect orientation.

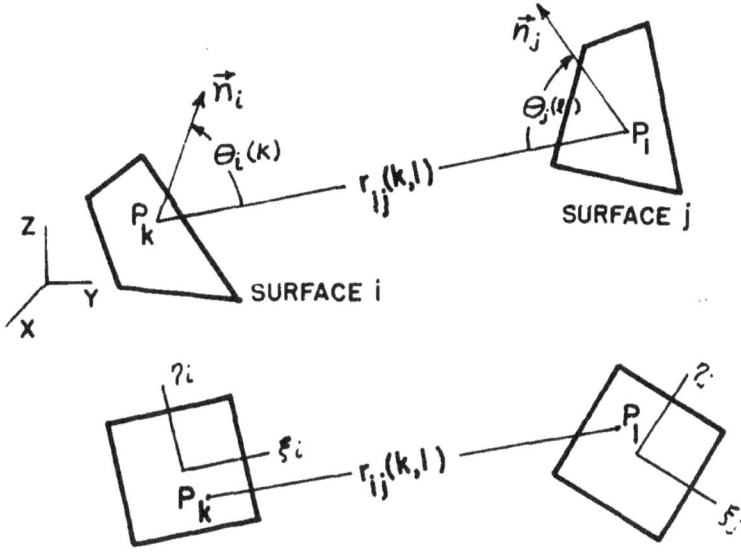

Figure 5.- Double area integration (showing mapping onto a unit square).

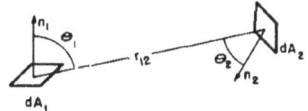

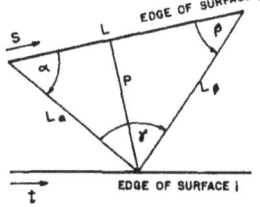

Figure 6.- Definition of angles and lengths for contour integration.

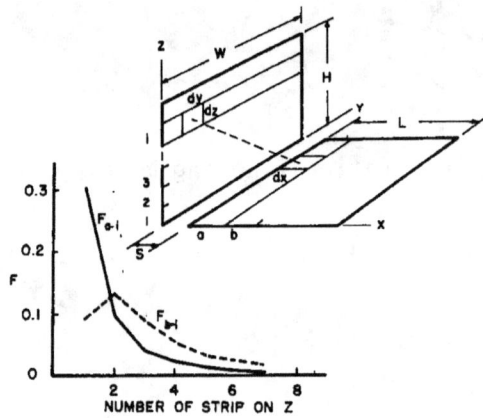

Figure 7.- View factors between surfaces at right angles to each other and separated by the distance S.

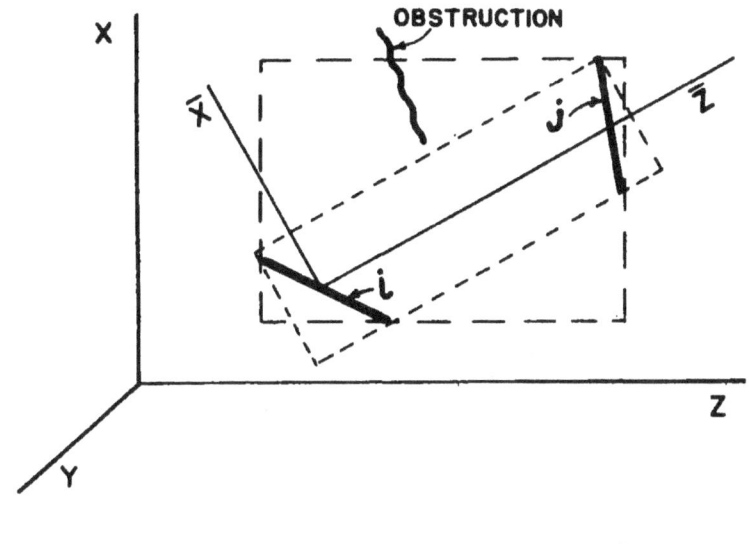

(a) Projected view of a view prism between surfaces i and j.

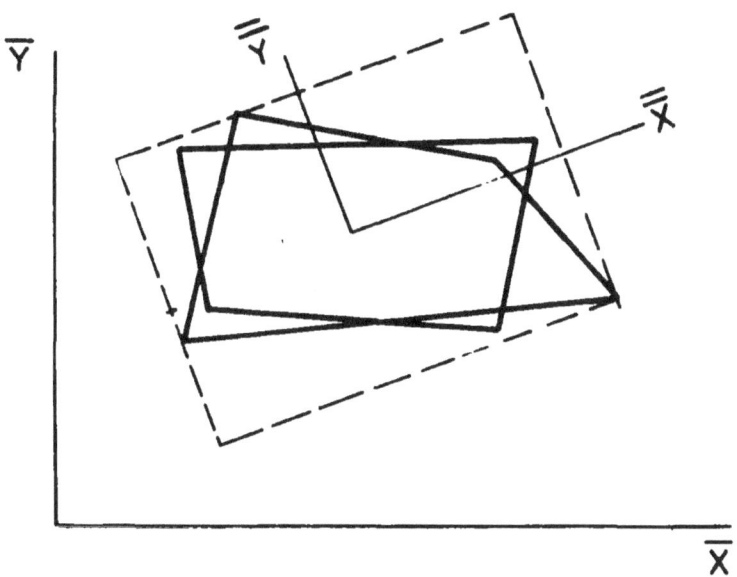

(b) Projected end view of the prism showing the rotation to achieve minimum cross section.

Figure 8.- Views of prisms between surfaces i and j.

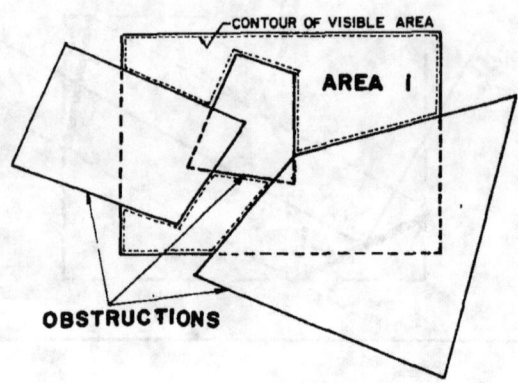

Figure 9.- View of surface i with interposed obstructions showing the multiple segmented contours.

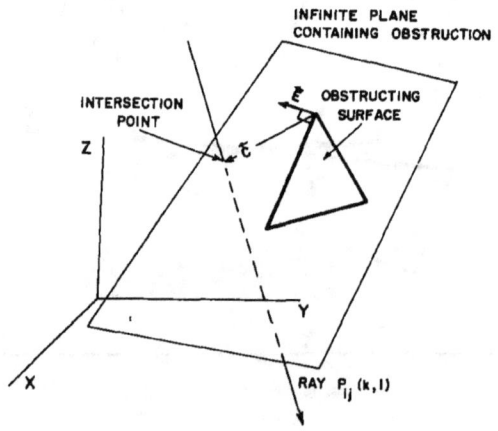

Figure 10.- Geometry for determining if the ray P(k,l) intercepts an obstructing surface.

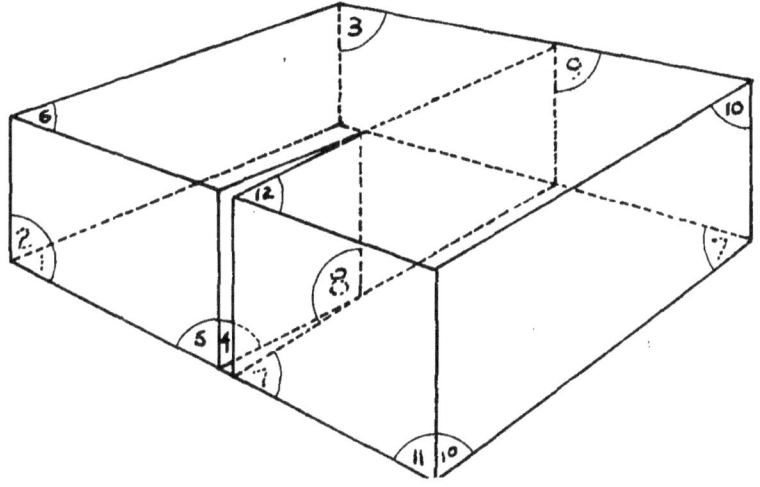

Figure 11.- The adjoint problem in which two contiguous surfaces have an obstructed view of each other.

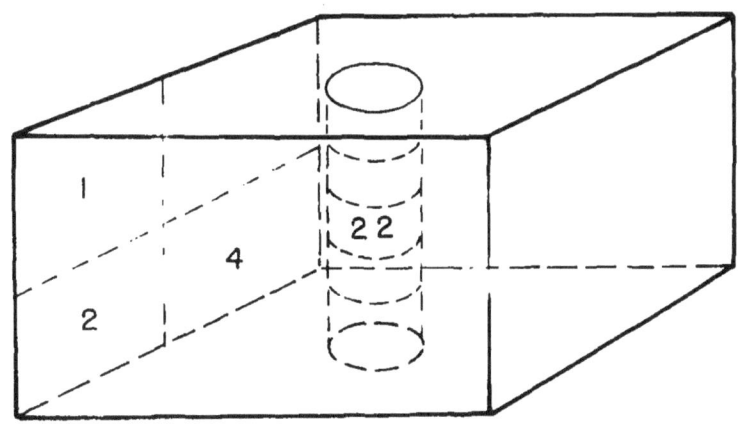

Figure 12.- The enclosed cylinder.

RECENT DEVELOPMENTS IN THERMAL RADIATION SYSTEM ANALYZER (TRASYS)

Robert A. Vogt
NASA Johnson Space Center
Houston, Texas

SUMMARY

The Thermal Radiation Analyzer System (TRASYS) computer program remains a dynamic program. Many changes have been made in the last few years. Because of the program modularized structure it has been a building experience of adding new capabilities while keeping intact the same data input structure. The overview of the program structure and general capabilities should be sufficient background to grasp a discussion of recent developments showing the progress in the last year. Where appropriate, assessments are made of the new features. The last section discusses the application of TRASYS peripheral programs and the importance they have in developing a totally integrated thermal analysis system.

INTRODUCTION

Large and complex configurations such as the Shuttle Orbiter and its payloads have established the need for more exact definition of the thermal radiation environment for on-orbit thermal analysis. For example, the Shuttle being larger, having more extensive self-shadowing, and presenting areas with greater sensitivity to the extreme environmental conditions than found in the previous manned spacecraft programs made it evident that improved analytical tools and modeling methods would be required. The cavities created by the open payload bay doors/radiators and the payload bay with payloads contribute to very steep thermal gradients because of the trapped edge and self-shadowing effects inherent in the configuration. Other locations, which place stringent requirements on the radiation analysis tools and methods, are the unpressurized internal equipment compartments, especially the uninsulated ones such as Orbiter aft section. These areas are radiation dominated and the geometry requires greater modeling detail to predict accurate temperature levels and thermal gradients. These situations coupled with the large size and long mission scenarios have made unprecedented demands for improvements in the computational and storage efficiency for thermal radiation analyzer computer programs and for more effective utilization of these tools.

Realizing in 1970 that unique requirements would be imposed by the next generation post-Apollo spacecraft, and that existing tools would be inadequate, NASA Johnson Space Center (JSC) began preliminary design and planning for what eventually evolved as the thermal radiation analyzer TRASYS computer program (ref. 1). This computer program has been actively developed since 1972. Although the TRASYS computer program presently meets JSC needs, development continues on further improving its' capabilities and performance. Continual studies have also made substantial improvements in efficiency by identifying and educating users on more optimum methods in the application of TRASYS.

This paper will initially give a review of the program structure for those unfamiliar with the program. With this as background, basic features, recent development and support programs will be discussed.

PROGRAM STRUCTURE

When TRASYS is executed, generally two subprograms are used; a preprocessor and processor. The preprocessor performs two basic functions. First, it reads and converts the user defined model geometry in the form required by the processor. Secondly, it interprets the TRASYS psuedo Fortran code that the user specifies in the input data to define what computations are desired and the sequence in which they should be performed. Based upon this data the preprocessor generates driving logic using dynamic storage techniques with only those program segments required to obtain the requested solution. The implication of the second function is that the user may readily customize the desired solutions, thus having a very definite influence on the accuracy and computational time.

The processor is the work horse. Its function is to obtain the desired solution and output the data computed by the processor in one or more optional formats. This is accomplished by executing the code the preprocessor created.

PROGRAM CAPABILITIES AND FEATURES

TRASYS, the Thermal Radiation Analyzer System is a modularized computer program system designed to compute the total thermal radiation environment for a spacecraft in orbit. The principal end products are the radiation conductors, and total heating as function of time or averaged. The output is a lumped parameter nodal representation formatted for direct interface with a thermal analyzer. The radiation conductors account for the radiation interchange between a network of nodes that make up the geometric model defined by the user. The radiation interchange includes the direct contribution from the sun, albedo, and planet plus the intra-network reflections of this energy. Self-shadowing can be considered for the direct and reflected heat loads.

The program's major attribute is its flexible structure and margin for growth which has allowed the program to keep pace with requirements while maintaining the basic input structure.

The program includes, but is not limited to, the following features:

- 1000 node capability with extended core.

- 9 different surface types to describe geometry.

- 15 user called segments that perform specific functions, e.g., compute form factors, compute grey bodies, plot geometric model, etc.

- The user can write his own executive to customize the desired solution with numerous program segments, subroutines, and variables to choose from.

- An efficient easy to use restart capability that minimizes loss of output.

- Convenient thermal analyzer interface easily tailored to other thermal analyzers.

- Choice of form factor solution techniques: Nusselt Unit Sphere or Hybrid which automatically chooses between double summation or Nusselt Unit Sphere.

- Form factor imaging for symmetrical configurations.

- Macroinstructions include optimized application of previous flux computations.

- Self-shadowing of external flux on a discrete element basis and/or with precomputed shadow tables generated by the program.

- Accepts trajectory tape input to define orbit position and attitude.

- 3 plot segments which will plot surface node data, sun-planet-spacecraft relationship, and heating rates vs time.

- Geometric and optical properties and orbit/attitude may be a function of time.

- Pure diffuse or mixed diffuse-specular radiation property model for infrared and/or solar waveband.

The JSC TRASYS program is operational on the UNIVAC 1100 series computer with central memory that varies depending upon the model size and the largest instructional bank of the various segments mapped. The minimum core is approximately 40K decimal words.

RECENT DEVELOPMENTS

The following paragraphs spotlight the more significant changes to the JSC TRASYS program in the last year and describe their key features and/or overtones.

Ray Tracing Segment

A new infrared (IR)/solar radiation interchange segment (RTCAL) was developed for mixed diffuse-specular surfaces. The segment uses a ray tracing procedure that is conveniently integrated into the overall program structure so it has an interface with the grey body calculation (GBCAL) link similar to the real body calculation (RBCAL) link. Unlike the RBCAL link though there are no restrictions on surface type, number of nodes per surface, and number of specular reflections.

More time will be needed to evaluate and optimize the TRASYS ray tracing segment before it can be considered a viable analytical tool.

Application of Direct Incident Flux Shadow Tables

Without the use of shadow tables the direct fluxes are computed with shadowing inherently considered on a element basis. Previous timing studies have shown that up to ninety percent of the CPU time is expended in the TRASYS shadow routines when computing orbital heating rates. The application of shadow tables is one way that a significant reduction in computer time can be made because it bypasses the time consuming subroutines. Shadow tables are precomputed at specific clock and cone angles. They can be used repeatedly on subsequent runs as long as the configuration is not changed.

Shadow factors are applied in TRASYS in the following manner: For the direct solar portion of the heat flux computations the position of the sun is determined based upon the orbit and attitude input parameters. An unshadowed solar flux, after being computed, is multiplied by the shadow factor obtained via table look-up of the precomputed shadows tables. The total direct flux is the product of the unshadowed direct incident flux and the shadow factor. Similarly, the albedo and planetary flux are computed with each planet node becoming a point source and the total albedo or total planetary flux being a summation from all planet nodes.

To minimize the error associated with interpolation of shadow tables with large step functions, the program tests to ensure that the tabulated dependent values interpolated between do not exceed the tolerance for the test. If the tolerance is exceeded at any time the shadow tables are temporarily not used and the program reverts back to calculating the total incident flux to the node on an element by element basis. A separate tolerance may be specified; one for solar and one for albedo/planetary heating.

As an example of how the program executes, suppose the shadow factor solar tolerance is 0.5 and for a given sun position, the interpolation would occur between values of 0.0 and 1.0. The tolerance is exceeded so shadow tables will not be used to compute the total solar flux to that node. Opposingly, if the interpolation had been between tabulated shadow factors of 1.0 and 0.62, the shadow tables would be used.

The albedo/planetary flux computations work similarly except the table is entered and the test made using the shadow factor albedo/planetary tolerance for however many nodes the planet is divided into for a particular spacecraft node. Decreasing the tolerances will improve accuracy while increasing computer time. When shadow tables are used extensively there is the risk that some of the nodes will have excessive errors. Additional controls allow the user, with a feel for the problem, to basically eliminate any significant errors without penalizing the approach as a whole. All or selected nodes may be excluded from using shadow tables for solar and/or albedo and planetary when accuracy requirements and their sensitivity to shadowing dictate it.

As of this writing, NASA/JSC has obtained very favorable performance with the control parameters selected to never use shadow tables to compute the solar fluxes and to use them 100% of the time in albedo/planetary calculations. Typically the computer charges may be reduced by 50% while comparison of predicted temperatures showed better than 90% were less than 3°C and the maximum was 5.5°C difference.

Hybrid Form Factor Segment

A new form factor segment was developed to replace the previous double summation form factor solution. The new form factor calculation (FFCAL) link automatically chooses between a double summation method and the unit sphere method to calculate form factors. The choice between the two methods is based on a criteria involving the nearness of the node pairs. For closely adjacent nodes the unit sphere method is used for its superior accuracy in this condition. For more distant nodes the double summation method is selected because it is faster and does not suffer from the inherent accuracy problem which occurs with this method when the nodes are closely spaced. The user still retains direct control of relative accuracy with input accuracy parameters.

At NASA/JSC the new FFCAL link is becoming the primary segment to compute form factors replacing the pure unit sphere method as the mainstay, the reasons being an approximate 40% reduction in computer time, with no noticeable degradation in accuracy.

Form Factors to Space

The way in which radiation conductors are computed to space has created accuracy problems in certain situations. Normally after computing form factors and node to node interchange factors the interchange factor to space is computed implicitly utilizing the residual for conservation of energy. The screening out of small form factors, and inaccuracies in the form factors themselves affect the accuracy of the radiation conductors to space. The error will have a greater relative effect on the temperature predictions for nodes that have high form factor sums. An alternate solution is to compute form factors to space explicitly. This will eliminate the accummulative error in interchange factors that gets dumped into the space conductor. On the other hand, the fact that the form factor to space is explicitly computed does not necessarily mean the conductor to space will be better. It will be better only if the form factor to space is more accurate. Because of the sprawling nature of a space node this will not always be possible and/or practical with limited computer resources.

The procedure utilized is to generate 100 rays from the center of each element evenly distributed outward in the half space. Each ray is checked to see if it is blocked by one of the possible shadowing surfaces. With all the form factors to space known the radiation interchange to space is computed as part of the network by the GBCAL link.

Currently the form factor to space capability needs to have its characteristics evaluated to determine when it is practical to use and whether further improvements can be made.

Identical Form Factor Request Matrix

A new capability has been added which allows more than one configuration to share the same form factor request matrix. Previously, even when there was no difference in the request matrix, it was necessary to repeat it under the proper current configuration name. This change is basically a potential time saver to the user.

Restart Tape Form Factor Updates

Frequently, after a restart tape has form factors stored, it is necessary to make model changes. This makes the tape incompatible with the new model unless the node array generated with the updates is identical to the node array stored on the restart tape. This means the same node numbers, and the same sequence of nodes. Recent changes have been made which allow a model to be reduced in size and still retrieve form factors from the larger model's restart tape. The program will automatically reduce the size of the form factor file by deleting the factors to the non-existing nodes.

Nodes can also be added to a model and still read form factors from the restart tape. The new node numbers must be unique from any on the restart tape. The new nodes may be added anywhere in the Surface Data Block. The program will utilize all of the values stored on the tape and create a program request matrix to compute all the form factors required because of the additional nodes. A combination of additions and deletions is also possible. Similar requirements apply for this to be accomplished.

The above capability will allow greater utilization of the binary restart tape which is preferred over the other alternative of manually selecting applicable form factors from previous models.

Frequently there is useful form factor data on more than one incompatible restart tape. A capability is being developed that will allow up to three restart tapes to be used from which to retrieve form factors. Another aspect of restart tapes is that as models have grown and mission simulation time and complexity have increased, tape overflow problems occasionally arise. One measure taken to reduce this risk is to no longer write two complete sets of form factors to the restart tape if they are not a function of the optical properties. This occurs if there are surfaces with transmissive or specular properties. As a result the majority of models require only one set of form factors. The program has been changed to automatically read/write one or two sets as required. Provisions are made to allow restart tapes to be read from previous program versions or if property changes are made which have impacted the previous program choice.

Trajectory Tape Input

Two new macroinstructions have been developed to generate the proper executive code to read the NASA/JSC common formatted trajectory tapes. The trajectory tape input capability was developed to assist in better preflight predictions and postflight data correlation, when the usual method of approximating the mission timeline is not sufficiently detailed to resolve critical issues. The preprocessor reads the position and attitude data from the tape and expands the code for the Operational Data Block. Consequently, it does not have to have the trajectory tape on subsequent runs and the user can customize the Operational Data Block further beyond the standard trajectory tape options. A trajectory tape used in its purest form would be very costly in computer time. This problem has been addressed to some extent but will probably warrant additional study and changes as more experience is gained.

Currently the following flexibility and degree of optimization have been implemented for a given time segment. A nominal time between positions (steps) can be specified by the user. This time interval will be adhered to unless there is a meaningful step function in the position vector to the sun or earth. The user may specify what direction cosine value qualifies as a step function. The program will recognize valid sun or earth attitude hold periods and consequently make optimum use of similar fluxes and/or planetary form factors available from previous computations. When the Shuttle is in the earth's shadow and in a earth hold attitude the program will extend the elapsed time between points to characterize the constant nature of the heating with just two points.

Extended Orbit Generator Capabilities

The orbit generator macroinstruction capability has been extended with the addition of two new arguments. One of the arguments will allow the initial time for the initial true anomaly to be specified. Previously the program assumed the time to be equal to the time since periapsis passage. This would not allow the initial true anomaly to be greater than zero without fudging the time.

The other new change will permit the user to specify the initial step number. This provides a greater flexibility to mix orbit and trajectory macroinstructions and to add new ones on subsequent runs without step number conflicts.

Source Editing

Previously orbit generation and other macroinstructions in the Operational Data Block were expanded after the source editing file was created by the preprocessor. A modification was made to include all card images generated by the macroinstructions in the source edit file. They will also be listed with edit numbers when a source listing is printed. This change will allow the user to make customized edits to these standard routines.

Possible Shadowers

Form factor blockage factors between each node pair are printed by the form factor segments and stored on the restart tape for subsequent printing. This is done because experience has shown that frequently it is easier to eyeball a blockage factor to judge the reasonableness of a suspect form factor than the form factor itself. An additional diagnostic aid is now available. A list of all possible form factor shadowing surfaces for all node pairs can be obtained with or independent of the form factor link.

Extended Core

The TRASYS program at JSC was initially developed to operate in 65K memory. As the program and models have grown there have been occasional problems mapping some of the larger links. This occurs with approximately 600-700 nodes. The newest version is an extended core version where all model size dependent variables in common will be mapped into extended core.

TRASYS ANCILLARY PROGRAMS

The major TRASYS support program available at JSC plays an important role. Several programs use the TRASYS restart tape which was designed to function not only as a restart tape but as a TRASYS interface to other programs. The programs listed are only a start in developing the full potential of utilizing the restart tape to perform tasks more efficiently external of TRASYS.

Restart Tape Print

The program will list data from selected psuedo files. It will also read and list all the header records on the restart tape. The program is used to inspect specific data, validate what is on the tape, or see whether it can be read or not.

Trajectory Print

There are two trajectory print programs; one for preflight and one for postflight. They list all the information that is normally of interest to the Shuttle thermal analyst and in particular the TRASYS trajectory tape user.

Thermal Analyzer Total Heat

Programs are available that collect the total heat rates as a function of time from a series of restart tapes generated with orbit and/or trajectory runs. A tape/file is created with proper format for a thermal analyzer flux read routine. This approach is preferred to the alternate method provided by TRASYS of using arrays and cyclic interpolation routines. It has more flexibility and requires less storage, a critical factor in large RC network Shuttle models.

Interactive Graphics

An Adage 340 minicomputer is utilized to validate TRASYS input data, and plot the surface/node geometry. It will also plot attitude and orbit relationships. The user may interact with such functions as zoom, translation, transform and hidden lines. The value of this system can not be overestimated in time and errors saved.

CONCLUDING REMARKS

The TRASYS computer program has made significant improvements over the last year. Form factor computations times have been reduced approximately 40% and the longer flux runs have been decreased 50% when shadow tables are used. Trajectory and ray tracing capability will require further development but both have received a significant start. The basic structure of TRASYS will allow it to grow in whatever direction it needs to.

REFERENCES

1. Jensen, Carl L.; Goble, Richard G.: Thermal Radiation Analysis, TRASYS II User's Manual Revision 3 June 1981. NASA CR 161077

ROLE OF IAC
IN LARGE SPACE SYSTEMS THERMAL ANALYSIS

G.K. Jones, J.T. Skladany, and J.P. Young
Goddard Space Flight Center
Greenbelt, Maryland

INTEGRATED ANALYSIS CAPABILITY (IAC)

To produce practical, alternative large space structure configurations, design analysts must have highly adaptable and efficient computer analysis programs to evaluate critical coupling effects that can significantly influence spacecraft system performance. These coupling effects arise from the varied parameters of the spacecraft systems, environments, and forcing functions associated with disciplines such as thermal, structures, and controls. Adverse effects can be expected to significantly impact system design aspects such as structural integrity, controllability, and mission performance.

One such neeeded design analysis capability is a software system that can "integrate" individual discipline computer codes into a highly user-oriented/interactive-graphics-based analysis capability. This "integration" of computer codes must be done in a manner that will greatly accelerate interdisciplinary data flow by maximizing use of modern data handling techniques and new generation computer systems. By providing this type of computer-assisted interdisciplinary design analysis capability, the analyst will be afforded a rapid and efficient system to minimize solution turnaround time as well as having basic solution capabilities hitherto unavailable.

Therefore, the purpose of the integrated analysis capability is to provide new system analysis capability wherein coupling effects of multidisciplinary design drivers can be rapidly evaluated and design alternatives assessed.

The IAC system can be viewed as being the following two products (see fig. 1):

a. Core framework system which serves as an "integrating base" whereby users can readily add desired analysis modules. The IAC is explicitly being designed so as to greatly ease the task of interfacing new analysis modules.

b. Self-contained interdisciplinary system analysis capability having a specific set of fully integrated multidisciplinary analysis programs that deal with the coupling of thermal, structures, controls, antenna radiation performance, and instrument optical performance disciplines.

Use of the IAC will be adaptable to the full range of design process stages starting at the definition phase and progressing to the final design verification stage.

INTEGRATED ANALYSIS CAPABILITY

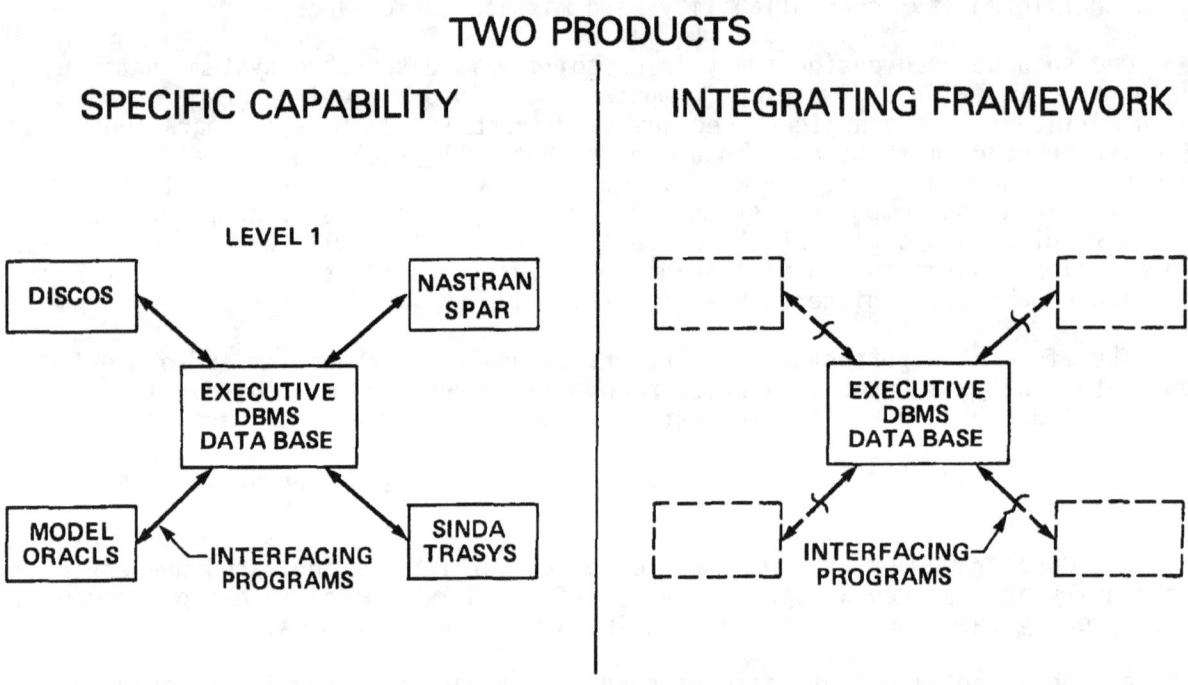

Figure 1

IAC SOLUTION PATHS

Much of the required technical capability of the IAC can be described as being part of one or more distinct "solution paths" (fig. 2). Each path is actually a class of solutions which consists of a number of selectable options and variations, rather than a rigidly predefined and automated process. An engineer-in-the-loop mode of operation is therefore possible and, in fact, is emphasized. Currently, five such solution paths have been defined. The standalone (uncoupled) operation of each technology or major technical module is defined to be Solution Path I. Paths II through V involve an increasing degree of interdisciplinary coupling and correspondingly greater complexity. Solution Path II provides thermal deformations via the coupling of a thermal analyzer such as SINDA, NASTRAN, or SPAR with a structural analyzer such as NASTRAN or SPAR. Obviously, a major coupling task is to handle the generally incompatible thermal and structural models. Path III accomplishes a structural/control analysis in either the frequency or time domain by providing required modal data from a structural analyzer to a system dynamics analyzer module such as DISCOS. Solution Path IV provides a time domain structural/control analysis, including a time-varying but quasi-static thermal loading; i.e., thermal loads are unaffected by the dynamic motions. Paths I-IV are to be fully implemented within the Level 1 program. Finally, Path V provides a fully coupled analysis in the frequency domain, and is directed at problems such as thermal flutter of long spacecraft members. This last solution path requires development and verification of new analysis technology such as thermal mode solution technique.

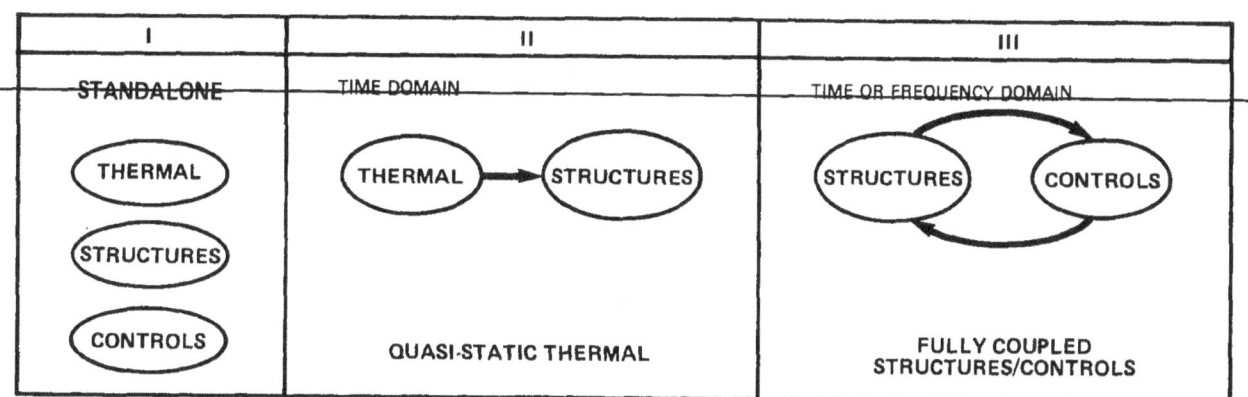

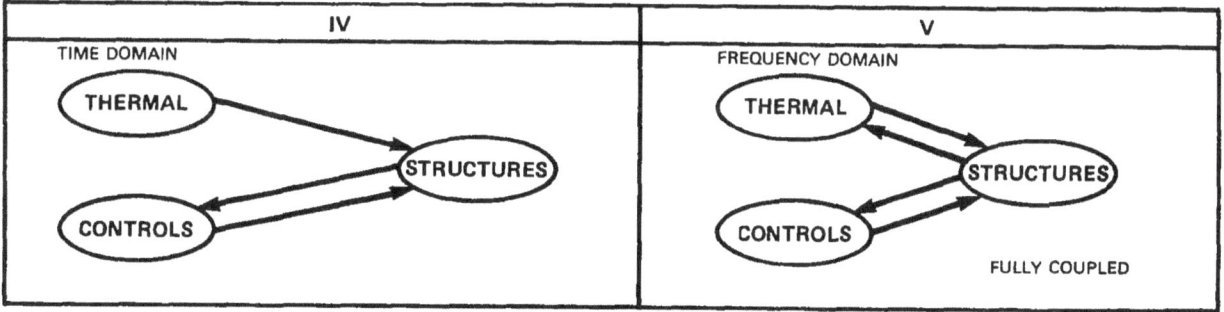

Figure 2

IAC ARCHITECTURE

The IAC design has an architectural plan not too unlike any common database-dependent software system consisting of data-handling capability and unified system executive encircled by application programs and a key supportive interactive graphics module (fig. 3). The diagram also shows the required interface to IPAD via use of the RIM database manager. A very important aspect of the IAC system, as indicated by the "OTHER" module block, is that specific attention is being given to making the system "open-ended" by facilitating the effort necessary to add other analysis capabilities. One design criterion for the early IAC release levels is to incorporate, where possible, analysis modules that are considered "industry standard." The most notable exceptions to this criterion are the SAMSAN and MODEL control system analysis-related programs which are currently being developed at the Goddard Space Flight Center. Creation of the interface programs, shown as the broad arrows A - G, constitutes a major part of the total IAC development activity. The bulk of the remaining task has centered around building up the total data handling and executive systems.

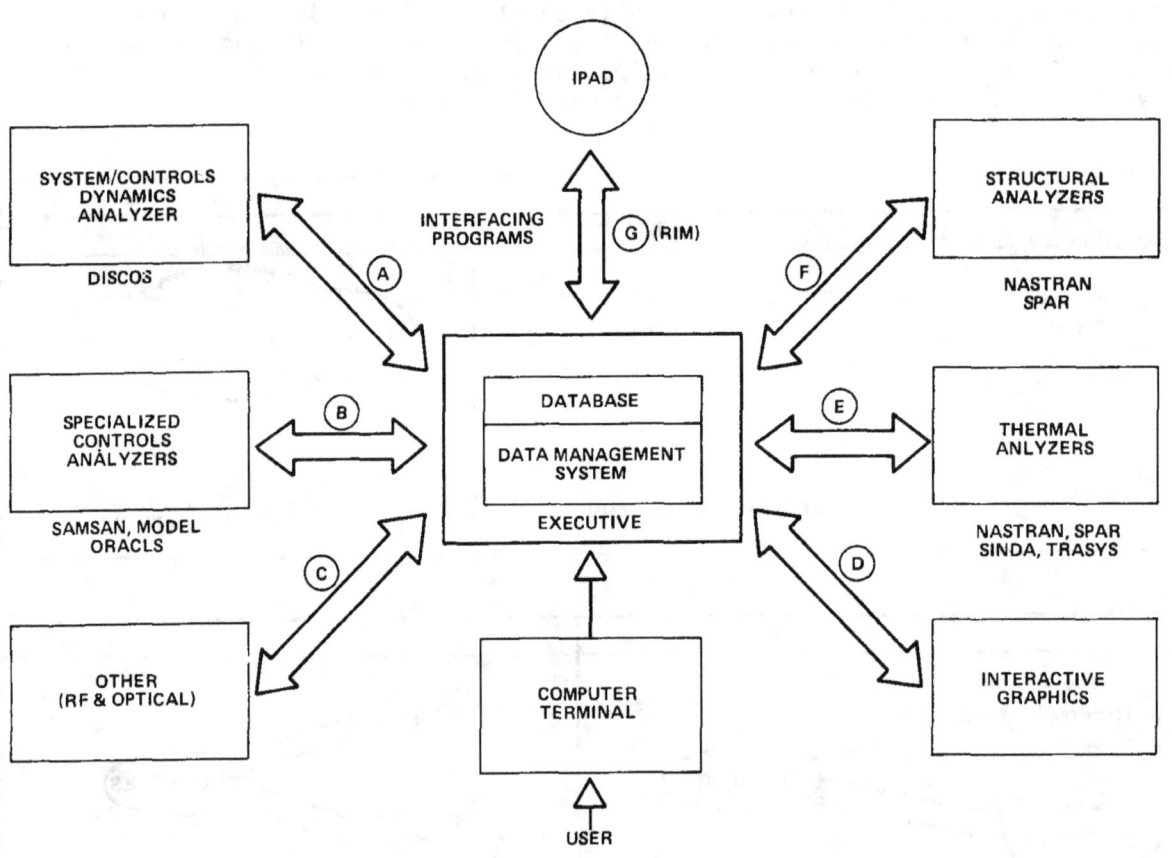

Figure 3

IAC DEVELOPMENT PLAN

Figure 4 gives a picture of the projected staged level delivery schedule of the IAC through FY 1985. In addition, a FY 1980 accomplishment is shown as the completion of Phase I and delivery of a pilot program and a detail system definition. The Level 1 through Level 4 IAC systems are shown as being completed on approximately 1-year intervals starting in early FY 83. Each level will successively incorporate additional capability as briefly noted in the chart. For definition of the solution paths (S/P) I-V as shown, refer to figure 2. The first host computer (H/C) will be the DEC VAX 11/780 super minicomputer manufactured by Digital Equipment Corporation. The second H/C has not yet been selected. Selection will be delayed as long as possible to allow the current-generation computer user market to develop further. Since a significant class of large space structures appears to be of a complex tension-stiffened (T/S) member type of construction, the Level 3 IAC is projected to contain solution algorithms unique to such structures. The need for an improved capability to analyze for geometric nonlinearities is anticipated and is projected for incorporation into the Level 4 IAC. After several years of usage, more effective ways of integrating the technical analysis modules will undoubtedly become known. In addition, after such usage experience it may well be advantageous to "pause" and re-evaluate the total IAC design concept from a top-down software design point of view. Therefore, provision for these tasks is shown during the Level 4 development period.

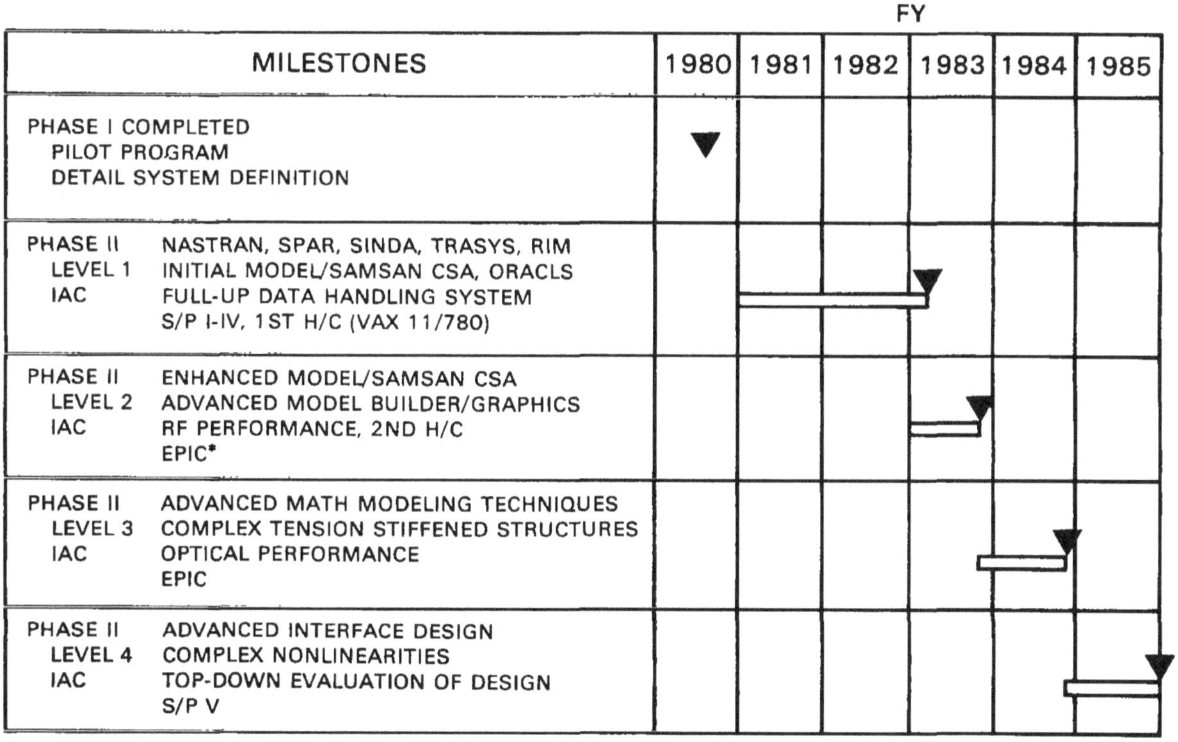

MILESTONES		1980	1981	1982	1983	1984	1985
PHASE I COMPLETED PILOT PROGRAM DETAIL SYSTEM DEFINITION		▼					
PHASE II LEVEL 1 IAC	NASTRAN, SPAR, SINDA, TRASYS, RIM INITIAL MODEL/SAMSAN CSA, ORACLS FULL-UP DATA HANDLING SYSTEM S/P I-IV, 1ST H/C (VAX 11/780)				▼		
PHASE II LEVEL 2 IAC	ENHANCED MODEL/SAMSAN CSA ADVANCED MODEL BUILDER/GRAPHICS RF PERFORMANCE, 2ND H/C EPIC*					▼	
PHASE II LEVEL 3 IAC	ADVANCED MATH MODELING TECHNIQUES COMPLEX TENSION STIFFENED STRUCTURES OPTICAL PERFORMANCE EPIC					▼	
PHASE II LEVEL 4 IAC	ADVANCED INTERFACE DESIGN COMPLEX NONLINEARITIES TOP-DOWN EVALUATION OF DESIGN S/P V						▼

*Enhanced Plug-In Capability

Figure 4

MAJOR THERMAL MODULES

The four major thermal analyses modules are presented in figure 5. The NASTRAN thermal analyzer (NTA) and SPAR thermal analyzer perform steady-state and transient thermal analyses. Both use the finite-element solution technique. Their main advantage is that they compute temperatures at points which are completely analogous to structural grid points. Their major drawback is their limited acceptance by the thermal community.

The SINDA program is a finite-difference thermal analyzer. It has a wide range of options and capabilities, and unlike the NTA it is widely accepted by the thermal community. The major drawback is the lack of direct compatibility with the structural model.

TRASYS is probably the most widely used radiation analysis system. It calculates all the parameters dealing with radiation, including black-body view factors, interchange factors, and complete absorbed fluxes from the Sun and planets throughout an orbit.

- NTA — NASTRAN THERMAL ANALYZER

- SINDA — SYSTEMS IMPROVED NUMERICAL DIFFERENTIAL ANALYZER

- TRASYS — THERMAL RADIATION ANALYSIS SYSTEM

- STA — SPAR THERMAL ANALYZER

Figure 5

THERMAL INTERFACES REQUIRED

The existence of a number of thermal programs establishes the need for some thermal interfaces (fig. 6). The NASTRAN-to-SINDA link is basically a conversion of the structural model to a SINDA thermal model.

The TRASYS/NASTRAN link is envisioned as a two-way link. First TRASYS must be able to accept model data in NASTRAN format, and secondly the output from TRASYS must be converted to NTA format. A link from TRASYS to SINDA exists as part of standard TRASYS output.

Finally, a link from SINDA output back to the structural model must be provided in order to perform a structural analysis with thermal data included. MIMIC (model integration via mesh interpolation coefficients), which is a program that derives point-wise spatial interpolation coefficients, will transform temperature data from one set of nodes to another.

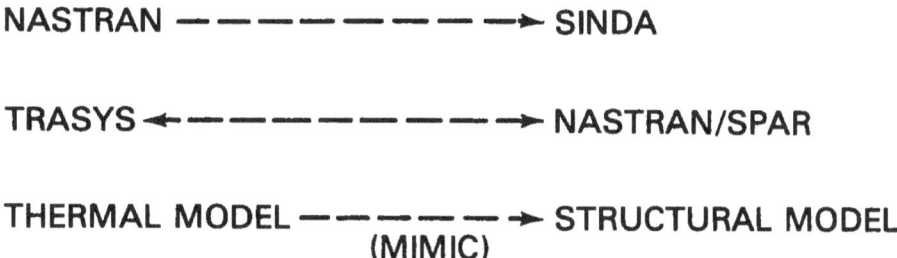

Figure 6

THERMAL MODULE CAPABILITIES

The combination of several thermal programs is required because no one program exists which performs all the required functions (fig. 7). The analyzers (e.g., SINDA, NTA) generally do not calculate the radiant inputs to the thermal model. That is, in general, a special-purpose program such as TRASYS is required to provide the radiant energy to the model from the Sun and planets. In addition, most of the analyzers contain (at best) only a limited capability to calculate radiant interchange factors. Some additional capability in this area has recently been added to the MSC NASTRAN by the addition of the VIEW program. The SPAR program is shown open-ended because currently there is significant on-going activity to extend SPAR capabilities.

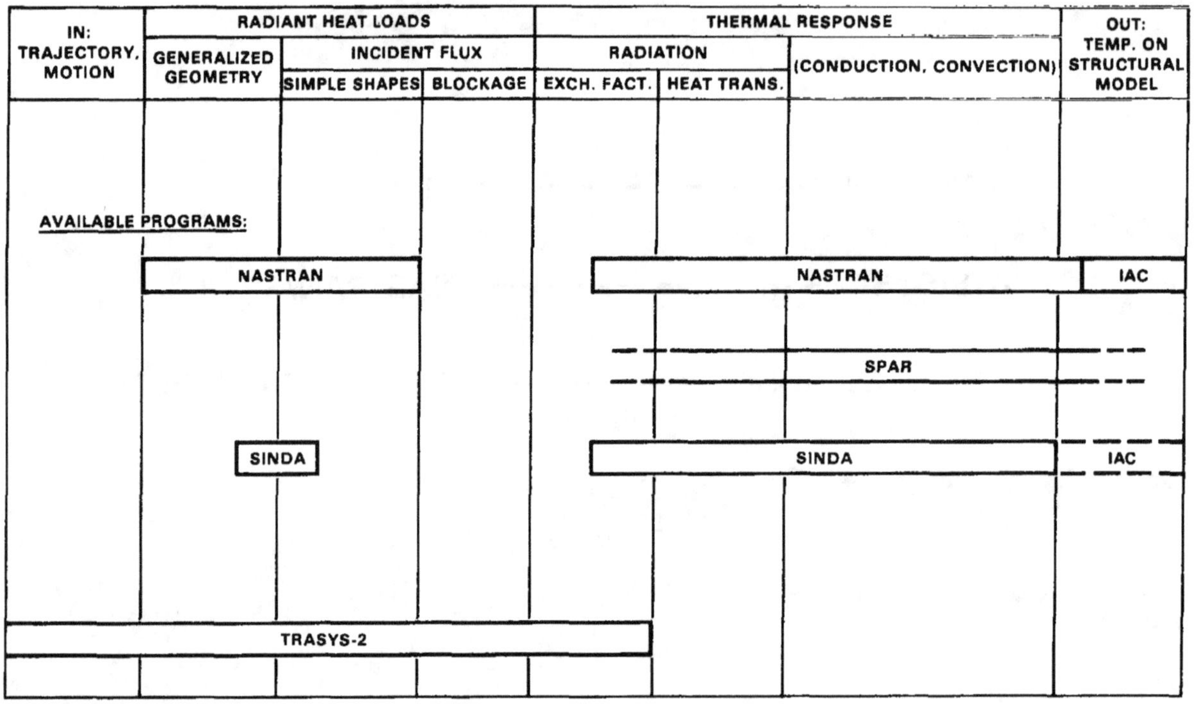

IN: TRAJECTORY, MOTION	RADIANT HEAT LOADS			THERMAL RESPONSE			OUT: TEMP. ON STRUCTURAL MODEL
	GENERALIZED GEOMETRY	INCIDENT FLUX		RADIATION		(CONDUCTION, CONVECTION)	
		SIMPLE SHAPES	BLOCKAGE	EXCH. FACT.	HEAT TRANS.		
AVAILABLE PROGRAMS:							
	NASTRAN					NASTRAN	IAC
						SPAR	
	SINDA					SINDA	IAC
TRASYS-2							

Figure 7

THERMAL ANALYSIS MODULES AND DATA FLOW

The thermal process is envisioned as beginning with a basic finite-element structural model (fig. 8). Engineering intervention is required to add some thermal data at this point. From this point, part of the data is transferred to the region where the radiant interchanges are computed. Meanwhile the user now has a choice as to which thermal analyzer he wishes to run. If he chooses SINDA, the finite-element data must be converted to a finite-difference format. Once the thermal analyzer has been run, its output can be converted to temperatures at the structural model grid points so that further processing can occur.

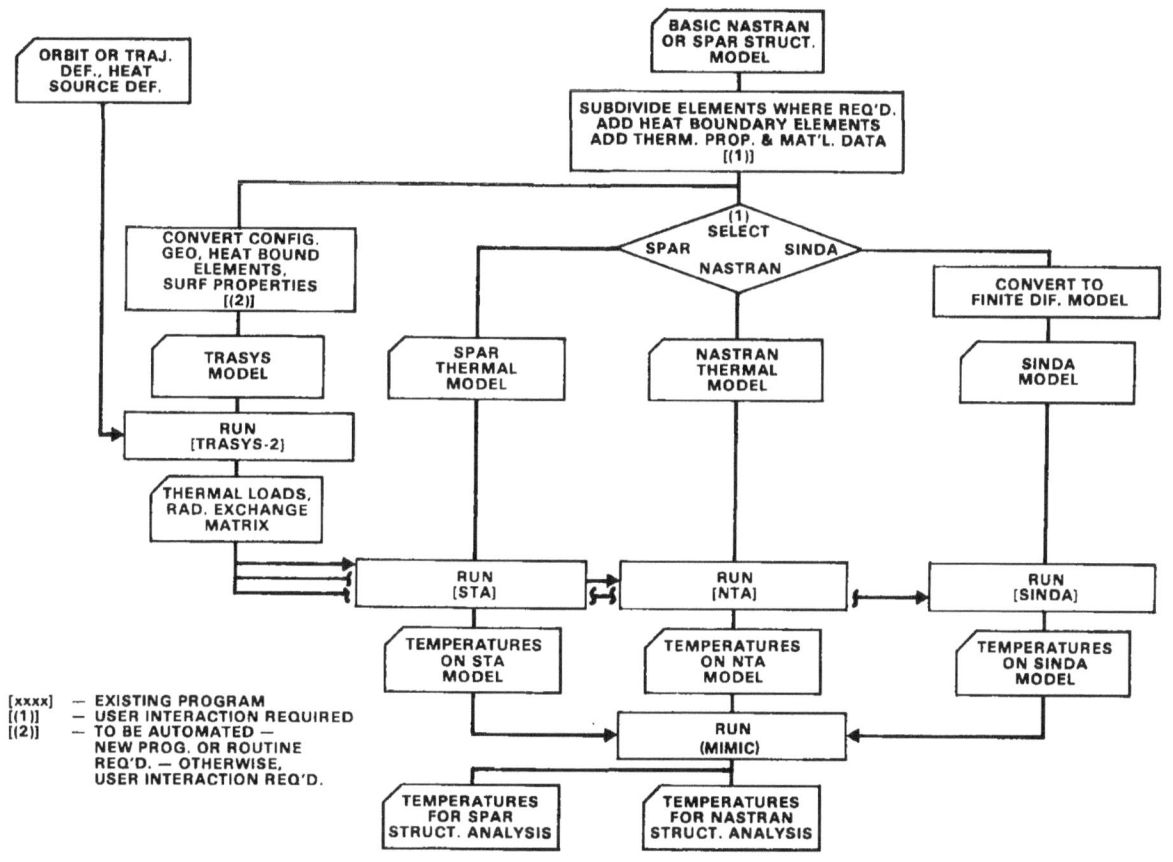

Figure 8

THERMAL ANALYSIS VIA THERMAL MODES

The investigation of the concept of thermal modes was intially motivated by a desire to be able to include thermal effects in coupled-structure thermal-controls stability analyses. Conventional thermal analysis methods are ill-suited for such coupled-system dynamic analysis. This technique also appeared to have considerable potential for problem size truncation in transient thermal analyses. In this application, it was spectulated that a large number of thermal modes could be discarded prior to performing the transient analysis with little effect on the results (fig. 9).

WHY :

1. POTENTIAL FOR PROBLEM SIZE TRUNCATION

2. APPLICATION TO COUPLED THERMAL-STRUCTURE-CONTROLS PROBLEMS

Figure 9

ACTIVITY TO DATE

Much of the work on thermal modes presented herein was performed by a contractor, John Anderes, of Swales and Associates. Papers and reports were reviewed and several different thermal mode formulations were investigated. The CAVE (conduction analysis via eigenvalues) formulation was used in the CAVE III thermal analysis code developed for LaRC by Grumman Aerospace Corportion. In a recent paper (ref. 1), H. P. Frisch of GSFC presented a formulation of the thermal heat balance equation for thermal mode solution that included linearized radiation. This reported activity has resulted in the formulation of a more generalized thermal mode technique, as described in figure 10, that was implemented using NASTRAN combined with DMAP to obtain the desired matrix data. Then a general-purpose matrix code, titled FLAME, was used to solve the matrix equation for nodal tempertures. Two test problems were defined and solved:

1. A 10-node slab problem was selected from the CAVE III report. This one-dimensional problem had only conduction coupling.

2. A 55-node parabolic dish antenna model was available from previous IAC studies. This 3-dimensional problem had both conduction and nonlinear radiation coupling.

- REVIEW OF SEVERAL DIFFERENT THERMAL MODE FORMULATIONS
    - CAVE (LaRC)
    - FRISCH (GSFC)

- IMPLEMENTATION OF CODE TO FIND THERMAL MODES AND TO SOLVE THE RESULTING MODAL EQUATIONS
    - NASTRAN THERMAL ANALYZER, DMAP
    - FLAME

- TRIAL SOLUTIONS AND COMPARISONS WITH CONVENTIONAL TECHNIQUE
    - SLAB, 10 NODE, ONE DIMENSION, CONDUCTION
    - PARABOLIC DISH, 55 NODE, 3-D, CONDUCTION & RADIATION

Figure 10

TECHNIQUE

The basic thermal equation used in this investigation was the formulation developed for the NASTRAN thermal analyzer (NTA) (fig. 11). This formulation supports both linear and nonlinear radiation coupling. In this formulation the nonlinear radiation terms are included on the right side of the equation in a forcing function role. The right side of the equation was set equal to zero and the eigenvalue and eigenvectors were determined using the EISPAK eigensolvers in FLAME. These eigenvalues and temperature patterns (eigenvector) may be physically interpreted by considering a structural analogy. If an undamped structure is physically deformed into one of its structural mode shapes and then released, it will vibrate indefinitely with that shape at the frequency of that mode. Likewise, if a structure is given an initial thermal pattern (distribution) of one of its thermal modes, referenced to some mean temperature, the transient nodal temperatures will decay to the mean temperature with a time constant equal to the reciprocal of the eigenvalue. During this transient period, the thermal pattern will maintain its original shape and only its magnitude will change.

- BASIC EQUATION:

$$B\dot{T} + KT = P + N(T)$$

  T — NODAL TEMPERATURES
  B — HEAT CAPACITANCE MATRIX
  K — HEAT CONDUCTION PLUS LINEARIZED RADIATION MATRIX
  P — THERMAL INPUT MATRIX
  N — NON LINEAR RADIATION TERMS

- SET RIGHT SIDE EQUAL TO ZERO AND ASSUME A SOLUTION:

$$T = \phi e^{-\lambda t}$$

- THE REDUCED EQUATION IS THEN:

$$\left\{ B^{-1} K - \lambda I \right\} \phi = 0$$

- USE EIGENSOLVER TO GET

  λ_i — EIGENVALUES (DECAY CONSTANTS)
  ϕ_i — EIGENVECTORS (TEMPERATURE PATTERNS)

Figure 11

PARABOLIC DISH MODEL

Two thermal transient problems were solved via thermal modes. The more complex analysis used an NTA model of a parabolic dish (fig. 12). The model contained 55 grid points and the analysis included both conduction and nonlinear radiation coupling. Topologically the model consisted of a parabolic dish, a feed assembly, and a pedestal which is partially obscured in this hidden-line plot.

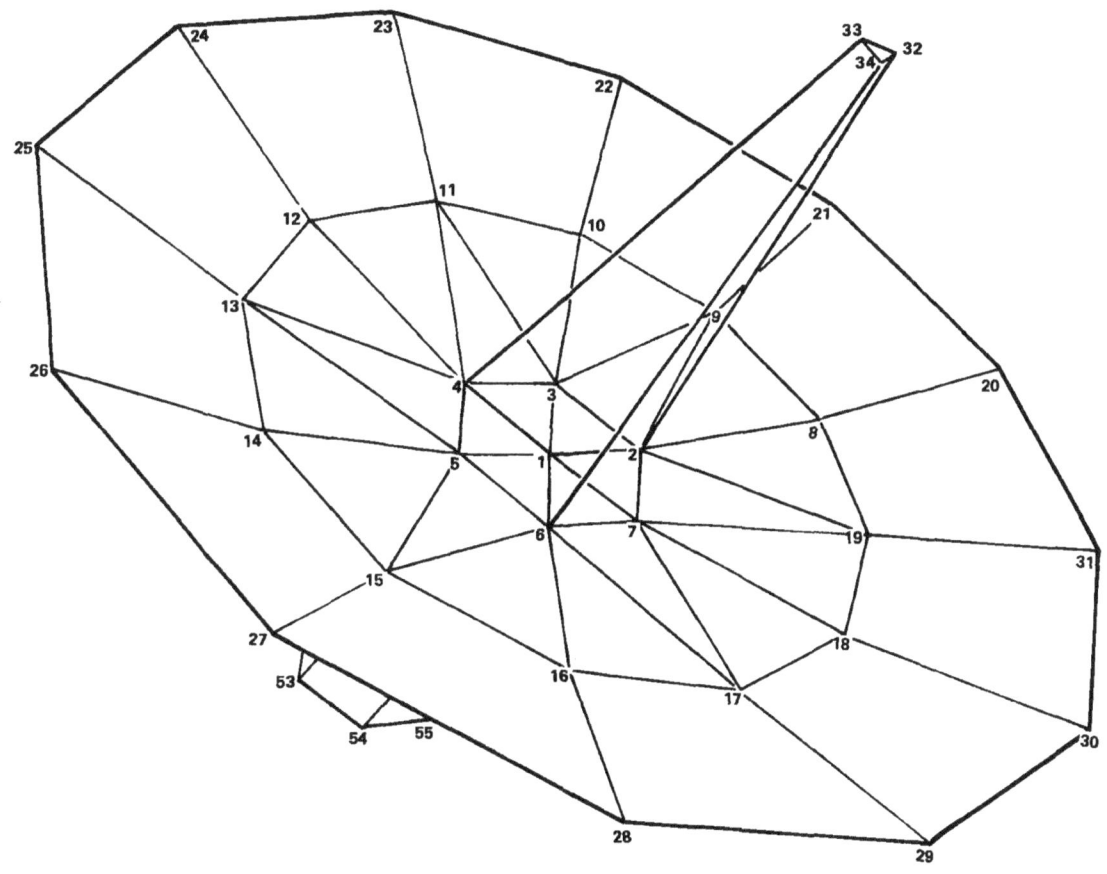

Figure 12

PARABOLIC DISH-THERMAL MODE 1

The visualization of thermal modes on a conventional black-and-white graphic terminal would be difficult. For this study we used the MOVIE.BYU graphics software together with a 512 x 512 color raster display system. The first thermal mode of the parabolic dish is shown in figure 13. The lightest shade (maximum yellow) represents +1° and the darkest shade (maximum blue) represents -1°. The black lines on the plot are contour lines representing 0.5° increments. The decay constant for this mode is 1.6×10^5 sec. The decay constants for the other 54 modes range from 0.59×10^5 sec to 0.0039×10^5 sec.

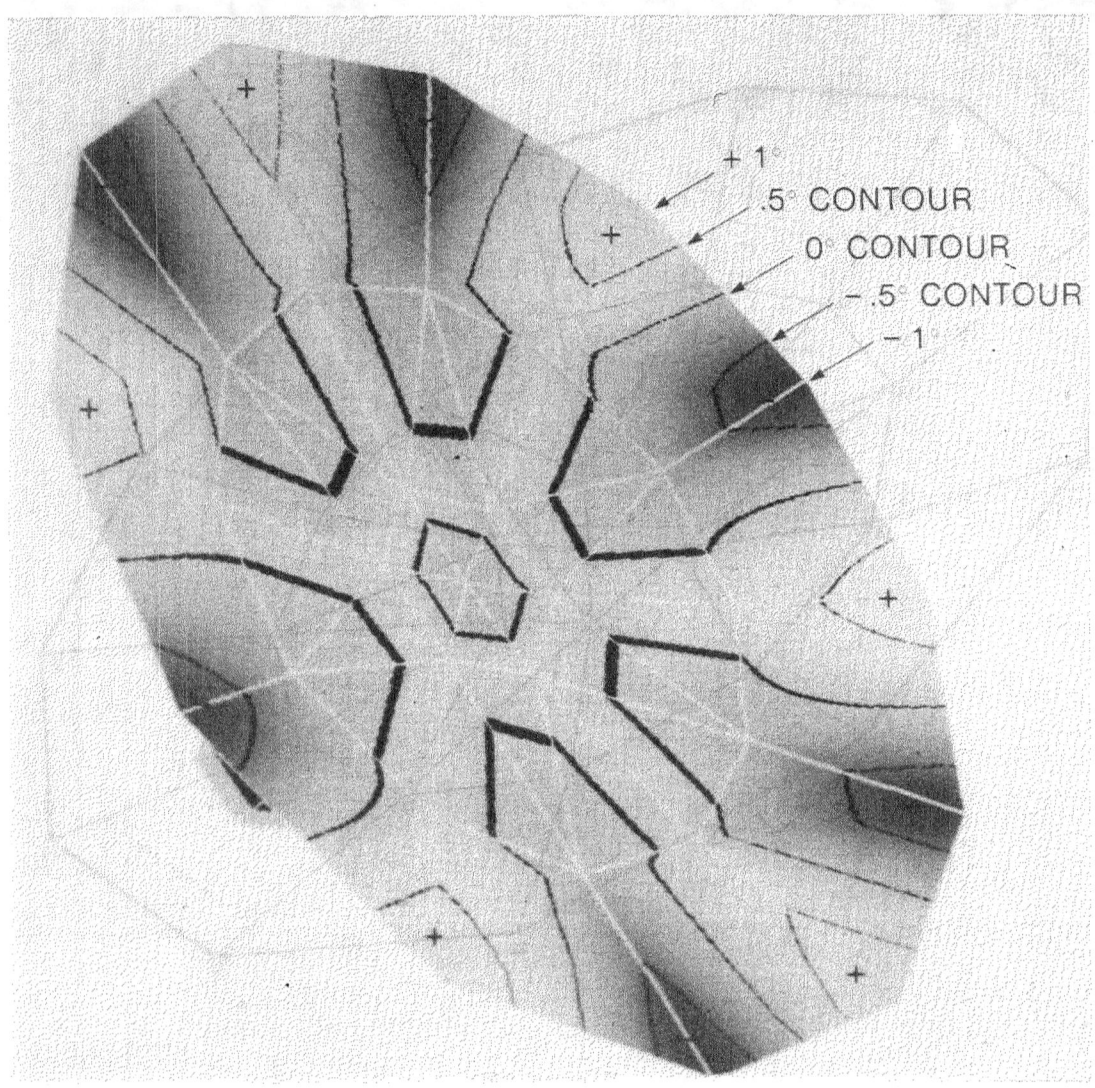

Figure 13

THERMAL TRANSIENT RESPONSE-PARABOLIC DISH

The antenna dish transient problem consisted of calculating the temperature changes of the dish from an initial temperature of 23.5° with in-orbit flux input and nonlinear radiation (fig. 14). The modal solution using all thermal modes (55) was found to match exactly the solution found via conventional analysis. Computer runs investigating the effects of truncating the modal set are in progress. For example, the predicted response of the pedestal end points using the lower 36 modes almost exactly matches the results using all 55 thermal modes. At this point in time, however, we have developed no firm criteria to select modes to be retained.

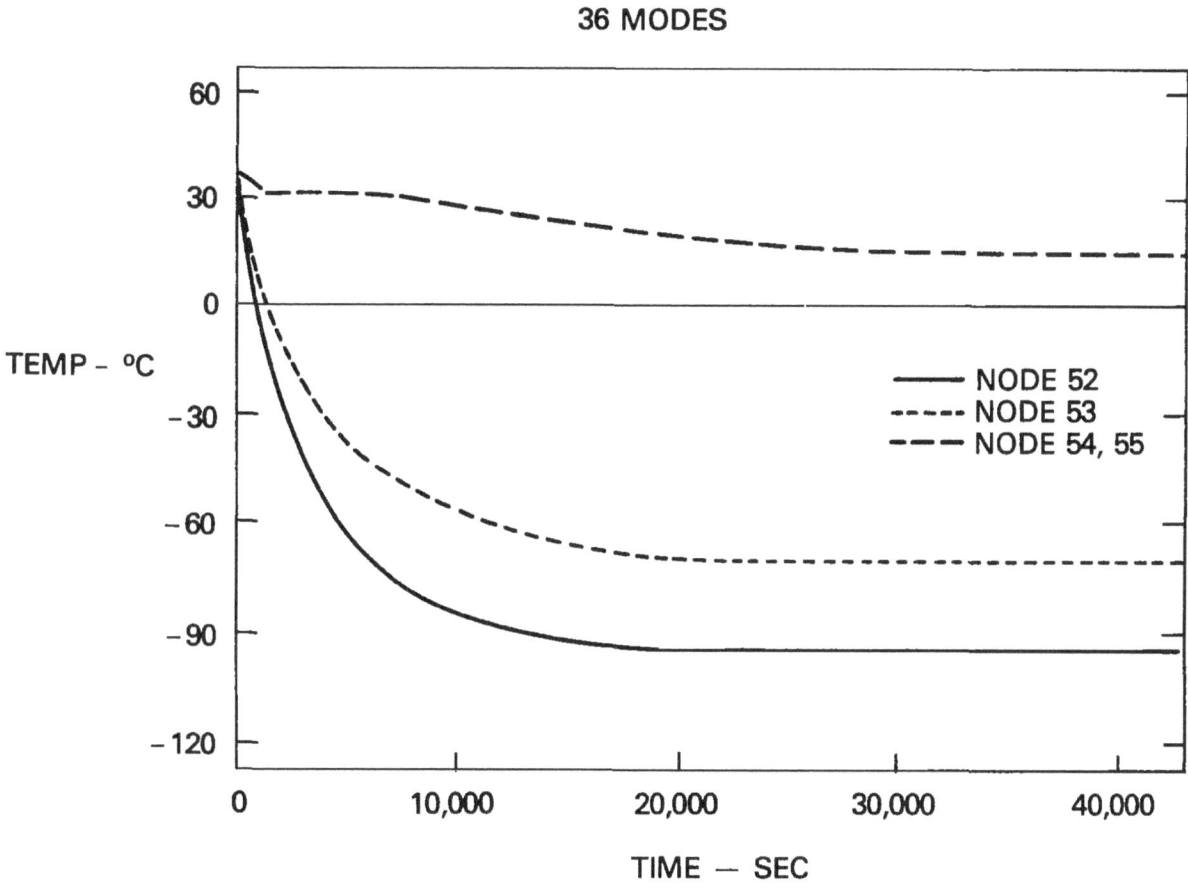

Figure 14

RUN TIME PERFORMANCE-PARABOLIC DISH

Presented in figure 15 is a comparison of the CPU time involved in three different analyses of the antenna dish (that is, conventional method, thermal mode using all 55 modes, and thermal mode using the first 36 thermal modes). As was expected, truncation of the modal set reduces the CPU run time. Surprisingly, the modal solution using all 55 modes was somewhat quicker than the conventional direct solution. At this point in time the thermal mode technique shows promise of being more efficient than the conventional direct solution, but additional experience using thermal modes must be obtained before any firm conclusions can be reached.

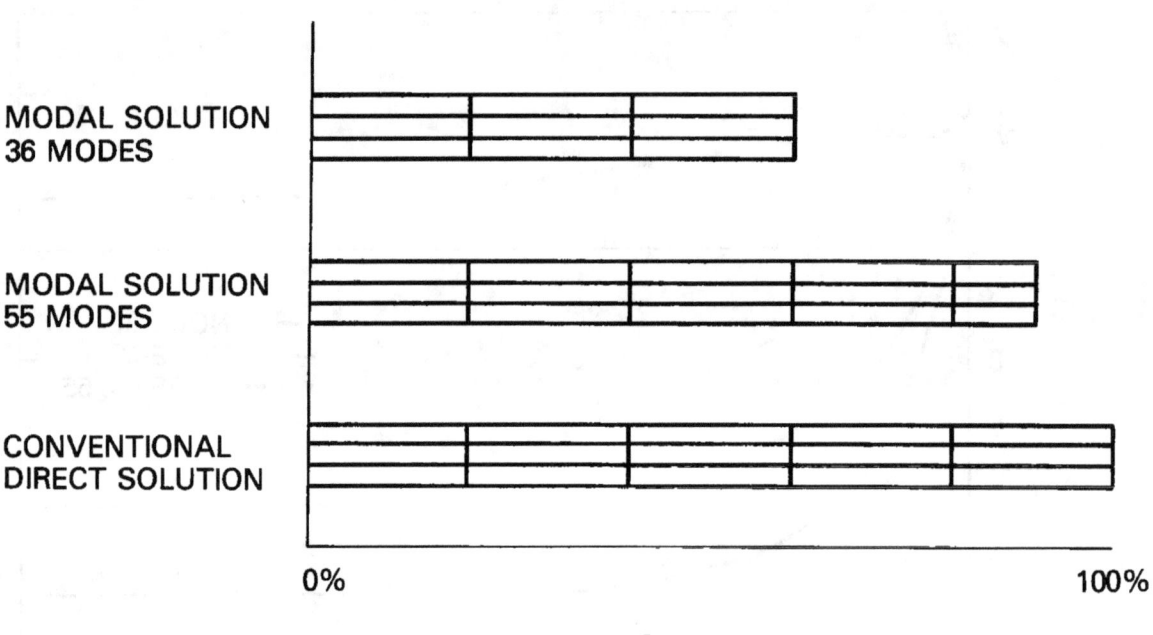

Figure 15

ANTICIPATED FUTURE ACTIVITY

With the completion of this study, the fundamental validity, practicality, and means for implementation of the thermal mode process have been demonstrated. Projected future activity will consist of evaluating the technique on larger real-world problems, developing an understanding of the mode selection process, and developing interfaces to other thermal analysis codes, i.e., SPAR and SINDA (fig. 16).

- EVALUATE TECHNIQUE FOR LARGER REAL WORLD PROBLEMS

- DEVELOP UNDERSTANDING TO ENABLE EFFICIENT MODE SELECTION

- DEVELOP INTERFACES TO OTHER THERMAL ANALYSIS CODES (I.E., SPAR, SINDA)

Figure 16

REFERENCE

1. Frisch, H. P.: Thermally Induced Response of Flexible Structures: A Method for Analysis. J. Guidance and Control, Vol. 3, No. 1, Jan. - Feb. 1980.

REENTRY HEATING ANALYSIS OF SPACE SHUTTLE
WITH COMPARISON OF FLIGHT DATA

Leslie Gong, Robert D. Quinn, and William L. Ko
Dryden Flight Research Facility
NASA Ames Research Center
Edwards, California

SUMMARY

Surface heating rates and surface temperatures for a space shuttle reentry profile were calculated for two wing cross sections and one fuselage cross section. Heating rates and temperatures at 12 locations on the wing and 6 locations on the fuselage are presented. The heating on the lower wing was most severe, with peak temperatures reaching values of 1240°C for turbulent flow and 900°C for laminar flow. For the fuselage, the most severe heating occured on the lower glove surface where peak temperatures of 910°C and 700°C were calculated for turbulent flow and laminar flow, respectively.

Aluminum structural temperatures were calculated using a finite difference thermal analyzer computer program, and the predicted temperatures are compared to measured flight data. Skin temperatures measured on the lower surface of the wing and bay 1 of the upper surface of the wing agreed best with temperatures calculated assuming laminar flow. The measured temperatures at bays 2 and 4 on the upper surface of the wing were in quite good agreement with the temperatures calculated assuming separated flow. The measured temperatures on the lower forward spar cap of bay 4 were in good agreement with values predicted assuming laminar flow. However, temperatures measured on the aft spar cap were higher than the values calculated for laminar flow. The upper spar cap temperatures computed for separated flow were in fairly good agreement with the flight data.

INTRODUCTION

The space shuttle orbiter is designed to be used for approximately one hundred flights. During each flight the vehicle must withstand severe aerodynamic heating during reentry through the atmosphere. The space shuttle skin and substructure are constructed primarily of aluminum which must be protected during reentry with a thermal protection system (TPS) from being overheated beyond the design temperature limit of 177°C so that the integrity of the structure is maintained for subsequent flights.

In addition to the temperature limit, the temperature gradient within the structure must not be too severe or the resulting thermal stress will buckle the skin and cause possible bondline failure.

In order to gain confidence in the thermal protection system, reentry heating analysis of the shuttle must be made and the temperature distribution within the structure must be estimated to assure that the design temperature is not exceeded and the thermal stresses are not excessive.

This paper presents calculated surface (aerodynamic) heating rates and surface temperatures for the wing at wing station (WS) 240 and wing station (WS) 328. Also presented are the aerodynamic heating rates and surface temperatures computed for six locations on the fuselage at fuselage station (FS) 877. Wing stations 240 and 328, and FS 877 were chosen for analyses because these locations were most heavily instrumented. In addition, calculated aluminum skin and spar cap temperatures for the wing at WS 240 are shown and compared to available measured flight data.

SYMBOLS

H   altitude, m

Q   heating rate, $\frac{kw}{m^2\text{-sec}}$

T   temperature, °C

V   velocity, $\frac{m}{sec}$

α   angle of attack, degrees

SPACE SHUTTLE

A planform view and side view of the space shuttle showing the two wing locations and one fuselage location for which aerodynamic heating analyses were made are shown in figure 1. As shown, calculations were made for WS 240, WS 328, and FS 877. A thermal model was also made for the wing cross section at WS 240 and structural temperatures were calculated for this location. Cross sections of the wing at WS 240 and WS 328, and the fuselage at FS 877 showing the general moldline geometry are presented in figure 2.

FLIGHT CONDITIONS

In order to make aerodynamic heating calculations, time histories of altitude, Mach number or velocity, and angle of attack must be determined. For the present investigation the nominal STS-1 (Space Transportation System) reentry flight trajectory shown in figure 3 was used. Also shown in this figure are the actual reentry flight time histories of angle of attack, velocity, and altitude. As can be seen the two trajectories are in excellent agreement and, therefore, calculations made using the nominal trajectory can be used to compare with flight data.

DESCRIPTION OF WING STRUCTURE

The geometry of the wing section at WS 240 for which structural temperatures were calculated is shown in figure 4. Both the upper and lower skins and forward spar web of bay 1 are made of aluminum honeycomb sandwich plates. The skins for bays 2, 3, and 4 are made of spanwise "hat" stringer reinforced aluminum skins. The remaining spar webs are made of corrugated aluminum. The lower wing skin is covered with high temperature reusable surface insulation (HRSI), with the strain isolation pad (SIP) lying between the wing skin and the HRSI. Most of the upper skin of bay 1 is protected by low temperature reusable surface insulation (LRSI) under which lies the SIP layer. The HRSI and LRSI are bonded to the SIP with room temperature vulcanized (RTV) silicone rubber and the SIP is bonded to the skin with RTV. The remainder of the upper skin of bay 1 and all of the upper skin of bays 2, 3, and 4 are covered with felt reusable surface insulation (FRSI). The FRSI is bonded directly to the skin, with RTV, and there is no SIP layer.

CALCULATING METHODS

Aerodynamic Heating

External heating rates and surface temperatures were computed by the DFRC computer program "THEOSKN". This program solves the one-dimensional thin skin heating equation and computes time-histories of heating rates, temperatures, heat transfer coefficients, skin friction, etc. At present this program can compute turbulent heat transfer by the theory of van Driest (reference 1) and Eckert's reference enthalpy method (reference 2), and laminar heat transfer by Eckert's reference enthalpy method (reference 2). Also, 3-D stagnation point laminar heat transfer and 2-D stagnation point laminar heat transfer with and without sweep can be computed by the theory of Fay and Riddell (reference 3). The swept cylinder theory of Beckwith and Gallagher (reference 4) is used to compute turbulent stagnation point heating.

The local flow conditions used in the heating equations can be computed by the program or can be input to the program from some other source. At present, the program can calculate normal shock local flow with or without sweep, and local flow conditions based on the oblique shock theory, the Prandtl-Meyer expansion theory, and/or the tangent cone theory. Also, heating rates can be arbitrarily changed by user input, and a transition number based on Reynolds number and/or Mach number may be input to the program to change heating calculations (e.g. laminar to turbulent flow). All calculations are based on real gas properties of air.

In the present investigation, heating rates and surface temperatures were calculated for the upper and lower surfaces of the wing at WS 240 and WS 328, and for the fuselage at FS 877. Three cases of heating rates and surface temperatures were calculated for the wing at WS 240. Namely, (i) turbulent flow for both the lower and upper surfaces, (ii) laminar flow for both the lower and upper surfaces, and (iii) laminar flow for the lower surface and bay 1 of the upper surface, and separated flow for the aft bays of the upper surface. For WS 328, heating rates and surface temperatures were computed only for case iii. The laminar heating rates were computed by Eckert's reference enthalpy method, and the turbulent heating rates were computed by the theory of van Driest. For both cases a Reynolds analogy factor of 1.12 was used. The local flow conditions were computed by the oblique shock theory and the Prandtl-Meyer expansion theory. The initial wedge angle was taken to be 30 degrees. The flow distance was the chordwise distance measured from the leading edge of the wing. The laminar and turbulent heating rates for the upper surface were computed by the same procedure used to calculate the lower surface heating rates except that the flow expansion was limited in such a way that the local static pressures did not go below three-tenths of the free stream static pressure. Finally, the separated flow heating rates for the upper surface were estimated. For this analysis, the separated flow heating rates were taken to be one-half of the attached laminar flow calculated heating rates.

Heating rates and surface temperatures were calculated for the lower surface of the fuselage by the turbulent swept cylinder theory of Beckwith and Gallagher and the laminar swept cylinder theory of Fay and Riddell. Calculations for the lower glove surface were also made using the above theories. However, the resulting heating rates were increased by 20 percent as suggested by results from wind tunnel tests (references 5 and 6). For the leading edge of the "glove," only laminar calculations were made using the swept cylinder theory of Fay and Riddell. The upper "glove" surface was known to be in a low heating separated flow region. Measured results for similar geometry on the X-15 airplane during a reentry flight (reference 7) showed the lower surface heating to be about thirty times the upper surface heating. Therefore, the heating rates for the upper glove surface were estimated by taking one-thirtieth of the heating rates calculated for the lower "glove" surface. For the side of the fuselage, it was assumed that the flow was separated from the fuselage glove junction to a point (attachment point) on the fuselage where the TPS changed from

FRSI to LRSI. This location is about halfway up the side of the fuselage. Heating rates and surface temperatures were calculated for the side of the fuselage using Eckert's reference enthalpy method for laminar flow and assuming that the local attached flow conditions were equal to free stream values. The separated flow heating rates on the side of the fuselage were assumed to be equal to one-tenth the attached flow calculations. The heating rates and surface temperatures for the upper fuselage surface were computed by Eckert's reference enthalpy method for laminar flow with and without transition at a local Reynolds number of 5×10^5. The local flow conditions were calculated using the Prandtl-Meyer expansion theory with initial flow conditions equal to free-stream values. The flow distance was measured from the attachment point on the side of the fuselage. Two calculations were made. In the first calculation, the local flow was allowed to expand until the local static pressure was equal to one-half the free stream pressure. In the second calculation, the local static pressure was limited to one-fourth the free stream value.

Structural Temperature Calculations

The structural temperatures were computed using the Lockheed Thermal Analysis program (reference 8). This program computes transient temperature distributions in configurations of arbitrary complexity, using the electrical resistance capacity analogy. Solutions are obtained by converting the physical system into one consisting of lumped thermal capacitors connected by the thermal resistors and then using the lumped-parameter, or finite-difference approach, to solve for the temperature history of the system. This program permits direct solutions of complex transient problems involving conduction, convection, radiation, and heat storage. Furthermore, since it is also possible to specify any quantity as an arbitrary function of any other, it is also possible to solve such problems as change of state, variable thermodynamic properties, arbitrary variable boundary conditions (such as aerodynamic heating) and other non-linear effects. Input format is not restricted to any particular geometry, but is such that resistors and capacitors can be connected in any manner desired.

When using this program to compute temperatures for a large thermal model, it is desirable to make the computing interval as large as possible so that the computational time is not excessive. The computing interval for this program is determined by multiplying the minimum RC product by a given factor. The default value of this multiplying factor is 0.25. The RC product is the product of the capacity of a lump times the equivalent resistance of that lump, and the equivalent resistance is the parallel combination of all the resistors connected to the lump. The multiplying factor of 0.25 can be changed to any desired value. However, care must be exercised, as too large a computing interval will result in unreliable results. In the present investigation, a multiplying factor of 0.9 was used which resulted in a computing interval of approximately 1.0 second.

The thermal properties of the TPS and aluminum structure are functions of pressure and/or temperature and therefore must be updated at frequent intervals if good results are to be obtained. In the present analysis, the thermal properties were updated at each computing interval (1.0 second). Also when calculating the conduction resistors, the temperature used to update the conductivity was the average of the temperature of the two lumps connected by the resistor.

THERMAL MODEL

The third bay of the wing cross section at WS 240 was first modeled by a one-dimensional thermal model. One of the primary purposes for making the one-dimensional calculations was to determine how many layers[1] (lumps) the HRSI should be divided into in order to get a good solution with minimum computer running time. Consequently, the thermal model was made with the HRSI divided into 5, 10 and 15 layers. The thermal model with the HRSI divided into 10 layers is shown in figure 5. The circled numbers are the lump numbers, and numbers on the right are the resistors connecting the lumps. Resistors 1 through 14 and 16 through 19 are the conduction resistors. Resistor 15 is the internal radiation resistor and 26 and 27 are the external radiation resistors which radiate to the free-stream temperature T_∞. The results from the one-dimensional calculations are presented in figure 6 which shows a plot of the maximum calculated lower skin temperatures versus the number of HRSI layers. As shown, the temperature difference between the 5 layer model and 15 layer model is 1.67°C. However, the difference between the 10 and 15 layer model is only 0.27°C and is insignificant. Based on these calculations, it was decided that for all subsequent thermal models, the HRSI would be divided into approximately 10 layers.

The two dimensional thermal model for the entire wing cross section, excluding the leading edge and elevon, at WS 240 is shown in figures 7, 8, and 9. As shown in figure 7, the TPS and aluminum structure were divided into 410 lumps (capacitors). Figure 8 shows the conduction resistors and the external radiation resistors. There are a total of 486 conduction resistors and 45 external radiation resistors. Also shown in figure 8 are the aerodynamic heating inputs which are denoted by the arrows labeled Q20 to Q41, and Q50 to Q71. Because of the gaps in the reusable surface insulation (RSI), heat conduction was allowed only in the RSI thickness direction. As shown in figure 8, each external lump radiated to the ambient temperature T_∞. The emissivity used to compute the radiation heat flux was 0.85 for the lower surface and 0.80 for the upper surface. Also the view factor used in the radiation calculations

---

[1] In this discussion, the word layers refers to lumps in the TPS thickness direction. Therefore, for the one-dimensional thermal model, the two words are synonymous.

was 1.0. Typical internal radiation resistors are shown in figure 9. There are a total of 553 internal radiation resistors. View factors were computed for each internal radiating lump, and the emissivity used in the heat flux calculations was 0.667.

RESULTS AND DISCUSSION

Heating Rates and Surface Temperatures

<u>Wing</u> - The surface heating rates calculated for wing station 240 are shown in figure 10. Turbulent and laminar heating rates were computed for the lower surface and bay 1 of the upper surface. For bays 2, 3, and 4 of the upper surface, turbulent, laminar, and separated heating rates were calculated. Details of the methods used to compute the heating rates and surface temperatures are presented in the previous section of this paper called CALCULATING METHODS. The turbulent heating rates are shown by the solid lines. The laminar heating rates are shown by the dashed lines, and the heating rates computed for separated flow are shown by the long and short dashed lines. For the lower surface, the maximum heating rate calculated, assuming turbulent flow, varied from 225 kw/m^2-sec at bay 1 to 175 kw/m^2-sec at bay 4. For laminar flow, the maximum heating rate varied from 75 kw/m^2-sec at bay 1 to 40 kw/m^2-sec at bay 4. The maximum upper surface heating rate computed for turbulent flow was 5.9 kw/m^2-sec, and for laminar flow was 5.8 kw/m^2-sec. The heating rates calculated assuming separated flow varied from a maximum of 2.2 kw/m^2-sec at bay 2 to 1.6 kw/m^2-sec at bay 4. It may be noted that the laminar heating rates computed for the upper surface are higher than the turbulent heating rates for the first 1000 seconds of the reentry trajectory, and the total heating rates calculated for turbulent flow are only slightly higher than the total laminar heating rates. This apparent abnormality is due to the very low Reynolds numbers on the upper surface of the wing.

The calculated surface temperatures for the lower surface at WS 240 are shown in figure 11. The maximum temperatures calculated for turbulent flow vary from 1240°C at bay 1 to 1090°C at bay 4. For laminar flow, the maximum temperatures vary from 900°C at bay 1 to 690°C at bay 4. The upper surface temperatures are shown in figure 12. The maximum surface temperature occurs at bay 1 and is 315°C. It may also be noted that the maximum calculated temperatures for turbulent flow and laminar flow are nearly the same. The surface temperatures computed for separated flow range from 165°C at bay 2 to 120°C at bay 4.

Calculated heating rates for WS 328 are shown in figure 13. The laminar flow heating rates computed for the lower surface of the wing reached maximum values of 85 kw/m^2-sec at bay 1 and 53 kw/m^2-sec at bay 3. For the upper surface, the heating rates were computed assuming laminar flow for bay 1 and separated flow for bays 2 and 3. The maximum heating rate at bay 1 is 6 kw/m^2-sec, and for bays 2 and 3 the maximum heating rates are slightly above 2 kw/m^2-sec.

The surface temperatures at WS 328 are shown in figure 14 for the lower surface and figure 15 for the upper surface. All of the lower surface temperatures and the temperatures for bay 1 of the upper surface were calculated assuming laminar flow. The temperatures for bay 2 and bay 3 of the upper surface were computed assuming separated flow. The lower surface reaches maximum temperatures of 880°C, 800°C, and 740°C at bays 1, 2, and 3 respectively. For the upper surface, the maximum temperature at bay 1 is 370°C, and the peak temperatures at bays 2 and 3 are 230°C and 208°C, respectively.

Fuselage - Heating rates at six locations on the fuselage at FS 877 are presented in figure 16. Calculations are shown for location 1 (lower fuselage centerline), location 2 (lower surface of the glove), location 3 (leading edge of the glove), location 4 (upper surface of the glove), location 5 (side of the fuselage), and location 6 (top centerline of the fuselage). Details of the methods used to make these calculations are discussed in the previous section called CALCULATING METHODS. At location 1, heating rates are shown for both laminar and turbulent flow. The maximum value obtained for turbulent flow was 100 kw/m^2-sec and the peak laminar heating rate calculated was 48 kw/m^2-sec. Turbulent and laminar heating rates are also shown for location 2. The peak heating rate for turbulent flow is 115 kw/m^2-sec and for laminar flow is 45 kw/m^2-sec. It should be mentioned that the turbulent calculated heating rates at location 2 were empirically increased by 20 percent as discussed in the previous section. However, this empirical factor was not applied to the laminar calculations. At location 3, only laminar flow heating rates were calculated and as shown, the peak value at this location is 80 kw/m^2-sec. Two curves of calculated heating rates assuming separated flow are shown for location 4. The lower curve represents the estimated heating rates that were expected at this location. However, because of the uncertainty of the heating at this location, due to the complex flow field on the upper glove, conservative estimates of the heating rates were also made and are shown by the upper curve. At location 5, heating rates were generated assuming separated flow and assuming attached laminar flow with transition to turbulent flow at 1350 seconds. The attached flow calculations produced a peak heating rate of 14 kw/m^2-sec. However, if the flow is assumed separated at location 5, the maximum heating rate is 1.4 kw/m^2-sec. Two curves of calculated heating rates are shown for location 6. The lower curve was computed assuming laminar flow and also assuming that the local static pressures were equal to one-forth of the free-stream values. As shown, this curve reaches a peak value of 4 kw/m^2-sec. The upper curve was computed assuming laminar flow with transition to turbulent flow at a Reynolds number of 5 x 10^5 (time = 1050 seconds) and also assuming that the local static pressures were equal to one-half the free-stream values. The maximum heating rate shown by this curve is 8 kw/m^2-sec.

Surface temperatures calculated for the same six locations at which the heating rates were computed are shown in figure 17. As shown, the heating at locations 1, 2, and 3 is quite severe with peak temperature at location 1 and location 2 reaching approximately 910°C for turbulent flow and 700°C for laminar flow. At location 3 the peak temperature is 850°C. The temperatures at location 4 reached peak values of 190°C and 275°C depending on which heating rate curve was used (see figure 16). The maximum surface temperature calculated at location 5 was 157°C if separated flow was assumed and 450°C if the flow is assumed to be attached. The surface temperatures at location 6 reached a peak value of 325°C when it was assumed that the flow transitioned to turbulent flow at 1050 seconds.

Structural Temperatures

Predicted aluminum skin temperature time histories at four locations on the lower surface of the wing at WS 240 are shown in figure 18 and figure 19. Also shown for comparison are STS-1 measured flight temperatures. Figure 18 shows the skin temperatures for bays 1 and 2, and figure 19 shows the skin temperatures for bays 3 and 4. Except for bay 3 (figure 19), flight data for the time interval 0 to 1178 seconds were not available due to telemetering "blackout." Calculated temperatures are shown for turbulent flow using 80 percent TPS thickness, laminar flow using 80 percent TPS thickness, and laminar flow using 100 percent TPS thickness. Eighty (80) percent of the TPS thickness was used to account for gap heating effects[2]. As shown, the measured data falls about halfway between the two laminar curves up to 1650 seconds of the flight profile, and it is apparent that the flow on the lower surface of the wing was laminar. The fact that data falls midway between the two laminar curves indicates that the effect of gap heating may not be as severe as that imposed on the calculations by using 80 percent TPS thickness. It is obvious from

---

[2]The 80 percent TPS thickness was the design criterion used by space shuttle manufacturer to account for the effects of gap heating.

the comparisons shown in figure 18 and 19 that good agreement between the measured and calculated data up to 1650 seconds of the flight profile would have been obtained if 90 percent TPS thickness had been used in the calculations. After 1650 seconds, the flight data shows an increasing deviation from the calculated values and the agreement between the measured and calculated data is poor. The result was expected since the forced convection cooling from 1550 seconds to touchdown at 1916 seconds was not accounted for in the calculations (see figure 10). Also after touchdown the free convection external cooling and the free convection internal heating were neglected in the calculations.

Comparisons between measured and calculated skin temperatures on the upper surface of the wing are shown in figures 20 and 21. Skin temperature time-histories for bays 1 and 2 are shown in figure 20, and skin temperature time-histories for bay 3 and bay 4 are shown in figure 21. Calculated temperatures are shown for turbulent flow using 80 percent TPS thickness and laminar flow using 80 percent TPS thickness. Also shown are temperatures computed assuming laminar flow and 100 percent TPS thickness for bay 1 and temperatures calculated for separated flow with 100 percent TPS thickness for bays 2, 3 and 4. For bay 1, it can be seen that the measured flight data are in quite good agreement with the temperatures calculated assuming laminar flow and 100 percent TPS thickness. At bays 2 and 4, the temperatures calculated assuming separated flow and 100 percent TPS thickness are in fairly good agreement with the measured flight data. It may be noted that the measured temperatures at bays 2 and 4 continue to increase after touchdown. This increase in temperature of the upper skins is due to convection and radiation heating from the hotter lower skins. The upper skin of bay 1 does not show this increase in temperature after touchdown because the skins of bay 1 are made of aluminum honeycomb core sandwich plates which insulates the thermocouple, located on the outer skin, from the internal heating effects.

Comparisons between measured and calculated temperatures on the lower spar caps are shown in figure 22. The flight data for the lower forward spar cap of bay 4 are in good agreement with the laminar flow curve for 80 percent TPS thickness up to 1800 seconds. It may be noted that the measured skin temperatures at bay 4 (see figure 19) do not agree as well with the calculated values as do the measured temperatures of the forward spar cap. This somewhat poorer agreement between the measured and calculated skin temperatures may result from the fact that the skin was actually made of "hat" stringer reinforced aluminum, whereas, the skin used in the thermal model was an equivalent flat plate. The measured temperatures for the rear spar cap are higher than the laminar flow curve for 80 percent TPS thickness. The lower predicted values for the aft spar cap are probably due to the assumption of total insulation of the aft side of the rear spar web. Like the lower skin data, the flight data for the lower spar caps level off and remain virtually constant after 1800 seconds.

Comparisons between measured and calculated temperatures on the upper spar caps are shown in figure 23. As was the case for the upper skin temperatures, the measured flight data for the spar caps agree best with the values calculated assuming separated flow and 100 percent TPS thickness. The measured data show a higher rate of increase after touchdown than predicted by the calculated curve, and this higher heating rate is probably due to the effects of internal convection which were neglected in the calculations.

CONCLUDING REMARKS

A transient aerodynamic heating program was used to compute time-histories of surface heating rates and surface temperatures for two wing cross sections and one fuselage cross section. The heating on the lower surface of the wing was most severe, with peak temperatures reaching values of 1240°C for turbulent flow and 900°C for laminar flow. For the fuselage, heating was most severe at the lower glove surface where the peak temperatures were 910°C for turbulent flow and 700°C for laminar flow.

A finite-difference thermal analyzer computer program was used to compute structural temperatures for a wing cross section at WS 240. The predicted structural temperature time-histories were compared with measured flight data. These comparisons showed that, for the first 1650 seconds of the reentry trajectory, the temperatures measured on the lower surface of the wing were in fair agreement with values calculated assuming laminar flow. After 1650 seconds the flight data deviates from the predicted values due primarily to the fact that the external convection cooling and the internal convection heating were neglected in the calculations. The temperatures measured on the upper surface at bay 1 were in quite good agreement with values computed assuming laminar flow, and the upper surface temperatures measured aft of bay 1 were in fairly good agreement with values calculated assuming separated flow. The differences that do exist between the measured and calculated temperatures on the upper surface of the wing and the lower surface of the wing prior to time 1650 seconds could be caused by the following assumptions made in the thermal model: (1) the use of effective thickness for the TPS to account for gap heating; (2) initial temperatures and emissivities; (3) total insulation of the aft and forward spar webs; (4) the use of effective thickness for stiffened skin, corrugated spar webs, and honeycomb core skins; (5) no internal convection; and (6) the two-dimensional nature of the thermal model.

REFERENCES

1. van Driest, E. R.: The Problem of Aerodynamic Heating, Aeronautical Engineering Review, Vol. 15, No. 10, pp. 26-41, Oct. 1956.

2. Eckert, E. R. G.: Survey of Boundary Layer Heat Transfer at High Velocities and High Temperatures, WADC Technical Report 59-624, Wright Air Development Center, U.S. Air Force, April 1960.

3. Fay, J. A.; and Riddell, F. R.: Theory of Stagnation Point Heat Transfer in Disassociated Air, Journal of Aeronautical Science, Vol. 25, No. 2, pp. 73-85, 121, Feb. 1958.

4. Beckwith, I. E.; and Gallagher, J. J.: Local Heat Transfer and Recovery Temperatures on a Yawed Cylinder at a Mach Number of 4.15 and High Reynolds Numbers, NASA TR R-104, 1961 (Superceeds NACA Memo 2-27-59L).

5. Space Shuttle Aerothermodynamics Technology Conference, Volume I - Flow Fields, NASA TM X-2506, Feb. 1972.

6. Space Shuttle Aerothermodynamics Technology Conference, Volume II - Heating, NASA TM X-2507, Feb. 1972.

7. Quinn, R. D.; and Olinger, F. V.: Heat Transfer-Measurements Obtained on the X-15 Airplane with Correlation with Wind Tunnel Results, NASA TM X-1905, 1969.

8. Thermal Analyzer Computer Program for the Solution of General Heat Transfer Problems, LR 18902, Lockheed California Company, July 1965.

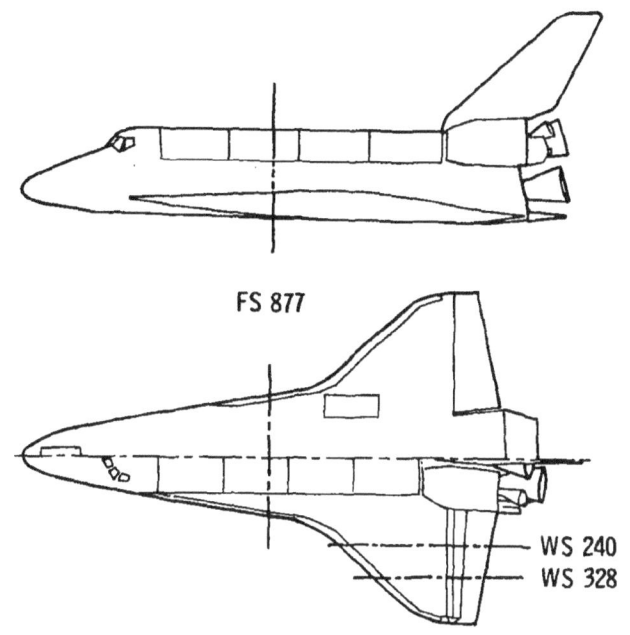

Figure 1.- Wing and fuselage locations analyzed.

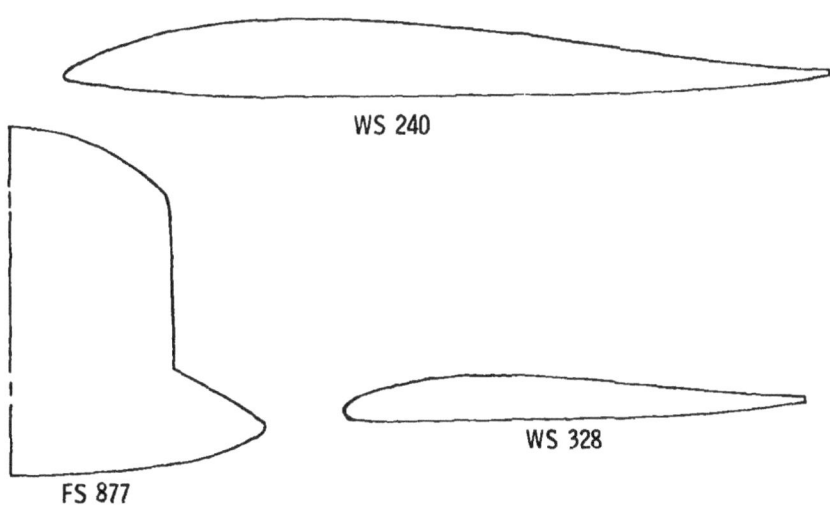

Figure 2.- Wing and fuselage cross sections.

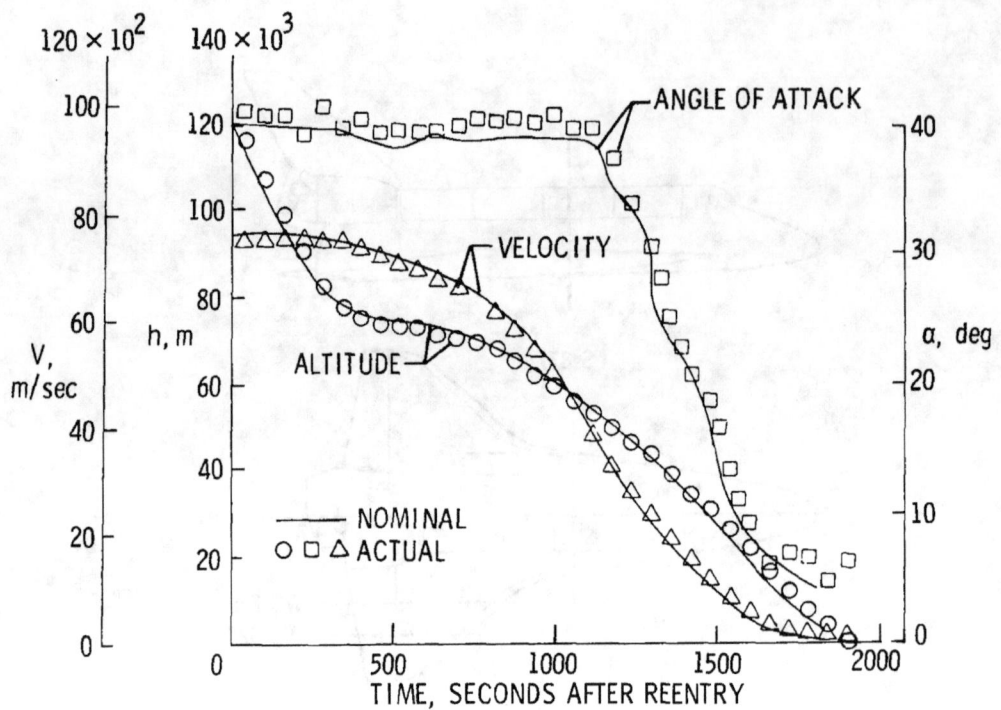

Figure 3.- Flight time histories.

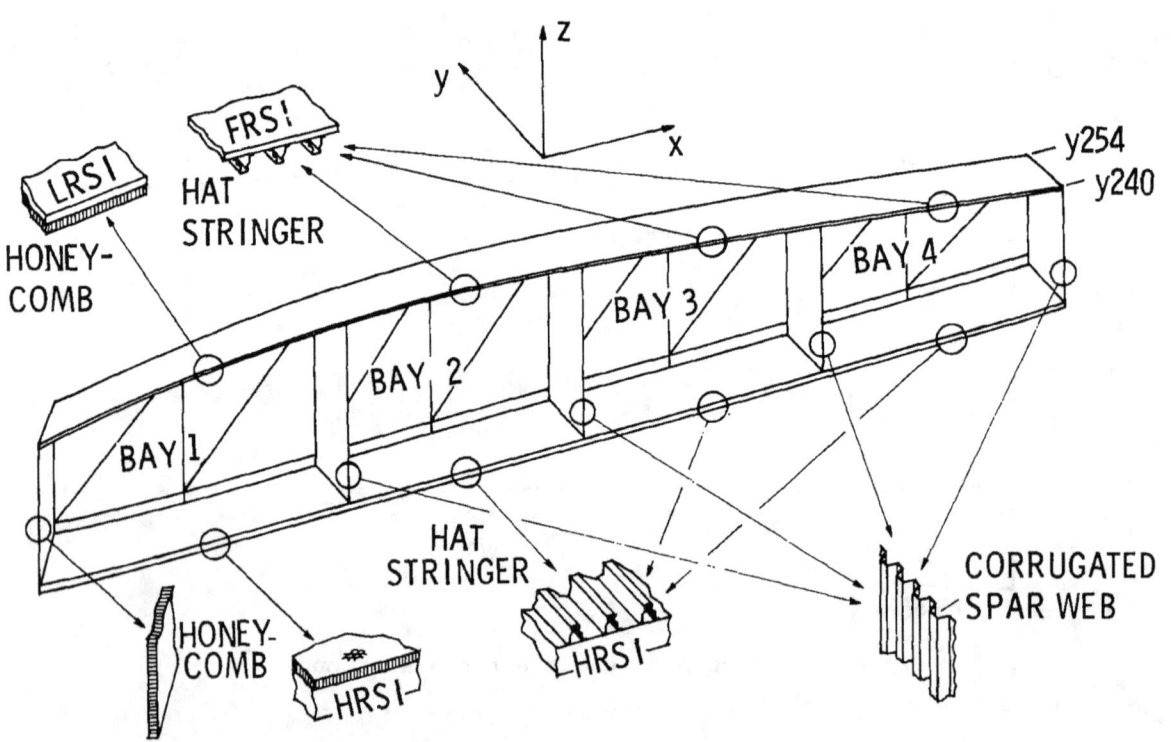

Figure 4.- Geometry of wing at WS 240.

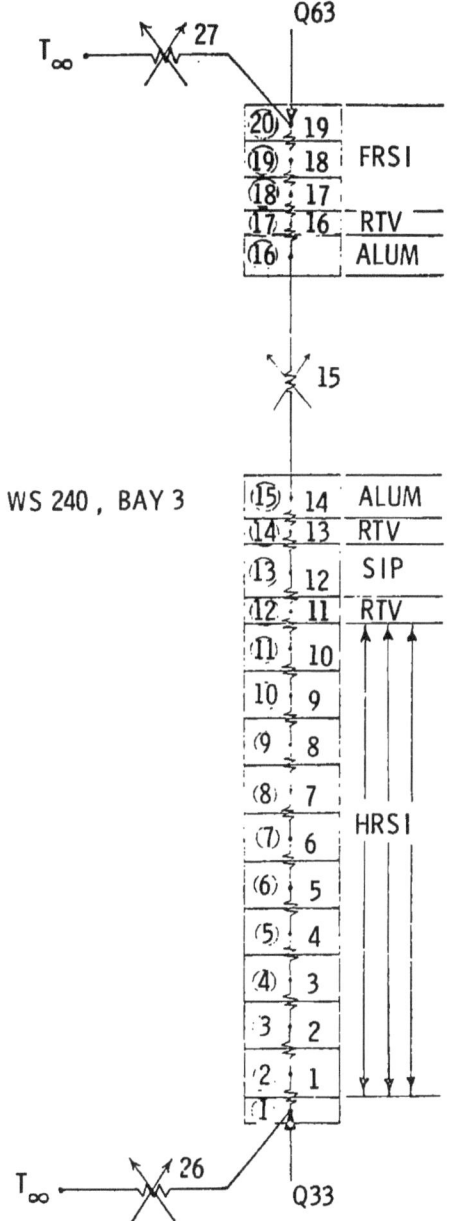

Figure 5.- One-dimensional thermal model.

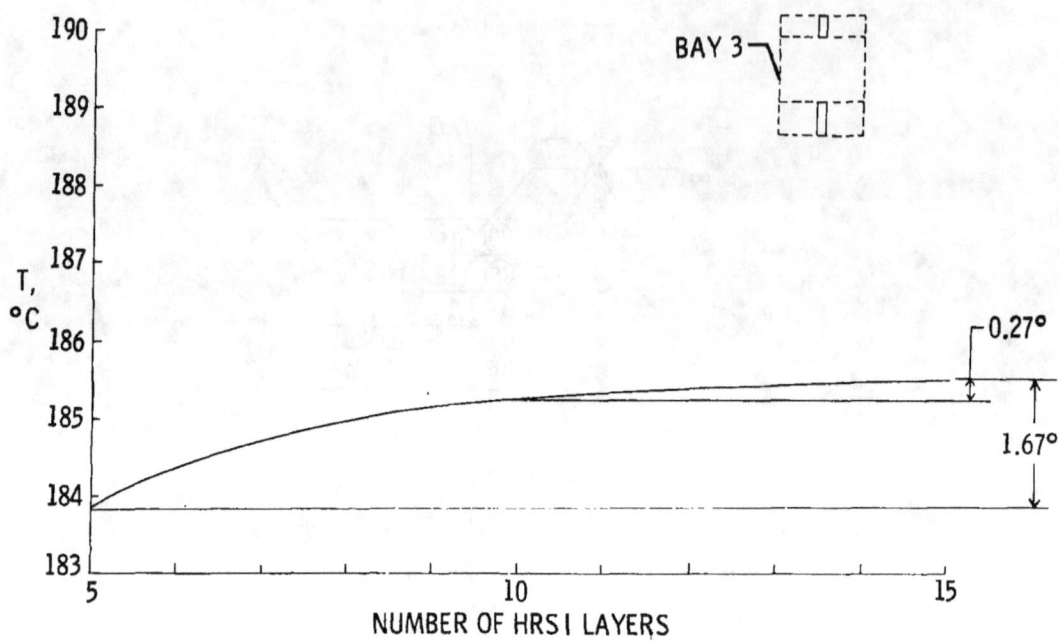

Figure 6.- Structural temperature variation, one-dimensional model.

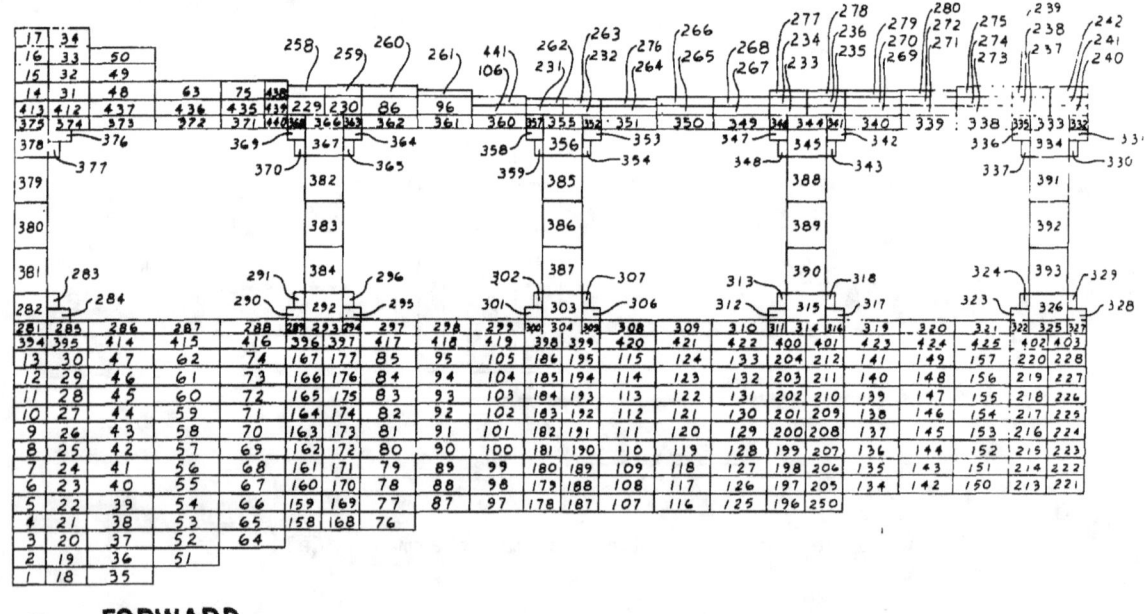

← FORWARD

Figure 7.- Thermal model, capacitors at WS 240.

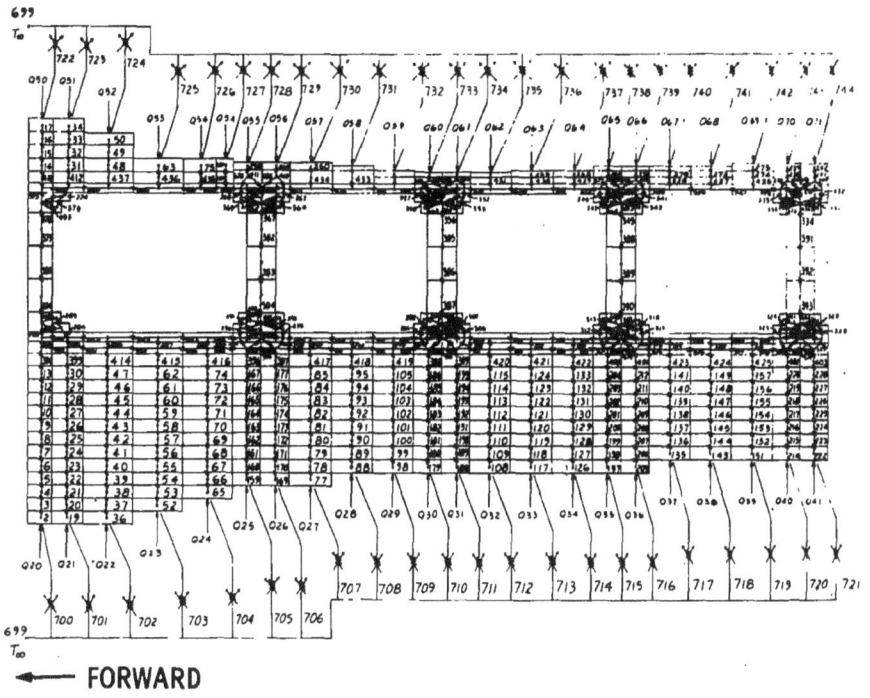

Figure 8.- Thermal model, resistors at WS 240.

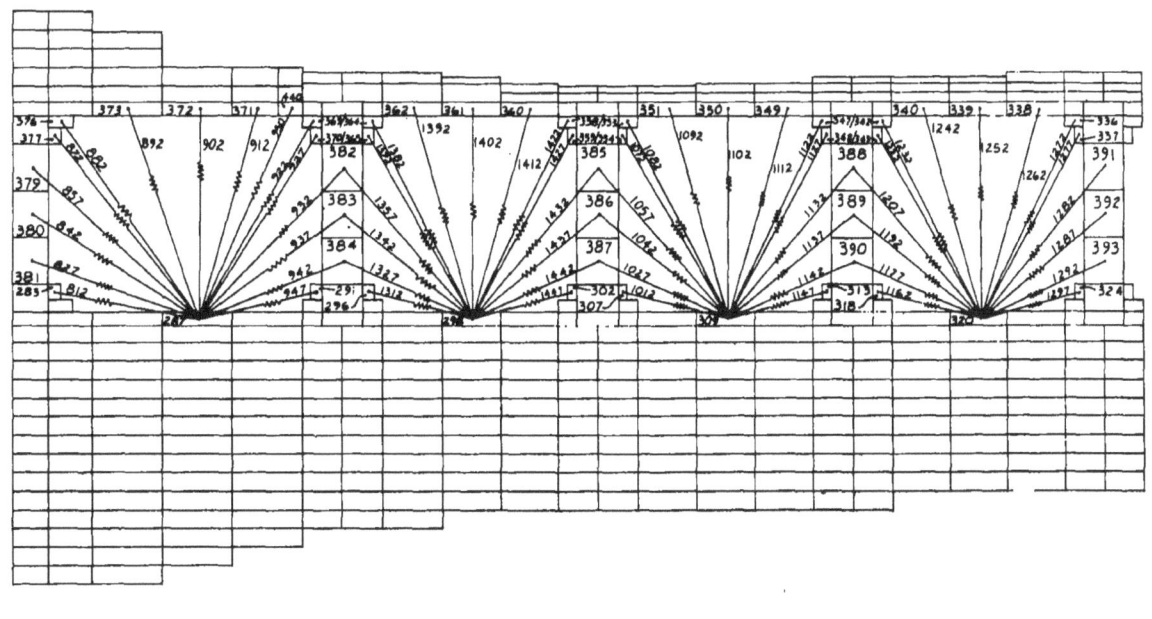

← FORWARD

Figure 9.- Thermal model, internal radiation at WS 240.

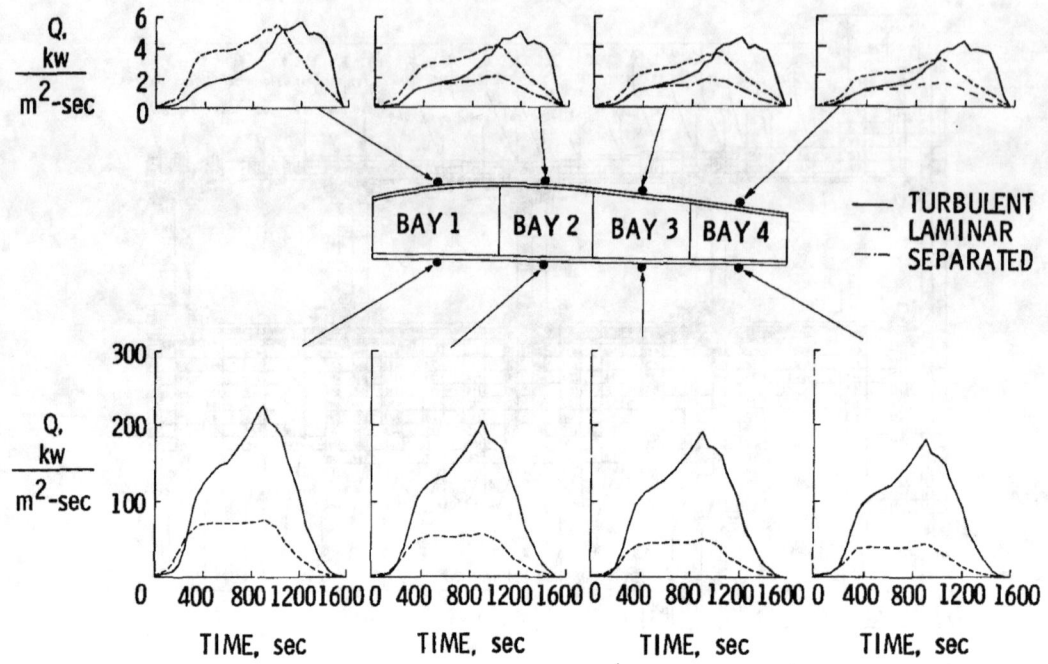

Figure 10.- Calculated heating rates, WS 240.

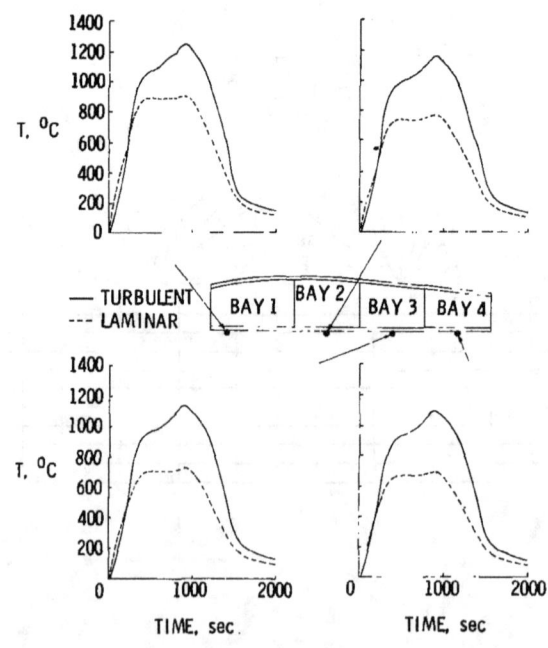

Figure 11.- Calculated surface temperatures of lower surface, WS 240.

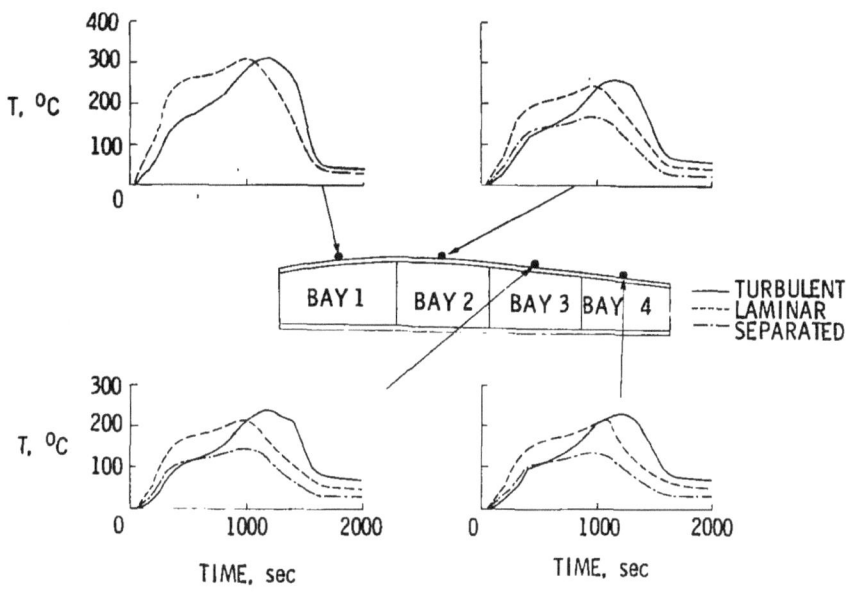

Figure 12.- Calculated surface temperatures of upper surface, WS 240.

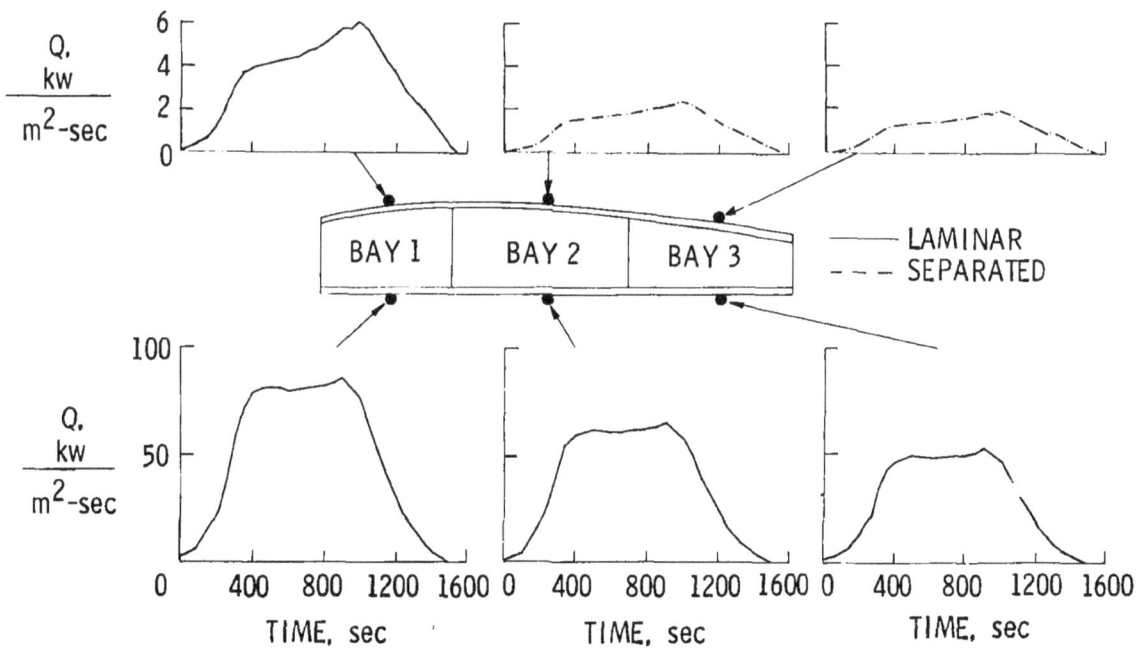

Figure 13.- Calculated heating rates, WS 328.

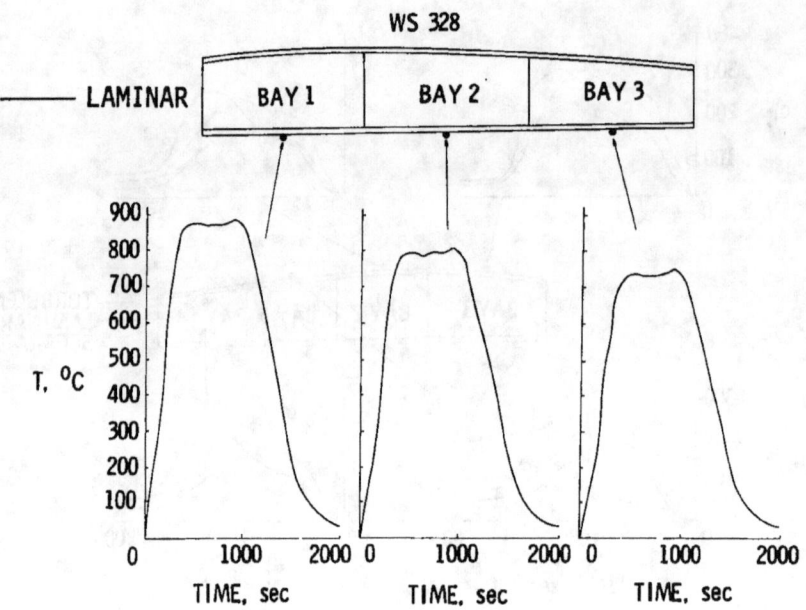

Figure 14.- Calculated surface temperatures of lower surface, WS 328.

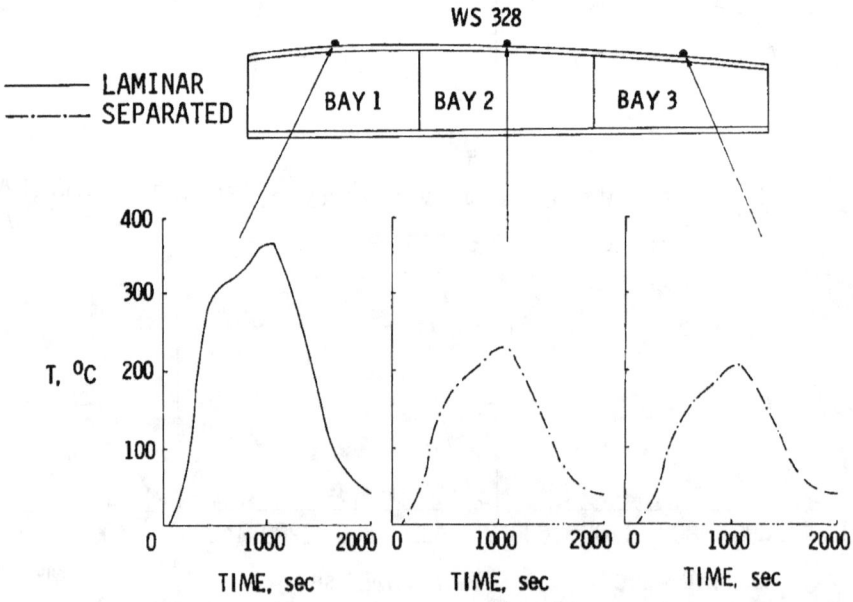

Figure 15.- Calculated surface temperatures of upper surface, WS 328

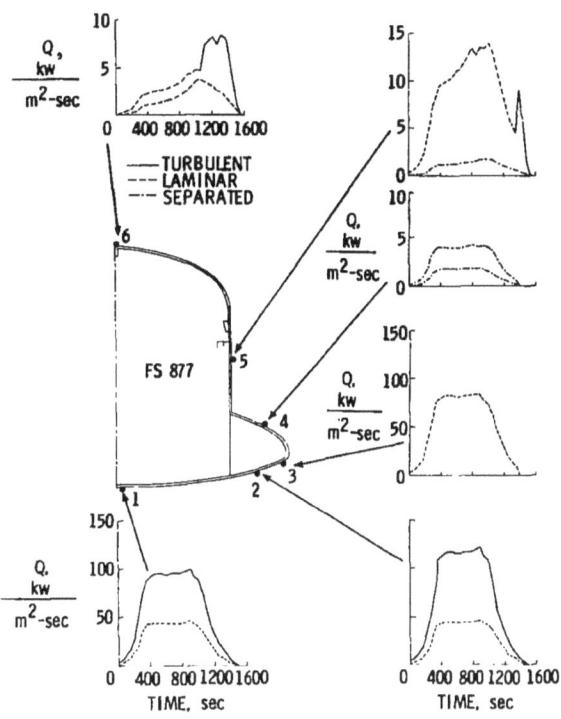

Figure 16.- Calculated heating rates, FS 877.

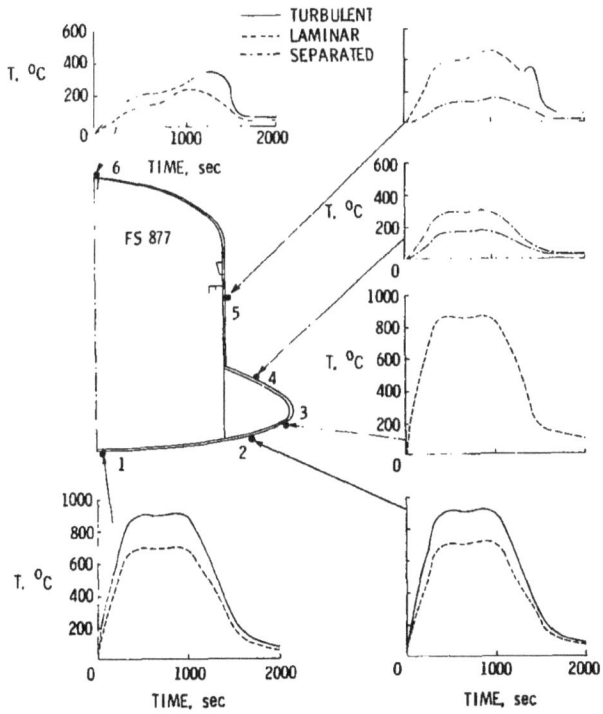

Figure 17.- Calculated surface temperatures at FS 877.

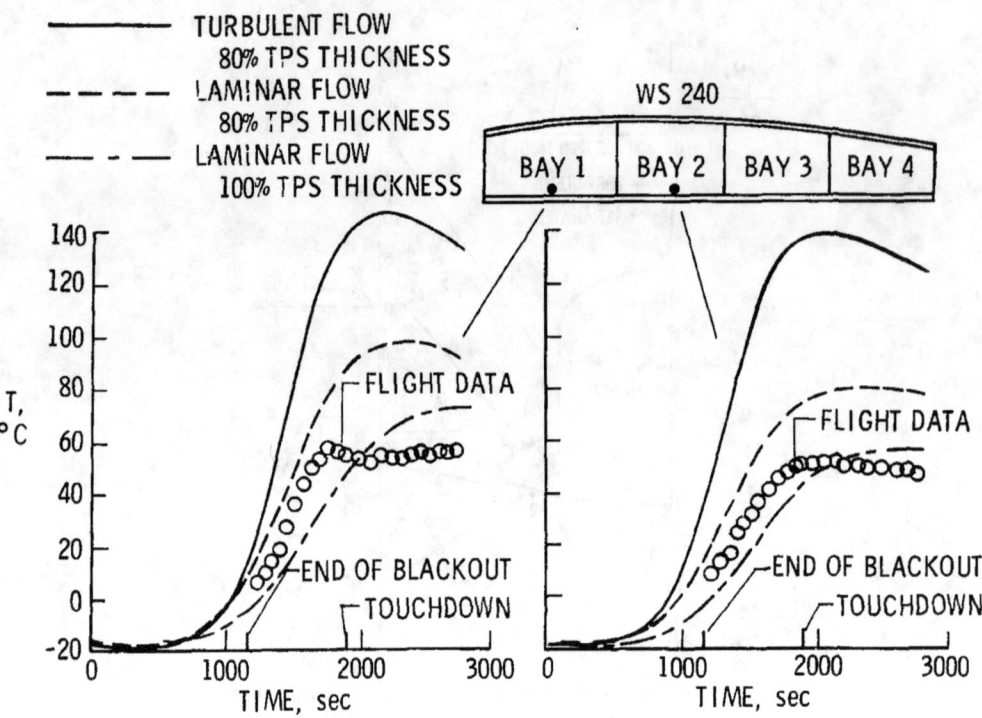

Figure 18.- Lower skin temperatures of bays 1 and 2, WS 240.

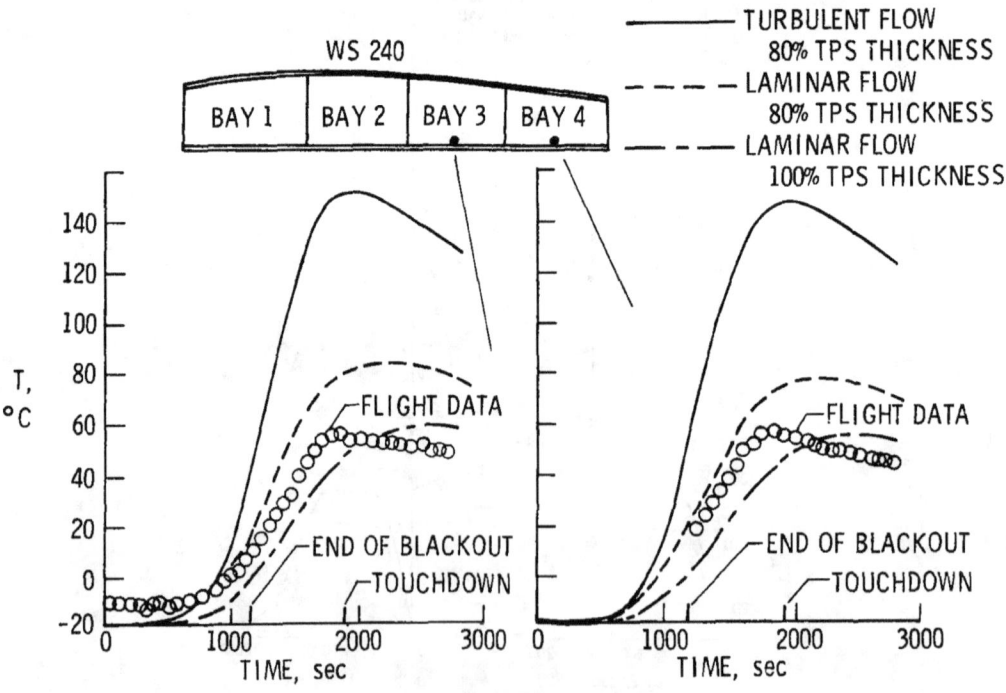

Figure 19.- Lower skin temperatures of bays 3 and 4, WS 240.

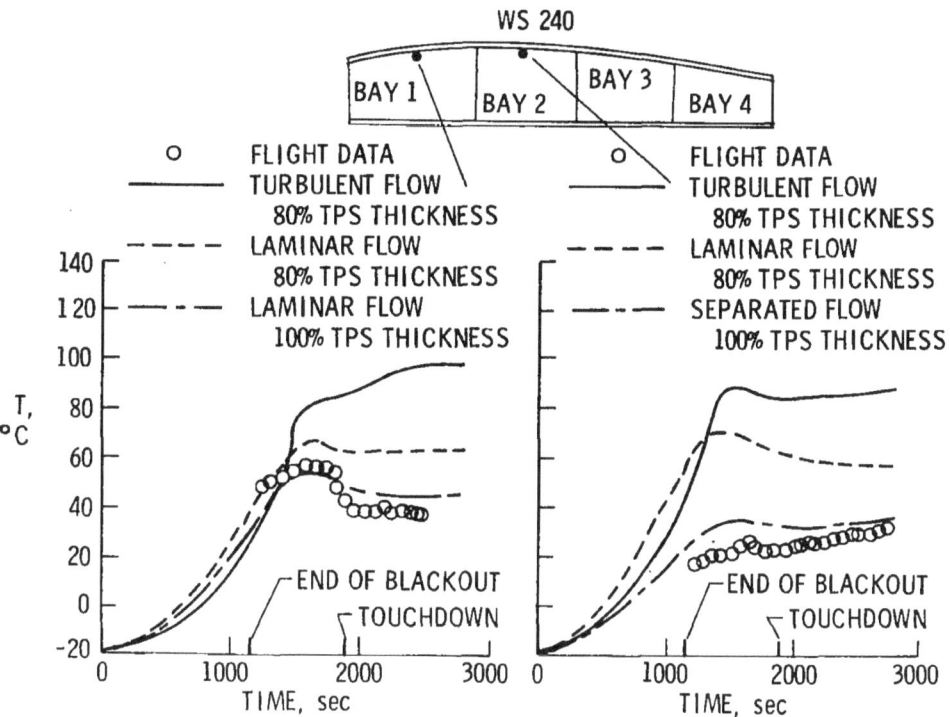

Figure 20.- Upper skin temperatures of bays 1 and 2, WS 240.

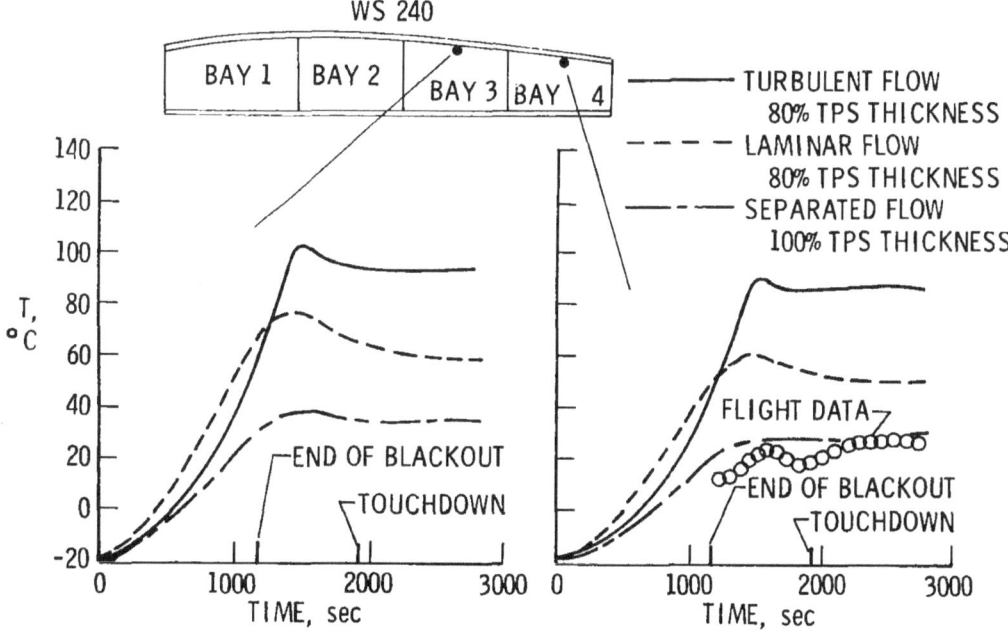

Figure 21.- Upper skin temperatures of bays 3 and 4, WS 240.

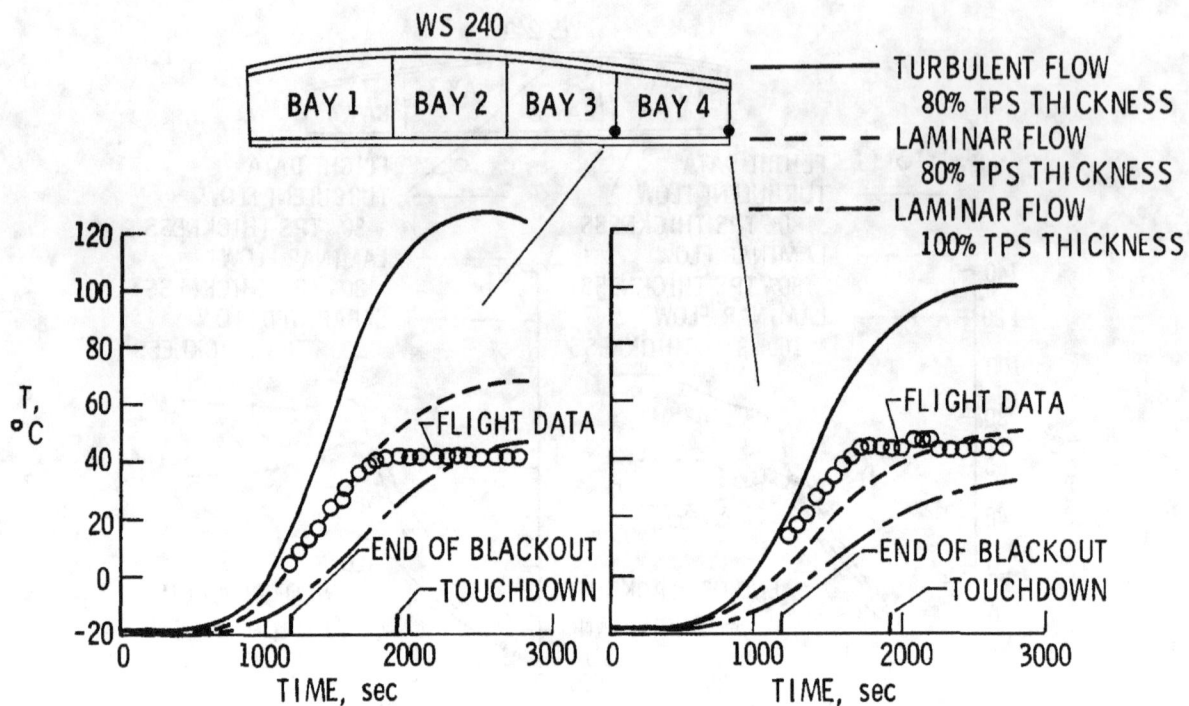

Figure 22.- Lower spar cap temperatures.

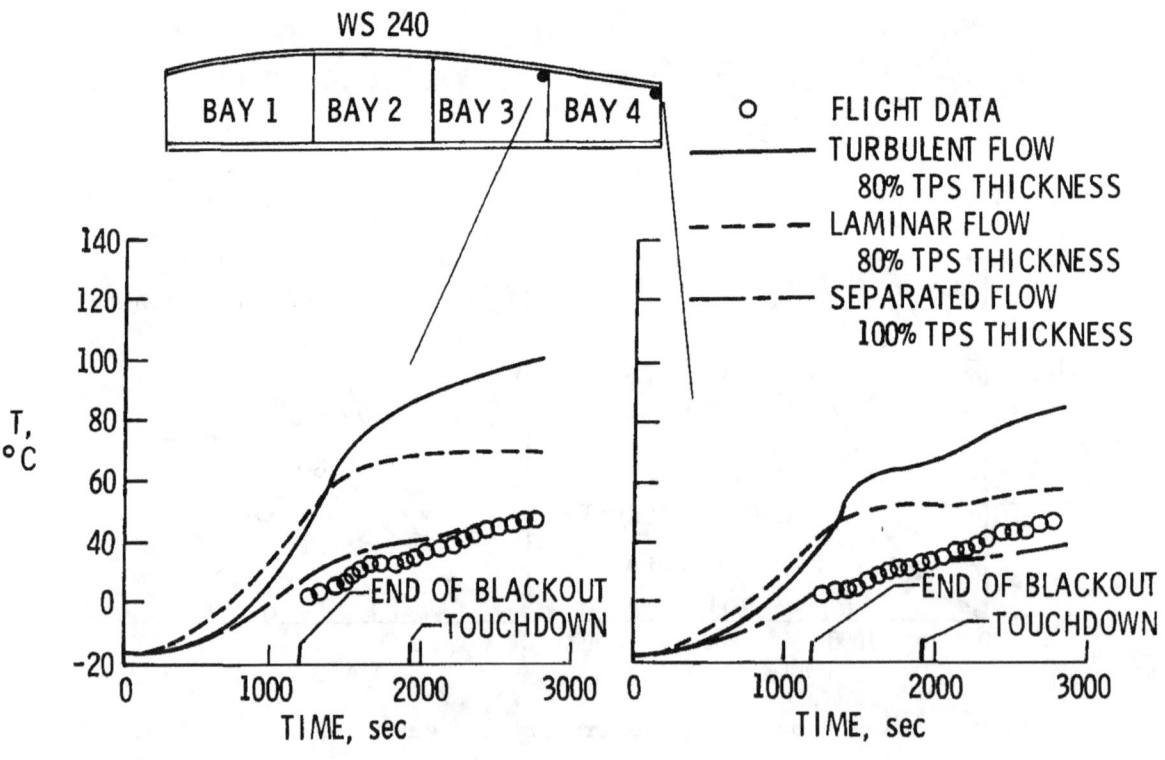

Figure 23.- Upper spar cap temperatures.

REENTRY HEAT TRANSFER ANALYSIS
OF THE SPACE SHUTTLE ORBITER

William L. Ko, Robert D. Quinn, Leslie Gong,
Lawrence S. Schuster, and David Gonzales
Dryden Flight Research Center
Edwards, California

SUMMARY

A SPAR (structural performance and resizing) finite element thermal analysis computer program was used in the reentry heat transfer analysis of the space shuttle. Two typical wing cross sections and a midfuselage cross section were selected for the analysis. The surface heat inputs to the thermal models were obtained from aerodynamic heating analyses, which assumed (1) a purely turbulent boundary layer, (2) a purely laminar boundary layer, (3) separated flow, and (4) transition from laminar to turbulent flow.

The effect of internal radiation was found to be quite significant. With the effect of the internal radiation considered, the wing lower skin temperature became about 39° C (70° F) lower.

The results were compared with flight data for STS-1 (space transportation system, trajectory 1). The calculated and measured temperatures compared well for the wing if laminar flow was assumed for the lower surface and bay 1 upper surface and if separated flow was assumed for the upper surfaces of bays other than bay 1. For the fuselage, good agreement between the calculated and measured data was obtained if laminar flow was assumed for the bottom surface. The structural temperatures were found to reach their peak values shortly before touchdown.

In addition, the finite element solutions were compared with those obtained from the conventional finite difference solutions.

INTRODUCTION

The space shuttle orbiter is designed to be used for at least 100 missions. During each flight cycle, it must withstand the vibrations of lift-off and survive severe aerodynamic heating as it reenters the atmosphere at extremely high velocity (approximately Mach 25 at the start of reentry) and high angle of attack (approximately 40° during the early phase of reentry). The space shuttle skins are constructed primarily of aluminum and/or graphite/epoxy composites, which must be protected from being heated beyond the design limit (176° C, 350° F). Overheating would result in the loss of structural integrity required for subsequent flights. Thus, a thermal protection system (TPS) is bonded to the skins. Another area of great concern is the thermal stresses. The strain isolation pad (SIP) is designed to absorb the skin buckling effect on the TPS. If skin buckling is too severe, the TPS and the shuttle skins may debond, resulting in partial or total loss of the TPS function.

In order to gain confidence in the thermal performance of the shuttle under a prescribed reentry flight profile (time histories of velocity, altitude, and angle of attack), a preflight reentry heating thermal analysis of the space shuttle was conducted. The structural temperature distribution obtained from the thermal analysis can then be used to calculate thermal stresses for studying structural performance. The present paper is limited to the prediction of structural temperature distribution. The results are compared with flight measured data from the first flight.

ANALYSIS

For the analysis, it was assumed that the space shuttle enters the atmosphere under the nominal (or design) STS-1 (space transportation system, trajectory 1) trajectory (shown in fig. 1 (solid lines)). The flight data shown in the figure closely follow the nominal trajectory. Time zero in the analysis corresponds to the start of reentry, which is defined to occur at an altitude of 121,920 meters (400,000 feet). Based on the nominal STS-1 trajectory, the preflight aerodynamic heating was calculated using the classical aerodynamic theories and assuming several types of flow conditions (turbulent, laminar, and separated). Two typical wing cross sections, wing stations (WS's) 240 and 328, and a midfuselage cross section, fuselage station (FS) 877, were selected for the thermal analysis. The locations of WS 240, WS 328, and FS 877 are shown in figure 2.

DESCRIPTION OF STRUCTURES

The geometry of the wing segment bound by WS 240 and WS 254 is shown in figure 3. Both the upper and lower skins of bay 1 are made of aluminum honeycomb core sandwich plates. The skins of bays 2, 3, and 4 are made of hat-stringer-reinforced aluminum. The spar webs are made of corrugated aluminum plates, except for the forward spar web of bay 1, which is made of aluminum honeycomb core sandwich plates. The entire lower wing skin is covered with HRSI (high temperature reusable surface insulation), with SIP lying between the wing skin and the HRSI. Most of the upper skin of bay 1 is protected by LRSI (low temperature reusable surface insulation), under which lies the SIP layer. A small portion of the upper skin of bay 1 and the upper skins of bays 2, 3, and 4 are covered only with FRSI (felt reusable surface insulation).

The wing segment bound by WS 328 and WS 342.5 (fig. 4) has only three bays. The forward spar web of bay 1 is made of honeycomb core sandwich plate, and the rest corrugated plates. All the lower and the upper skins are hat-stringer stiffened. The lower skin is covered with HRSI, and the upper skin LRSI. No FRSI appears on the upper surface of WS 328.

The geometry of the fuselage cross section FS 877 is shown in figure 5. Both the fuselage belly and the sidewall are made of T-stiffener-reinforced aluminum. The lower and upper skins (except for the leading edge region) are made of hat-stringer-reinforced aluminum. The leading edge region of the glove skin is made of aluminum honeycomb core sandwich structure. A small portion of the bay door inner surface is covered by an RTV (room-temperature-vulcanized) layer, which serves as a heat sink. The fuselage belly and the lower glove are protected with HRSI. The upper glove is covered partly with HRSI (region of the honeycomb core sandwich structure) and partly with LRSI. The lower portion of the sidewall is covered with FRSI, and the upper portion with LRSI. The outside surface of the bay door is protected by FRSI.

THERMAL MODELING

To account for the spanwise heat flow and the effect of the rib trusses at WS 254, the wing segment shown in figure 3 was modeled in three dimensions using a SPAR (structural performance and resizing) finite element thermal analysis computer program (ref. 1) rather than the conventional finite difference method. The entire finite element thermal model for the wing segment, shown in figure 6, has 920 joint locations (JLOC's or nodes). The modeling was limited to the major load-carrying portion (bays 1 to 4) of WS 240. In modeling, all the wing skins and the spar webs were replaced by solid plates having corresponding effective thicknesses. The equivalent wing skins and spar webs, rib cap shear webs, TPS surface coatings, and RTV layers (lying on both sides of the SIP) were then modeled using SPAR K41 elements (4-node, two-dimensional heat conduction elements). The spar caps, rib caps, and rib trusses were modeled using SPAR K21 elements (2-node, one-dimensional heat conduction elements). The TPS was modeled in 10 layers on the lower surface and 3 to 4 layers on the upper surface using SPAR K81 elements (8-node, three-dimensional heat conduction elements) and K61 elements (6-node, three-dimensional heat conduction elements). The K61 elements were used only in the region where the modeled TPS layers changed from four to three layers on the upper surface (see fig. 6). The SIP was modeled by one layer of SPAR K81 elements.

A small one-dimensional thermal model for WS 240 bay 3 was used to determine the effect of the number of modeled TPS layers on the solutions. The lower skin peak temperature difference between 10 and 15 layer modeling was found to be only 0.8 percent; thus, modeling the lower surface TPS in 10 layers was considered quite adequate.

Because of the existence of gaps in the reusable surface insulation (RSI), heat conduction in both HRSI and LRSI was permitted only in the RSI thickness direction. To account for gap heating between the RSI tiles, the thickness of RSI was reduced to 80 percent of its original thickness. Aerodynamic surfaces were modeled using one layer of K41 elements of unit thickness to provide source heat generation. For the external and internal radiant heat energy exchanges, a layer of SPAR R41 elements (4-node thermal radiation exchange elements) was attached to the outer surface of the TPS and the exposed aluminum surfaces. For the spar webs with two sides exposed (lying between bays 1 and 2, 2 and 3, and 3 and 4), one layer of R41 elements was used for each exposed surface. Only one SPAR R41 element was used to represent radiation into space (see fig. 6). The size of the space R41 element was conveniently chosen so that it had unit length in the vertical direction and a width equal to the spanwise width of the wing segment. Because the surface areas were so small, thermal radiation exchanges were ignored at the rib cap shear webs and rib trusses. Outer surfaces of the forward and rear spar webs were totally insulated. Internal convection and external convective cooling (negligible during reentry) were ignored.

Modeling of wing segment bound by WS 328 and WS 342.5 (fig. 4) was quite similar to the previous case. For this wing segment, TPS was modeled in 13 layers on the lower surface and 5 layers on the upper surface. Both the lower and upper TPS thicknesses were reduced to 80 percent of their original thicknesses to account for the gap heating. The three-dimensional thermal model for the wing segment shown in figure 7 has a total of 915 joint locations.

Modeling of FS 877 was two dimensional. Because it is symmetrical, only one-half of the fuselage cross section was modeled. The SPAR thermal model for FS 877 is shown in figure 8. The model has a total of 605 joint locations. The T-stiffener and hat-stringer-reinforced skins were represented by equivalent smooth skins with proper effective thicknesses. The equivalent skins, glove honeycomb core sandwich plate skins, bay door composite skins, longerons, vertical wall between the two lower longerons, torque box, and top centerline beam were modeled using SPAR K21 elements. The glove honeycomb core was modeled using SPAR K41 elements. The bay door honeycomb core was modeled using both SPAR K41 and K31 elements (3-node, two-dimensional heat conduction elements). The TPS was modeled in 10 layers using SPAR K41 elements.

To account for gap heating, the TPS thickness was reduced to 80 percent of its original thickness. Heat conduction in both HRSI and LRSI was allowed only in the RSI thickness direction to account for the existence of gaps between the RSI tiles. Heat conduction in FRSI was two dimensional. Both the SIP and the RTV heat sinks were modeled using SPAR K41 elements. To provide aerodynamic heating, the aerosurface was modeled using SPAR K21 elements for source heat generation. SPAR R21 elements (2 node thermal radiation exchange elements) were attached to the TPS outer surface and the exposed structural surfaces to handle both external and internal radiation heat exchange. Radiation to space was modeled using just one SPAR R21 element of unit length. Radiation exchange inside the torque box was neglected, and the cargo bay was assumed to be empty. Internal convection and external convective cooling (negligible during reentry) were not taken into account.

AERODYNAMIC HEATING

The external heat inputs to the thermal models were computed by a NASA computer program called THEOSKN using the velocity-altitude-angle-of-attack time history of the nominal STS-1 trajectory given in figure 1. The THEOSKN program solves the one-dimensional thin skin heating equation and computes time histories for surface temperature, heat transfer coefficients, heating rates, and skin friction. Real gas properties of air were used in all calculations.

Representative heating rate calculations at WS 240 are shown in figure 9. Three cases of heat inputs were generated: turbulent flow for both the lower and upper surfaces, laminar flow for both the lower and upper surfaces, laminar flow for the lower surface and the upper surface of bay 1, and separated flow for the upper surfaces of bays other than bay 1. The laminar heating rates (dashed lines in fig. 9) were computed by Eckert's reference enthalpy method (ref. 2) and the turbulent heating rates (solid lines in fig. 9) were computed by the theory of van Driest (ref. 3). For both cases, a Reynolds analogy factor of 1.12 was used. The local flow conditions were computed by the oblique shock theory and the Prandtl-Meyer expansion theory. The initial wedge angle was taken to be the angle between the tangent to the wing skin (TPS surface) and the horizontal line passing through the stagnation point, which is parallel to the centerline of the shuttle. The flow distance was taken to be the chordwise distance as measured from the leading edge of the wing. The laminar and turbulent heating rates shown for the upper surface were computed by the same procedure used to calculate the lower surface heating rates, except that the flow expansion was limited in such a way that the local static pressure did not go below three-tenths of the free-stream static pressures. Finally, the separated flow heating rates for the upper surface (broken lines in fig. 9) were the estimated heating rates. For this analysis, the separated flow heating rates were taken to be one-half of the calculated heating rates of the attached laminar flow.

For WS 328, only one case of heat input was generated (fig. 10), namely, laminar flow for the lower surface and the upper surface of bay 1, and separated flow for the upper surfaces of bays 2 and 3. Calculation of the heating rates followed the same procedures used in the calculation of the laminar and separated flow heating rates for WS 240.

Heating rates calculated at FS 877 are shown in figure 11 for five typical locations: the lower centerline (point 1), the glove lower surface (point 2), the upper glove surface (point 3), the sidewall surface (point 4), and the upper centerline (point 5). Two cases of heat inputs were generated for the fuselage thermal model: turbulent flow on the fuselage belly surface, separated flow on the glove upper surface, and laminar/turbulent transition flow on both the sidewall surface and the bay door surface (80 percent TPS thickness); and laminar flow on the fuselage belly and the glove lower surfaces, separated flow on both the glove upper surface and the sidewall surface, and laminar flow on the bay door surface (80 percent TPS thickness).

The heating rates shown by the solid line for the lower centerline (point 1) were computed by the turbulent swept-cylinder theory of Beckwith and Gallagher (ref. 4), and the values shown by the dashed curves were computed by the laminar swept-cylinder theory of Fay and Riddell (ref. 5). The heating rates for the glove lower surface (point 2) were calculated using the method described above; however, the computed values were increased by 20 percent as suggested by results from shuttle wind tunnel tests. The upper glove surface (point 3) was known to be in a low heating separated flow region. Measured results for the similar geometry of the X-15 airplane during a reentry flight (ref. 6) showed the lower surface heating to be about 30 times the upper surface heating. Therefore, the heating rates for the upper glove surface were estimated by taking one-thirtieth of the heating rates calculated for the lower glove surface.

For the sidewall surface (point 4), it was not known if the flow was attached or separated; consequently, two calculations were made. The upper curve represents values calculated assuming attached laminar flow with transition to attached turbulent flow at a Reynolds number of 5×10^5 (at time = 1400 seconds from reentry). These heating rates were calculated using the flat-plate theories of Eckert's reference enthalpy method and van Driest for laminar and turbulent flow, respectively. A flow distance of 0.61 meter (2 feet) was used, and the flow conditions were assumed to be equal to free-stream values. The separated flow heating rates were assumed to be one-tenth of the heating rates of the laminar attached flow.

The heating rates represented by the upper curve for the upper centerline (point 5) were calculated assuming attached laminar flow with transition to turbulent flow at a Reynolds number of 5×10^5 (at time $\approx$ 1000 seconds from reentry). For these calculations the local static pressures were assumed to be equal to one-half of the free-stream static pressures and the flow length was measured from the attachment point on the sidewall of the fuselage. The heating rates shown by the lower curve were computed assuming laminar flow with local pressures equal to one-fourth of the free-stream values. For both cases the flat-plate heating theories of Eckert's reference enthalpy and van Driest were used for laminar and turbulent calculations, respectively.

For more detailed discussion on the calculation of heating rates for the wing and the fuselage, see reference 7.

RADIATION EXCHANGE

For both the external and the internal thermal radiation exchanges, all the view factors were calculated from the equation (ref. 8)

$$A_i F_{ij} = A_j F_{ji} \tag{1}$$

where A_i is surface area of radiation exchange element i and F_{ji} is view factor, defined as the fraction of radiant heat leaving element j incident on element i. In the calculation of view factors for the external radiation exchanges (considering that element i represents the space element and element j any radiation exchange element on the wing or fuselage surface), F_{ji} was taken to be unity; therefore, $F_{ij} = \frac{A_j}{A_i}$ according to equation (1). In the view factor calculations for the fuselage internal radiation exchanges, each radiation exchange element was set to receive not only from the other elements but also from mirror images of all elements. In other words, the entire fuselage cross section was used to compute the fuselage internal radiation view factors. Values of emissivity and reflectivity used to compute radiant heat fluxes were:

Surface	Emissivity	Reflectivity
Windward	0.85	0.15
Leeward	0.80	0.20
Internal Structure	0.667	0.333
Space	1.0	0

The initial temperature distribution used in the analysis was obtained from the shuttle manufacturer. The effect of neglecting internal radiation was investigated using a small, three-dimensional thermal model for WS 240 bay 3, assuming total insulation from neighboring bays.

TRANSIENT THERMAL SOLUTIONS

The SPAR thermal analysis finite element computer program was used in the calculation of temperature time histories at all joint locations of the thermal models. The SPAR program used the approach described below to obtain transient thermal solutions.

The transient heat transfer matrix equation

$$(K_k + K_r)T + C\dot{T} = Q + R \tag{2}$$

where

K_k = conduction matrix
K_r = radiation matrix
T = temperature
C = capacitance matrix
$[\dot{\ }]$ = time derivative
Q = source load vector
R = radiation load vector

was integrated by assuming that the temperature vector T_{i+1} at time step t_{i+1} can be expressed as

$$T_{i+1} = T_i + \dot{T}_i \Delta t + \frac{1}{2!}\ddot{T}_i \Delta t^2 + \frac{1}{3!}\dddot{T}_i \Delta t^3 + \cdots \tag{3}$$

where T_i is the temperature vector at time step t_i, and Δt is the time increment. The vector $\dot{T}$ is determined directly from equation (2); i.e.,

$$\dot{T} = -C^{-1}(K_k + K_r)T + C^{-1}(Q + R) \tag{4}$$

Higher order derivatives are obtained by differentiating equation (2) according to the assumption that (1) material properties are constant, (2) Q varies linearly with time, and (3) R is constant over Δt:

$$\ddot{T} = -C^{-1}(K_k + 4K_r)\dot{T} + C^{-1}\dot{Q} \tag{5}$$

$$\dddot{T} = -C^{-1}(K_k + 4K_r)\ddot{T} + 4\dot{K}_r\dot{T} \tag{6}$$

etc.

In the present computations, the Taylor series expansion (eq. (3)) was cut off after the third term. The pressure dependency of the TPS and SIP properties was converted into time dependency based on the nominal trajectory given in figure 1.

Time dependent properties were then averaged over "time intervals," which were taken to be 2 seconds. Temperature dependent properties were evaluated at the temperatures computed at the beginning of each time interval. Q, $\dot{Q}$, and R were computed every 2 seconds.

RESULTS AND DISCUSSION

TPS Surface Temperatures

Computed TPS surface temperature time histories at eight typical locations (four locations on the lower surface and four locations on the upper surface) at WS 240 are shown in figure 12. For the lower surface, the peak heating for all four bays took place at about 900 seconds from the reentry for both turbulent and laminar flow. Heating at bay 1 was most severe, with peak temperatures reaching 1243° C (2269° F) for turbulent flow and 899° C (1650° F) for laminar flow. For the upper surface, the maximum heating for all four bays occurred at 1200 seconds for turbulent flow and at 1000 seconds for both laminar and separated flow. Again, the upper surface of bay 1 experienced approximately 310° C (590° F) for both turbulent and laminar flow.

The calculated TPS surface temperature time histories for WS 328 are shown in figure 13. The temperatures for the lower and the upper surfaces reached their peak values at t = 900 seconds and t = 1000 seconds respectively. Bay 1 was heated most severely and the peak temperatures there reached 873° C (1604° F) for the lower surface and 368° C (695° F) for the upper surface.

Calculated TPS surface temperature time histories at five locations at FS 877 are shown in figure 14. Except for the fuselage sidewall (point 5), the curves for all locations (points 1 to 4) have plateaus lying between 400 seconds and 1000 seconds. Heating was most severe at the lower glove surface (point 2) where the peak temperature was 919° C (1687° F) for turbulent flow and 707° C (1304° F) for laminar flow.

Structural Temperatures

Predicted aluminum skin temperature time histories at eight locations at WS 240 are shown in figure 15 for the three heating cases mentioned earlier. The STS-1 flight data are also shown for comparison. Flight data for the time interval 0 to 1178 seconds were not available because of the telemetering "blackout." Only data for the bay 3 lower skin were available, which were obtained from on-board instrumentation. For the entire lower skin and the bay 1 upper skin, the flight data correlate better with the laminar flow curves for 80 percent TPS thickness than with the laminar flow curves for 100 percent TPS thickness and turbulent flow curves, reflecting the effect of TPS gap heating and the fact that the actual flow was laminar. The measured results for the upper skins of bays 2 and 4 compare better with the separated flow curves (100 percent FRSI thickness, without gap heating) than with the laminar and turbulent flow curves, showing flow separation in the actual flight. No acceptable flight data were obtained for the upper skin of bay 3. The increasing deviation of the flight data for the lower skins from the laminar flow curves for 80 percent TPS thickness may be due to internal convection and to external convective cooling, which were neglected in the analysis. The continuous increase of the measured temperatures for the upper skins of bays 2 and 4, starting at 100 seconds before touchdown, may also be due to internal convective heating from the lower skin.

Temperature time histories of the lower and upper spar caps at WS 240 and a rib cap at WS 254 are shown in figure 16. The flight data for the lower spar and rib caps of bay 3 closely follow the laminar flow curves for 80 percent TPS thickness up to 1800 seconds; beyond that point, like the lower skin data, they deviate from the curves. The measured temperatures for the rear lower spar cap (top right plot) are higher than the laminar flow curve for 80 percent TPS thickness. The lower predicted values may be due to the assumption of total insulation on the aft side of the rear spar web. Like the upper skin data, the flight data for the upper spar and rib caps correlate better with the separated flow curves (100 percent TPS thickness) than with the laminar and the turbulent flow curves. Again, the data show continuous heating after touchdown, suggesting possible convective heat transfer from the lower skin.

The chordwise temperature distributions of the aluminum skins at WS 240 are shown in figure 17 for profile time t = 1600 seconds. The "scalloped" shape reflects the drop in temperature at the spar caps. For the lower surface, the measured temperatures follow the laminar flow curve for 80 percent TPS thickness quite well except at the rear spar (see also top right plot, fig. 16). For the upper surface, the plot shows laminar flow on bay 1 and the effect of separated flow on bays 2, 3, and 4.

The time histories of the predicted and the flight measured WS 328 skin temperatures are shown in figure 18. The flight data were available for only three locations (i.e., lower skin at bays 1 and 3, upper skin at bay 2) and only for the time after t = 1178 seconds. For the lower skin, the flight data agree fairly well with the laminar flow curves (80 percent TPS thickness) after touchdown. However, before the touchdown the flight data show higher temperatures than predicted. The cause of difference for bay 3 may be that part of the flight profile might be turbulent or transitional flow, and the cause of difference for bay 1 may be due to the higher initial temperature in the actual flight. Unlike bay 3 data, which show continued cooling after landing, the bay 1 data show no cooling at all. This is due to heating from the upper skin, which is hotter than the lower skin (see fig. 19). For the upper skin (fig. 18), the flight data for bay 2 agree very nicely with the separated flow curve (80 percent TPS thickness) up to t = 1600 seconds. After that the measured temperature continues to increase, whereas the calculated temperature increases at a very low rate, reflecting possible convective heat transfer from the hotter lower skin. In figure 18, the laminar flow curves for 100 percent TPS thickness are also plotted to show the effect of gap heating. The chordwise distribution of WS 328 skin temperatures is shown in figure 19 for the profile time of 1600 seconds. For the lower surface, the measured temperatures are slightly higher than the calculated curve. For the upper surface the flight data point falls right on the calculated curve (80 percent TPS thickness).

The temperature time histories of the fuselage skins at five locations are shown in figure 20. The measured belly temperatures agreed with the laminar flow curve for 80 percent TPS thickness until about 1800 seconds; then the data show rapid cooling, which suggests possible convective heat transfer effect of the air mass inside the cargo bay and the effect of external convective cooling, which were neglected in the calculations. The lower glove measured value was far below the laminar flow curve for 80 percent TPS thickness, which should be the theoretical lower bound for the attached flow. It is possible that the data are incorrect or that the flow conditions in this area are substantially different from what one would assume. The predicted temperatures for the sidewall and the upper longeron are rather poor because of uncertainty in the flow field from which the heat inputs were calculated (e.g., location of flow re-attachment point, etc.). No acceptable data were obtained for the upper glove skin temperatures.

The circumferential distribution of the fuselage structural temperatures is shown in figure 21 for profile time 1600 seconds. The "valleys" of the curves indicate the temperature drop at the heat sinks (structural junction points). The cargo bay outer skin was heated more than the aluminum skin even though the heat input there was relatively low, indicating poor conductivities of the bay door materials.

Effect of Internal Radiation

The effect of internal radiation was investigated for bay 3 of WS 240 assuming total insulation from the neighboring bays. The results shown in figure 22 are for Mission 3 heat input. With the effect of internal radiation taken into account, the temperatures of the lower and the upper skins were brought closer (especially after landing), and the peak temperature of the lower skin was reduced by approximately 39° C (70° F).

COMPARISON WITH FINITE DIFFERENCE SOLUTIONS

The conventional finite difference method was also used in the thermal analysis in order to verify the solutions obtained from the SPAR finite element method. For this purpose, two-dimensional finite element and finite difference thermal models were set up for WS 240. The two models are shown in figure 23. For the SPAR finite element model, the lower and the upper TPS were modeled in 10 and 3 layers respectively. For the finite difference model, the divided TPS sublayers changed in number stepwise along the chordwise direction. For the lower TPS, the number of sublayers changed between 13 and 8, while for the upper TPS, the number of sublayers changed between 4 and 1.

The time histories of the WS 240 aluminum skin temperatures predicted by the two methods are given in figure 24. The two methods predicted fairly close temperatures except for bay 1. For the present modeling, the SPAR finite element method gives slightly higher lower skin temperatures and lower slopes for the upper skin temperature curves. The chordwise distributions of aluminum skin temperatures at t = 1600 seconds predicted by the two methods are shown in figure 25. The two types of solutions agree fairly well except for bay 1. The relatively large differences between the two solutions at bay 1 may be attributed to the difference in TPS modelings in the two models. By making the two thermal models as identical as possible, the two types of solutions should converge.

For those who are familiar with the finite element structural analysis, SPAR is attractive because the finite element thermal modeling and the finite element structural modeling are quite similar.

CONCLUSIONS

A finite element thermal analysis computer program was used in the reentry heat transfer analysis of the space shuttle. Thermal models were set up for two typical wing cross sections and for a midfuselage cross section.

The comparison between calculated and measured temperatures for the wing was quite good if laminar flow was assumed for the lower surface and the bay 1 upper surface and if separated flow was assumed for the upper surfaces of bays other than bay 1. The differences that did exist could be caused by the following assumptions made in the thermal modeling: (1) the use of effective thicknesses for the TPS, stiffened skins, corrugated spar webs, and honeycomb core; (2) no internal convection and no external convective cooling; (3) initial temperatures and emissivities; and (4) total insulation on the outboard sides of WS 240 and WS 328, on the inboard sides of WS 254 and WS 242.5, and on the outer surfaces of both forward and rear spar webs.

For the fuselage, good agreement between the calculated and measured temperatures was obtained for the lower surface if laminar flow was assumed. Why the calculated and measured temperatures for the lower glove differed so greatly is not known. It is possible that the flight data were inaccurate or that the flow conditions in this area are substantially different from what one would assume. The difference between the prediction and the flight data for the upper fuselage surface is probably due to the very complex flow in this region— exactly where the flow is attached and/or separated is not known. Other causes of data/prediction discrepancies could be the following assumptions made in the thermal modeling of the fuselage: (1) the use of effective thicknesses for the TPS and stiffened skins; (2) no internal convection and no external convective cooling; and (3) initial temperatures, emissivities, and two-dimensional nature of the thermal model.

REFERENCES

1. Marlowe, M. B.; Moore, R. A.; and Whetstone, W. D.: SPAR Thermal Analysis Processors Manual, System Level 16. Volume 1: Program Execution. NASA CR-159162, Oct. 1979.

2. Eckert, E. R. G.: Survey of Boundary Layer Heat Transfer at High Velocities and High Temperatures. WADC-TR-59-624, Wright Air Development Center, Wright-Patterson AFB, Ohio, Apr. 1960.

3. van Driest, E. R.: The Problem of Aerodynamic Heating. Aeronaut. Engr. Rev., vol. 15, no. 10, Oct. 1956, pp. 26-41.

4. Beckwith, I. E.; and Gallagher, J. J.: Local Heat Transfer and Recovery Temperatures on a Yawed Cylinder at a Mach Number of 4.15 and High Reynolds Numbers. NASA TR R-104, 1961.

5. Fay, J. A.; and Riddell, F. R.: Theory of Stagnation Point Heat Transfer in Disassociated Air. J. Aeronaut. Sci., vol. 25, no. 2, Feb. 1958, pp. 73-85, 121.

6. Quinn, R. D.; and Olinger, F. V.: Heat-Transfer Measurements Obtained on the X-15 Airplane Including Correlations With Wind-Tunnel Results. NASA TM X-1705, 1969.

7. Gong, L.; Quinn, R. D.; and Ko, W. L.: Reentry Heating Analysis of Space Shuttle with Comparison of Flight Data. Computational Aspects of Heat Transfer in Structures, NASA CP-2216, 1982. (Paper no. 17 of this compilation.)

8. Sparrow, E. M.; and Cess, R. D.: Radiation Heat Transfer. McGraw-Hill Book Co., Inc., New York, 1978.

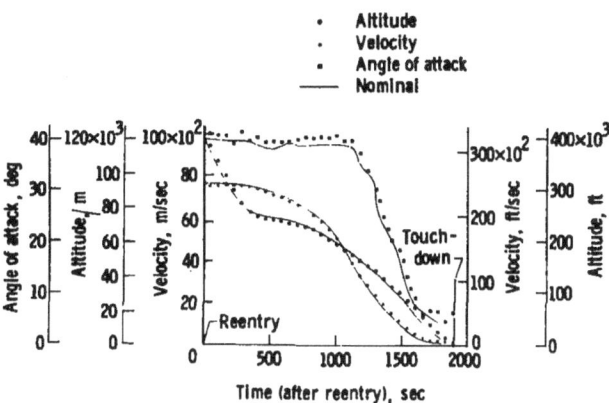

Figure 1. Nominal versus actual trajectory time history.

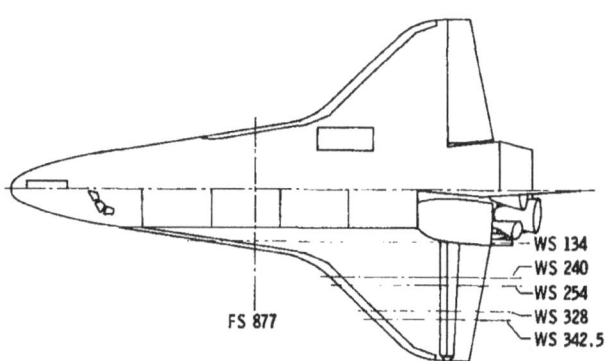

Figure 2. Locations of wing stations and fuselage stations.

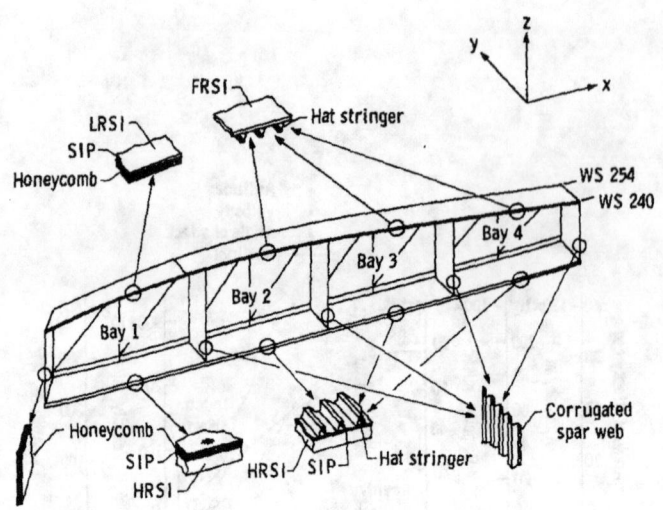

Figure 3. Geometry of wing segment between WS 240 and WS 254.

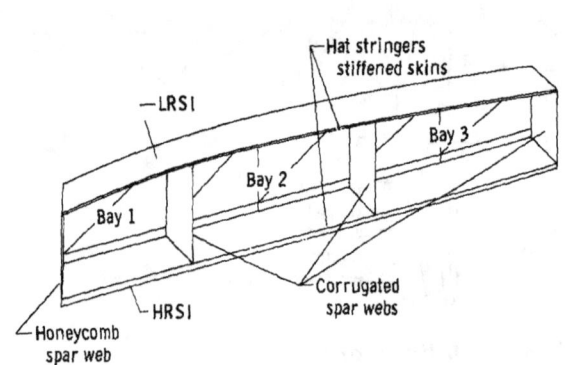

Figure 4. Geometry of wing segment between WS 328 and WS 342.5.

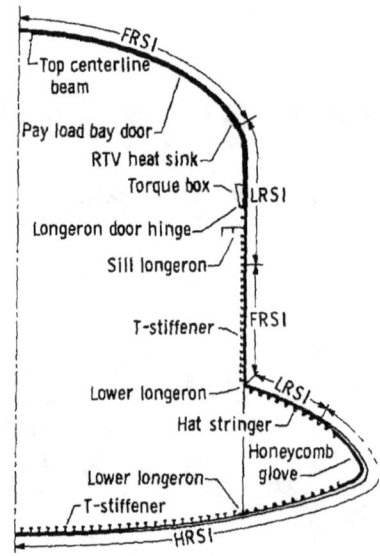

Figure 5. Geometry of fuselage cross section at FS 877.

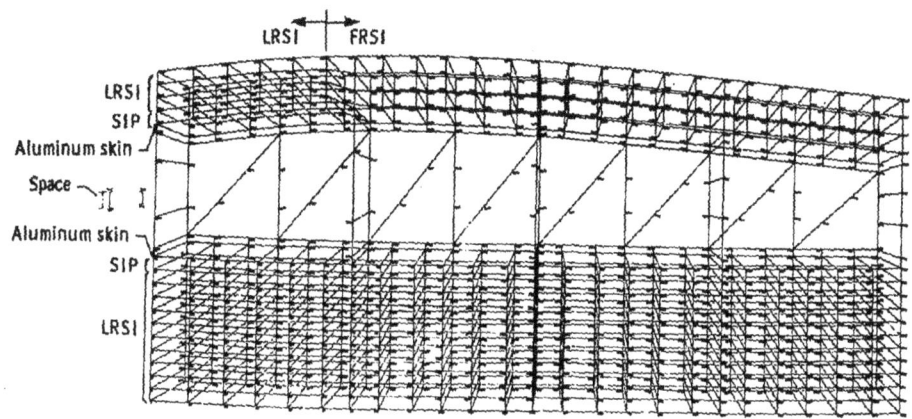

Figure 6. SPAR finite element thermal model for WS 240 (three-dimensional).

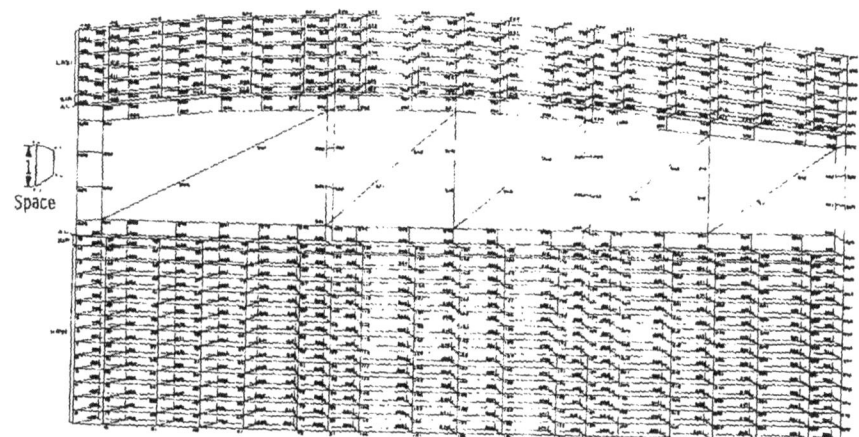

Figure 7. SPAR finite element thermal model for WS 328 (three-dimensional).

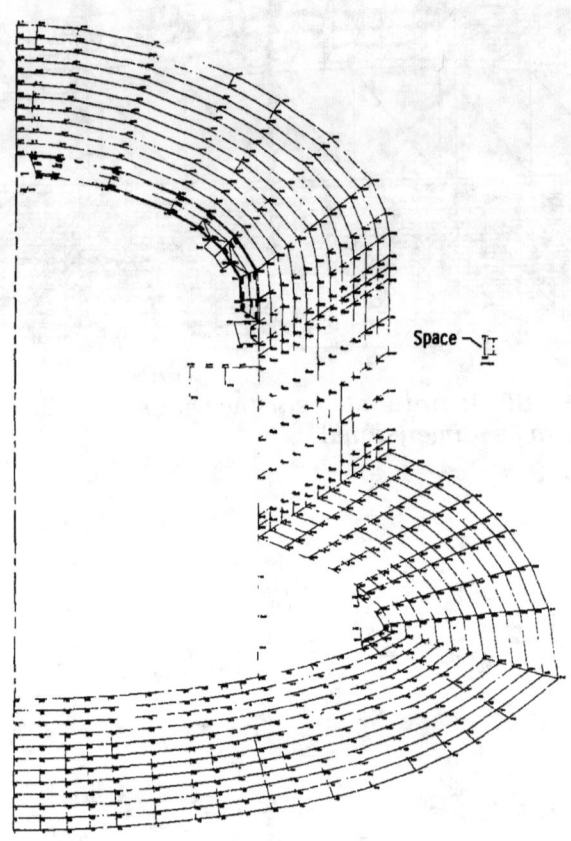

Figure 8. SPAR finite element thermal model for FS 877 (two-dimensional).

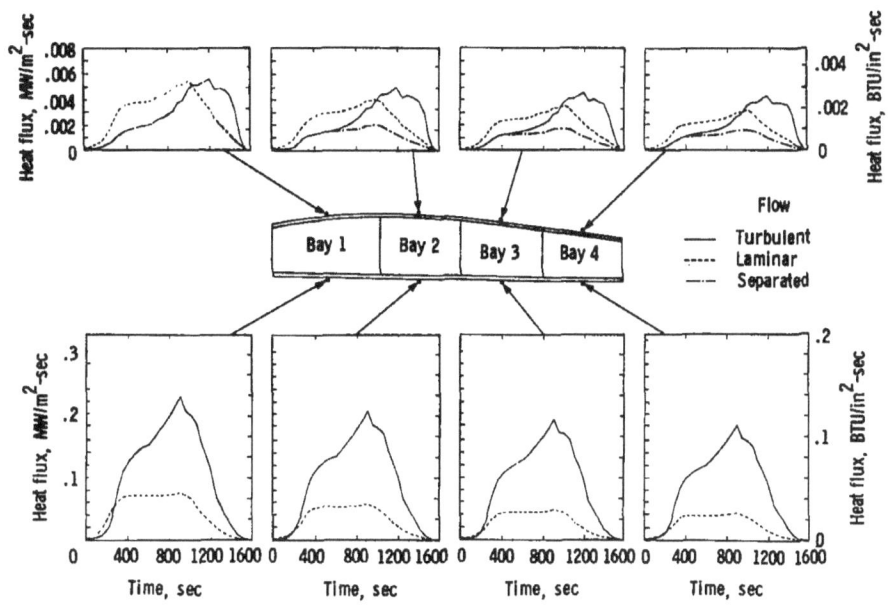

Figure 9. Surface heating rates at WS 240. STS-1 flight.

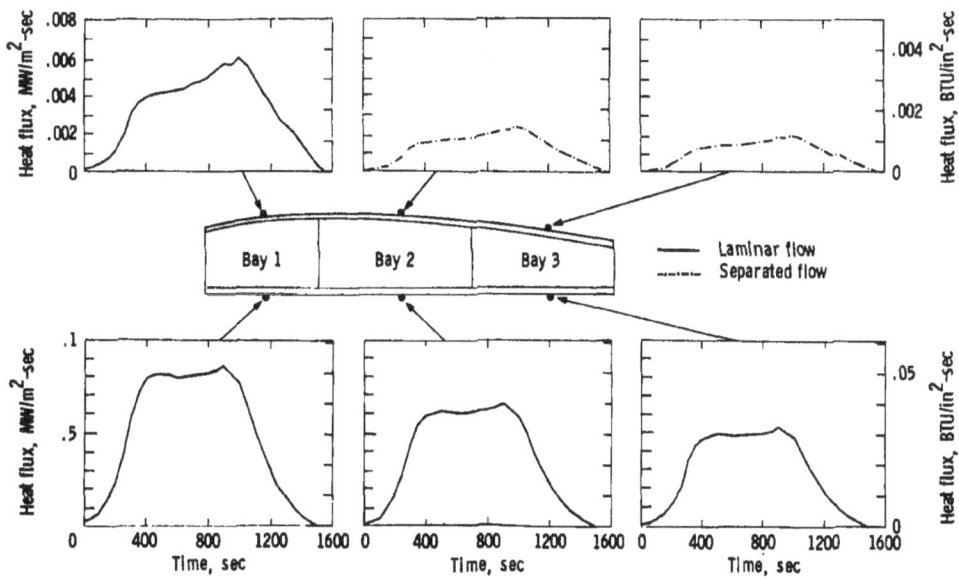

Figure 10. Surface heating rates at WS 328. STS-1 flight.

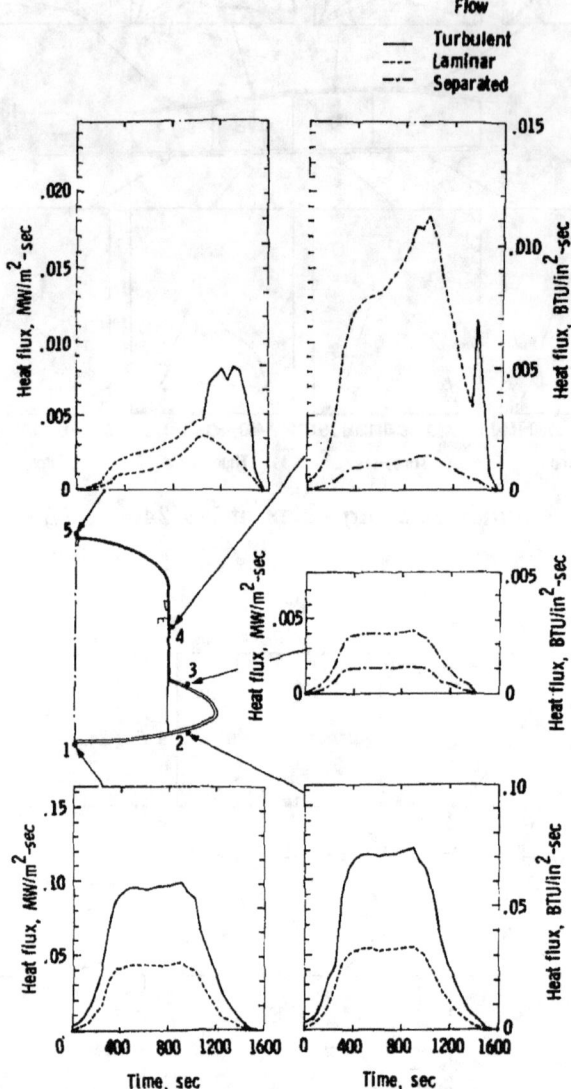

Figure 11. Surface heating rates at FS 877. STS-1 flight.

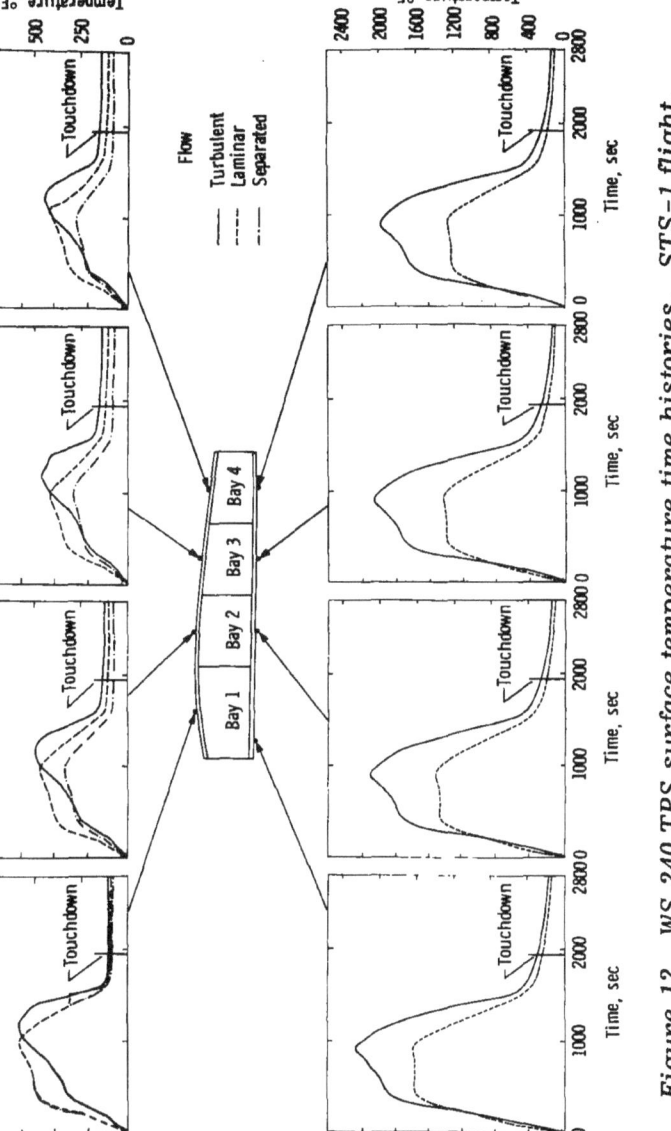

Figure 12. WS 240 TPS surface temperature time histories. STS-1 flight.

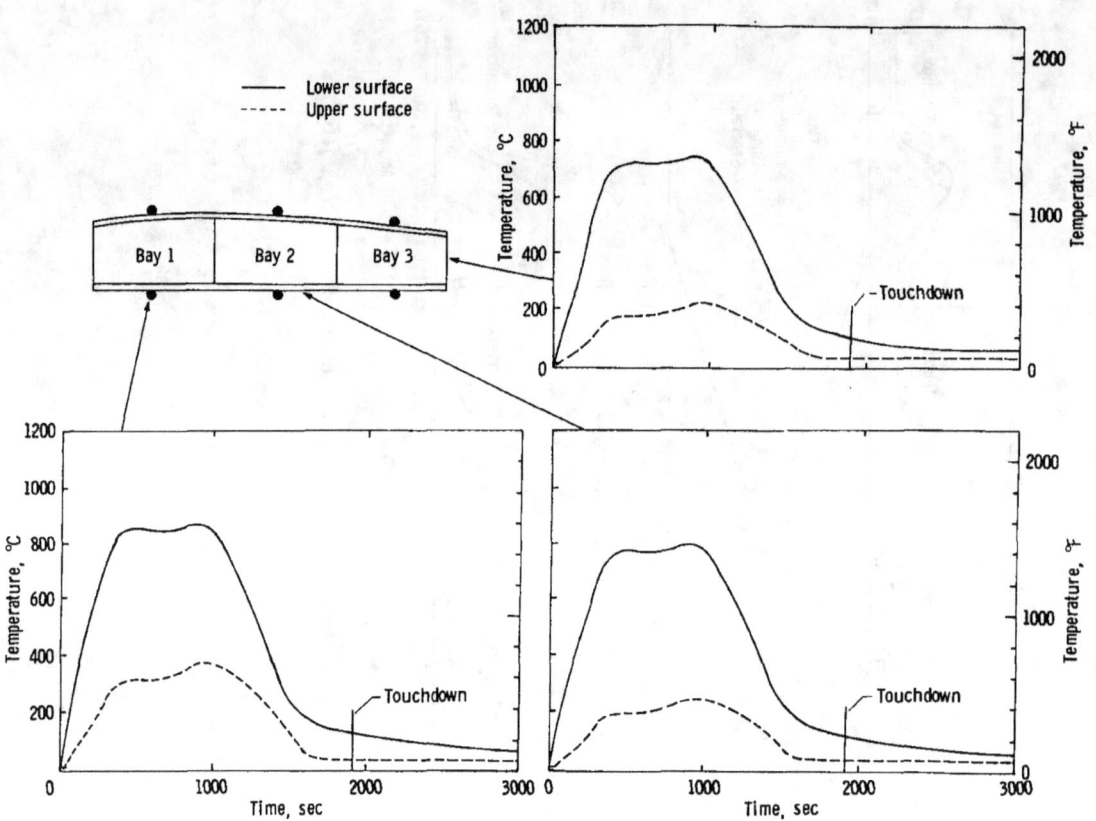

Figure 13. WS 328 TPS surface temperature time histories. STS-1 flight.

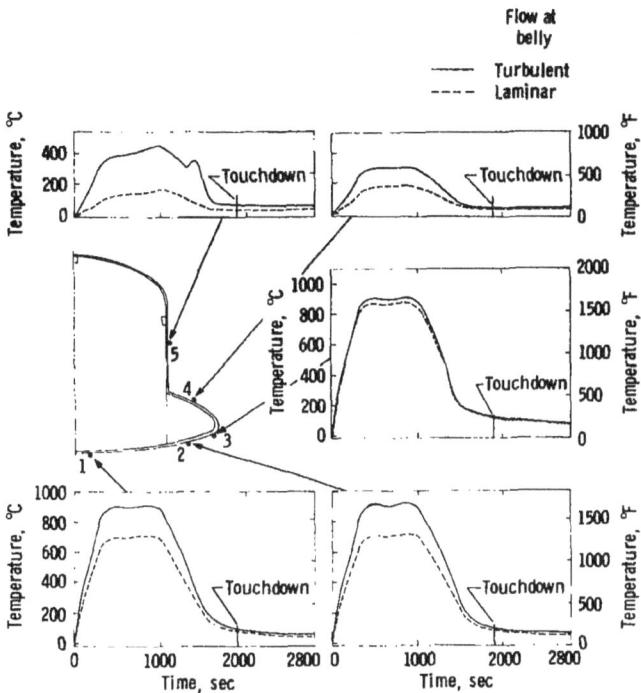

Figure 14. FS 877 TPS surface temperature time histories. STS-1 flight.

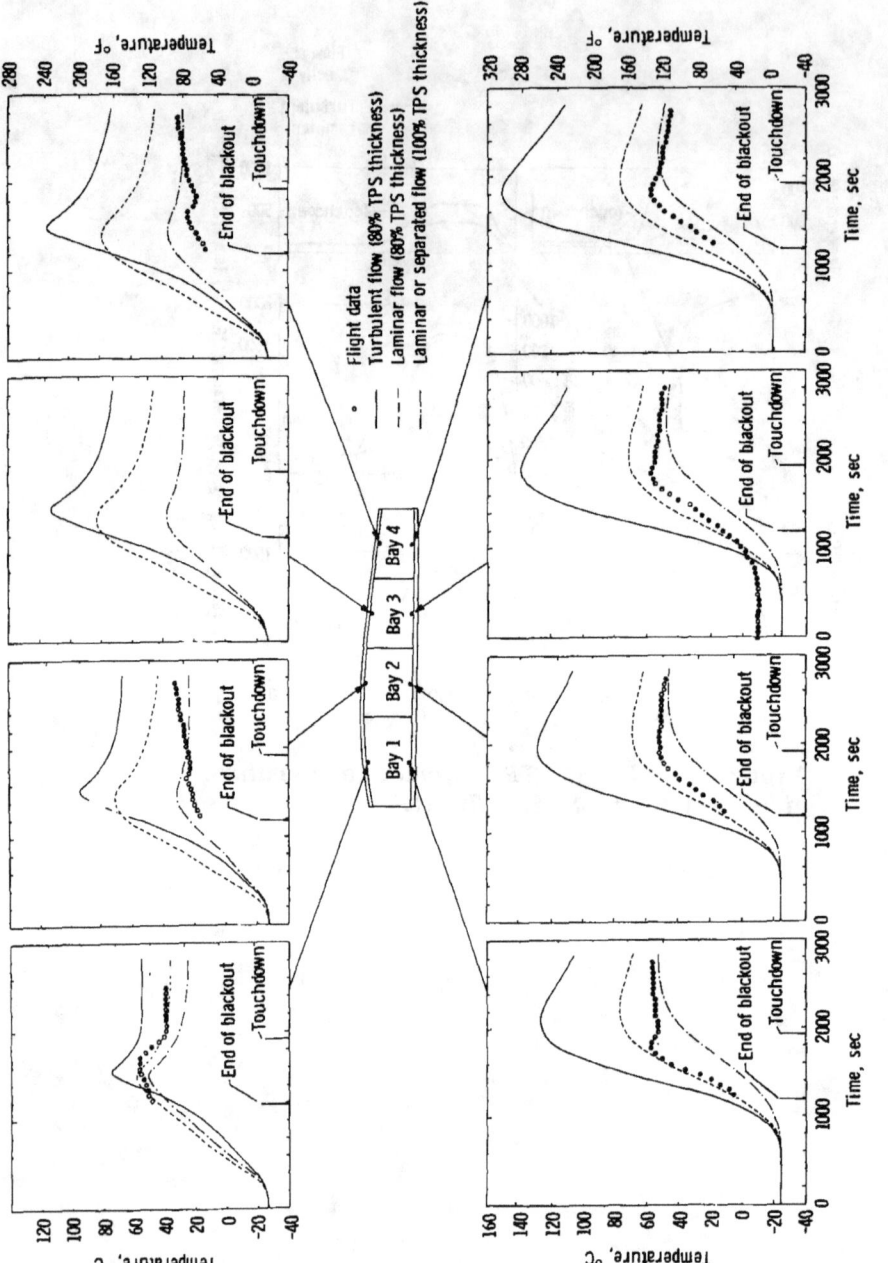

Figure 15. WS 240 aluminum skin temperature time histories. STS-1 flight.

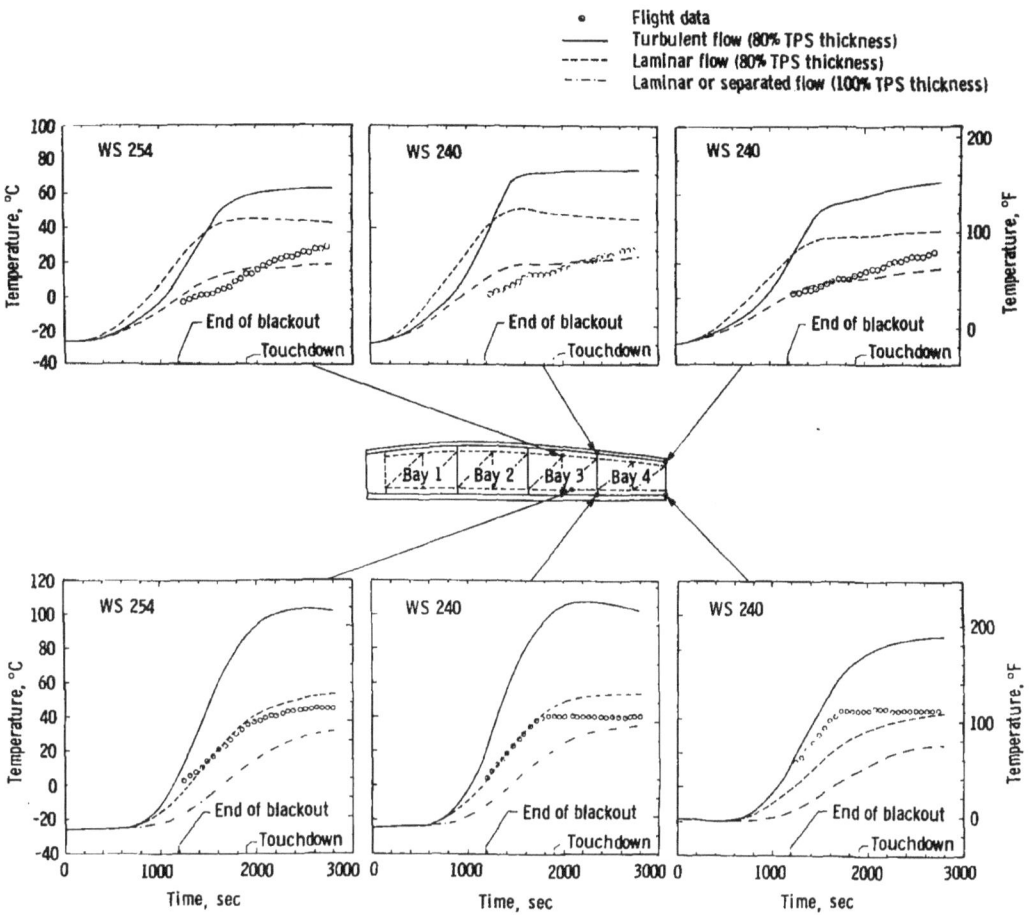

Figure 16. SPAR and rib caps temperature time histories. STS-1 flight.

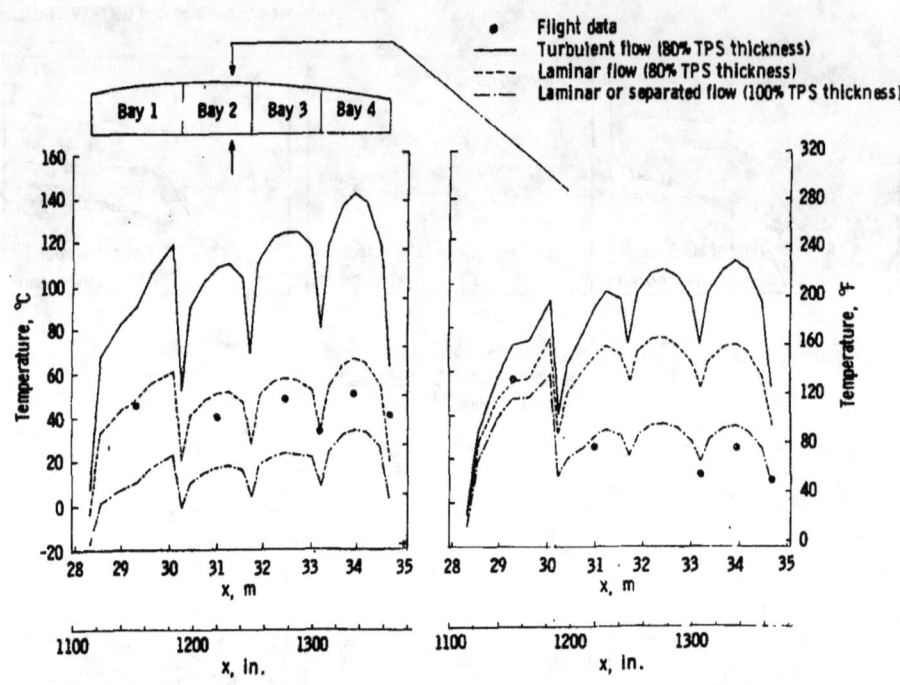

Figure 17. WS 240 chordwise distribution of the aluminum skin temperatures based on different surface heatings.
Time = 1600 seconds.

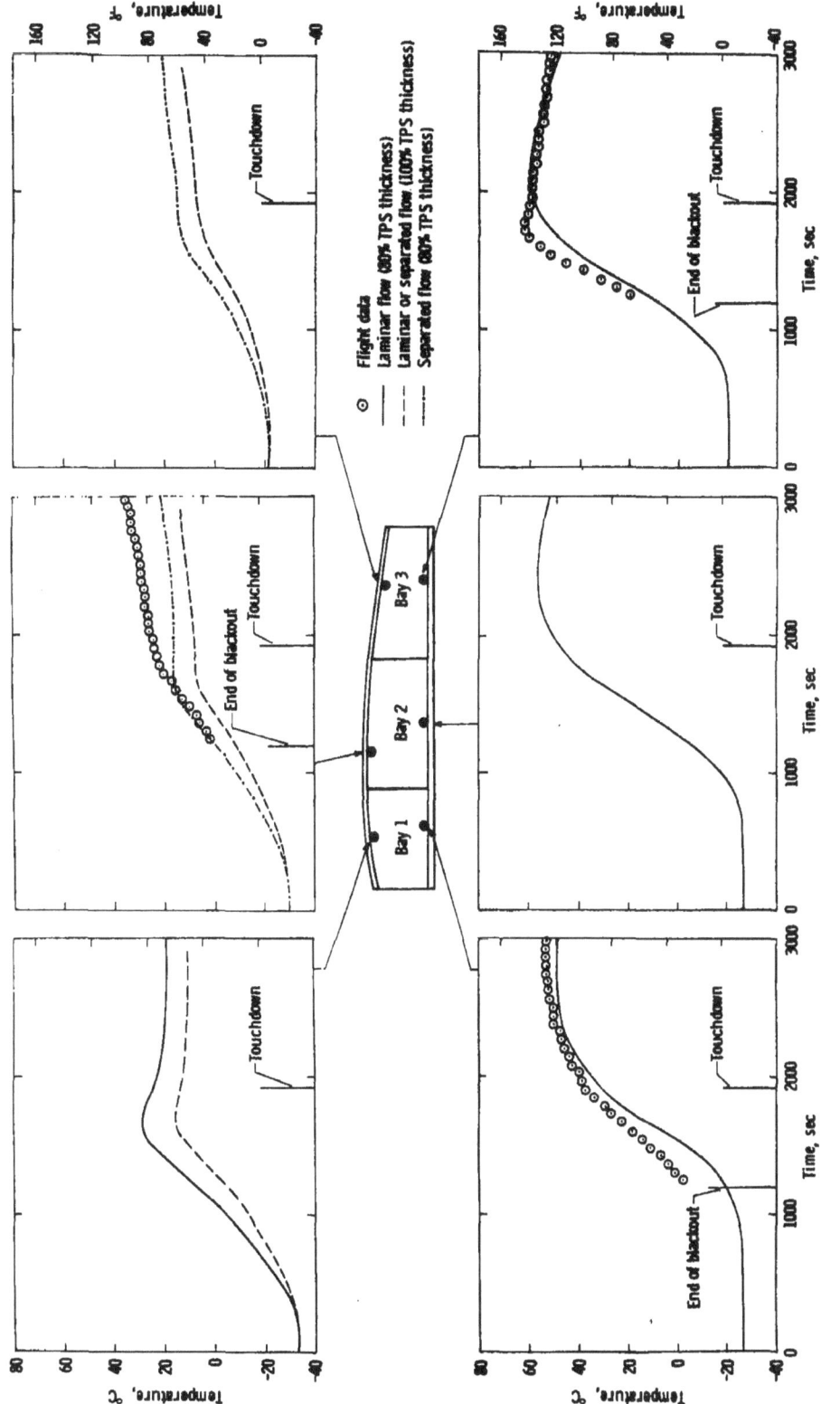

Figure 18. WS 328 aluminum skin temperature time histories. STS-1 flight.

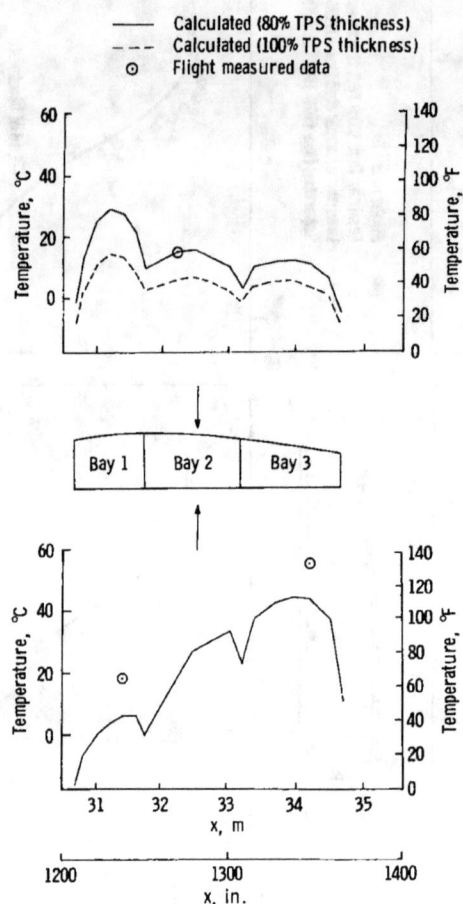

Figure 19. WS 328 chordwise distribution of the aluminum skin temperatures. STS-1 flight. Time = 1600 seconds.

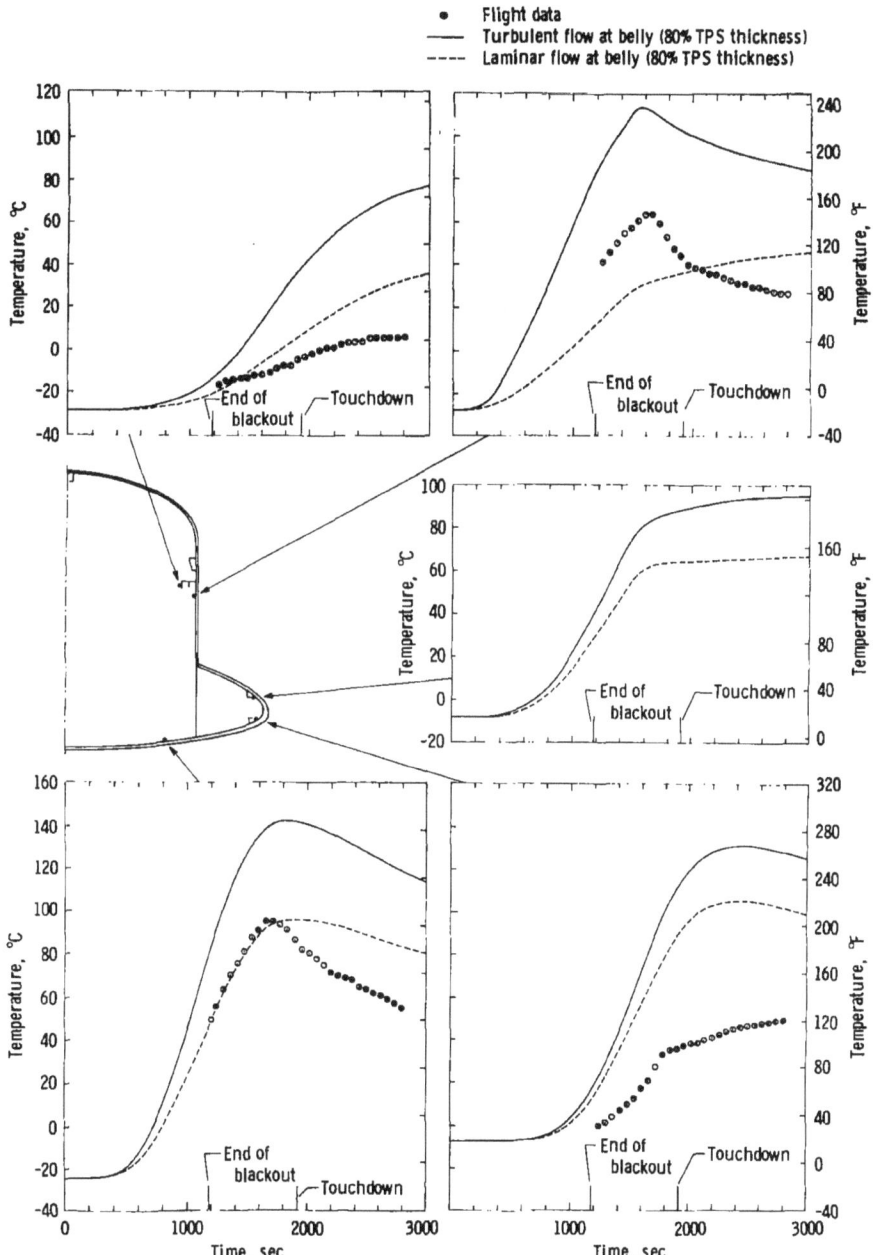

Figure 20. FS 877 aluminum skins temperature time histories. STS-1 flight.

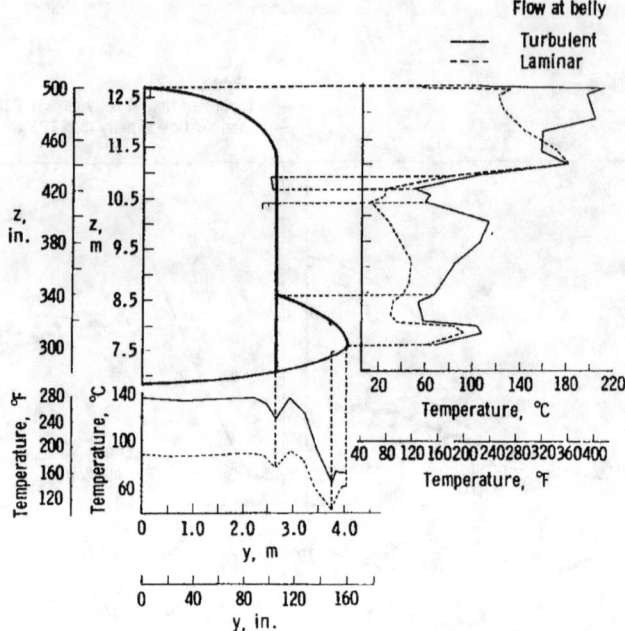

Figure 21. FS 877 circumferential distribution of the structural temperature. STS-1 flight. Time = 1600 seconds.

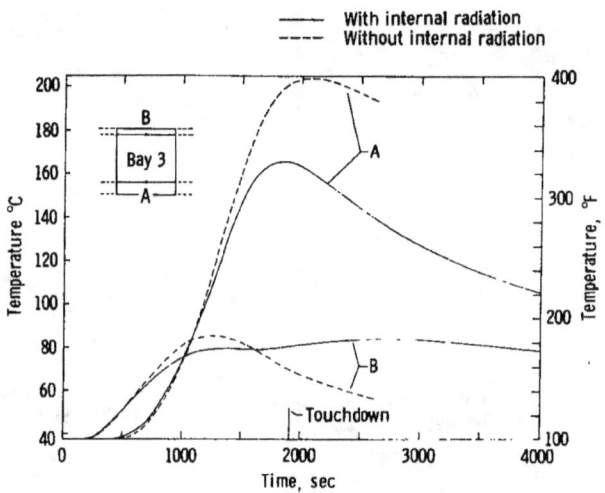

Figure 22. Effect of internal radiation on WS 240 bay 3 aluminum skin temperatures. Mission 3 heating.

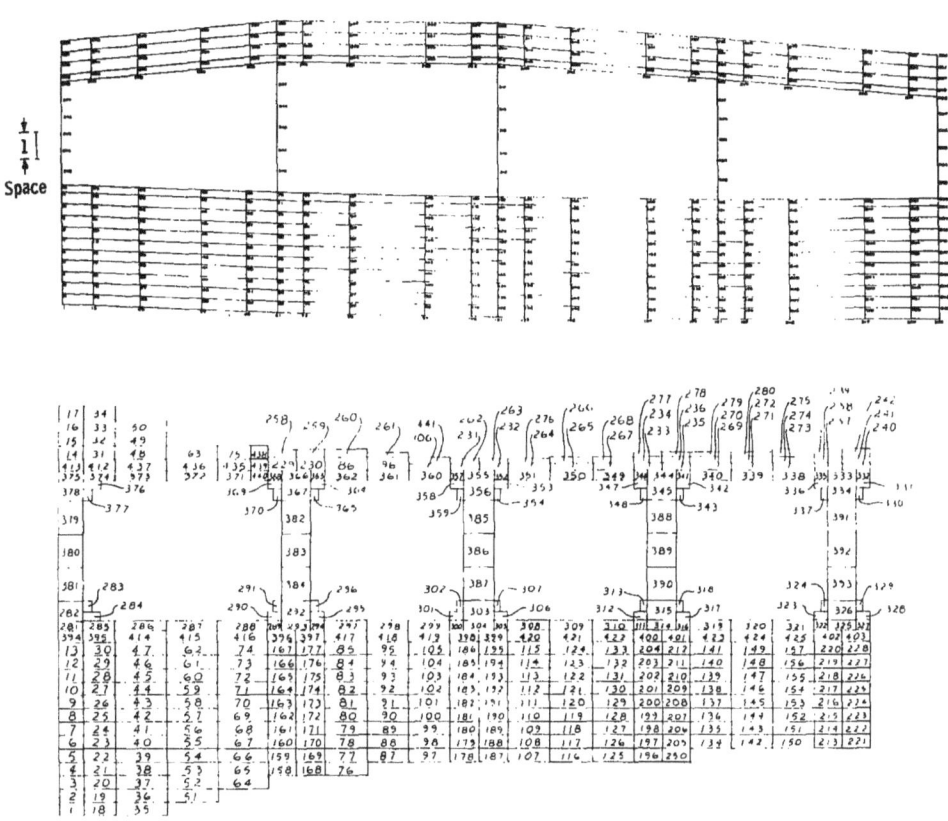

Figure 23. SPAR finite element thermal model and finite difference model for WS 240 (two-dimensional).

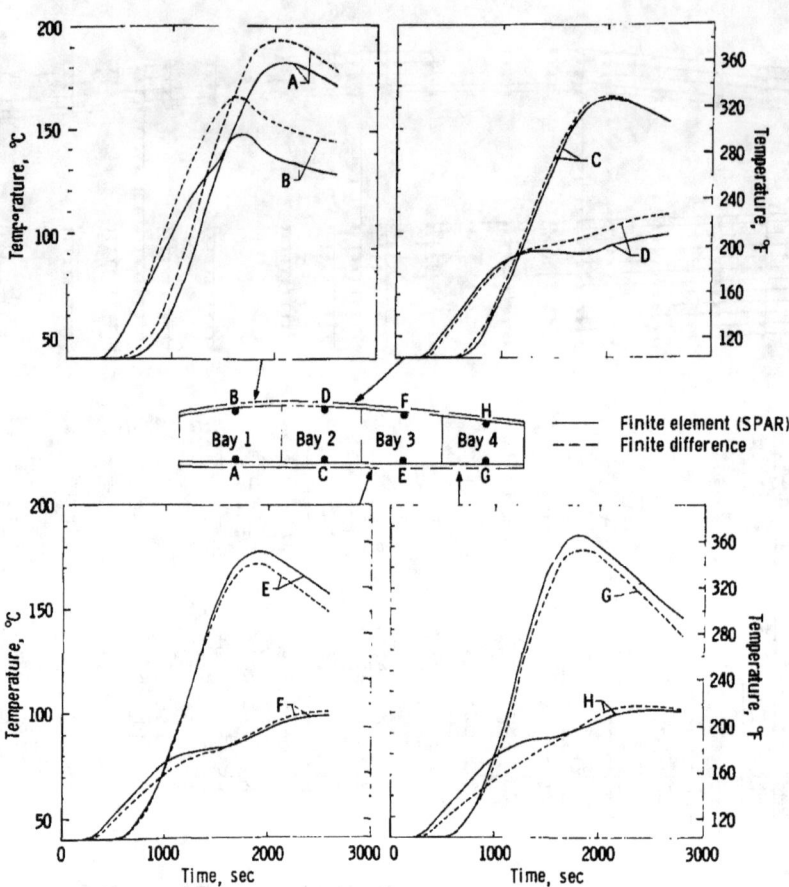

Figure 24. WS 240 aluminum skin temperature time histories predicted by finite element and finite difference methods. Mission 3 heating.

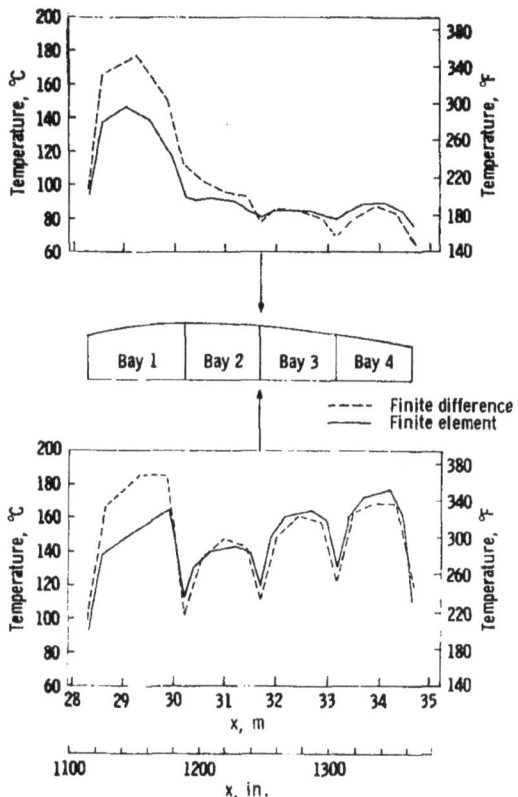

Figure 25. WS 240 chordwise aluminum skin temperature distributions predicted by finite element and finite difference methods. Time = 1600 seconds. Mission 3 heating.

SPACE SHUTTLE ORBITER ENTRY
HEATING AND TPS RESPONSE: STS-1
PREDICTIONS AND FLIGHT DATA

Robert C. Ried, Winston D. Goodrich,
Chien P. Li, Carl D. Scott, Stephen M. Derry,
and Robert J. Maraia
NASA Lyndon B. Johnson Space Center
Houston, Texas

SUMMARY

The first Orbital flight test of the Space Transportation System, STS-1, was a highly successful demonstration of the technology associated with reusable manned spacecraft. In particular, this paper addresses aerothermodynamic development flight test data, transmitted after entry blackout, which confirm engineering predictions of boundary layer transition, numerical simulations of the Orbiter flow field and tend to substantiate preflight predictions of surface catalysis phenomena. The thermal response of the thermal protection system was as expected. The only exception is that internal free convection was found to be significant in limiting the peak temperature of the structure in areas which do not have internal insulation.

INTRODUCTION

The Space Shuttle Orbiter is the first reusable entry spacecraft built on a foundation of technology and experience gained from the Apollo, Gemini, and Mercury programs (fig. 1). One of the most critical elements to the development of this capabiltiy is the reusable thermal protection system (TPS) mounted on the aluminum structure. The TPS has to be reusable to reduce operational costs and of low mass to achieve necessary and desirable vehicle performance. Experience indicated that the untaxed potential and significant unnecessary mass in a TPS can be attributed to limited understanding and the associated compounded conservatism due to uncertainties in trajectory, environment, system properties, system performance, and system requirements. This program incurred the risk associated with the development of the first reusable TPS and simultaneously the risk associated with significant reductions in conservatism, e.g., the entry design environment was based on the use of nominal heating obtained from the state-of-the-art methodology. The first risk was overcome by a substantial investment in the TPS development while the second risk was treated by the development of understanding for preflight confidence, e.g., aerothermodynamic technology, the subject of this paper. The second risk was also tempered by selecting initial flight trajectories which are not quite as severe as the design entry trajectory. The products of this approach to the Orbiter design and development are an efficient TPS and an aluminum structure which experiences signficant thermal strain and stress as a result of the entry heating. The thermal heat load to the structure is a minute fraction of the aerodynamic heating to the TPS, which is in turn a small fraction of the energy dissipated by atmospheric braking.

In this paper, representative aerothermodynamic and TPS thermal response data obtained on the first atmospheric entry test flight (STS-1) of the Space Shuttle Orbiter are compared to preflight predictions. These predictions are based on rather sophisticated computational and experimental investigations which have complemented the design and development activities as "benchmark" information. Although much of this information was utilized in the design, development and preflight assessment process, the predictions presented here are not the design values nor have they been extended over all regions of the vehicle as required for the design heating rates and TPS response characteristics (refs. 1 & 2). The prediction methodology presented here has also been used to evaluate the sensitivity of preflight predictions to uncertainties in independent parameters and establish system performance uncertainties (ref. 3).

ENTRY HEATING PREDICTIONS

Two features of the Shuttle present particular challenges to the aerothermodynamicist: the temperature limits of reusable TPS materials and the complex geometry of the Orbiter vehicle. The combination of these features in particular pushed the state-of-the-art beyond previous experience even though the definition of the aerothermodynamic environment associated with entry from low earth orbit has been addressed for about thirty years.

Design Approach

In the beginning of the Shuttle program considerable debate ensued as to the most appropriate aerothermodynamic methodology to use for defining the entry heating. The practical state-of-the-art in flow field modeling was limited to two-dimensional flows used in conjunction with ground test facilities which were not capable of simulating all of the significant parameters. NASA/JSC and the prime contractor, Rockwell International (RI) agreed to place a heavy reliance on hypersonic wind tunnel testing of geometrically scaled models to simulate the three-dimensional features of the flow dynamics while using appropriate two-dimensional flow models calibrated by wind tunnel data to simulate flight-related high velocity, real gas phenomena. This methodology was subsequently applied to flight with equilibrium air thermodynamic and transport properties. This approach is schematically illustrated in figure 2 and is the foundation for the design heating methodology.

Orbiter Flow Field Simulations

It is obvious that the flow of air around the Orbiter during entry is three-dimensional and therefore the use of two-dimensional flow models calibrated with wind tunnel data is questionable when extrapolated for use at flight conditions. As such, complementary computational fluid mechanics activities at NASA/ARC and JSC were applied to the development of Orbiter Flow Field Simulations (OFFS) to obtain more reliable techniques for extrapolating wind tunnel data to flight. This rather extensive effort is documented in references 4 through 10. The wind tunnel data on the Orbiter served as good verification for computations performed at wind tunnel conditions (ref. 9). This paper is the first comparison between flight predictions based on these numerical simulations and STS-1 flight data. The results presented here are based on three-dimensional inviscid computations with two-dimensional boundary layer solutions applied along a surface streamline as illustrated in figure 3 (ref. 10). A "coupled" three-dimensional flow field capability which is based on numerical solutions to the "Parabolized Navier-Stokes" equations (refs. 11 & 12) enables computation of flow around the Orbiter chine, wing fillet and lee side. All flow field computations performed to date are either for flight conditions corresponding to the design trajectory or wind tunnel tests.

Surface Catalysis

At entry flight conditions, heat transfer to the Orbiter is realized not only for kinetic thermal energy of the air (as at wind tunnel conditions) but also potential energy stored in chemical changes such as latent heat of dissociation. In general, the air, processed by a hypersonic shock, is not in chemical equilibrium. Since it is necessary to know the chemical composition, finite rate air chemistry flow field computations have been performed for select design trajectory conditions (ref. 13). These results had been incorporated as boundary conditions for finite rate boundary layer computations (ref. 8) to obtain heat transfer. Heat transfer to a surface for a real gas out of chemical and thermodynamic equilibrium also depend on properties of the surface such as the surface catalytic recombination and chemical energy accommodation rates. Surface catalytic recombination rates for the Orbiter TPS have been determined from arc jet testing and analysis (ref. 14) and applied as boundary conditions to finite rate boundary layer computations (ref. 15) coupled to the finite rate inviscid computations (refs. 6, 7, & 13) to obtain more accurate predictions of flight heating. This process is illustrated schematically in figure 4. Data obtained on STS-1 began after peak heating and after the major significance of finite catalysis effects. Hopefully STS-2 data, particularly with an experiment dedicated to this phenomenon, will provide much more useful information.

BOUNDARY LAYER TRANSITION

Quantitative studies of turbulent phenomena and transition from laminar to turbulent flow have been underway for over a century. However, the only approach for defining turbulent heating and boundary layer transition for the Orbiter was and is empirical correlation (fig. 5). In spite of the large dimensions of the Orbiter, the Reynolds numbers (associated with the high heating portion of atmospheric entry) are low enough to permit laminar flow if the configuration is properly controlled (ref. 16). Discontinuous surface radii of curvature are to be avoided (ref. 17). Once the entry configuration is picked, given all of the desired operational entry constraints for the Orbiter, the minimum TPS requirements are obtained by flying just outside the boundary layer transition flight conditions (ref. 18). Selecting a proper configuration and restricting the trajectory to a laminar flow regime has eliminated on the order of 1000 kg of mass from the Orbiter TPS.

Smooth Body

Parametric wind tunnel testing of the Orbiter configuration led Rockwell to correlate boundary layer transition data with a local momentum thickness Reynolds number Re_θ divided by the local Mach number M_1. This correlation parameter varied with location on the vehicle but surprisingly only slightly with angle-of-attack in the range of interest. This parameter was very effective in correlating the available data on this configuration over the range of hypersonic wind tunnel test conditions. These tests were also capable of simulating the predicted values of this parameter at flight conditions for the appropriate angle-of-attack. Because this approach has been shown to be in virtual agreement with the use of a simplistic normal shock Reynolds number (ref. 16)--which worked quite well for the Apollo configuration--Re_θ/M_1 was agreed upon as a suitable parameter for correlating smooth body boundary layer transition. It should be emphasized that the main requirement for smooth body boundary layer transition was geometric similitude (including angle-of-attack) and shock layer flow Reynolds number and Mach number simulation. Wall-to-total temperature ratio was found to have no discernible effect on smooth body transition.

Real Body

The major portion of the Orbiter windward surface is covered with TPS tiles, nominally 15 cm (6") square with nominally a 1 mm (.045") gap. Since it was not clear how to analytically account for the counteracting influences of a "cold" and "rough" surface on boundary layer transition, a parametric experimental program, as close to similitude as possible, was pursued (refs. 19-21). It was, and is still, not clear whether distributed or single point roughnesses dominate the boundary layer transition. A 0.0175 scale Orbiter heat transfer model, for which an existing smooth body transition data base existed, was modified to include as much detailed geometric simulation as possible. Randomly distributed protruding tiles were formed in the model to provide a realistic simulation of misaligned TPS tile heights k. The tile gaps were beyond the simulation capability of this model. Also the wall-to-total temperature ratio T_w/T_o was varied throughout the range of interest by cooling the model prior to testing. By varying both the height of the randomly distributed tiles and the temperature ratio, at the same Reynolds number and angles-of-attack used during the smooth body tests, the effects of roughness and cooling could be established.

The combined effects of tile height and T_w/T_o on boundary layer transition location can be seen in figure 6. Note that Re, M and angle-of-attack remain constant during the tests shown in this figure. Even so, the location of transition moves forward as k is increased and as T_w/T_o is decreased.

In the final analysis, this data was best correlated in terms of a departure of Re_θ/M_1 from smooth body transition as a function of Re_k (Reynolds number based on step height k, and conditions at the height of the step for a smooth surface flow.) The results of this transition correlation are shown in figure 7 as applied to the design trajectory. The tile step height data on the Orbiter was never obtained directly. However, the RMS step height, measured on a number of vibro-acoustic test simulation panels (before and after launch simulation), was on the order of .8 mm (.030"). This value was used for preflight predictions although as can be seen in figure 7, step heights below 1.2 mm (.05") do not significantly alter the boundary layer transition from smooth body correlations. The overall logic for predicting boundary layer transition on the Shuttle Orbiter is illustrated in figure 8.

TPS THERMAL PERFORMANCE

The windward surface (bottom) of the Orbiter is protected by a high temperature, low density ceramic tile TPS. These brittle tiles require a thin (0.406 cm) strain isolation pad (SIP), composed of Nomex nylon felt. The system re-radiates most (>95%) of the incident convective heating by maintaining a high coating (reaction-cured glass) temperature. The low density ceramic (5.6 g/cc) is approximately 90% porous; therefore, a very effective insulation. Its thermal diffusion properties are temperature and pressure dependent, but this diffusion is predominately one-dimensional.

The gap between the tiles represents a significant, local departure from one-dimensionality, due to the complex coupling of sidewall coating conduction, radiation interchange, and gap-flow convection from non-adiabatic air (ref. 22). In general, the gap convection is not proportional to the surface heating. Further, the importance of gap heating increases with Reynolds number, particularly as the external flow becomes turbulent. The gaps are designed to be held to a small enough dimension (1.14 mm width), however, that their contribution to the total thermal diffusion to the Orbiter structure is small ($\sim$ 25%).

The TPS thermal analysis is characteristically treated as a one-dimensional diffusion with suitable modification for the gap and radiant contribution. This thermal analysis is calibrated to arc jet test simulations of local heating histories. The basic TPS thermal analysis logic is illustrated schematically in figure 9.

FLIGHT PREDICTIONS AND RESULTS

Predictions

All of the extensive numerical Orbiter flow field simulations described above have been performed either at wind tunnel conditions or at select points along the design trajectory for benchmark purposes. Although the Orbiter entry flight trajectories are all within a relatively narrow band, it was necessary to develop techniques for extrapolating this information to the STS-1 flight test conditions. To achieve this end, state-of-the-art two-dimensional flow models have been (and are being) calibrated to the benchmark simulations. The procedure is similar to the design methodology presented in figure 2 with the exception that the wind tunnel test data is replaced with flow field simulations that are quite close to the flight conditions of interest. If the Shuttle Orbiter design was initiated today, the design methodology would be performed in this manner. The turbulent heating was calculated in the same manner as the design methodology, i.e., with the Spalding and Chi theory calibrated to wind tunnel data.

Surface Temperatures

The primary surface environment information obtained from the Orbiter flight test program is through "surface thermocouples" which essentially measure the temperature of the TPS tile coating. Since the dominant heat transfer processes are aerodynamic heating and re-radiation, the surface temperature measurements are virtually heat transfer measurements. However, to properly account for conduction and thermal capacity, the measured temperatures are compared directly with predicted temperatures. Figure 10 shows this comparison for data obtained along the windward pitch plane of the Orbiter. These predictions are based on the extrapolation of three-dimensional flow field computations for the design trajectory through the use of two-dimensional oblique shock flow models. The forward region exhibits the response to laminar as well as turbulent heating and presents a clear indication of a boundary layer transition. Aft of the midfuselage boundary layer transition had occurred prior to the available data. The agreement between predictions and data is quite good.

Predicted and inferred heating rates are shown in figure 11 for a representative mid-fuselage location. Here it can be seen that only tenuous conclusions can be drawn concerning the anticipated finite catalysis phenomenon. Confirmation of this phenomenon must await additional flights with complete data. It should be noted that state-of-the-art methodologies normalized to wind tunnel data are significantly above these predictions at flight conditions. The obvious transition in the state-of-the-art methodologies presented in figure 2 was used only for preflight assessment, whereas the TPS design assumed smooth body transition.

Boundary Layer Transition

The boundary layer transition data obtained on the instrumented half of the STS-1 Orbiter vehicle are fantastic. Most surface temperatures show a clear indication of the onset and completion of the transition process. In select regions of the vehicle the measurements show an incipient transition, a reversal toward laminar values and then a final transition process. These are generally not isolated measurements but rather this effect can be seen as a definite flow pattern in regions of the vehicle. Since the Orbiter is undergoing a decrease in angle-of-attack as well as changing flight conditions, it is not clear whether this behavior is inherent to the transition phenomena or reflects the vehicle behavior. Time contours for the incipient and final boundary layer transition times are illustrated in figure 12.

TPS Thermal Response

In general, the TPS thermal response and aluminum temperatures were consistent with predictions. Figure 13 depicts the temperature transients at three body points on the bottom of the vehicle. At each Development Flight Instrumentation (DFI) plug location, thermocouples were located in the tile coating, several locations within the tile, at the tile-SIP interface, and on the aluminum skin. Most of the flight data were available only for the time after 1100 seconds of entry, due to a data recorder malfunction.

In figure 14, thermal response for two locations where the total aluminum bondline temperature transients are available, were simulated by the numerical model. Note that the adiabatic backwall analysis (see BP 1600) does not follow the flight data. Using backface radiation and free convection, however, the model matches the two data curves perfectly. This was a post-flight modification to the math model. These models were driven (at the surface) by analytically-derived convection coefficients h and local recovery temperatures.

Figure 15 provides a means of comparing bondline temperature-response predictions for STS-1 and the design trajectory as well as STS-1 data for a typical mid-fuselage station (i.e., BP 1500). At this location the major temperature difference between STS-1 and the design trajectory is due to the initial temperature. The data and predictions (labeled JSC predictions) indicate that peak allowable bondline temperatures would not be exceeded for the design trajectory. However, bondline temperature predictions at the time of maximum stress (between TAEM and landing) do not provide a conservative outlook.

CONCLUDING REMARKS

The general agreement between these preflight predictions and the temperature measurements obtained on STS-1 leads to the conclusions summarized in figure 16. The use of computational fluid mechanics as a valuable tool in the design and development process has been demonstrated here. Although the STS-1 measurements point to the significance of finite surface catalysis, a firm conclusion requires a full set of flight data. Boundary layer transition on the windward side of the Orbiter occurred just as it was expected to. The quality of this data is excellent. The TPS thermal response was as predicted with the addition of internal free convection for un-insulated structure areas. The TPS appears to be generally adequate from a thermal response standpoint; which implies that the Orbiter has a warm structure.

REFERENCES

1. "Space Shuttle Orbiter Entry Aerodynamic Heating Data Book," Rockwell International Space Division, Downey, CA, SD73-SH-0184C, Oct. 1978.

2. "Space Shuttle Program Thermodynamic Design Data Book, Thermal Protection System," Rockwell International Space Division, Downey, CA, SD73-SH-0226, Vol. 2C, July 1977.

3. Goodrich, W. D,; Derry, S. M.; Maraia, R. J.: Effects of Aerodynamic Heating and TPS Thermal Performance Uncertainties on the Shuttle Orbiter, in Entry Heating and Thermal Protection, Vol. 69 of Progress in Astronautics and Aeronautics, Walter B. Olstad, ed. (1980) pp. 247-268.

4. Li, C. P.: "A Numerical Study of Laminar Flow Separation on Blunt Flared Cones at Angle-of-Attack," AIAA Paper 74-585, June 1974.

5. Kutler, P.; Reinhardt, W. A,; and Warning, R. F.: "Multishocked Three-Dimensional Supersonic Flow Fields with Real Gas Effects," AIAA Journal, Vol. 11, May 1973, pp. 657-664.

6. Rakich, John V,; and Mateer, G. G.: "Calculation of Metric Coefficients for Streamline Coordinates", AIAA Journal, Vol. 10, No. 11, Nov. 1972, pp. 1538-1540.

7. Rakich, John V.; and Lanfranco, Martin J.: "Numerical Computation of Space Shuttle Laminar Heating and Surface Streamlines", Journal of Spacecraft and Rockets, Vol. 14, No. 5, May 1977, pp. 265-272.

8. Tong, H.; Buckingham, A. C.; and Morse, H. L.: Nonequilibrium Chemistry Boundary Layer Integral Matrix Procedure, NASA CR-134039, July 1973.

9. Goodrich, W. D.; Li, C. P.; Houston, C. K.; Meyers, R. M.; and Olmedo, L.: "Scaling of Orbiter Aerothermodynamic Data Through Numerical Flow Field Simulations," NASA SP-347, Part 2, March 1975, pp. 1395-1410.

10. Goodrich, W. D.; Li, C. P.; Houston, C. K.; Chiu, P. B.; and Olmedo, L.: "Numercial Computations of Orbiter Flow Fields and Laminar Heating Rate," Journal of Spacecraft and Rockets, Vol. 14, May 1977, pp. 257-264.

11. Li, C. P.: "Application of An Implicit Technique to the Shock-Layer Flow Around General Bodies," To appear in AIAA Journal (1982).

12. Li, C. P.: "Numercial Simulation of Reentry Flow Around the Shuttle Orbiter Including Real Gas Effects," Paper Presented at the Symposium on Computers in Flow Predictions and Fluid Dynamics Experiments. ASME Winter Annual Meeting, Nov. 1981, Washington, D.C.

13. Rakich, John V.; Bailey, Harry E.; and Park, Chul: Computation of Nonequilibrium Three-Dimensional Inviscid Flow Over Blunt-Nosed Bodies Flying at Supersonic Speeds. AIAA Paper 75-835, June 1975.

14. Scott, Carl D.: Catalytic Recombination of Nitrogen and Oxygen on High Temperature Reusable Surface Insulation. AIAA Paper 80-1477, June 1981.

15. Scott, Carl D.: Space Shuttle Laminar Heating with Finite-Rate Catalytic Recombination, AIAA Paper 81-1144, June 1981.

16. Ried, R. C., Jr.; Goodrich, W. D.; Strouhal, G.; and Curry, D. M.: "The Importance of Boundary Layer Transition to the Space Shuttle Design," Proceedings of the Boundary Layer Transition Workshop held Nov. 3-5, 1971. Aerospace Report No. TOR-0172 (S2816-16) -5, Dec. 20, 1971.

17. Young, C. H.; Reda, D. C.; and Roberge, A. M.: "Hypersonic Transitional and Turbulent Flow Studies on a Lifting Entry Vehicle," AIAA Paper 71-100, Jan. 1971.

18. Curry, D. M.; Tolin, J. W., Jr.; and Goodrich, W. D.: "Effects of Selected Trajectory Parameters on Weight Trends in the Shuttle Thermal Protection System," NASA TMX-58113, Jan. 1974.

19. Goodrich, W. D.; and Stalmach, C., Jr.: "Effects of Scaled Heatshield Tile Misalignment on Orbiter Boundary-Layer Transition," Journal of Spacecraft and Rockets, Vol. 14, October 1977, pp. 638-640.

20. Bertin, J. J.; Idar, E. S., III; and Goodrich, W. D.; "Effect of Surface Cooling and Roughness on Transition for the Shuttle Orbiter," Journal of Spacecraft and Rockets, Vol. 15, March - April, 1978, pp. 113-119.

21. Bertin, J. J.; Hayden, T. E.; and Goodrich, W. D.: "Comparison of Correlations of Shuttle Boundary-Layer Transition Due to Distributed Roughness," AIAA Paper 81-0417, Jan. 1981.

22. Scott, C. D.; and Maraia, R. J.: Gap Heating with Pressure Gradients, in Entry Heating and Thermal Protection, Vol. 69 of Progress in Astronautics and Aeronautics, Walter B. Olstad, ed. (1980) pp. 269-286.

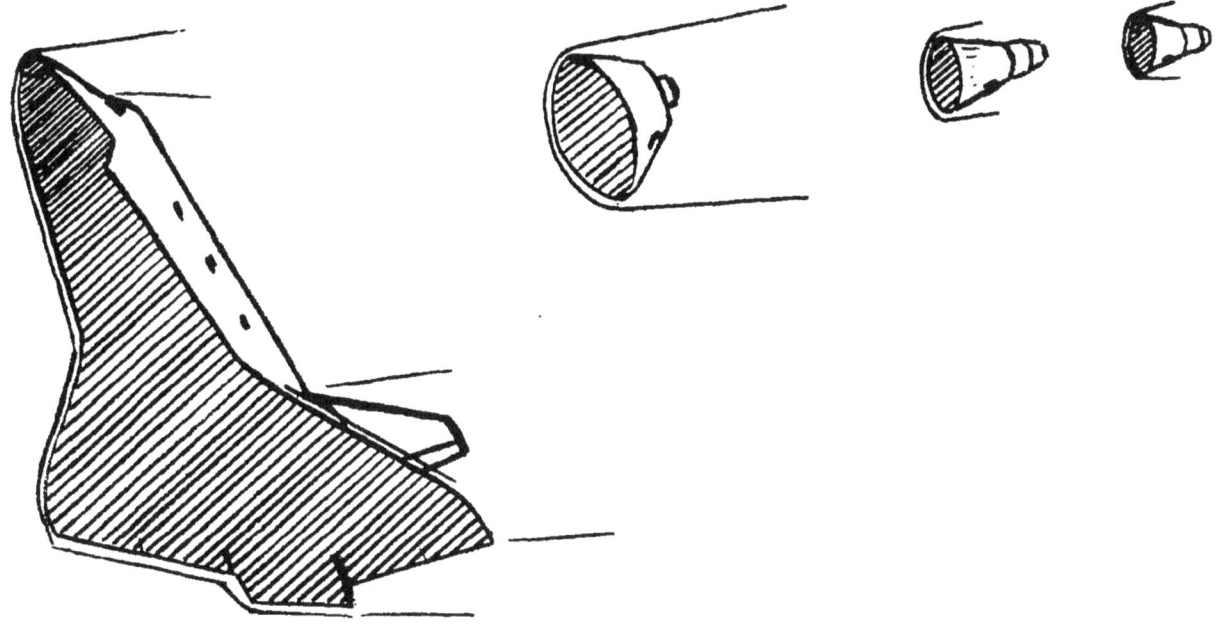

Figure 1.- Manned orbital entry spacecraft.

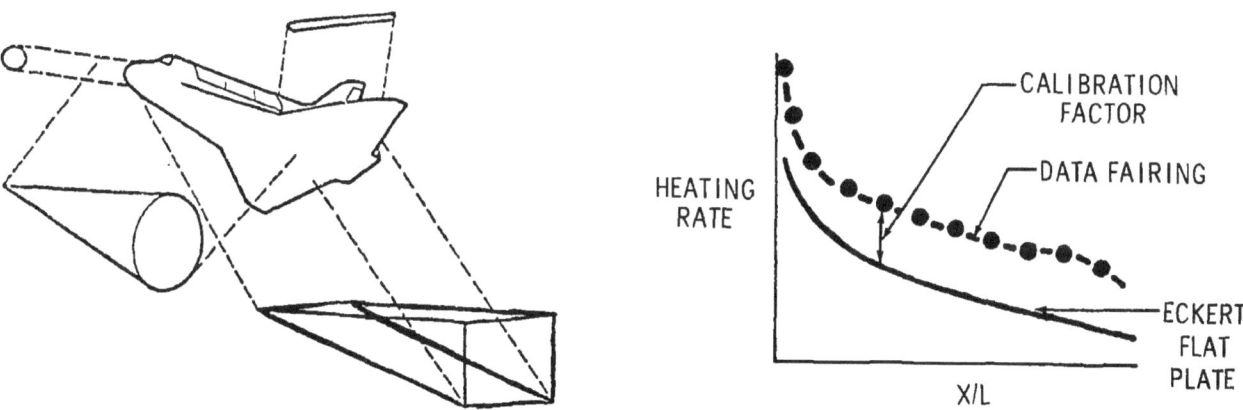

Figure 2.- Design heating methodology.

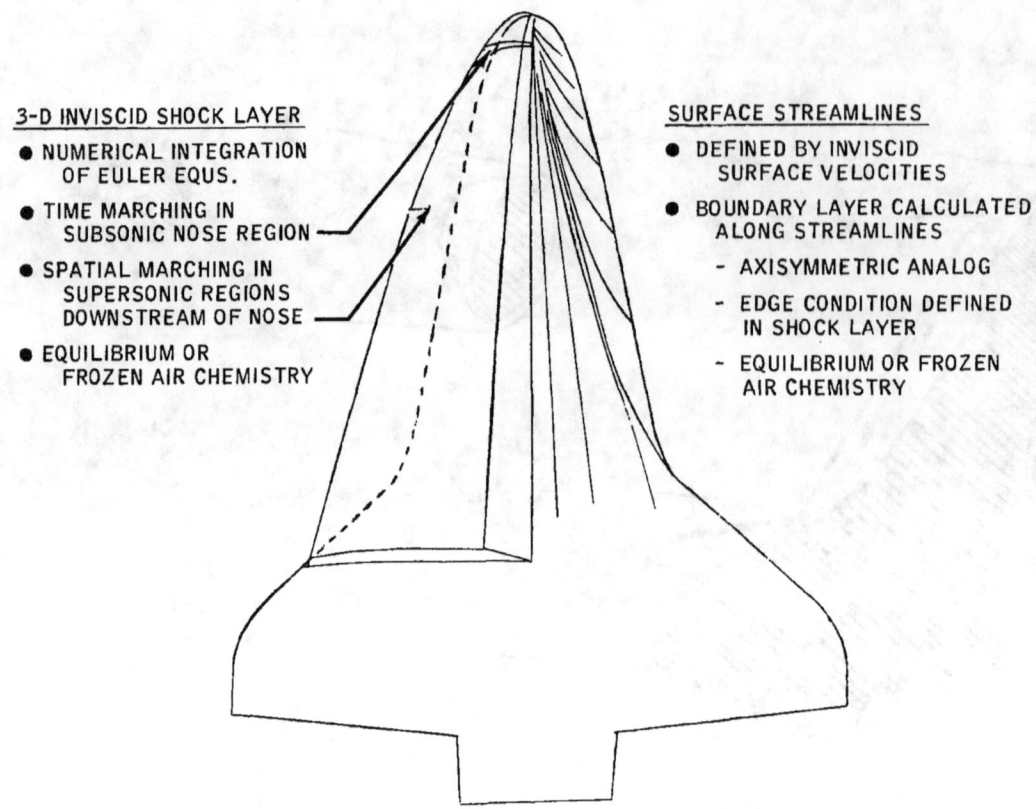

Figure 3.- Illustration of flow field simulation technology as applied to orbiter.

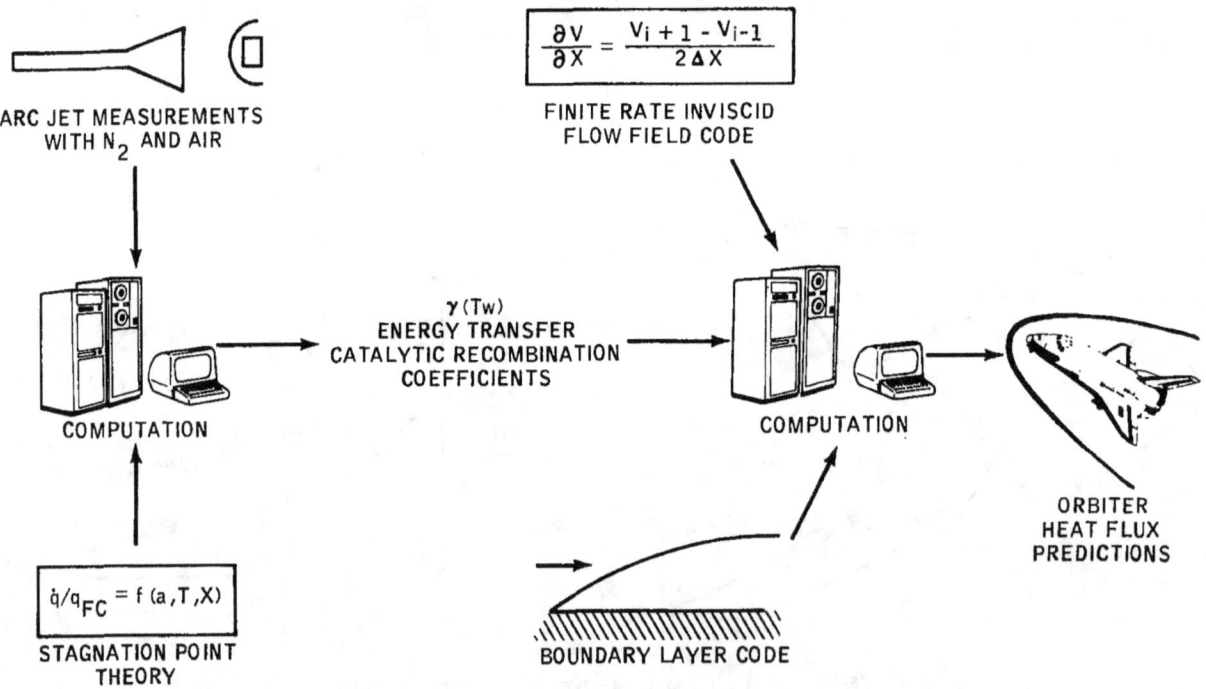

Figure 4.- Surface catalysis flight prediction process.

Figure 5.- State of the art in hypersonic B.L. transition.

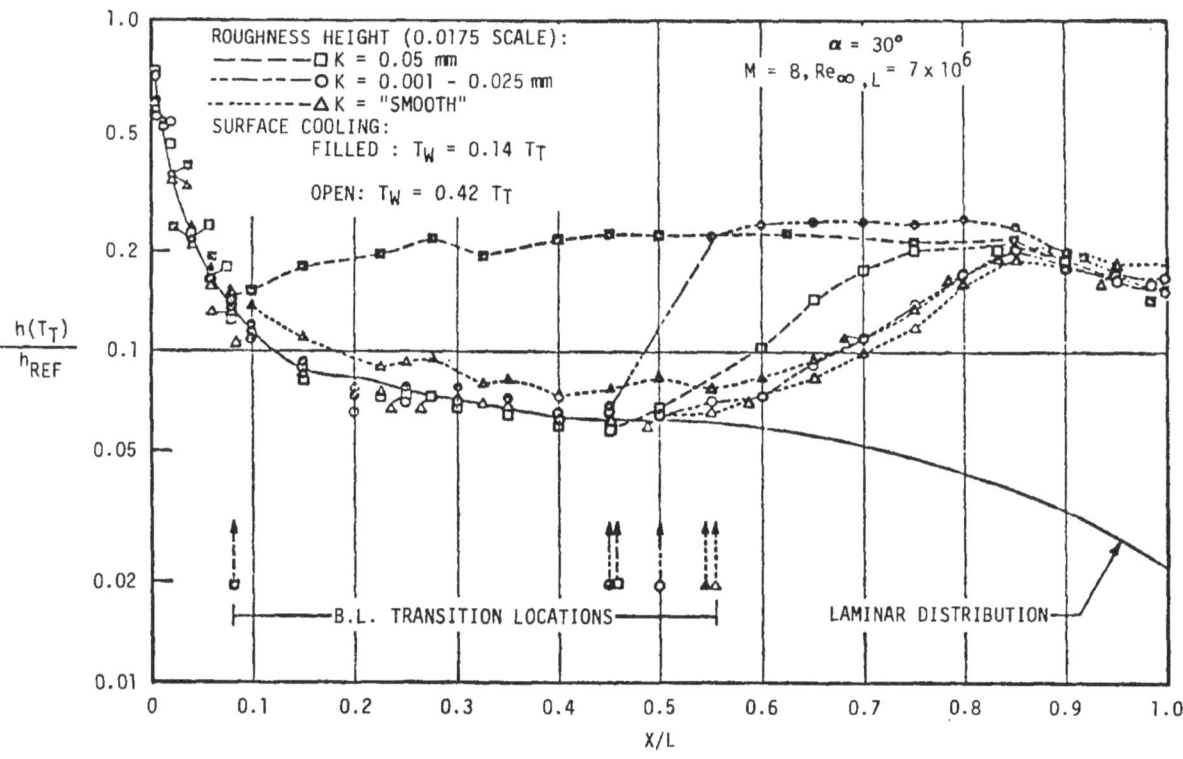

Figure 6.- Orbiter B.L. transition with scaled tile roughness.

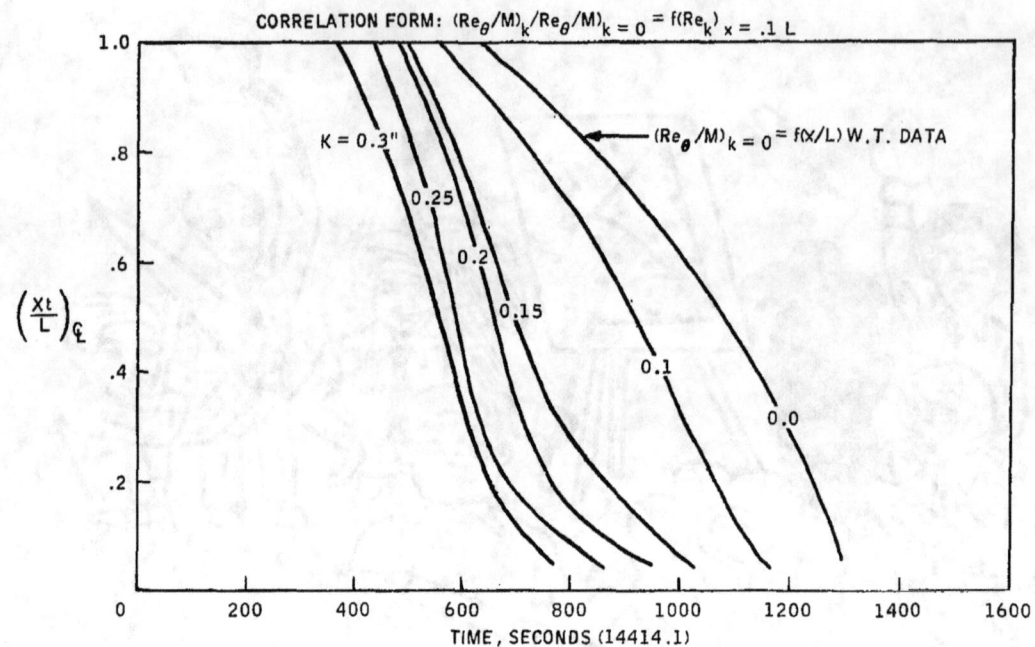

Figure 7.- History of roughness induced orbiter B.L. transition.

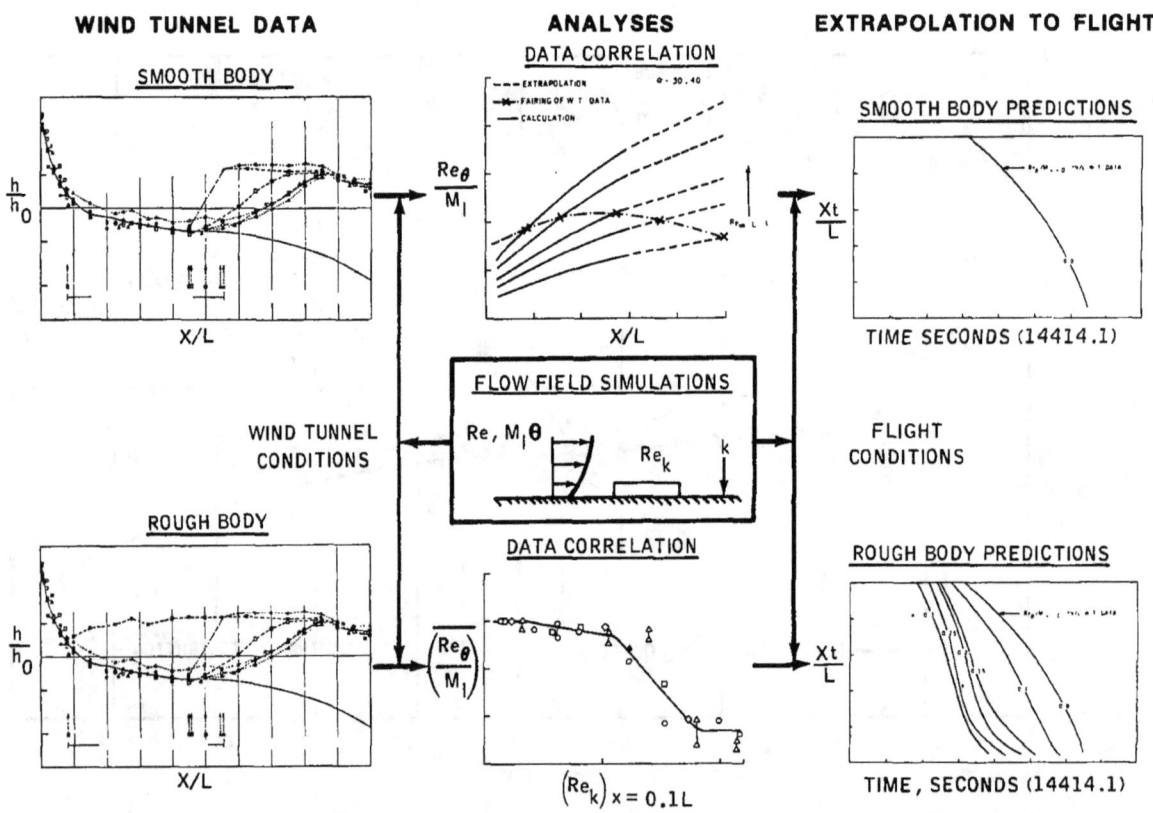

Figure 8.- Logic for predicting B.L. transition on orbiter.

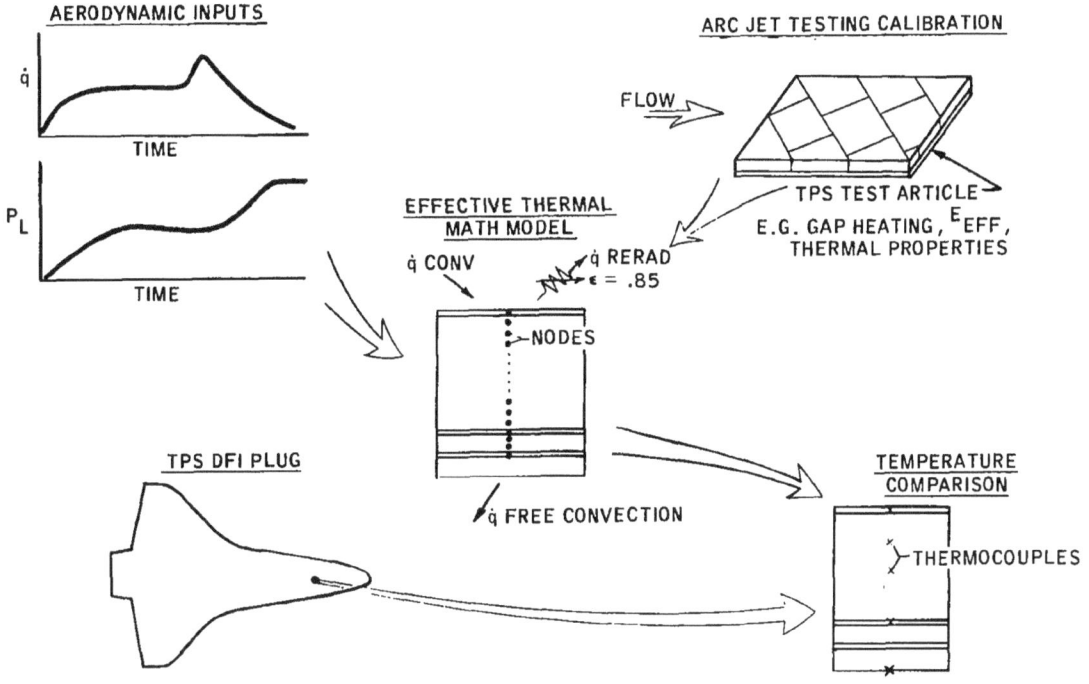

Figure 9.- TPS thermal analysis logic.

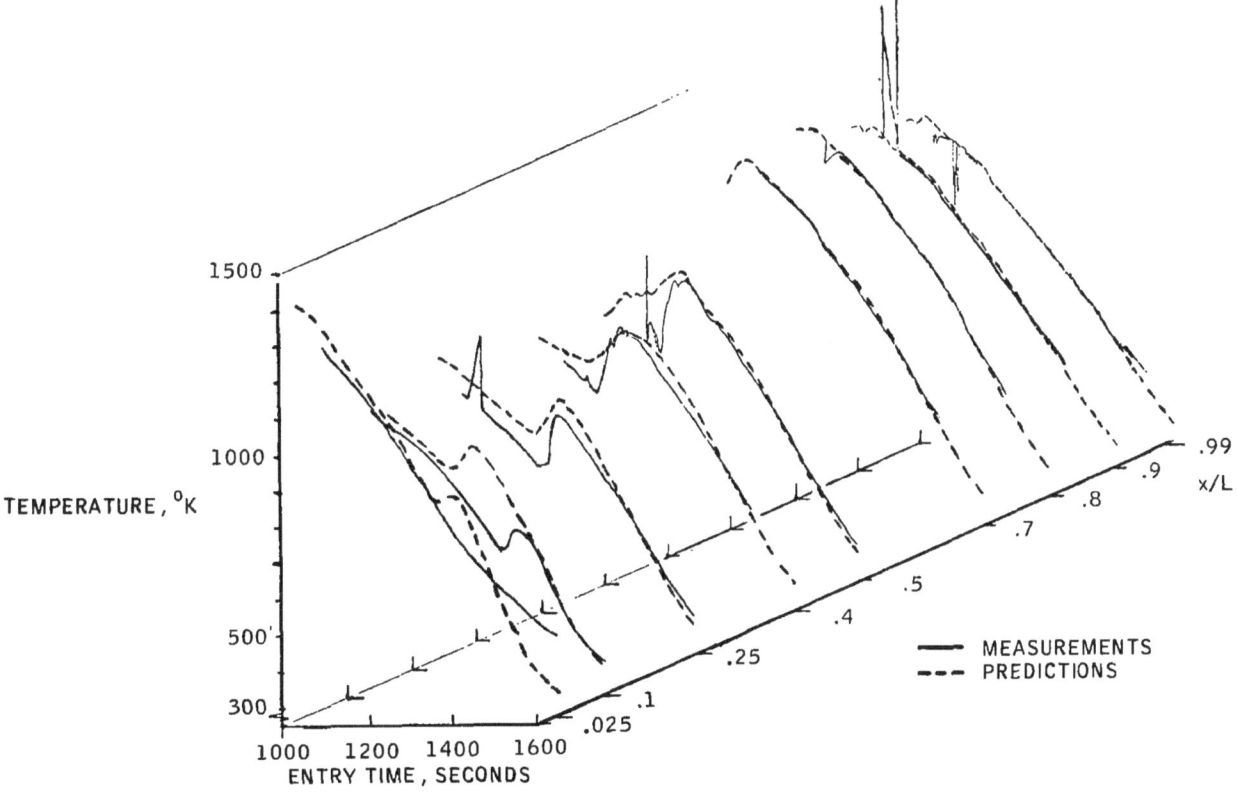

Figure 10.- STS-1 surface temperature measurements and predictions, windward centerline.

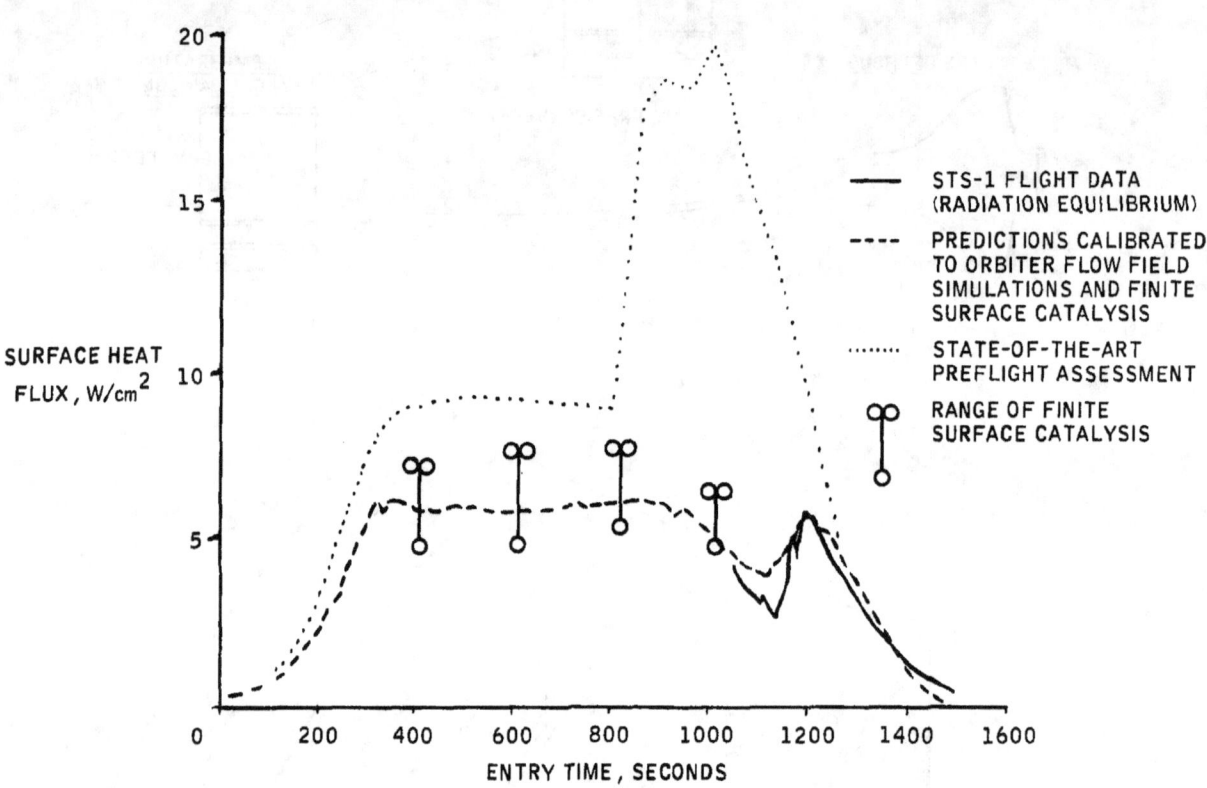

Figure 11.- STS-1 entry heating data and preflight predictions, windward centerline x/L = 0.4.

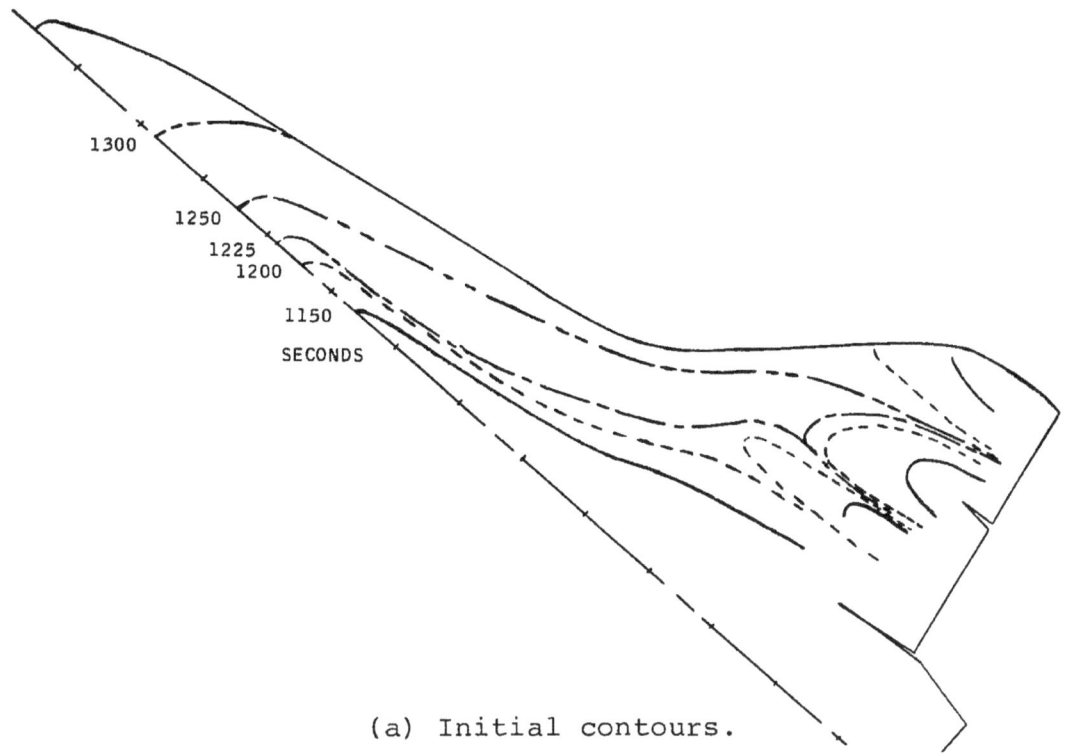

(a) Initial contours.

Figure 12.- Boundary layer transition tune contours, STS-1.

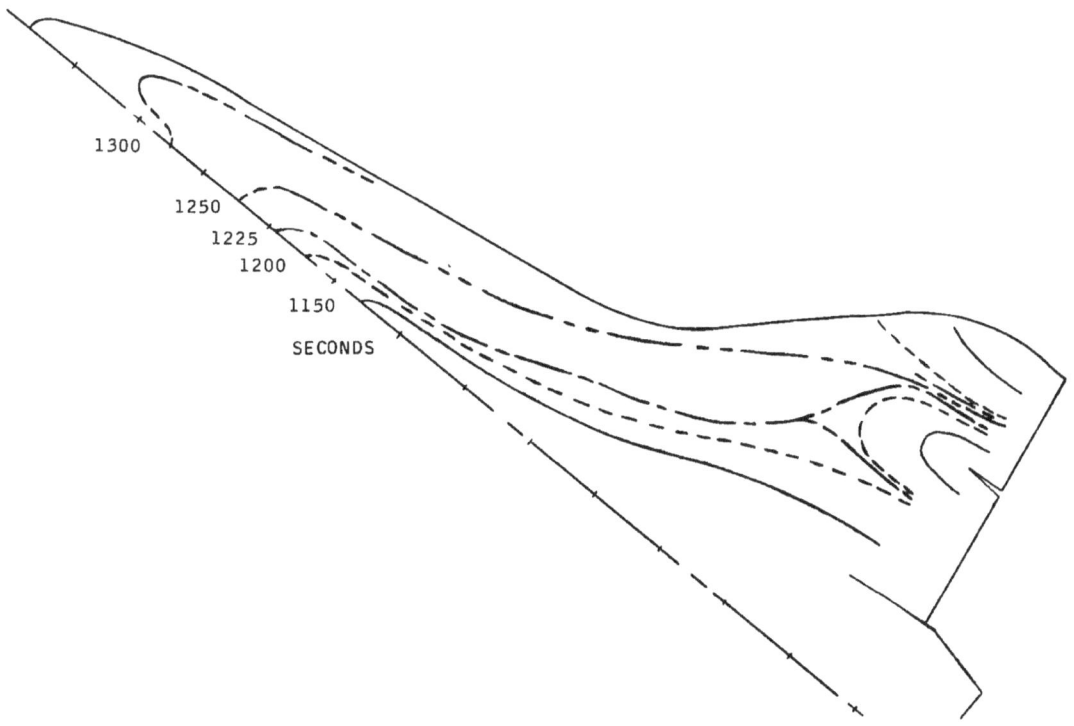

(b) Final contours.

Figure 12.- Concluded.

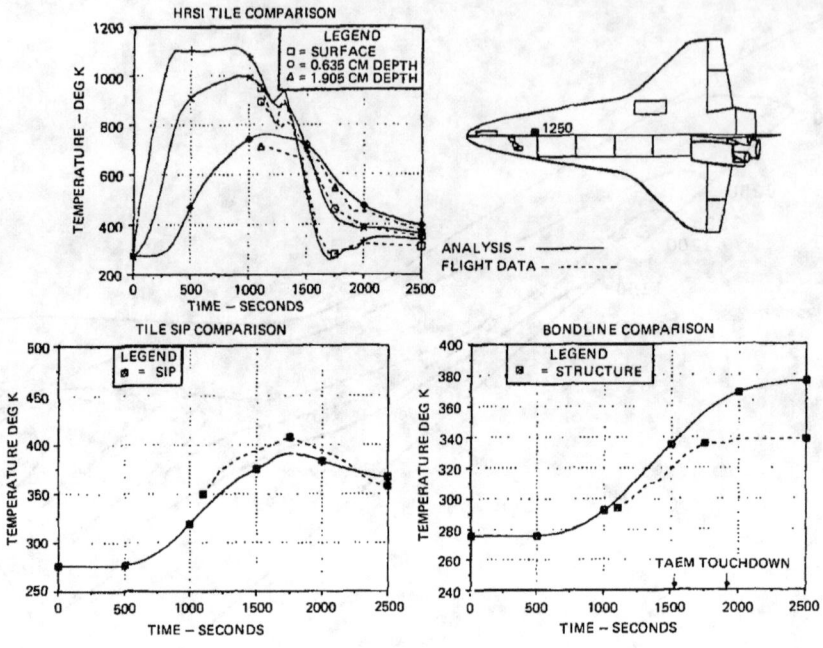

(a) At BP 1250.

Figure 13.- TPS plug temperature comparisons for STS-1.

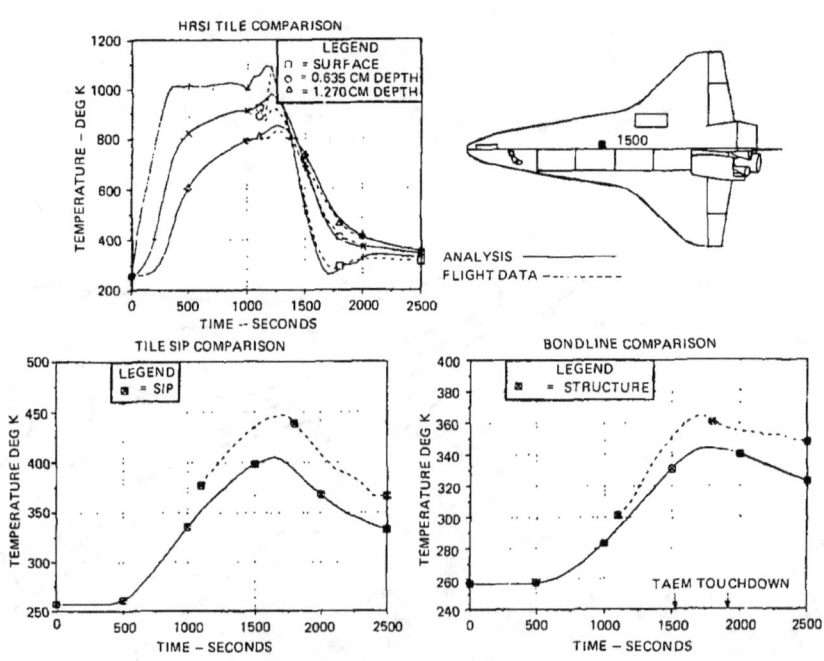

(b) At BP 1500.

Figure 13.- Continued.

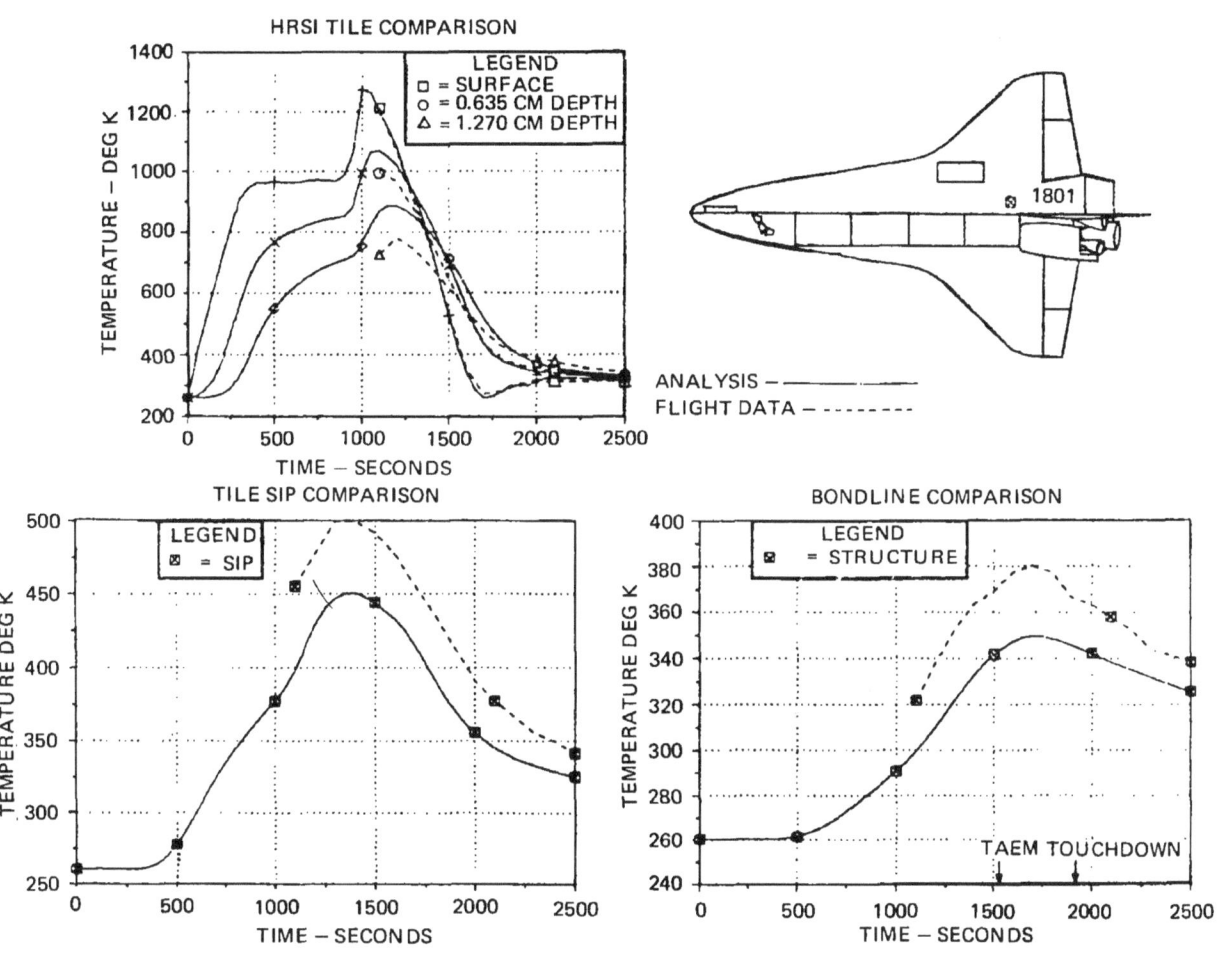

(c) At BP 1801.

Figure 13.- Concluded.

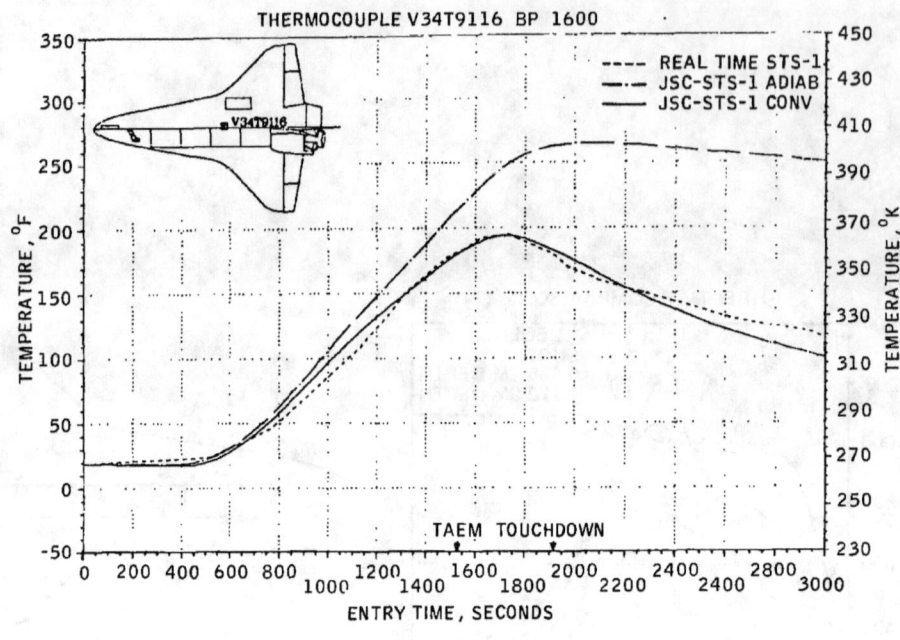

(a) At BP 1600.

Figure 14.- Bondline temperature comparison with STS-1 data.

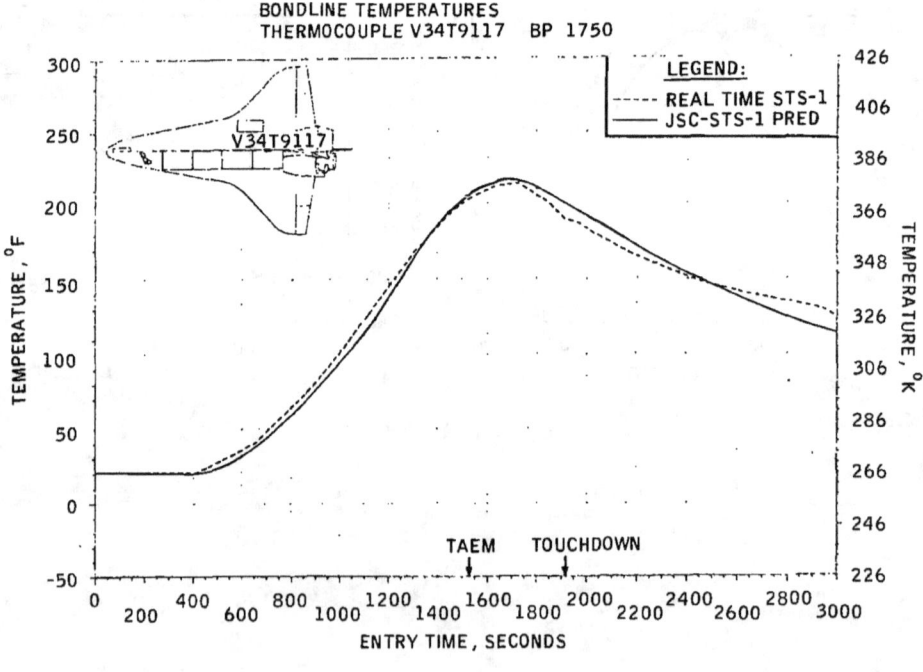

(b) At BP 1750.

Figure 14.- Concluded.

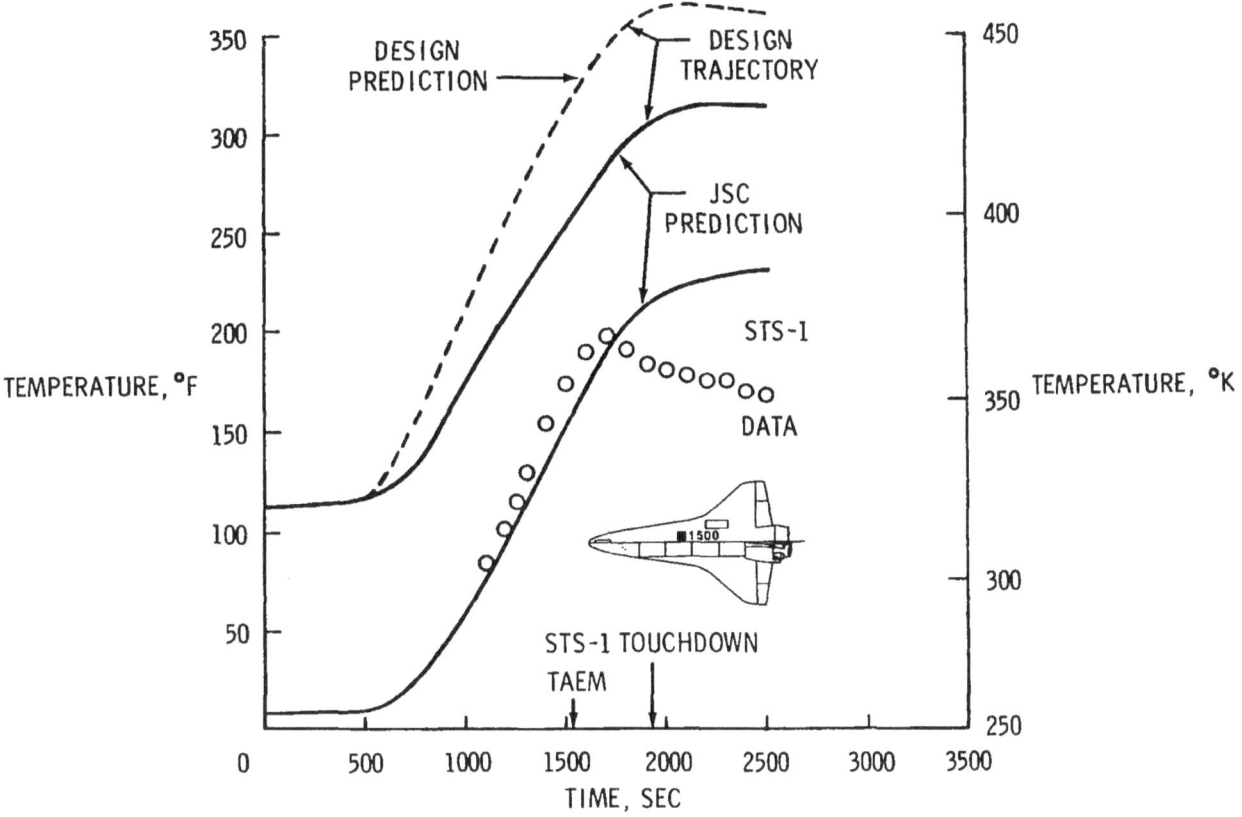

Figure 15.- Entry bondline thermal response, midbody windward.

- COMPUTATIONAL FLUID MECHANICS SIMULATIONS HAVE BEEN A VALUABLE TOOL FOR THE ORBITER DEVELOPMENT

- PREDICTED SURFACE TEMPERATURES AGREE WITH OR EXCEED MEASUREMENTS

- BOUNDARY LAYER TRANSITION OCCURRED AS PREDICTED

- FINITE SURFACE CATALYSIS IMPLIED BY MEASUREMENTS

- TPS THERMAL RESPONSE AS PREDICTED
    - INTERNAL FREE CONVECTION SIGNIFICANT

- THE ORBITER HAS A "WARM" STRUCTURE

Figure 16.- Conclusions.

TRANSIENT THERMAL ANALYSIS OF A TITANIUM MULTIWALL THERMAL PROTECTION SYSTEM

M. L. Blosser
NASA Langley Research Center
Hampton, Virginia

INTRODUCTION

(Figure 1)

This paper demonstrates the application of the SPAR Thermal Analyzer (ref. 1) to the thermal analysis of a thermal protection system concept. Thermal analysis is especially useful in the concept design and development stages to provide a basis for design and design modification decisions.

The titanium multiwall thermal protection system concept (ref. 2) consists of alternate flat and dimpled sheets which are joined together at the crests of the dimples and formed into 30 cm by 30 cm (12 in. by 12 in.) tiles as shown in the figure . The tiles are mechanically attached to the structure. The complex tile geometry complicates thermal analysis. Three modes of heat transfer must be considered: conduction through the gas inside the tile, conduction through the metal, and radiation between the various layers. The voids between the dimpled and flat sheets were designed to be small enough so that natural convection is insignificant (e.g., Grashof number < 1000).

A two step approach was used in the thermal analysis of the multiwall thermal protection system. First, an effective normal (through-the-thickness) thermal conductivity was obtained from a steady state analysis using a detailed SPAR finite element model of a small symmetric section of the multiwall tile. This effective conductivity was then used in simple one-dimensional finite element models for preliminary analysis of several transient heat transfer problems. The model used to determine the effective conductivity is shown on the next figure.

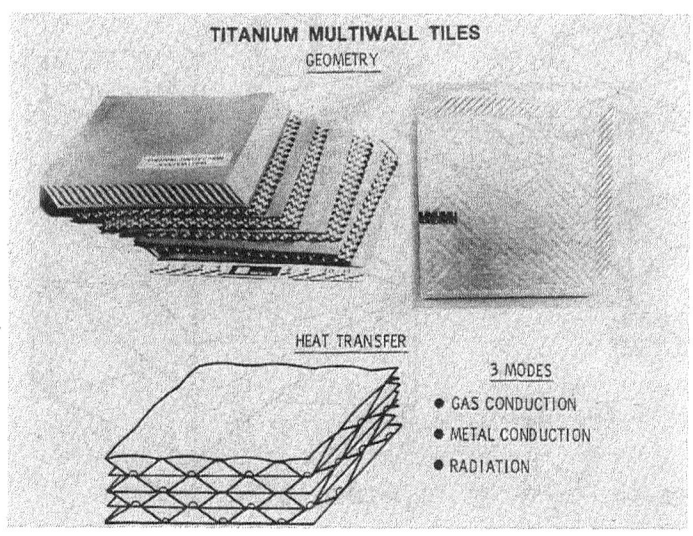

SPAR FINITE ELEMENT MODEL

(Figure 2)

An effective normal thermal conductivity for the 1.75 cm (0.688 in.) thick multiwall tile, shown in the figure, was calculated using a steady-state SPAR finite element analysis. Each dimple of the simplified model of a dimpled sheet, shown on the left of the figure, is represented by eight triangular areas. Each of the eight triangular areas is symmetrical with respect to the heat transfer through the tile; that is, each of the three sides is adiabatic. Therefore, heat transfer through the tile was analyzed with the prism shaped model shown on the right. The upper triangular surface of this model is only 0.20 cm^2 (0.031 in^2). The model contains 333 nodes, 288 metal conduction elements, 264 air conduction elements, and 512 radiation elements. On each of the horizontal and inclined planes of the model, which represent the flat and dimpled sheets respectively, 32 two-dimensional metal conduction elements are arranged as illustrated on the upper surface of the model. Three-dimensional air conduction elements fill the space between the planes of the model, as indicated by the typical element shown. Radiation elements are super-imposed on each side of the metal conduction elements. One element accounts for radiation to and from the upper surface of the metal sheet and the other accounts for radiation to and from the lower surface. Radiation view-factors for the radiation elements were calculated using the general purpose radiation computer program TRASYS II (ref. 3).

The temperature of the bottom surface of the model was held constant and a heating rate, q, was applied to the upper surface. The computed average temperature of the upper surface was used in the standard heat conduction formula, shown on the left of the figure, to calculate an effective conductivity. The results of the calculations are shown in the next figure.

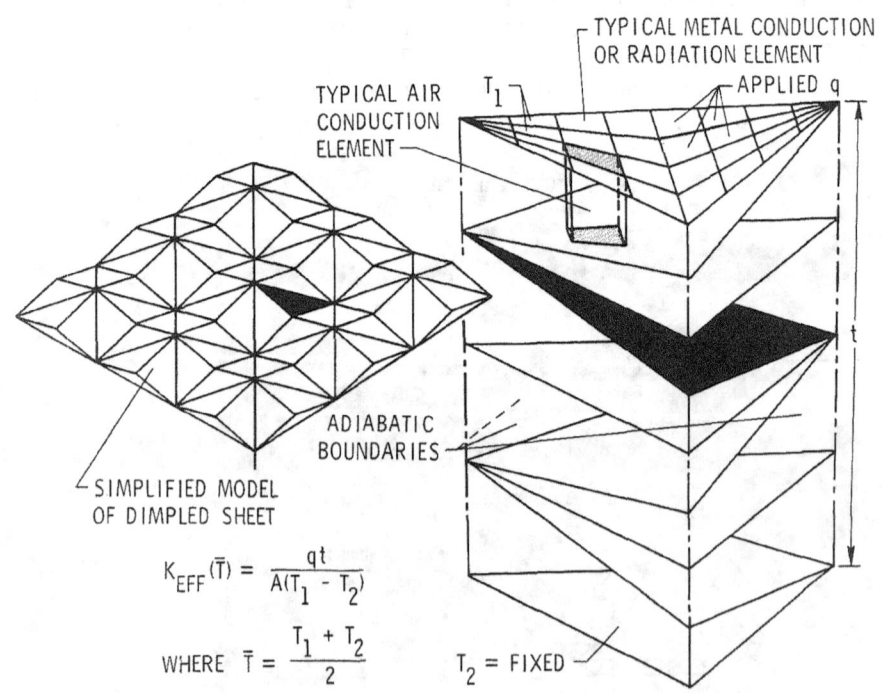

$$K_{EFF}(\bar{T}) = \frac{qt}{A(T_1 - T_2)}$$

$$\text{WHERE } \bar{T} = \frac{T_1 + T_2}{2}$$

EFFECTIVE THERMAL CONDUCTIVITY

(Figure 3)

This figure shows a comparison between the calculated effective thermal conductivity and the measured thermal conductivity taken from reference 4. The conductivity calculated using SPAR shows good agreement with test data for the same multiwall tile thickness.

The contribution of each mode of heat transfer is also shown. Radiation and gas conduction are the major modes of heat transfer, with radiation becoming the dominant mode at higher temperature. Metal conduction contributes relatively little to the total conductivity of the tile. Because each component was calculated independently, the coupling between the modes of heat transfer was not accounted for. Therefore, the sum of the components is slightly greater than the total conductivity.

The next figure shows an example of the use of this effective conductivity in a simplified transient finite element analysis.

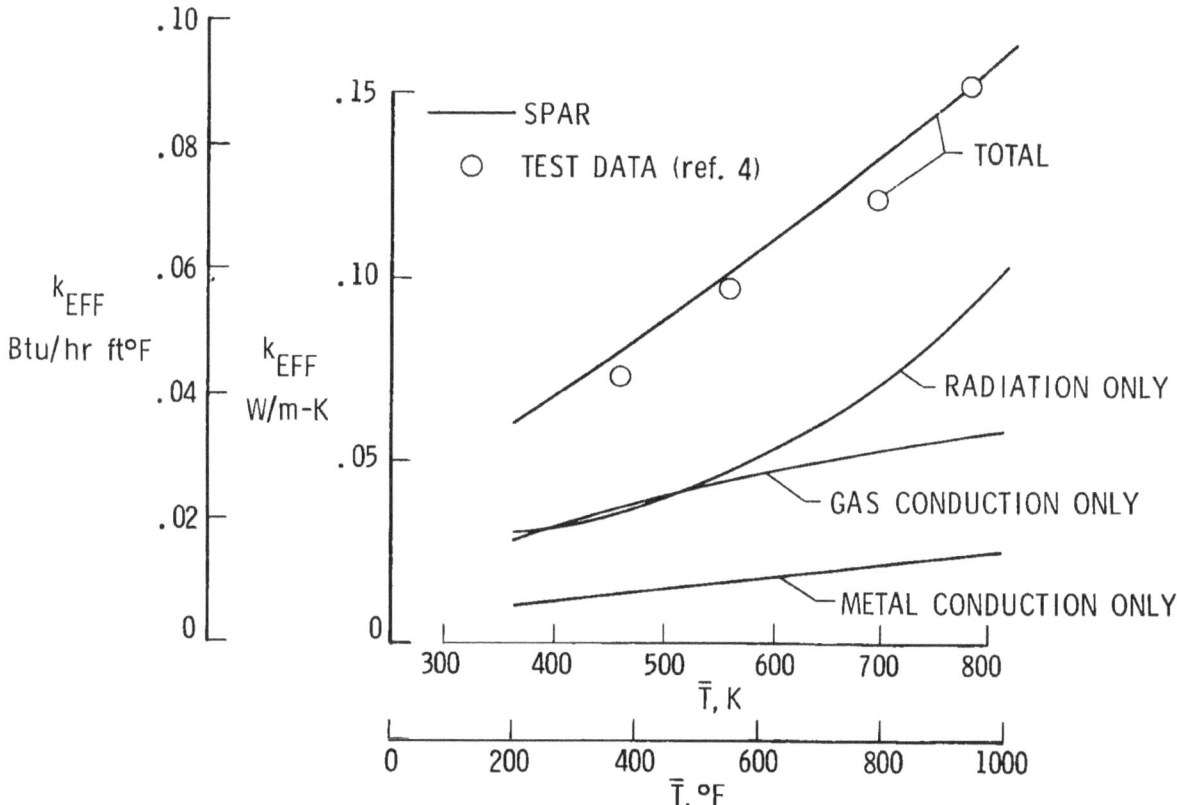

MODEL FOR TRANSIENT 1-D THERMAL ANALYSIS
(Figure 4)

The cross-section on the left of the figure represents a section through the center of a multiwall tile, including the underlying air gap and aluminum structure. The simple, one-dimensional, finite element model, shown to the right of the figure, was used in a SPAR preliminary transient thermal analysis. The model consisted of only 4 nodes and had 3 one-dimensional conduction elements and 2 point radiation elements. The total heat transfer through the multiwall tile was represented by a single 1-D conduction element which was assigned the temperature dependent effective conductivity shown in figure 3. Conduction through the air gap and aluminum was represented by 1-D conduction elements. Radiation across the air gap was accounted for by point radiation elements. Conduction across the air gap is dependent on the thickness of the gap. A temperature history, representative of the design entry thermal environment at Shuttle body point 3140 (a location on the upper center near the windows), was applied to the outer multiwall tile surface. No heat loss was allowed from the lower surface of the structure.

A temperature difference through the thickness of the multiwall tile will cause the tile to bow. The resulting change in air gap thickness was accounted for in the model by proportionately changing the conductivity of the air gap as a function of time. Thermal bowing of multiwall tiles is explained further in the next figure.

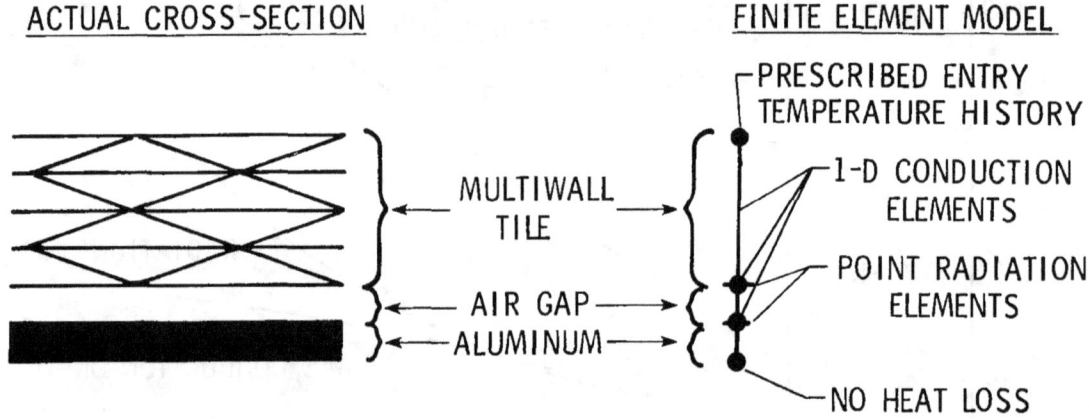

- 4 NODES
- EFFECTIVE CONDUCTIVITY ACCOUNTS FOR HEAT TRANSFER THROUGH MULTIWALL
- TIME-VARYING CONDUCTIVITY USED IN AIR GAP TO SIMULATE EFFECTS OF THERMAL BOWING OF TILE

THERMAL BOWING OF MULTIWALL TILE

(Figure 5)

This figure shows the amount of thermal bowing calculated using temperatures measured during transient heating tests. Two multiwall tiles, shown in the figure, were realistically attached to a well insulated aluminum plate and were subjected to radiant heating which simulated the entry thermal environment at body point 3140. The tiles were instrumented with thermocouples to measure the temperature histories at various locations on the tiles and underlying aluminum. A more complete description of the tests is given in reference 5.

Large differences between backface temperatures measured at the center of the tiles and those measured toward the edges suggested that thermal deformations may have had a significant effect on the thermal performance of the tiles. Temperatures measured at the center of the upper surface (T_1) and lower surface (T_2) of the tile were used to calculate the change in air gap thickness due to thermal bowing as a function of time. The calculated variation of the thickness of the air gap as a function of time is shown in the figure. As previously mentioned, the conductivity, rather than the length of the air gap conduction element, was varied to account for the effects of thermal bowing. The next figure shows the results of the transient analysis.

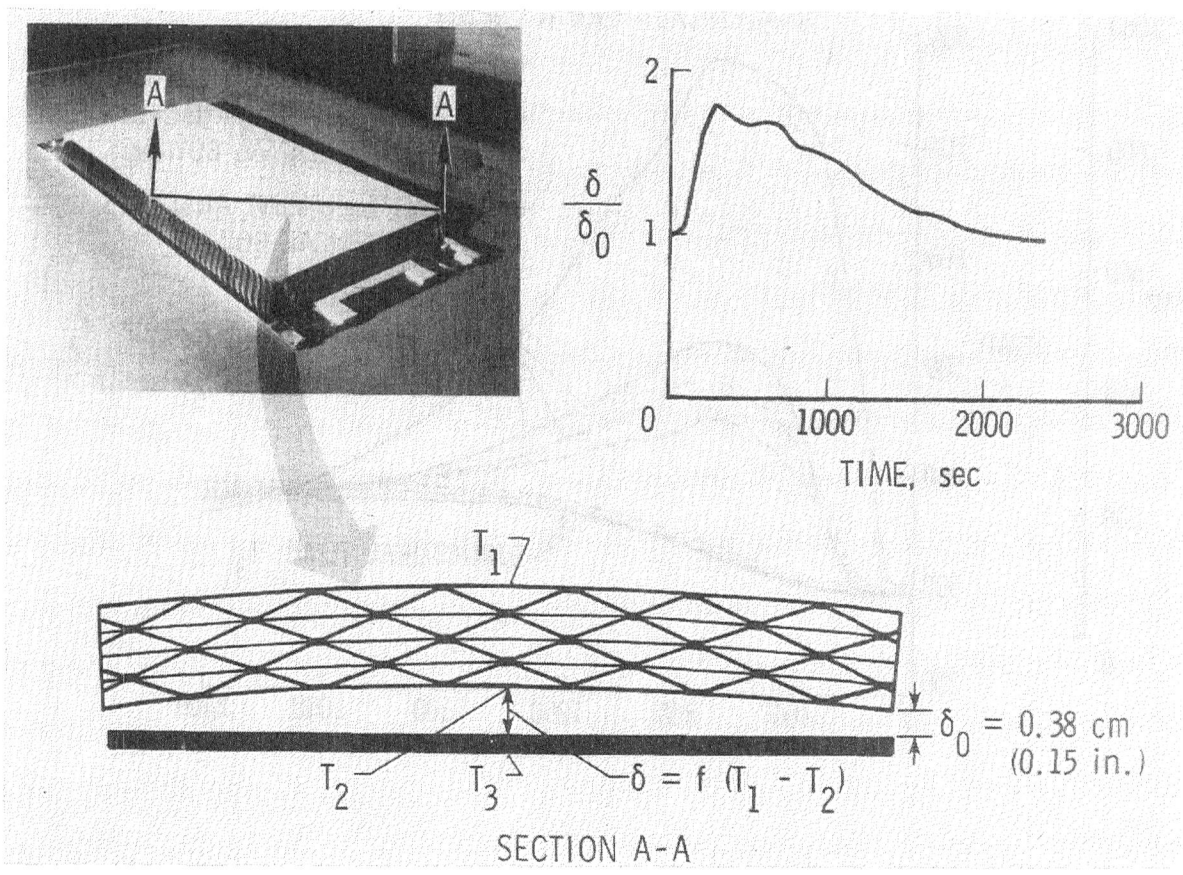

TRANSIENT RESPONSE OF MULTIWALL TILE - 1-D SPAR FINITE ELEMENT ANALYSIS

(Figure 6)

The figure shows a comparison between temperature histories calculated with the one-dimensional SPAR analysis and temperature histories measured during the two-tile radiant heating test. The measured surface temperature history, T_1, was applied to the surface of the finite element model. Temperatures were calculated both with and without accounting for the effect of thermal bowing. The analysis in which the effects of thermal bowing were neglected slightly overpredicted the structural temperature, T_3, and significantly underpredicted the temperature of the backface of the multiwall tile, T_2. When the effect of thermal bowing was included in the analysis the agreement was significantly improved. The temperature of the multiwall tile backface, T_2, is still underpredicted but there is good agreement between the calculated and measured structural temperatures.

The next figure introduces another problem for which this simple one-dimensional analysis was used.

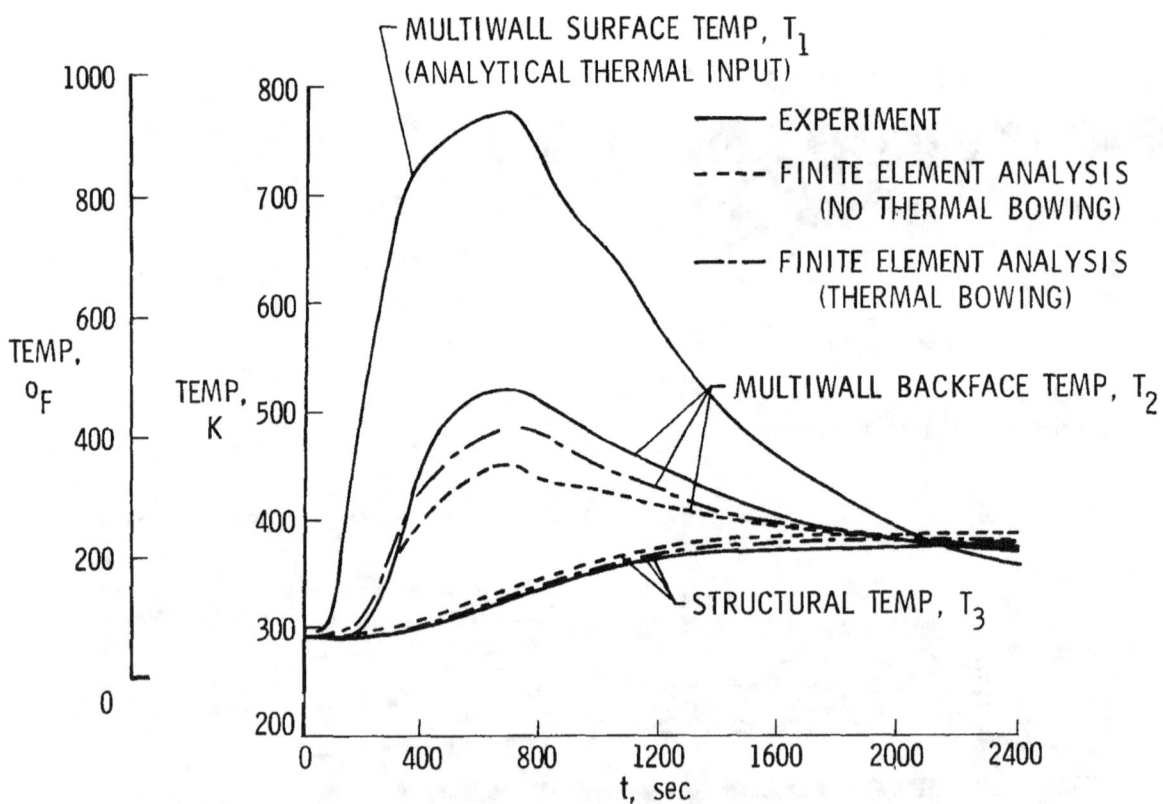

PRELIMINARY THERMAL ANALYSIS IN SUPPORT OF ORBITER EXPERIMENT OF TITANIUM MULTIWALL

(Figure 7)

As a part of a proposed Orbiter Experiments Program (OEX) the LRSI ceramic tiles on a 2.3 m^2 (25 ft^2) area on the Orbiter will be replaced by titanium multiwall tiles. Thermal analysis was required to determine if the present titanium multiwall tiles, designed for a different location on the Orbiter, would adequately protect the area being considered for the OEX equipment.

A simple one-dimensional SPAR finite element model, similar to the one previously described, was used for the transient thermal analysis at each of the six body points (BP) shown. Although several of these body points were located on the special RSI interface tiles, all body points were assumed to be located at the center of a multiwall tile so that the simple 1-D analysis could be used. For this analysis, thermal bowing was not considered and predicted heating rate histories were used as thermal inputs. A surface emissivity of 0.8 (representative of the surface coating on a multiwall tile) was used. The maximum structural temperature was calculated for each location. The temperatures which are shown on the figure were all below the maximum design temperature of 450 K (350°F). The calculated temperatures are considered to be conservative (high) since this analysis procedure has been shown to overpredict the structural temperature, especially when thermal bowing is not considered. Therefore, this preliminary analysis indicates that the present titanium multiwall tiles will adequately protect the Orbiter structure for this OEX experiment.

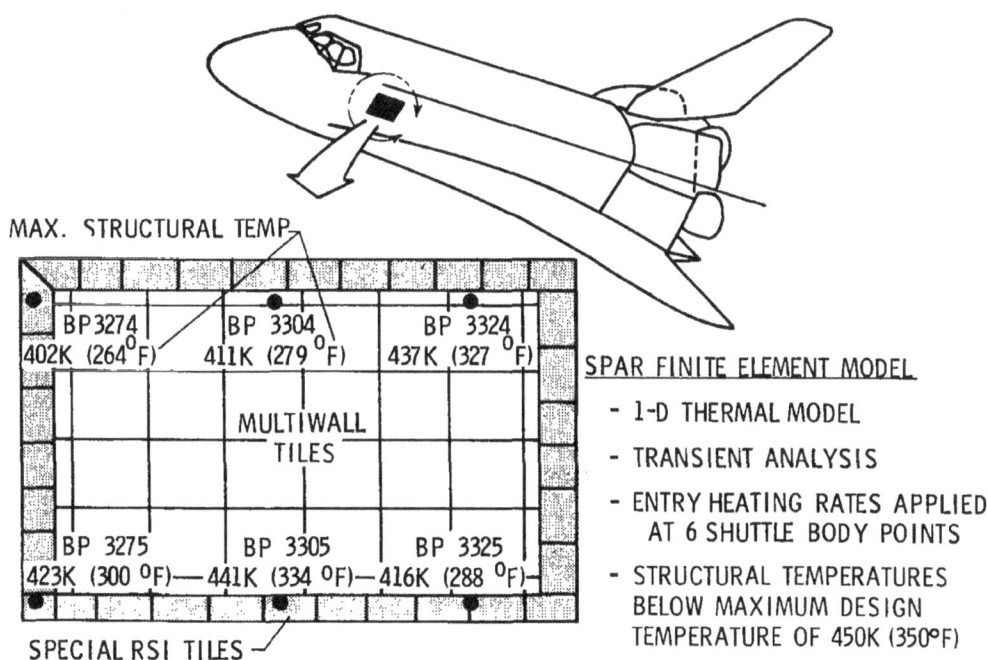

ANALYSIS OF EVACUATED MULTIWALL WITH 2-D SPAR FINITE ELEMENT MODEL

(Figure 8)

Since gas conduction is a major component of the heat transfer through multiwall tile (see fig. 3) the thermal performance of a multiwall tile can be greatly enhanced if the tile is evacuated. However, the pressure must be maintained at less than 10^{-4} mm of mercury to achieve this improvement, and the reliability of a multiwall tile to maintain such a high vacuum is a concern. To determine the effect of loss of vacuum, an evacuated multiwall tile array with a single tile which lost vacuum, as shown in the figure, was considered.

A simplified, two-dimensional SPAR finite element model was used to estimate the increase in structural temperature resulting from the loss of vacuum in one tile of an evacuated multiwall array. The model, shown in the figure, represents a wedge-shaped section with its sharp edge at the center of the unevacuated tile and extending the width of the neighboring evacuated tile. By modelling this wedge-shaped region, a 2-D model can be used to approximate a 3-D structure. The multiwall tile, air gap, and aluminum structure were modelled with 2D conduction elements. Element thicknesses were varied, as shown schematically, to account for 3-D heat diffusion. Radiation across the horizontal air gap was neglected because experience with the 1-D model, shown in figure 4, indicates that accounting for the radiation would have greatly complicated the analysis without significantly affecting the calculated structural temperatures. Radiation across the vertical air gap was neglected for simplicity.

In the normal direction the conductivity used for the evacuated multiwall tile was determined from figure 3 by subtracting the gas component from the total conductivity. The lateral conductance of the multiwall tile was assumed equivalent to that of the metal sheets because the contribution of air conduction was calculated to be negligible in comparison, and lateral heat transfer due to radiation was assumed to be negligible. The lateral conductance was approximately an order of magnitude higher than the transverse conductance.

Calculations were made for four different cases. In all cases the lower surface of the aluminum structure was assumed adiabatic. The first three cases had the prescribed entry temperature history of body point 3140 applied to the multiwall tile surface. For the first case all of the multiwall tiles were assumed unevacuated to determine the maximum structural temperature under an unevacuated array. In the second case the multiwall tiles were assumed to be evacuated to determine the maximum structural temperature under an evacuated array. For the third case, one multiwall tile was assumed unevacuated and the surrounding tiles were assumed evacuated to determine how the added energy absorbed due to vacuum loss in one tile diffused through the aluminum structure.

The purpose of the fourth case was to determine if a tile which had lost vacuum could be easily detected. Starting with the temperature distribution at landing, the surface of the multiwall was cooled by forced convection to ambient temperature, representative of a 5 km/hr (3 mph) wind, and the resulting surface temperature difference between evacuated and unevacuated tiles was computed. The results of these four cases are shown on the next figure.

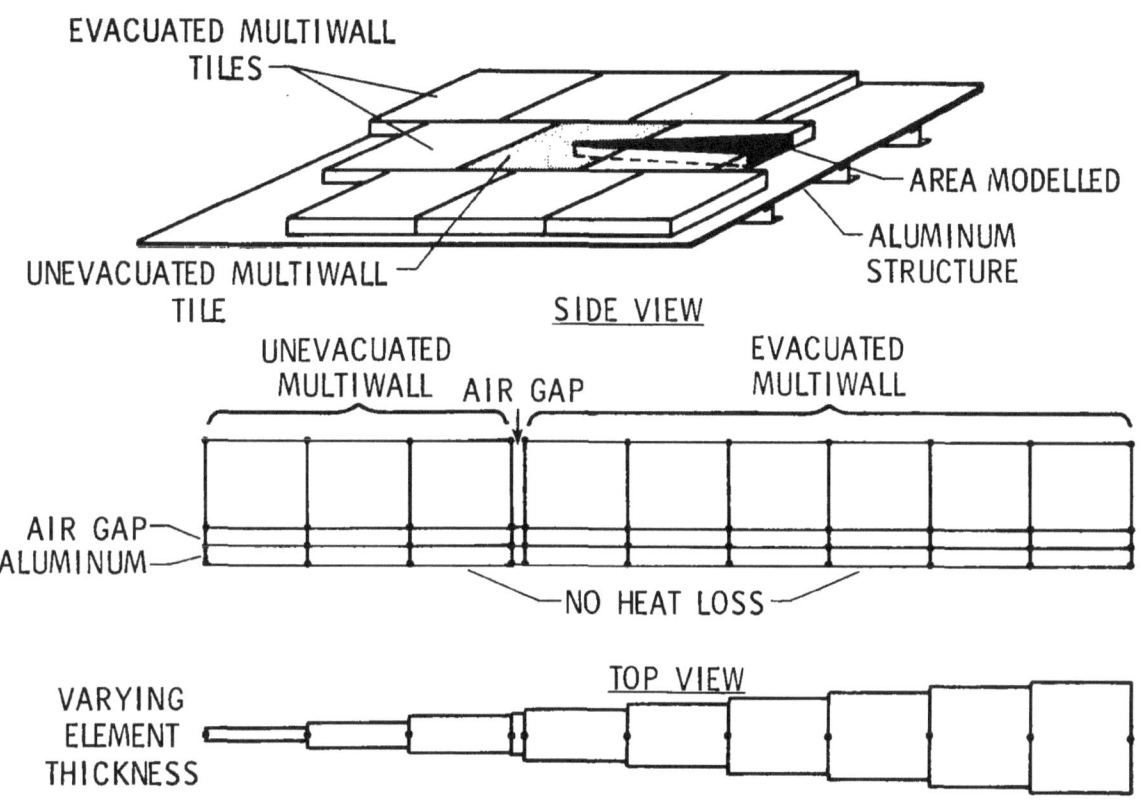

Figure 8

ANALYSIS OF EVACUATED MULTIWALL RESULTS

(Figure 9)

The maximum aluminum structural temperature, which was calculated to occur near landing, is shown on the left of the figure. The results from the first two cases are shown by the dashed lines. The upper dashed line represents the maximum structural temperature beneath an array containing all unevacuated tiles, and the lower dashed line represents the maximum structural temperature under an array containing all evacuated tiles. The 64 K (115°F) temperature difference is a measure of the improved thermal performance that results from using evacuated tiles. The solid line (case 3) represents the distribution of maximum temperatures in the structure underlying a single unevacuated tile in an evacuated array. The maximum temperature increase under the tile is only 20 K (35°F). As shown in the figure, the result of vacuum loss is not severe since the additional energy is diffused into the surrounding aluminum structure.

The surface temperatures of the evacuated and unevacuated tiles resulting from case 4 are compared on the right of the figure. Within five minutes after landing the surface temperature of an unevacuated tile, which was initially the same as that of an evacuated tile, exceeded that of the evacuated tile by approximately 11 K (20°F). Even after five hours, a 4 K (7°F) temperature difference remains as the structure slowly cools. These temperature differences could be easily detected by commercially available thermal scanning equipment. However, the structure may cool more rapidly since the analysis neglects heat loss from the backside of the structure, and consequently the surface temperature difference would diminish more rapidy. Further work would be necessary to quantitatively assess the effect of backside heat loss.

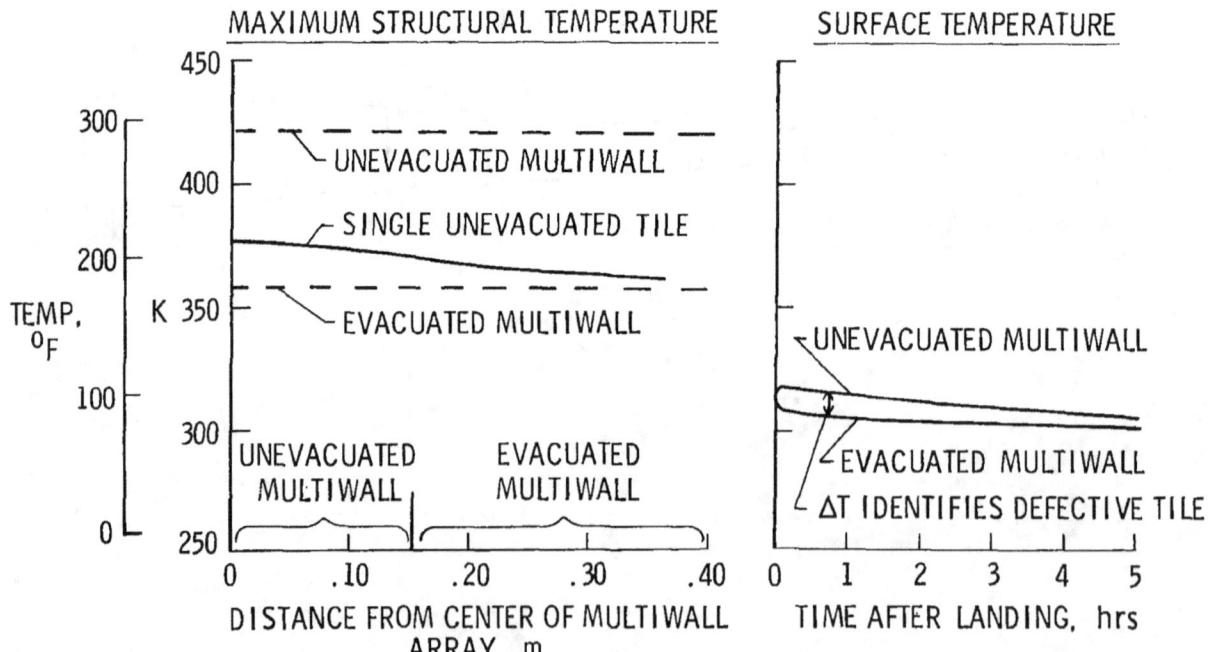

FUTURE THERMAL ANALYSIS OF TITANIUM MUTLIWALL

(Figure 10)

At some point in the development of a concept the simple preliminary analyses must be followed up by more detailed and complete analyses. The simple one-dimensional model which has been used to analyze the thermal performance of the titanium multiwall concept until now is only an approximation of the heat transfer at the center of a tile. The edge effects have been neglected. A more comprehensive analysis is planned which will incorporate the details of the titanium multiwall system shown in the figure. The heat transfer through the corrugated sidewall, the mechanical attachments, the gaps between and beneath the tiles, and the nomex felt, as well as the three-dimensional effects of thermal bowing will have to be considered in a more comprehensive analysis. The SPAR Thermal Analyzer will still be used for the analysis, but with a much more complex and detailed model.

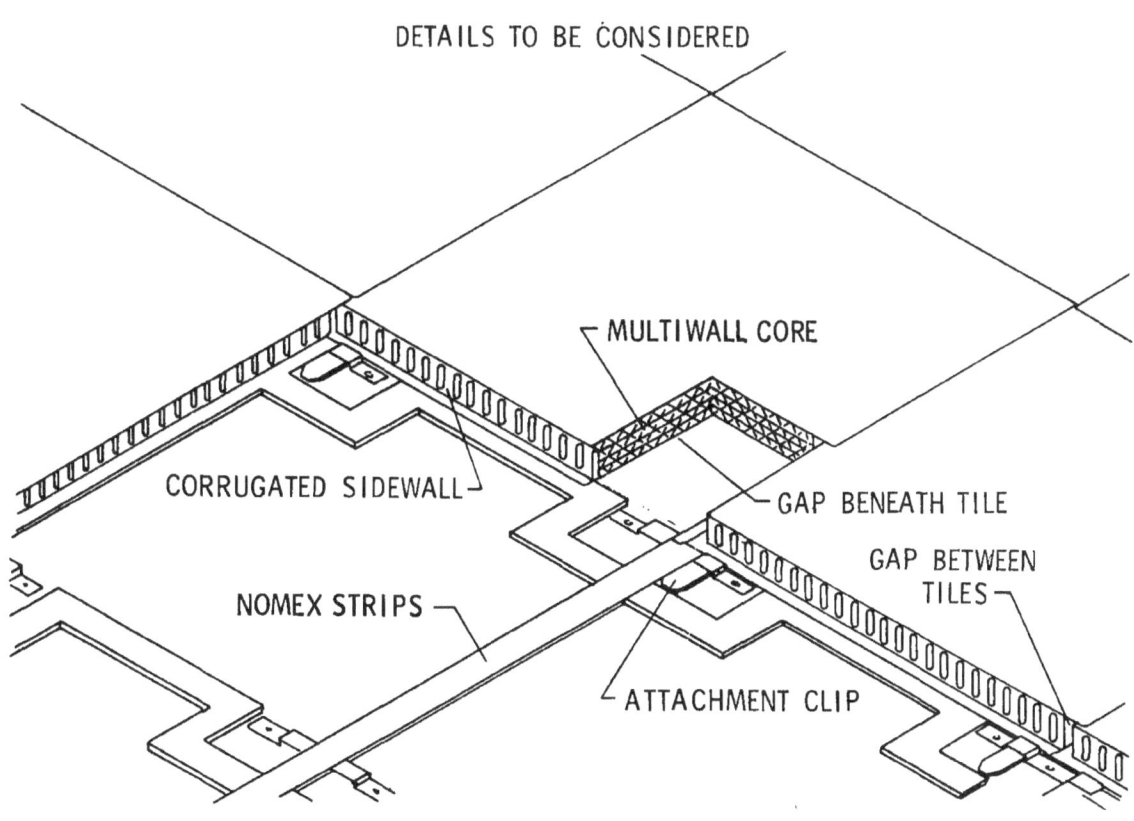

DETAILS TO BE CONSIDERED

SUMMARY

(Figure 11)

The SPAR Thermal Analyzer has been used for preliminary analysis of the titanium multiwall thermal protection system concept. First a steady state analysis was performed using a detailed finite element model of a small, representative region of a multiwall tile to obtain the effective conductivity of the tile. This effective conductivity was used with simple finite element models to determine the transient thermal performance for several preliminary design studies. A more comprehensive SPAR analysis which will incorporate details of the multiwall tiles and attachments will be necessary to more accurately predict the thermal performance of the titanium multiwall thermal protection concept for final design.

- DETAILED SPAR FINITE MODEL USED TO DETERMINE EFFECTIVE MULTIWALL CONDUCTIVITY

- SIMPLIFIED SPAR MODELS USED IN TRANSIENT THERMAL ANALYSES

- COMPREHENSIVE SPAR ANALYSIS REQUIRED TO ACCURATELY PREDICT THERMAL PERFORMANCE OF MULTIWALL THERMAL PROTECTION SYSTEM

REFERENCES

1. SPAR Thermal Analysis Processors Reference Manual, System Level 16. NASA CR-159162, 1978.

2. Jackson, L. Robert: Multiwall TPS. Recent Advances in Structures for Hypersonic Flight. NASA CP-2065, Part II, 1978, pp. 671-706.

3. Jenson, Carl L.; and Gable, Richard G.: Thermal Radiation Analysis System (TRASYS II) User's Manual. NASA CR-159273-1, 1980.

4. Blair, W.; Meaney, J. E.; and Rosenthal, H. A.: Fabrication of Titanium Multi-Wall Thermal Protection System (TPS) Test Panel Arrays. NASA CR-159383, 1980.

5. Avery, Don E.; Shideler, John L.; and Stuckey, Robert N.: Thermal and Aerothermal Performance of a Titanium Multiwall Thermal Protection System. NASA TP-1961, 1981.

HISTORICAL PERSPECTIVES ON THERMOSTRUCTURAL RESEARCH
AT THE NACA LANGLEY AERONAUTICAL LABORATORY FROM 1948 TO 1958

Richard R. Heldenfels
Distinguished Research Associate
NASA Langley Research Center
Hampton, Virginia

INTRODUCTION

This paper will describe some of the early research on structural problems produced by aerodynamic heating, conducted at the Langley Aeronautical Laboratory of the National Advisory Committee for Aeronautics from 1948 to 1958. That was the last decade of the NACA; in 1958 NACA became the nucleus of NASA.

I was one of the original investigators of these problems, became one of the leaders, and then managed such programs for the rest of my career at Langley. In this paper I will describe some activities in which I was personally involved using charts taken from papers published in those years. I have made a few literature searches to refresh my memory and locate suitable illustrations. I have not, however, approached this paper with the thoroughness of a historian; it is simply a personal recollection of some early research activities related to heat transfer in structures.

Figure 1 illustrates the organization of the NACA (ref. 1). The NACA was a committee established in 1915 to supervise and direct the scientific study of the problems of flight. The members were leaders of aeronautics in the United States and they represented government, industry, and universities. It was advised by committees and subcommittees composed of specialists in aeronautical technical areas. Only subcommittees under the Committee on Aircraft Construction are shown on figure 1 for simplicity. These committees determined policy and priorities for research. Often they focused on the urgent problems of the day, but some members were futurists who insured adequate research at the frontiers of flight. This particular type of committee organization was a significant factor in the attainment of world aeronautical superiority by the United States.

The NACA initially contracted for research but was aware that a well-equipped and suitably staffed laboratory was required to fulfill its obligations. Langley was established in 1920; the others listed were added during the NACA expansion in the World War II years.

Aircraft structures research in the NACA was concentrated at Langley, while Lewis conducted materials and structures research for propulsion systems.

PROGRESS OF THERMOSTRUCTURAL RESEARCH

Figure 2 shows the growth of research on structural effects of aerodynamic heating. The measure used is the number of papers presented at NACA conferences that had a session on structures (refs. 2-7). These conferences were held periodically to report significant research results to the aeronautical community in advance of the published reports. The proceedings were usually classified CONFIDENTIAL, a practice rarely used by NASA today.

Elevated temperature structures research, which had just begun in 1948, had become significant by 1951, and grew steadily thereafter with a significant increase between 1955 and 1957. These steps in growth correlate with recommendations of the Subcommittee on Aircraft Structures. In 1951, that subcommittee emphasized the need for more NACA research on current and future problems associated with elevated temperature of aircraft structures. In 1955 it became concerned that the number of people in this field had remained fairly constant and recommended that the effort be increased. This recommendation was approved by the NACA and the results were evident at the 1957 conference.

Most of the Langley structures research was done in the Structures Research Division. In 1948, 1.5 man-years of effort from 47 available professionals (3%) were devoted to high-temperature structures research. The numbers were 11 of 47 (23%) in 1952 and 50 of 62 (81%) in 1957. This was not a very large research effort by today's standards. These manpower percentages on heating problems are about the same as the conference paper percentages.

In the rest of this paper, some specific research activities will be described, starting with calculation of the temperature of the structure.

STRUCTURAL TEMPERATURE DISTRIBUTIONS

The basic principles of aerodynamic heating were known to early aeronautical scientists, but engineering data on heat transfer coefficients in supersonic flow was very limited. Figure 3 shows some results from the first NACA publication to calculate surface equilibrium temperatures in steady flight, reference 8. Results are given for Mach numbers from 2 to 10 for altitudes from 50 000 to 100 000 feet. Note that stagnation temperature was used as the maximum surface temperature instead of the adiabatic wall temperature. However, the recovery factor was discussed in the paper along with all other pertinent considerations. This 1946 paper concluded with a long list of areas needing further study.

The first transient skin temperature calculations are compared with those measured on a V-2 missile in figure 4 from a 1948 NACA publication, reference 9. This missile reached a maximum Mach number of about 5 just after 60 seconds and then coasted to 300 000 feet altitude at 100 seconds. The note concerning the basis of the calculations refers to the temperature used in the heat transfer coefficient equation.

Two NACA papers, references 9 and 10, were published at about the same time comparing calculations with the V-2 data. These papers differed in methods for calculating the heat transfer coefficients and the numerical time integration procedures used. In those days before the electronic digital computer, such calculations could be rather tedious. Our computing machine was an electric-powered mechanical calculator.

Figure 5 shows measurements made on an NACA rocket-powered model reported at the 1951 conference, reference 3. This data was used to determine recovery factors and heat transfer coefficients which were found to be in good agreement with the available theories. Confidence was thus established in our ability to calculate thin-skin temperatures at supersonic speeds.

We turned then to the more complex problem of calculating internal structural temperature and explored the numerical solution of problems involving heat conduction within the structure. Figure 6, from the 1953 conference, reference 4, shows the methods that were evaluated by comparison with wing structural temperatures we had measured in a hot supersonic jet.

Figure 7 shows a comparison of results from two calculation methods and the data for a skin and web combination. The agreement is reasonably good. Adiabatic wall temperature and heat transfer coefficients were determined from the thin-skin temperature histories to define the conditions in the test facility.

Figure 8 shows a similar comparison along the centerline of a cross section of the wing with all important internal conduction included in the calculations. Two-dimensional conduction was required to analyze the solid leading and trailing edges.

We were pleased with these results, so our research did not emphasize techniques for calculating temperature distributions until more complex methods were needed for ablation materials in the 60's. By that time much better computational facilities were available. We did, however, explore various other phenomena such as effects of internal radiation and conductivity of joints, both analytically and experimentally. In an attempt to simplify the computations, we used an analog computer to solve the Method III problem of figure 6, but the setup time required made that an unproductive endeavor.

Heat transfer research with rocket-powered models had produced data up to $M = 14$ by the time of the 1957 conference, reference 7, and much wind tunnel data was available to $M = 6.8$. The research airplanes had attained a maximum speed of $M = 3.2$. This speed was reached by the X-2 airplane in 1956. It went out of control later in that flight and crashed, ending the X-2 project. The research airplane program continued to collect structural heating data, however, with the X-1B and X-1E.

Skin temperature measurements were made on all high-speed research airplanes, but the X-1B, figure 9, was especially instrumented for extensive skin and internal structural temperature measurements. The airplane was brought to Langley in 1955 for instrumentation because of our experience with structural temperature measurements and was later flown at the High Speed Flight Station. This 1957 conference figure from reference 7 shows skin temperature measurement locations; many others were located in the interior to obtain about 300 total measurements.

This completes the discussion of structural temperature distributions. Their effect on the structure will be discussed next.

STRUCTURAL EFFECTS OF AERODYNAMIC HEATING

Figure 10 was used to introduce the session on structural effects of aerodynamic heating at the 1955 conference, reference 6. It was one of many such charts used in those days to educate structural engineers not yet involved in the design of supersonic airplanes. Papers on these effects were in demand for technical conferences as were papers, similar to one I presented at the 1955 conference, on some design implications of aerodynamic heating.

Temperatures in the airframe have been discussed in the previous section. The charts that follow will address some of the items under structures and touch briefly on alleviation. I will not address the items under materials for lack of time. The change in material properties with temperature is the most important effect of aerodynamic heating on structural design and was the subject of the earliest structures research. This effect, however, was relatively simple to incorporate into structural design because prediction of structural buckling and strength was based on the stress-strain characteristics of the material.

I came to work at Langley in 1947 after engineering jobs with an aircraft company and the U. S. Army Air Corps at Wright Field. My first assignment was to a team developing methods for structural analysis of a sweptback wing. Although airplanes were being built with sweptback wings, the structural design methods of the time could not predict accurately the stresses and deflections of this new wing configuration. Figure 11, from the 1948 conference, reference 2, shows the idealized structure we used to represent a wing structure we had tested. I show this chart to emphasize the limitations on our ability to calculate stresses in a complex redundant structure. With this idealization we were able to reduce the principal computation to reduction of a 9 x 9 matrix (ref. 11). Today computer programs are available to solve this problem in great detail very quickly.

Our analysis method for the sweptback wing was presented at the 1948 conference. That conference included, also, the first NACA paper on a structural problem produced by aerodynamic heating. It was a thermal stress analysis of a multiweb wing under an arbitrary temperature distribution. This preliminary analysis was not completed because the principal investigator left Langley. In August 1948, I was assigned to continue the development of methods for thermal stress analysis. I became one of a very few people at Langley who were working then on elevated temperature structural problems. Initially, I put thermal expansion terms into the current analytical and numerical methods and applied them to some illustrative examples (refs. 12 and 13). What we called numerical methods then were later called finite-element methods. However, in 1948 axially loaded rods and rectangular panels that carried only shear constituted our complete stable of finite elements. We did, however, create a special triangular element for our swept wing analysis (fig. 11).

We devised the simple experiment shown in figure 12 to obtain experimental verification of our thermal stress methods. Steady-state thermal stresses were induced in a large, thick plate by heating the center and cooling the edges. This is typical of the kind of experimental structures research we conducted. Tests were designed to be critical in nature and limited in scope to get to the crux of the problem quickly and economically.

Figure 13, presented at the 1951 conference, reference 3, shows the excellent agreement we obtained between the theory and experiment for the longitudinal direct stresses. Similar results were obtained for the shear stresses and the transverse direct stresses that occur because of the free ends (ref. 14).

The theoretical results were obtained from an approximate solution based on the principle of minimum complementary energy. To do that I had to derive the correct energy term; I did it by working backwards from the differential equations. In those days some theoreticians did not agree on a rational derivation of this term, but I was satisfied with one that worked.

The plate of figure 12 was set up on simple edge supports to conduct a thermal buckling experiment. Figure 14 shows a comparison of the results of those tests with calculated results, also from the 1951 conference. These calculations used the previously described thermal stress methods and the energy method to solve the large-deflection buckling problem (ref. 15). The plate contained initial curvature; therefore, it began to buckle as soon as thermal stresses were induced.

With adequate methods for analysis of thermal stress and thermal buckling, we left their refinement to others and began to investigate effects of rapid heating on strength and stiffness. A wide variety of tests and analyses were made of simple structures subjected to rapid and steady heating. Figure 15, from the 1957 conference, reference 7, presents some important results on the effect of thermal stress on the failure strength of beams. These square tubes were tested with and without thermal stress and the failure load was essentially the same. The lines on the figure are calculated failure loads based on material properties at temperature. Thermal stress, however, did reduce the buckling load for these beams. These results removed some concerns about the importance of thermal stresses because they did not affect certain modes of structural failure.

Figure 16, from the 1955 conference, reference 6, shows a cantilever plate, heated along the edges to investigate changes in structural stiffness produced by nonuniform temperature distributions and thermal stresses. The radiant heaters were turned off at 16.2 seconds and the plate began to cool at the edges. During the heating the plate was periodically struck to excite its fundamental bending and torsion modes.

Figure 17 (ref. 16) shows the change in the frequency of the first torsion mode (35%), the one most affected by this type of temperature distribution. The first bending frequency was reduced 21%. The plate twisted also, because the thermal stresses coupled with the initial twist in the plate.

The techniques for calculating thermal stress and buckling described with figures 13 and 14 were used with the addition of a frequency term to obtain the theoretical results which are seen to be in good agreement with the data. Measured temperature distributions were used in these calculations.

In the course of our stiffness reduction research we developed a system for following a resonant frequency as it changed during a heating test. We used it to test some wing structures; typical results are shown in figure 18 from the 1957 conference, reference 7. This solid, double wedge wing experienced small changes in the first five natural modes. The radiant heating used however, was not a good simulation of aerodynamic heating of this type of wing. Similar tests on multiweb wings showed frequency changes twice as large for the mode shapes characteristic of that type of structure.

We devoted much effort to the study of stiffness changes due to thermal stress, but I am not aware of this problem ever being important in the design of an airplane, missile, or space vehicle except with respect to panel flutter. Panel flutter is beyond the scope of this lecture because the primary investigations of the effects of aerodynamic heating on it were conducted in later NASA programs.

Our interest in changes of effective stiffness was not generated by any theoretical insight but by a 1952 experiment that produced startling and totally unexpected results. Figure 19, from the 1953 conference, reference 4, shows the test facility and one of the test specimens in the program. The test facility was a free jet, 27 x 27 inch size, with an exit Mach number of 2 and a stagnation temperature of 500°F. The model shown had a 20-inch chord and span, typical of most models tested. The first test was made on a model twice that size to obtain the temperature data shown in figures 7 and 8. Near the end of that test the model appeared to experience panel buckling and vibration that led to its destruction. Many additional tests were made on models like that shown here to identify the failure mode and methods for its prevention.

Figure 20 (ref. 4) shows the camber type flutter that resulted from stiffness changes produced by aerodynamic heating. The wings that fluttered had very low resistance to shear deformation of the cross section, and the fifth natural vibration mode, the one most affected by thermal stress, involving such deformations was predominant in the response.

The spectacular nature of these failures provided our program with high priority support but, again, I am not aware of such a failure mode being important in the design of any aerospace vehicle. In any event, this type flutter is easily prevented by the addition of a few ribs. A theoretical analysis of this type of flutter, that correlated well with our test results, was published in 1962 (ref. 17).

In addition to coping with aerodynamic heating, means to alleviate it were also of interest. Our initial analysis indicated that alleviation by insulation was of greatest interest at hypersonic speeds so we did not do much thermal protection research until the late 50's. Figure 21, from the 1957 conference, reference 7, shows some insulating panels that had been evaluated by a variety of tests. These panels were designed and constructed by the Bell Aircraft Company for lifting-entry vehicle applications; they called them double-wall construction. Research is still continuing on similar concepts but applications have been relatively few. Two that come to mind are the afterbodies of the Mercury and Gemini capsules.

Many other theoretical and experimental programs were undertaken in the years under discussion, but time does not permit comprehensive coverage. Equally important as the research planning and execution was the conception, construction, and operation of test facilities.

HIGH-TEMPERATURE STRUCTURAL TEST FACILITIES

Development of test equipment and facilities began along with the initiation of research projects and accelerated along with their expansion. Prior to the expansion in 1955, a presentation was made to several advisory groups on the NACA approach to high-temperature research facilities (ref. 18). Additional detail is given in reference 19 of some subsequent developments.

Figure 22, from reference 18, lists the types of facilities under development along with the general types of structures research testing that was needed. Combinations of furnaces and testing machines were the principal generators of data on materials and structural elements. Figure 23 shows a large furnace for strength and creep tests of structures. We did much short-time creep testing because it was thought to be an important design consideration for high-speed aircraft. However, when we related our results to design criteria, we concluded that airplanes would not be designed to operate in the creep range of the material (ref. 20). Therefore, we de-emphasized creep in our program starting in 1956. Subsequent events supported this decision.

Starting in 1951, we began to search for ways to simulate or duplicate aerodynamic heating in the laboratory. We evaluated a variety of devices for radiative and convective heating of structures. One of our goals was to achieve initial heating rates of 100 Btu per square foot per second. This was derived from calculations of the heat transfer rate to airplanes accelerating to $M = 3$ or $M = 4$ at 50 000 feet. That turned out to be a very valid long-range goal for airplanes because very few fly that fast even today.

The first device used extensively for rapid heating was the carbon-rod radiator shown in figure 24. It provided the desired heating rate but the high thermal inertia of the rods required that mechanical shields be used to control the heat radiated to the test specimen.

The tungsten filament lamp was a much better radiant heating device because it could be controlled adequately by the power input. But the available lamps were not sufficiently powerful to meet our goal. Fortunately, General Electric was developing a quartz-tube lamp with the desired characteristics. We acquired some development lamps, 5 inches long, in 1952 that were very promising. We requested that they make lamps with a 10 inch effective length. These lamps, shown in two double-row high-intensity heaters in figure 25, met our requirements and were the heat source used in most of our future heating tests. Coupled to an appropriate power supply and control system, this type of lamp, in lengths from 10 to 50 inches, became the principal method for rapidly heating structures in laboratories throughout the world. Numerous commercial applications were made also.

Convective heating to simulate or duplicate aerodynamic heating was investigated from the beginning of our facility development program and a variety of techniques were tried. The results were several supersonic jets and wind tunnels that provided a duplication of high-speed flight or a simulation with the stagnation temperature higher than that achieved in flight at the same Mach number. This was a consequence of the practical problems of duplicating hypersonic flight conditions in a wind tunnel.

Development of hot wind tunnels is a long and interesting story in itself, so I can only discuss a few highlights in this paper. Figure 26, which I used in a talk in April 1959, shows the operational (black) and planned facilities in the first year of NASA.

In March 1951 we had begun to plan an increase in our elevated temperature structures research. Langley management decided in June 1951 that we should plan also for large high-temperature structural research laboratory. Hot subsonic air flow and radiant heating panels were proposed to heat structures in a large test chamber. Further study and testing, however, revealed that a true-temperature, $M = 3$, blowdown wind tunnel was the best approach. This became the 9 x 6 Foot Thermal Structures Tunnel. Its basic characteristics were established in March 1952, the tunnel became operational in 1957, and research testing began in the summer of 1958. Construction was delayed when the funds initially appropriated were withdrawn by Congress in a federal budget reduction action. This facility was used to test a wide variety of structural models, many of which were evaluated for panel flutter. A structural failure in the air storage field, in September 1977, made further operations impractical.

The ethylene jet and the ceramic heaters were very high-temperature supersonic jets for testing materials and small models. The electric-arc powered jets subsequently carried this capability to extremely high temperatures. Their original development was motivated by the long-range ballistic missile program, but these Langley facilities made their major contribution later to the manned space flight programs, including the Space Shuttle.

The facility labeled 7' HTF is the initial concept of the facility now known as the 8-Foot High-Temperature Structures Tunnel. It is a true-temperature, $M = 7$ blowdown wind tunnel. Construction began in 1960 and high-temperature testing began in 1968. Although nearly 10 years elapsed between concept and research, this facility was on line long before the vehicles that benefited from its testing became a reality.

The rocket models listed on figure 22 have been discussed earlier. They made essential contributions to heat transfer data at very high speeds and did some structural testing also. Research airplanes were mentioned earlier, but that program received a new thrust when the NACA decided, in the spring of 1952, to initiate studies of problems likely to be encountered in space flight and of methods for exploring them.

THE X-15 RESEARCH AIRPLANE

A task group of five senior researchers was established at Langley in March 1954 to define the characteristics of an airplane to explore problems of hypersonic and space flight. The principal features of the vehicle they proposed are shown in Figure 27 from reference 21. It was a relatively small vehicle to be air-launched from a B-36 airplane, and then rocket-propelled to a maximum speed of 6 600 feet per second or to a maximum altitude of more than 250 000 feet.

The task group recommended a heat-sink type structure of Inconel X material. Their rationale is displayed in Figure 28 (ref. 21). Inconel X retains its strength well to 1200°F; this temperature established the heat-sink thickness required. However, much of the skin was strength critical so the heat sink criteria applied principally to secondary structure.

In December 1954, NACA, the Air Force, and the Navy agreed to sponsor this research airplane project with the Air Force managing the design and construction and NACA providing technical direction. The procurement process occupied most of 1955 with 4 of 10 interested companies submitting proposals. The winner is shown in figure 29 from reference 21. This airplane was very much like the results of the NACA study. If my memory serves me correctly, two of the other proposals presented a shielded structure and the third one recommended a magnesium heat sink. That rather novel approach raised some very valid concerns for the evaluation team since magnesium burns very intensely under certain conditions. We had great fun running a wide variety of tests using several different facilities to determine when a magnesium structure would ignite in flight. We found, for example, that a burning thin skin could be quenched by an adjacent, thicker spar cap.

Figure 30 (ref. 21) shows some of the early structural temperature calculations. In this case the wing-skin temperatures are much lower than the 1200°F limit because of strength requirements and mission characteristics.

Construction of the three X-15 airplanes was completed in 1959 with the first flight in June of that year. The flight program continued until December 1968 and provided much information on heat transfer and structural temperatures in high-speed high-altitude flight.

Support of the X-15 program was a high priority activity at Langley and we made many tests and analyses of potential problems. We made vibration tests of the horizontal tail under radiant heating and found that the resultant stiffness changes were not significant. Panel flutter, however, was a problem in several areas. Tests made in various wind tunnels included the horizontal and vertical tails in the 9 x 6 Foot Thermal Structures Tunnel. As a result, stiffeners were added to many thin-skin panels to prevent panel flutter within the flight envelope of the X-15.

NACA BECOMES NASA

The Soviet Union launched the first Earth satellite on October 4, 1957. This brought immediate changes in NACA programs as many people began to plan space research and flight programs. By December of 1957 I had prepared a plan for structures and materials research needed to rapidly advance manned space flight and we initiated some of these projects as people could be made available. The National Aeronautics and Space Act of 1958 (approved July 29, 1958) created NASA and at the close of business on September 30, 1958 the NACA ceased to exist. All of its property, facilities, and personnel were absorbed by NASA.

The NACA had excellent facilities and personnel that could get the space program off to a fast start. Much was accomplished in the year between Sputnik I and the official establishment of NASA. In fact, a bidder's briefing for a manned satellite capsule (Project Mercury) was held at Langley on November 7, 1958, just one year and five weeks after Sputnik I.

Although a new era in structures research had begun, we continued to support aircraft and missile needs along with the new emphasis on space. Our prior research experience, however, led us to concentrate much of our program on the technology required to return space vehicles to a safe landing on Earth.

Figure 31, from reference 19, which was prepared during the last days of the NACA, shows the flight regions in which our high-temperature structures research was focused. Charts like this were used with overlays to evaluate the capabilities of our test facilities relative to proposed flight systems. In addition to the airplanes and missiles that were the motivation of our initial research, we had supported the long-range ballistic missile program and the reentry glider of the USAF Dyna-Soar program for a manned orbital system. Dyna-Soar started in 1958 after preliminary studies called ROBO, BRASSBELL, BOMI, and HYWARDS. Less than a year later in 1959, I presented a similar chart that showed reentry vehicles at speeds twice orbital velocity and hypersonic airplanes at M = 6 to 9. The NASA years brought a greater scope and a faster pace to our research, but a decade of experience had prepared us well for this new challenge.

CONCLUDING REMARKS

In the foregoing, I have described briefly some of the research activities at Langley in the first decade of high-temperature structures research. Many other interesting activities could not be included.

Techniques for both experimental and analytical research have improved greatly in the last three decades with advances in electronics (instruments and computers) making the major contributions. Although much new knowledge is being acquired at a rapid rate, the search must always continue. My experience shows that the old problems are never completely solved; they just keep turning up in different situations and under other circumstances.

Our research began without a clear definition of the future vehicles to which it would apply. Therefore, we were concerned initially with generic research on potential problems. As a result, some of these problems were of little practical importance to the vehicles that were developed later. On the other hand, some vehicles that were proposed were never built or came into being much later than expected. For example, o   Few supersonic airplanes fly faster than M = 3 today.

o   The hypersonic airplane has not had a mission important enough to warrant its development.

o   A reusable orbital vehicle, the Space Shuttle, finally demonstrated that capability over twenty years after the Dyna-Soar project was started.

These examples lead to my principal message. Vehicle oriented research programs, which seem to be favored in today's environment, have the advantage of speeding the development of new technology for a specific mission or vehicle. An inherent danger in this approach, however, is that too much effort will be expended on developing technology that may not be used because the vehicle is never constructed. A healthy research program must provide freedom to explore new ideas that have no obvious applications at the time. These ideas may generate the technology that makes important, unanticipated flight or vehicle opportunities possible. Fortunately for the United States, this freedom of inquiry was fostered by the National Advisory Committee for Aeronautics, making possible our world leadership first in aeronautics and then in space.

REFERENCES

1. Forty-third Annual Report of the National Advisory Committee for Aeronautics - 1957. U.S. Government Printing Office, 1957.

2. NACA Conference on Aircraft Structures - A Compilation of the Papers Presented. NACA, May 1948.

3. NACA Conference on Aircraft Structures - A Compilation of the Papers Presented. NACA, Mar. 1951.

4. NACA Conference on Aircraft Loads, Flutter, and Structures - A Compilation of the Papers Presented. NACA, Mar. 1953. (Available as NASA TM X-57364.)

5. NACA-University Conference on Aerodynamics, Construction, and Propulsion. Volume I - Aircraft Structures and Materials. NACA, Oct. 1954. (Available as NASA TM X-57206.)

6. NACA Conference on Aircraft Loads, Flutter, and Structures - A Compilation of the Papers Presented. NACA, Mar. 1955. (Available as NASA TM X-57821.)

7. NACA Conference on Aircraft Loads, Structures, and Flutter - A Compilation of the Papers Presented. NACA, Mar. 1957. (Available as NASA TM X-67367.)

8. Wood, George P.: Calculation of Surface Temperatures in Steady Supersonic Flight. NACA TN 1114, 1946.

9. Huston, Wilber B.; Warfield, Calvin N.; and Stone Anna Z.: A Study of Skin Temperatures of Conical Bodies in Supersonic Flight. NACA TN 1724, 1948.

10. Lo, Hsu: Determination of Transient Skin Temperature of Conical Bodies During Short-Time, High-Speed Flight. NACA TN 1725, 1948.

11. Heldenfels, Richard R.; Zender, George W.; and Libove, Charles: Stress and Distortion Analysis of a Swept Box Beam Having Bulkheads Perpendicular to the Spars. NACA TN 2232, 1950.

12. Heldenfels, Richard R.: The Effect of Nonuniform Temperature Distributions on the Stresses and Distortions of Stiffened-Shell Structures. NACA TN 2240, 1950.

13. Heldenfels, Richard R.: A Numerical Method for the Stress Analysis of Stiffened-Shell Structures Under Nonuniform Temperature Distributions. NACA Rep. 1043, 1951. (Supersedes NACA TN 2241.)

14. Heldenfels, Richard R.; and Roberts, William M.: Experimental and Theoretical Determination of Thermal Stresses in a Flat Plate. NACA TN 2769, 1952.

15. Gossard, Myron L.; Seide, Paul; and Roberts, William M.: Thermal Buckling of Plates. NACA TN 2771, 1952.

16. Heldenfels, Richard R.; and Vosteen, Louis F.: Approximate Analysis of Effects of Large Deflections and Initial Twist on Torsional Stiffness of a Cantilever Plate Subjected to Thermal Stresses. NACA Rep. 1361, 1958. (Supersedes NACA TN 4067.)

17. Thomson, Robert G.; and Kruszewski, Edwin T.: Theoretical Study of Camber Flutter Characteristics of Monocoque and Multiweb Wings. NASA TR R-150, 1962.

18. Purser, Paul E.; and Heldenfels, Richard R.: Presentation on Facility Problems in High-Temperature Structures Research. NACA RM L56C24, 1956.

19. Heldenfels, Richard R.: High-Temperature Testing of Aircraft Structures. AGARD Rep. 205, 1958.

20. Heldenfels, Richard R.; and Mathauser, Eldon E.: A Summary of NACA Research on the Strength and Creep of Aircraft Structures at Elevated Temperatures. NACA RM L56D06, 1956.

21. Research-Airplane-Committee Report on Conference on the Progress of the X-15 Project - A Compilation of the Papers Presented. NACA, Oct. 1956. (Available as NASA TM X-57072.)

SYMBOLS

b	width
f_o	frequency of unheated structure
f/f_o	change in frequency due to heating
M	Mach number
T	temperature
T_{aw}	adiabatic-wall temperature
T_B	boundary-layer temperature
T_o	initial temperature
T_s	surface or skin temperature
T_T	stagnation temperature
t	thickness
W_{i_c}	initial center deflection of plate
W_c	center deflection of plate
ε	emissivity
τ	time
ω/ω_o	change in circular frequency due to heating

COMMITTEES

SUBCOMMITTEES

AERODYNAMICS
POWER PLANTS FOR AIRCRAFT
AIRCRAFT CONSTRUCTION
OPERATING PROBLEMS
INDUSTRY CONSULTING

AIRCRAFT STRUCTURES
AIRCRAFT LOADS
VIBRATION AND FLUTTER
AIRCRAFT STRUCTURAL MATERIALS

RESEARCH INSTALLATIONS

LANGLEY AERONAUTICAL LABORATORY	VIRGINIA
AMES AERONAUTICAL LABORATORY	CALIFORNIA
LEWIS FLIGHT PROPULSION LABORATORY	OHIO
HIGH SPEED FLIGHT STATION	CALIFORNIA
PILOTLESS AIRCRAFT RESEARCH STATION	VIRGINIA

Figure 1.- Organization of National Advisory Committee for Aeronautics (NACA).

DATE		PAPERS PRESENTED		PERCENT HEATING
		TOTAL	HEATING	
MAY	1948	17	1	6
MARCH	1951	15	4	27
MARCH	1953	16	5	31
OCTOBER	1954	5	2	40
MARCH	1955	16	8	50
MARCH	1957	19	15	79

Figure 2.- Structures papers presented at NACA conferences.

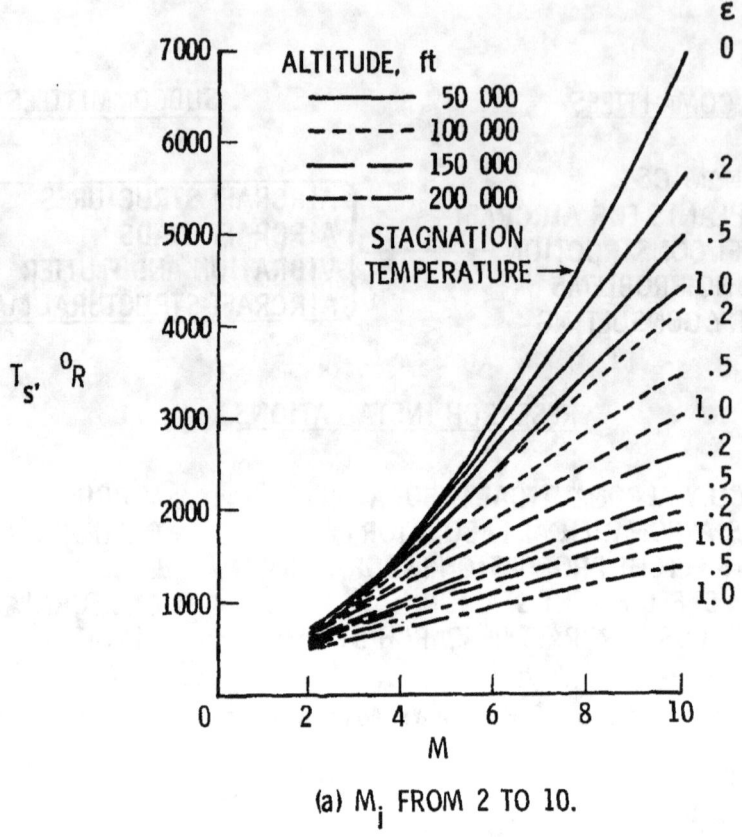

(a) M_i FROM 2 TO 10.

Figure 3.- Calculated surface equilibrium temperatures in steady flight.

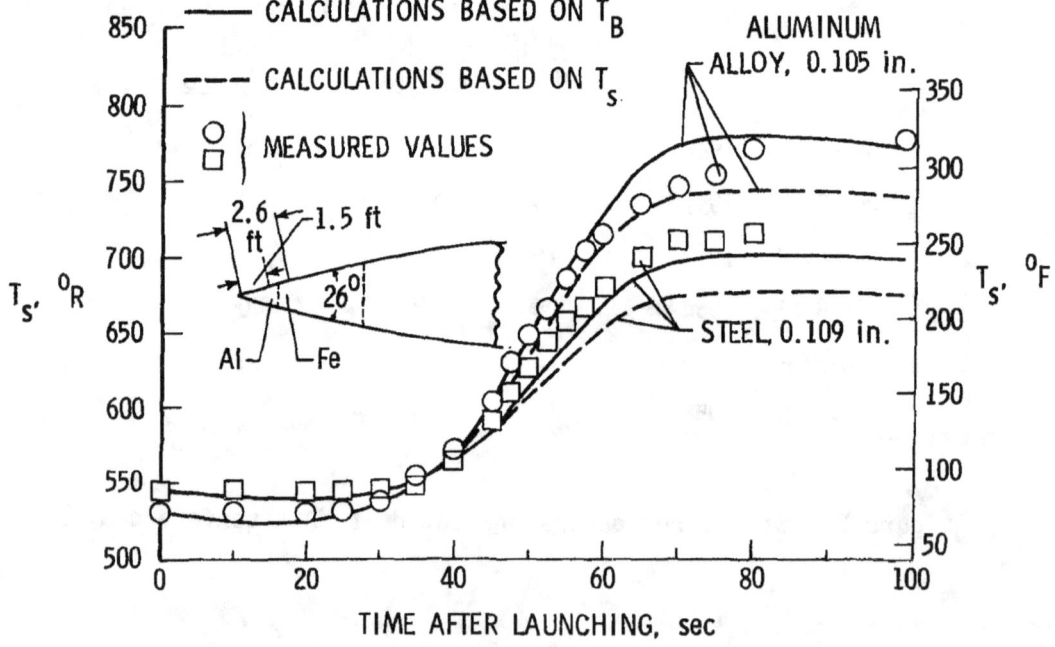

Figure 4.- Calculated transient skin temperatures compared with those measured on V-2 missile.

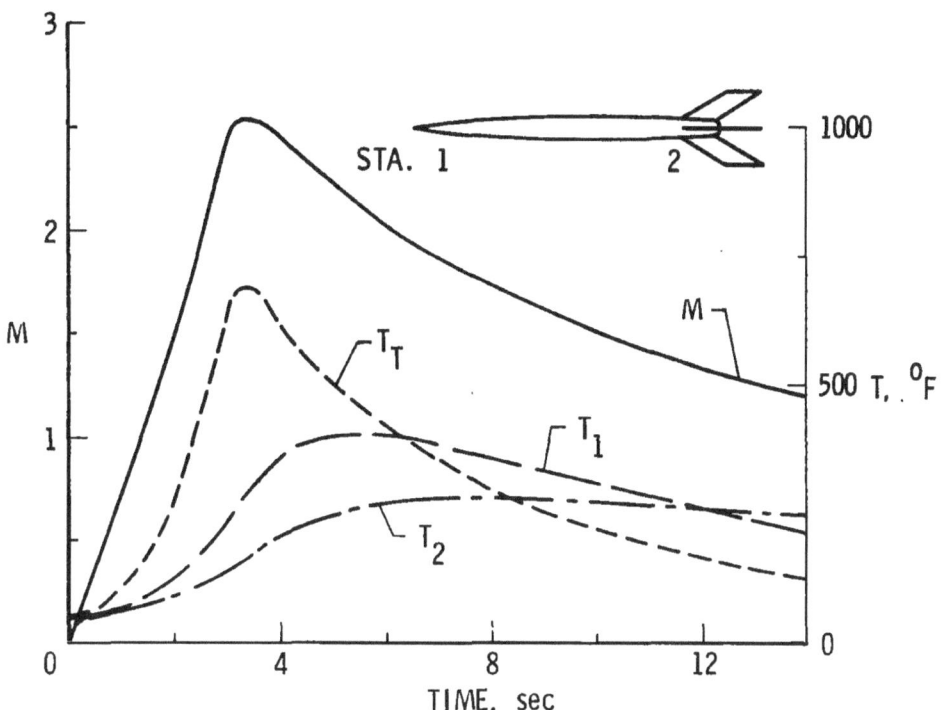

Figure 5.- Temperature history on NACA rocket-powered model.

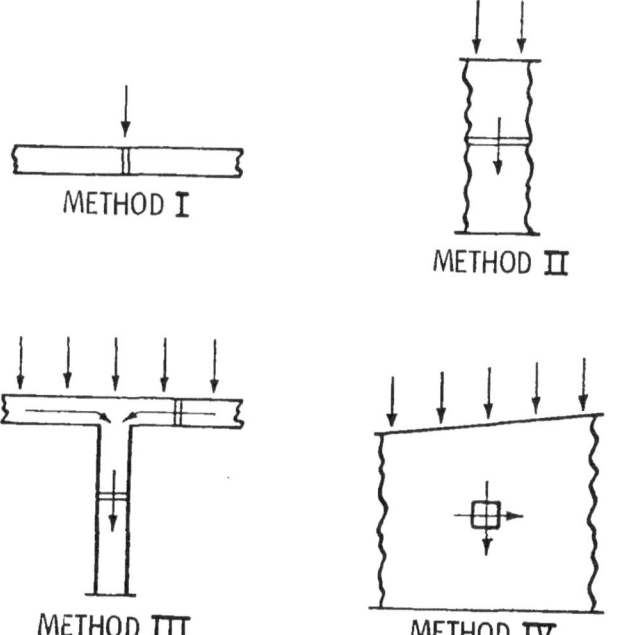

Figure 6.- Methods for calculating internal structural temperatures.

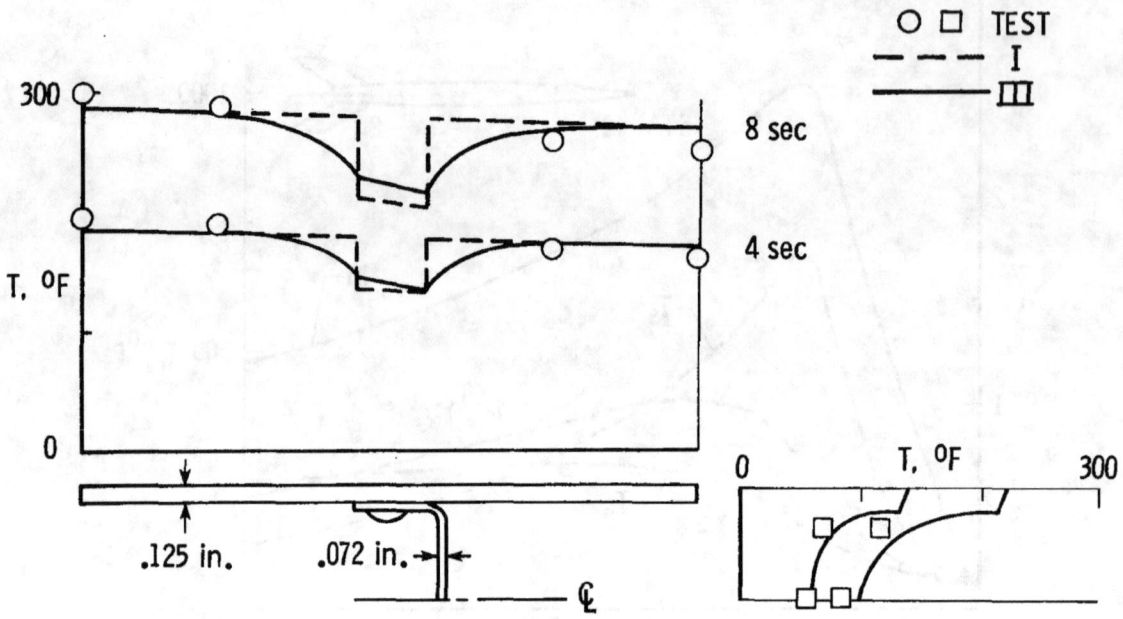

Figure 7.- Skin and web temperature distributions.

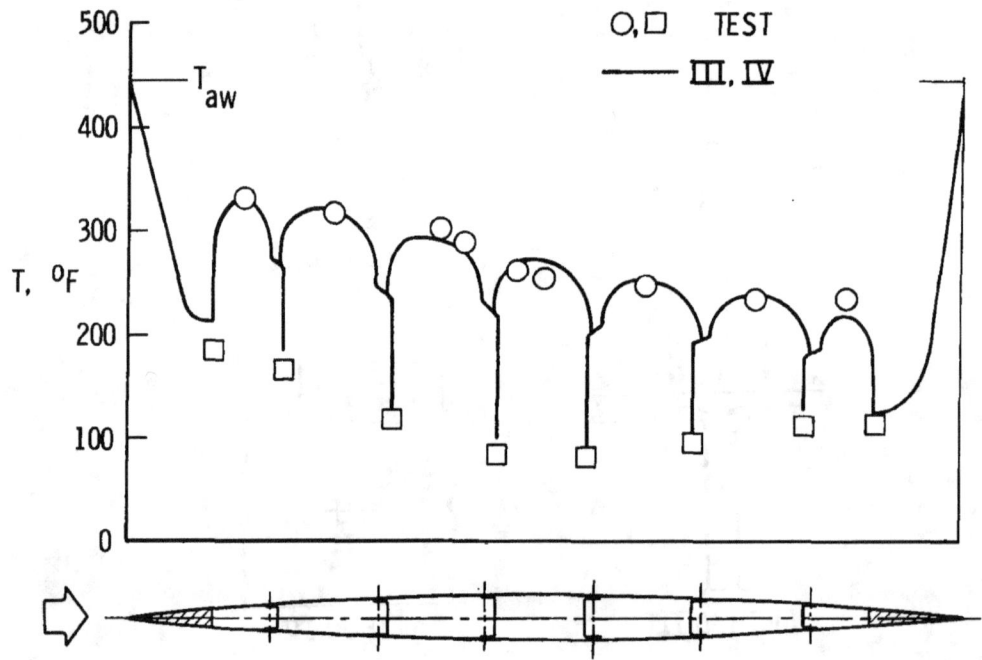

Figure 8.- Chordwise temperature distribution along centerline of wing cross section.

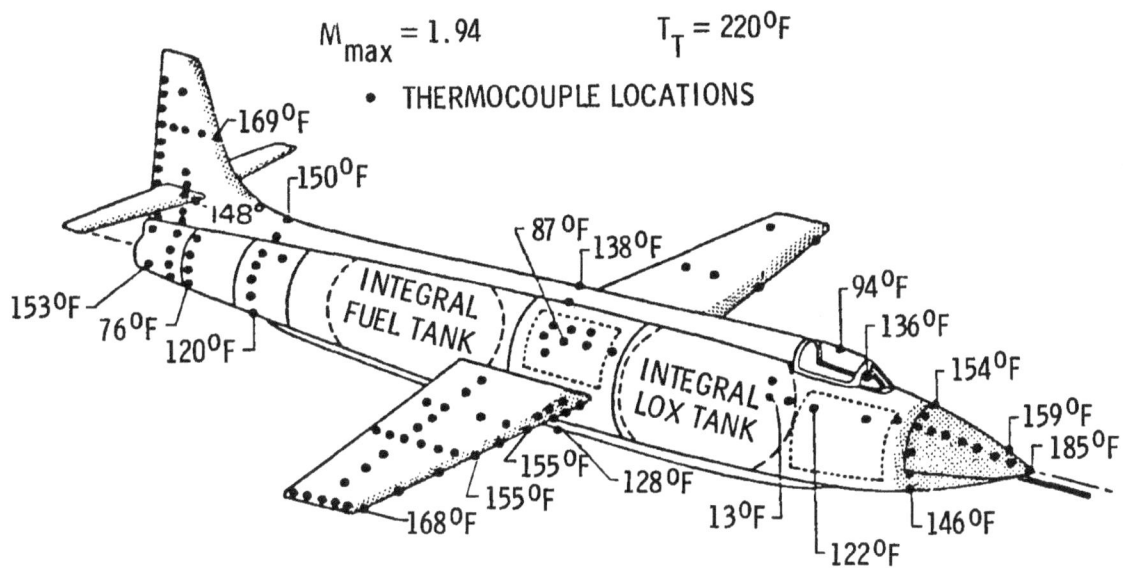

Figure 9.- Maximum measured temperatures on X-1B airplane.

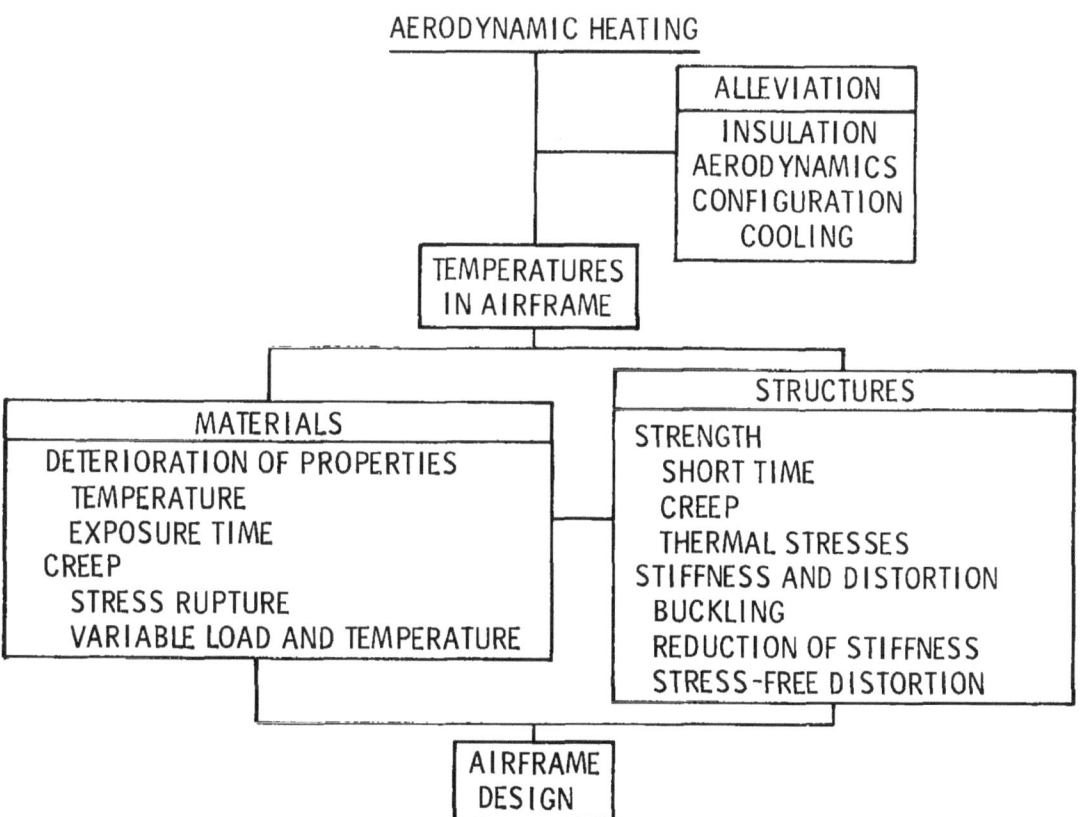

Figure 10.- Introductory chart at 1955 NACA Conference on Aircraft Loads, Flutter, and Structures.

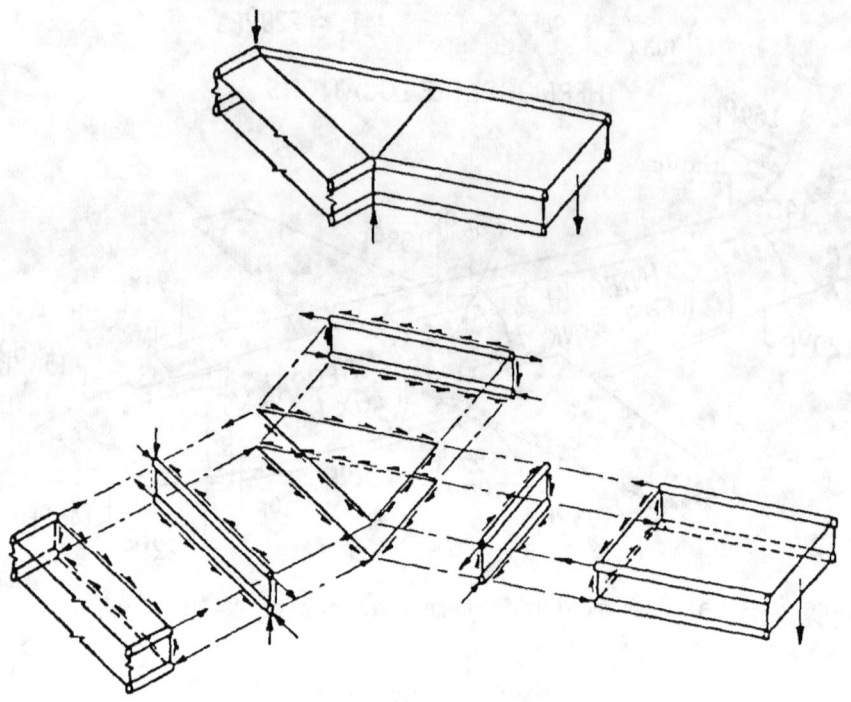

Figure 11.- Equivalent structure and breakdown used in structural analysis.

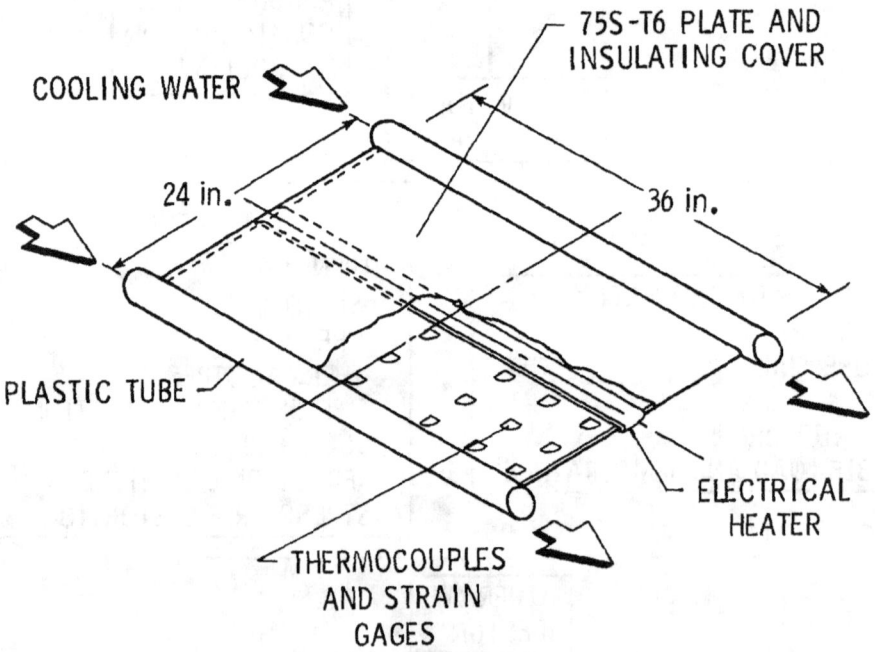

Figure 12.- Plate used for thermal stress test.

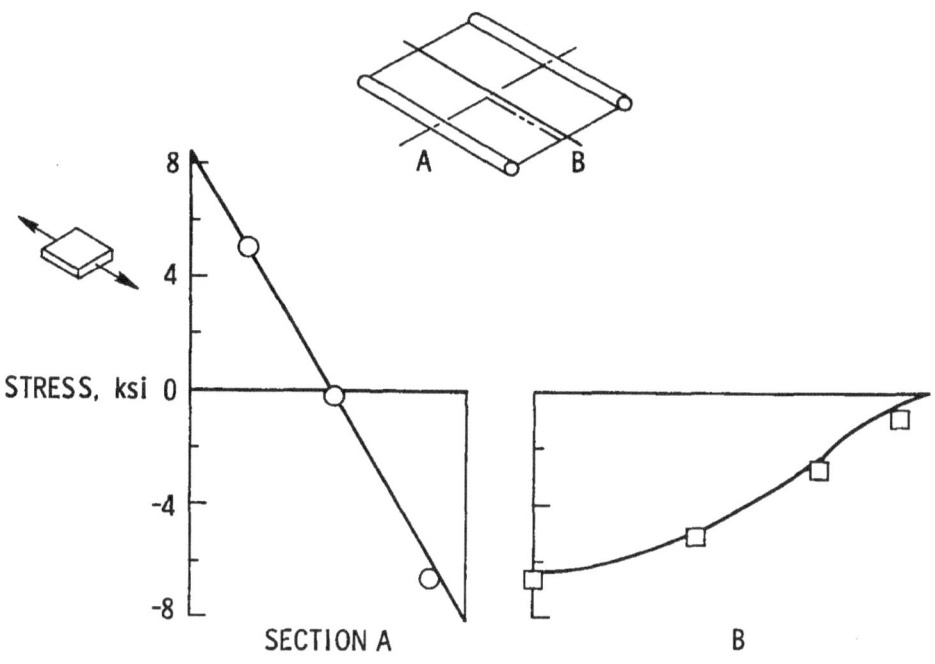

Figure 13.- Agreement between theory and experiment for longitudinal direct thermal stresses.

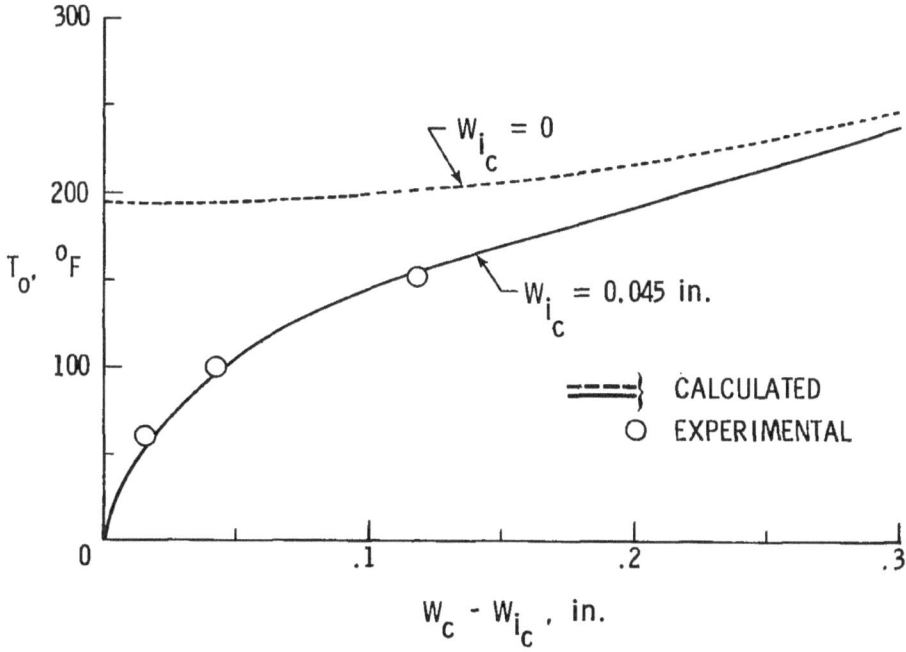

Figure 14.- Comparison between experiment and calculation for thermal buckling of a plate.

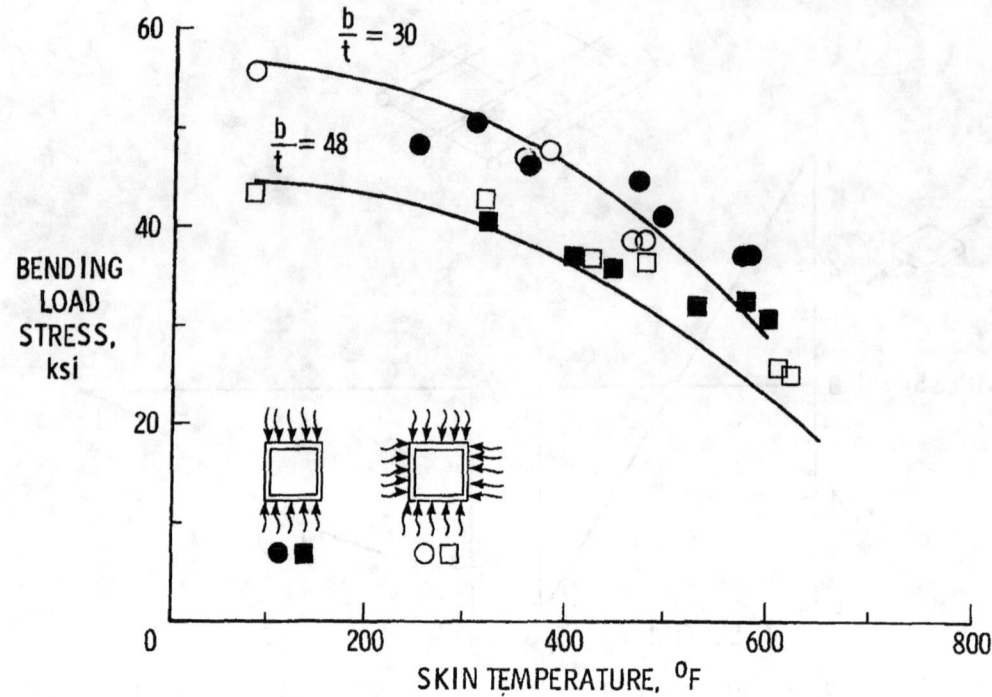

Figure 15. Rapid-heating effects on failure of 2014-T6 aluminum alloy beams. T = 100°F/sec.

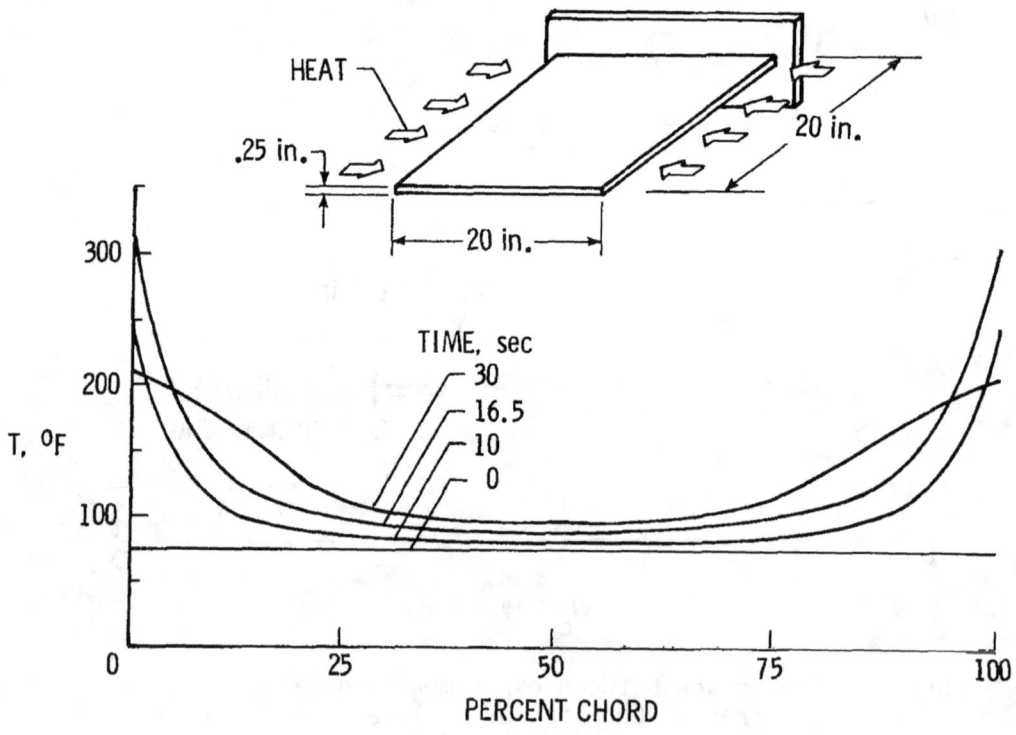

Figure 16.- Radiantly heated cantilever plate.

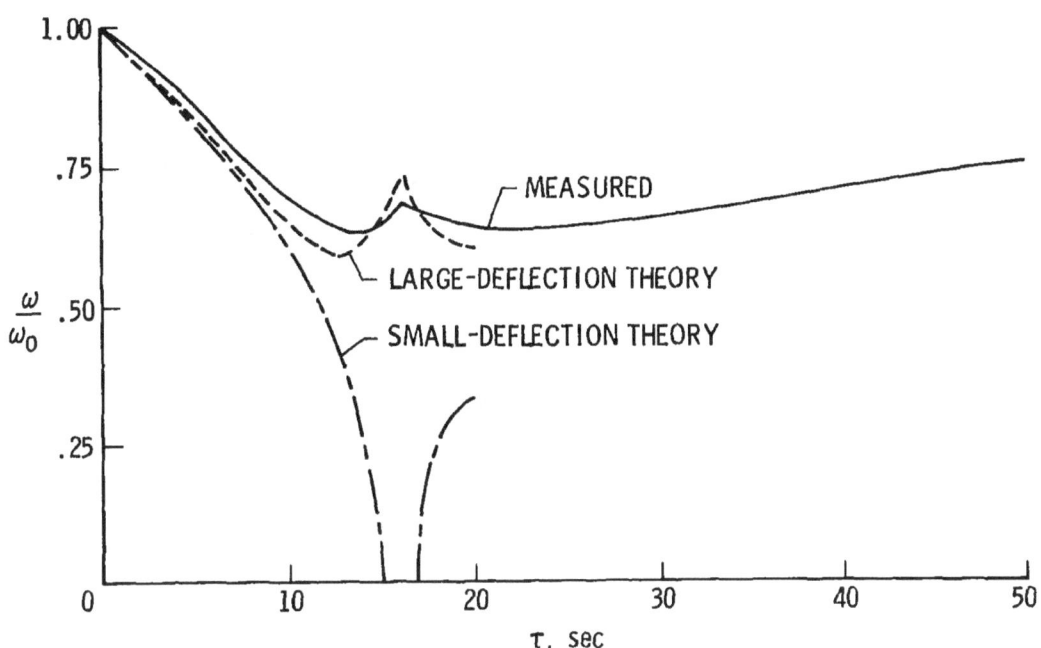

Figure 17.- Change in first torsion mode frequency of rapidly heated plate.

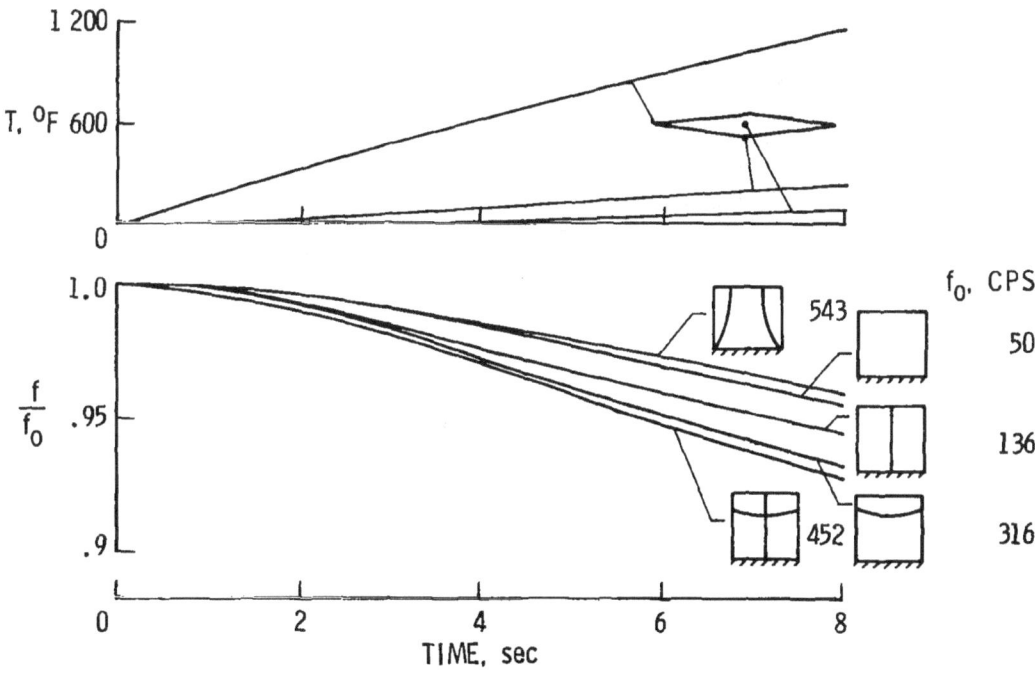

Figure 18.- Typical frequency history of double wedge wing.

Figure 19.- Jet test of wing structure.

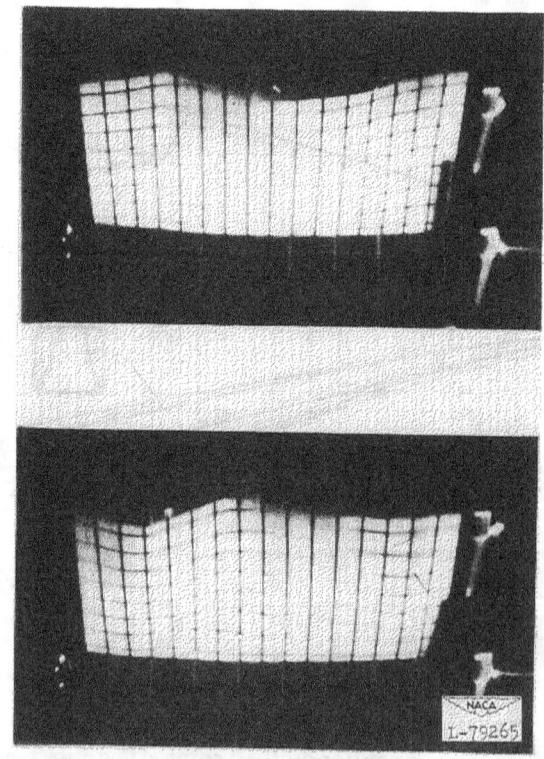

Figure 20.- Flutter resulting from stiffness changes due to aerodynamic heating.

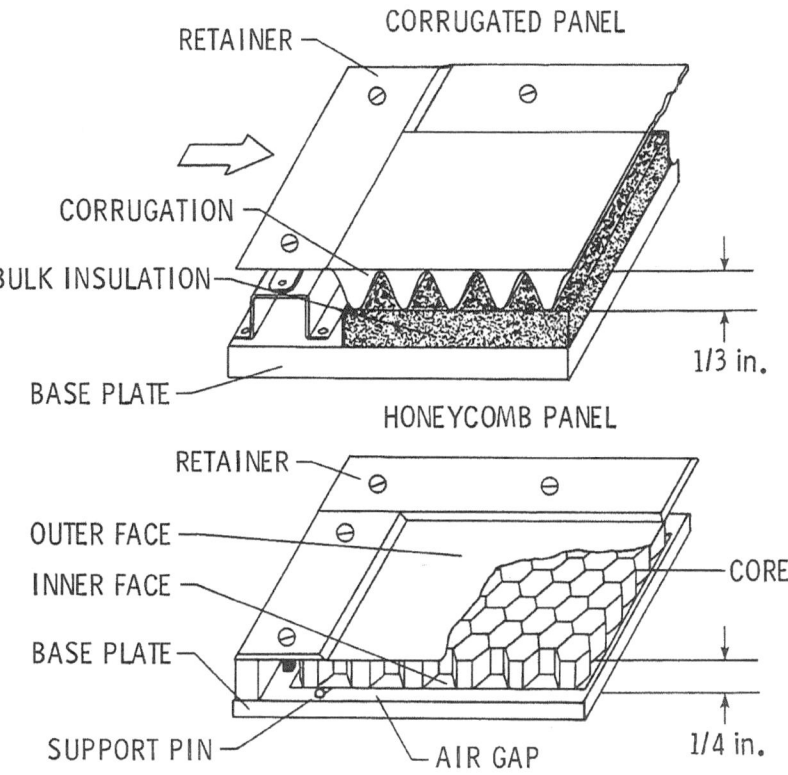

Figure 21.- Insulating panels.

- TYPES
  - LOADS
  - HIGH TEMPERATURES
  - RAPID HEATING
  - AERO-THERMO-ELASTICITY

- FACILITIES
  - TESTING MACHINES
  - OVENS AND FURNACES
  - RADIANT HEATERS
  - JETS AND WIND TUNNELS
  - ROCKET MODELS
  - RESEARCH AIRPLANES

Figure 22.- Structures research and facilities under development during 1955.

Figure 23.- Creep test equipment.

Figure 24.- Carbon-rod heat radiator.

Figure 25.- Quartz-lamp heat radiators.

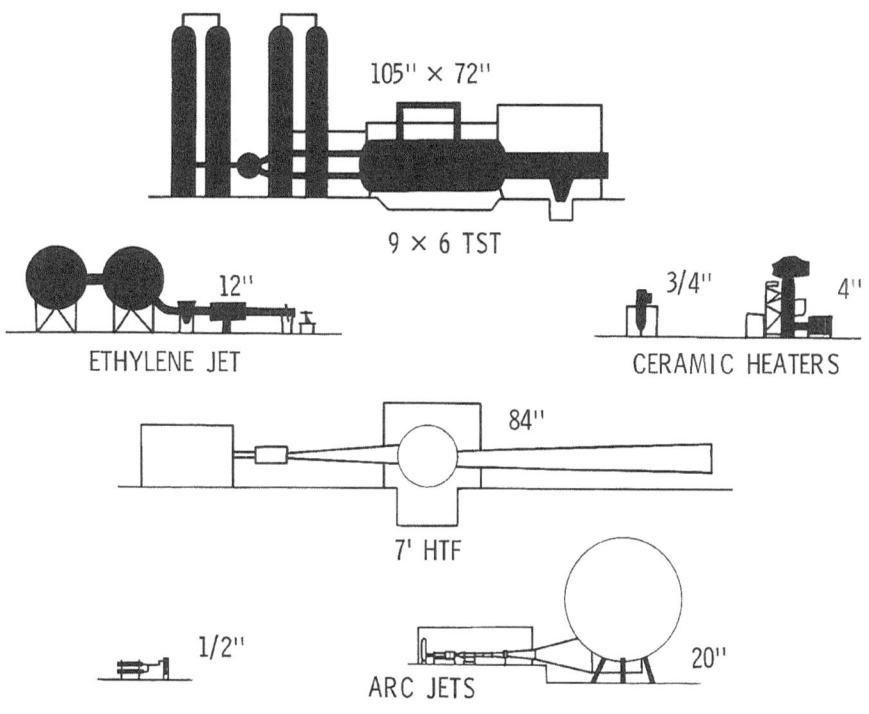

Figure 26.- Operational (black) and planned hot jets and tunnels in April 1959.

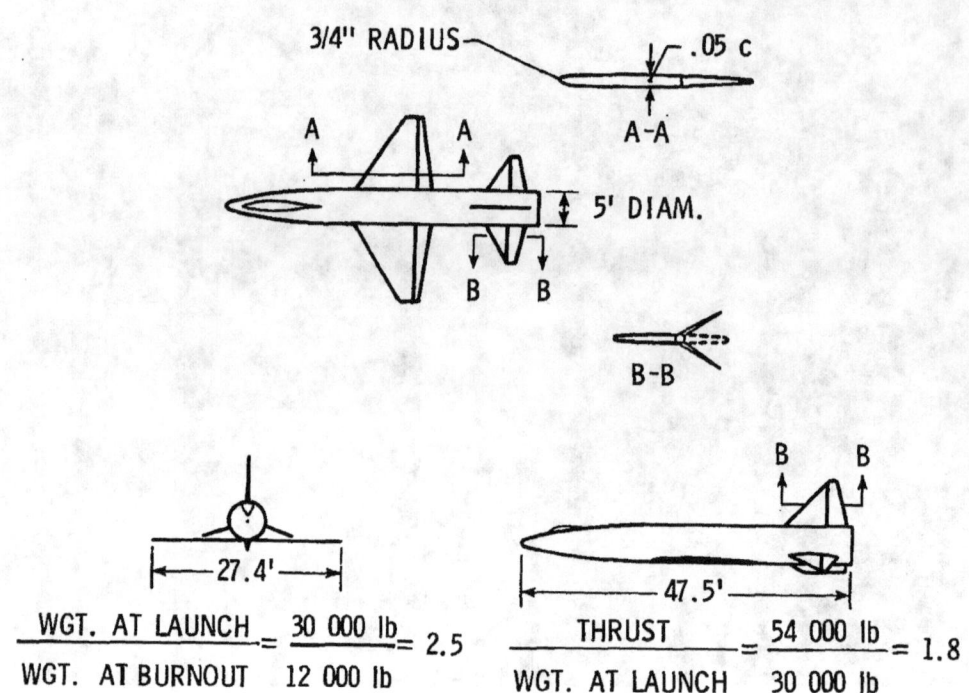

Figure 27.- Principal features of proposed X-15 research airplane defined during 1954.

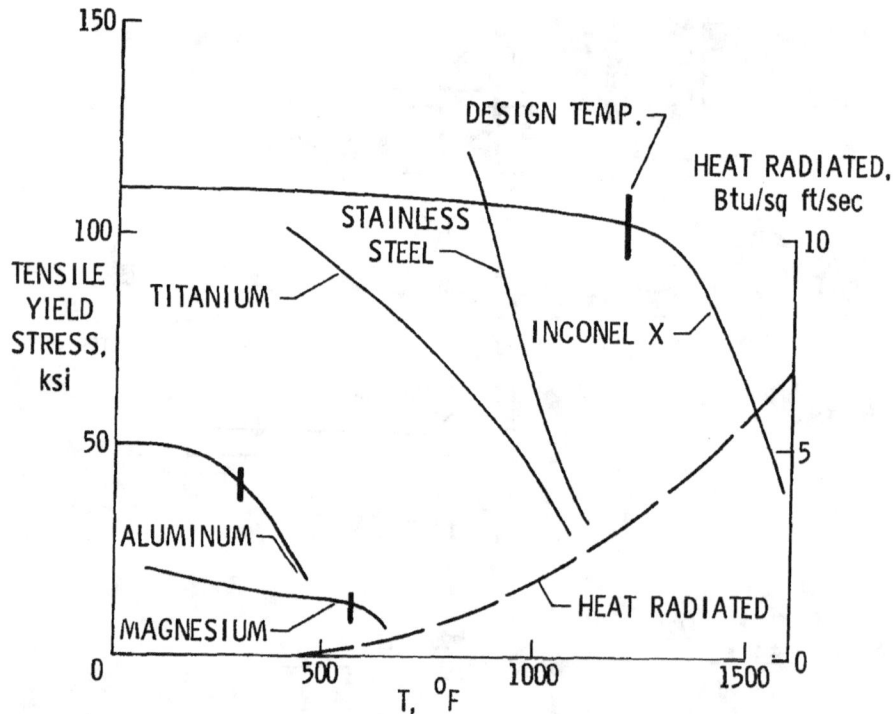

Figure 28.- Comparison of Inconel X with other candidates for X-15 applications.

Figure 29.- North American X-15 research airplane.

SPEED DESIGN MISSION

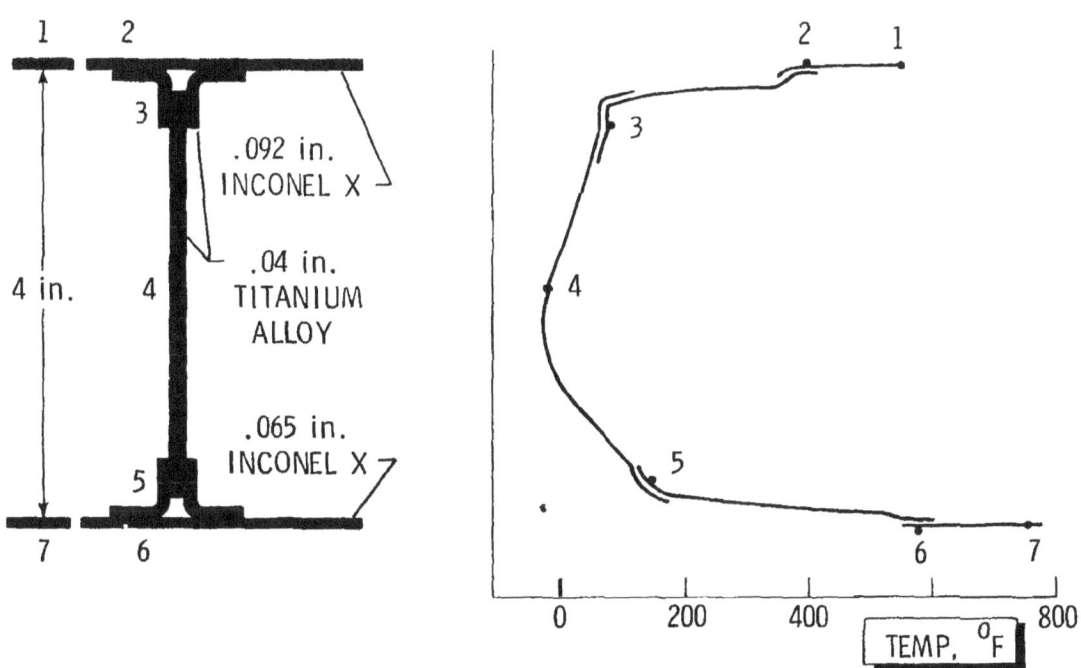

Figure 30.- Wing spar temperature calculations.

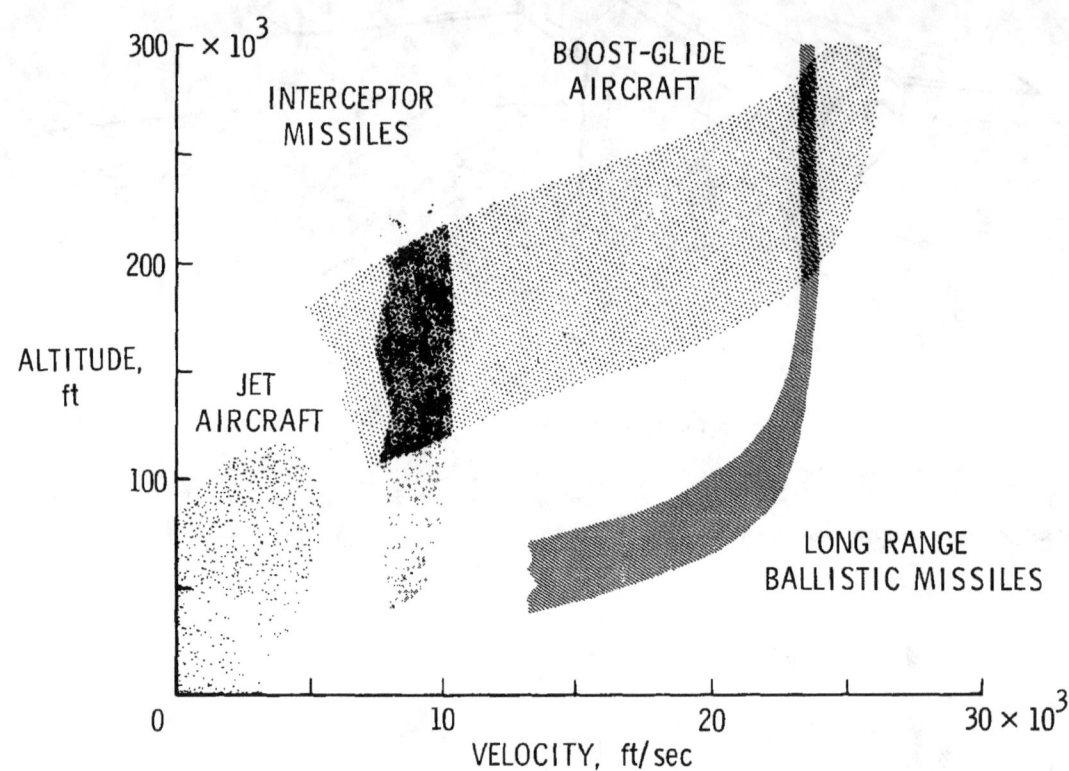

Figure 31.- Aircraft flight regions in which NACA high-temperature structures research was focused from 1948 to 1958.

GUIDELINES FOR DEVELOPING VECTORIZABLE COMPUTER PROGRAMS

E.W. Miner

Naval Research Laboratory
Washington, D.C. 20375

SUMMARY

This paper presents some fundamental principles for developing computer programs which are compatible with array-oriented computers. The emphasis is on basic techniques for structuring computer codes which are applicable in FORTRAN and do not require a special programming language or exact a significant penalty on a scalar computer. The intent is that researchers who are using numerical techniques to solve problems in engineering can apply these basic principles and thus develop transportable computer programs (in FORTRAN) which contain much vectorizable code. These principles are based primarily on the author's experience in running programs on the Texas Instruments Advanced Scientific Computer (TI-ASC), a vector processor, at the Naval Research Laboratory. The vector architecture of the ASC is discussed so that the requirements of array processing can be better appreciated. The "vectorization" of a finite-difference viscous shock-layer code is used as an example to illustrate the benefits and some of the difficulties involved. Increases in computing speed with vectorization are illustrated with results from the viscous shock-layer code and from a finite-element shock tube code. The applicability of these principles has been substantiated through running programs on other computers with array-associated computing characteristics, such as the Hewlett-Packard (H-P) 1000-F.

INTRODUCTION

The past decade has seen some considerable changes in the capabilities available to researchers involved in computational physics. Near the beginning of the last decade, scalar computers were the standard, but computers which would achieve higher computational speeds through parallelism or pipelining were already in the design stage and creating excitement among researchers. For example, the lead paper at the AIAA Computational Fluid Dynamics Conference in 1973 (ref. 1) was devoted to the future vector and parallel processors, their hardware, and their anticipated usefulness to the computational physics community. Since then, vector computers and other array oriented processors have become an actuality. The principal vector computers, for example, the CRAY-1, are still rather few in number and thus are available only to limited groups of researchers. However, other, more widely available computers have significant array-oriented features. In addition to the category of attached array processors, some mini-computers have array processing features. Specifically, the Hewlett-Packard (H-P) 1000-F series computers have what is called a Vector Instruction Set (VIS) implemented in firmware (microcode) which provides many of the benefits of vector programming and can increase computational speed by a factor of five or more. Even desktop computers (for example, the Tektronix 4050 series and the H-P System 35 and 45 desktops) have array oriented computational features in the BASIC language which they use. Thus, awareness of guidelines for developing computer codes which are compatible with array-oriented computing (i.e., vectorizable) should be advantageous to many researchers. Development of vectorizable codes should enhance code interchange and minimize reprogramming efforts when codes are exchanged between different computers.

Numerical techniques used by researchers to solve problems in computational physics typically are array oriented but common coding procedures sometimes reduce their compatibility with array processing. Such was the case with the particular program (a large boundary-layer type program) which is used as an illustration in this paper. By applying basic principles of "vectorization" as discussed in this paper, codes can be developed which are compatible with array processing and which require only a minimal, if any, increase in computing time on a scalar computer. In the case of the example program, the process of vectorization led to some code optimization and the scalar computing time was actually reduced.

Since FORTRAN is the most commonly used engineering programming language, attention is restricted in this paper to increasing the compatibility with array processing of programs coded in FORTRAN and is further restricted to FORTRAN code which is transportable. Although some of the examples used are specific to the Texas Instruments (TI) Advanced Scientific Computer (ASC), a vector processor, the examples illustrate considerations and techniques for developing codes which are compatible with array processing.

THE CONCEPT OF VECTORIZATION

In order to develop code which is vectorizable, it is necessary first to understand what vectorization is and how an array-oriented or vector computer differs from a scalar computer. The concept of vectorization is most easily introduced by example. Consider arrays A and B, each consisting of 100 numbers. Assume that one wishes to compute array C where $c_i = a_i + b_i$, $i = 1, 100$. The traditional scalar computer executes five assembly language instructions one hundred times. There are two memory fetches (a_i and b_i), one addition, one store to memory (for c_i), and an instruction that increments a counter, tests and branches back to load the next pair of input operands. Thus 500 scalar instructions are executed to add arrays A and B. A vector computer, or an array processor, has hardware which performs the 100 additions on the hundred pairs of input operands concurrently with the memory fetches and the stores to memory, greatly reducing the time required for computing array C. Such vector hardware may be available for virtually every arithmetic and logical operation.

The above example describes the singly subscripted FORTRAN DO loop:

```
        DO 100 I=1,100
        C(I) = A(I) + B(I)
100     CONTINUE
```

Doubly or triply subscripted arrays in loops nested 2 or 3 levels deep also may be collapsed on a vector machine into a single vector instruction.

Any FORTRAN DO loop representing one operation performed unconditionally on elements of one or two input arrays and producing elements of one output array is a candidate for a vector instruction and is thus vectorized. Vectorization may then be defined as getting as many operations as possible to vectorize. This requires designing, organizing, and writing programs so that the maximum possible number of arithmetic and logical operations can be executed as hardware vector instructions. Further, vectorization will be maximized when a programmer plans to operate on arrays of data instead of individual points of data. Such planning takes place at the program design level, at the subroutine level and at the line level within each subroutine.

ARRAY-ORIENTED PROGRAMMING

While vectorization is achieved by array-oriented programming, applications to specific computers may impose quite different constraints. For example, the attached support processors seem to be constrained by modest data transfer rates to and from the main computer's central memory. Thus the programmer would need to organize calculations such that many operations would be performed on the transferred data. In contrast, vector computers, such as the TI-ASC, can readily access large amounts of main memory and the vector architecture permits very rapid transfer of large arrays of data between the arithmetic units and main memory. In this case, the programmer needs to be less concerned with how the calculations are organized and can concentrate more fully on array-oriented programming.

For either hardware situation, array-oriented programming will require the following: choosing array-oriented algorithms which are vectorizable, planning program units which operate on arrays of data instead of points of data, minimizing the use of conditional operations, and planning array storage in memory for most rapid data transfer. These items are discussed more fully below.

Conditionality

A DO loop is the FORTRAN programmer's idiom for representing operations on arrays of data. The loop

```
      DO 100 I=1,50
      D(I)=A(I)*B(I)+C(I)
100   CONTINUE
```

represents two vector instructions; one which multiplies A and B, element-wise, and one which adds the elements of the product array to the elements of array C. The two instructions execute serially; the vector addition follows the vector multiplication. If this loop contains conditionality, e.g.

```
      DO 100 I=1,50
      IF(I.EQ.ITEST(I)) GO TO 100
      D(I)=A(I)*B(I)+C(I)
100   CONTINUE
```

it is no longer vectorizable. In this case the multiplication and addition may take place on some, but not all, of the array elements. An arithmetic or logical operation is vectorizable only if it is performed unconditionally on all elements of one or two input arrays to produce one resultant array.

Conditionality (the if-test) is often intrinsic to a computation, but significant vectorization may be achieved in the face of conditionality. A conditional operation can sometimes be transformed into one which is not conditional. Consider the loop

```
      DO 100 I=1,500
      IF(D(I).GT.DMAX) D(I)=DMAX
100   CONTINUE
```

which tests each element of array D and replaces only those values which pass the test. In the equivalent replacement loop

```
      DO 100 I=1,500
      D(I) = AMIN (D(I),DMAX)
100   CONTINUE
```

conditionality is eliminated and the operation is potentially vectorizable. In fact, vector machines may invoke a vector AMIN function which calculates the resultant vector D as a sequence of vector instructions on input vector D and scalar DMAX. Such vector functions are explained in a later section entitled "Vector Library Functions." Eliminating the "if-test" and consequently achieving array operations instead of scalar operations reduces the run time of the above example loop from 0.86×10^{-3} seconds to 0.4×10^{-4} seconds on the vector computer at NRL, a TI-ASC.

When using a vectorizing compiler an "if statement" within a loop will often inhibit vectorization of subsequent statements in the loop which are vector in character. Removing the "if-test" from the loop, or breaking the loop into several shorter loops, may result in significant vectorization. When the loop

```
          DO 100 I=1,500
          IF (X(I).GT.XMAX) X(I)=XMAX
          A(I)=C(I)*D(I)+X(I)
100       CONTINUE
```

is replaced by two loops

```
          DO 100 I=1,500
          IF (X(I).GT.XMAX) X(I)=XMAX
100       CONTINUE
          DO 110 I=1,500
          A(I)=C(I)*D(I)+X(I)
110       CONTINUE
```

its ASC execution time decreases from 0.17×10^{-2} seconds to 0.12×10^{-2} seconds. When conditionality is eliminated totally by using vector library function AMIN, the time drops to 0.86×10^{-4}. Minimizing the ill effects of conditionality is a central theme in the development of code which is compatible with vector computers and array processors.

Subroutine Organization for Array Operations

The fundamental principle for subroutine design is: plan, organize, and create subroutines which operate on arrays of data instead of points of data. For example, the program

```
          PROGRAM MAIN
          DIMENSON A(100),B(100),C(100)
          DO 20 I=1,100
          CALL SUB1 (A(I),B(I),C(I))
          CALL SUB2 (A(I),B(I),C(I))
          CALL SUB3 (A,(I),B(I),C(I))
20        CONTINUE
          END
```

locks the computation into scalar operations on points a_i, b_i, c_i and requires that the three subroutines be called 100 times each. The program above should be replaced by

```
          PROGRAM MAIN
          DIMENSION A(100),B(100),C(100)
          CALL SUB1 (A,B,C)
          CALL SUB2 (A,B,C)
          CALL SUB3 (A,B,C)
          END
```

where each subroutine operates on arrays A,B,C. This structure not only permits vectorized computation and but also minimizes costly subroutine linkage.

Vector Library Functions

Vectorizing compilers are built to apply the fundamental vectorization principle for subroutines. If a programmer codes

```
        DO 100 I=1,100
        B(I)=SIN(A(I))
100     CONTINUE
```

a vectorizing compiler can be expected to collapse the loop into a single call to a vector sine function with input vector A and resultant vector B.

Scalar computers have one system FORTRAN library. When a trigonometric function, square-root, maximum/minimum function, etc. is invoked, a point-wise (scalar) function is called with a scalar answer. Vector computers have such scalar functions, and, in addition, have a library of vector functions which operate on arrays of points. Vector functions are themselves vectorized. Five hundred sine calculations on the ASC take 0.11×10^{-1} seconds when done in scalar mode and 0.16×10^{-2} seconds in vector mode.

Algorithms and Mathematical Methods

Vectorization principles governing the choice of algorithms and mathematical methods may be deduced from the line level and subroutine level principles previously discussed. Methods chosen should involve significant unconditional computation on large arrays of data. Algorithms which entail more arithmetic operations may be preferred over those involving fewer arithmetic operations which do not vectorize.

On the ASC, recursive computations are intrinsically unvectorizable. Consider the loop

```
        DO 100 I=2,100
        A(I)=A(I-1)*B(I)
100     CONTINUE
```

where each element a_i of array A is computed from the element just previously computed, a_{i-1}. If this loop were performed in a vector mode, it would be equivalent to

```
        DO 100 I=1,100
        AA(I)=A(I)
100     CONTINUE
        DO 110 I=2,100
        A(I)=AA(I-1)*B(I)
110     CONTINUE
```

which yields different results from the original recursive code. The vectorizing compiler flags such loops as "vector hazards" and does not generate vector instructions for them.

When a recursive computation is required, it may be done in a loop by itself, isolated from other calculations. This prevents the vector hazard which it presents from inhibiting vectorization of subsequent calculations.

Memory Management

Vector instructions are most efficiently executed when the elements of operand arrays are stored, in central memory, contiguously with respect to the computation. The FORTRAN code

```
            DIMENSION A(10,50),B(50)
            DO 100 I=1,50
            A(K,I)=A(K,I)*B(I)
100         CONTINUE
```

exhibits non-contiguity for input operand A. The FORTRAN dimension statement declares that A is a 2-dimensional array and is stored column-wise in central memory. The multiplication occurs, element-wise, on a row or A. Thus every 10th value of A as it resides in memory is input and output to this computation. This substantially reduces the speed of the vector computation. A preferable coding for this situation would be

```
            DIMENSION A(50,10),B(50)
            DO 100 K=1,10
            DO 100 I=1,50
            A(I,K)=A(I,K)*B(I)
100         CONTINUE
```

With this arrangement (A is transposed) data streams from memory to the arithmetic unit quickly enough to ensure maximal execution speed.

Summary of Programming Principles for Vectorization

This list summarizes principles and guidelines already presented:

- Plan programs and subroutines which operate on arrays of data instead of points of data.
- Choose algorithms and mathematical methods which are array-oriented and vectorizable.
- Minimize and/or eliminate conditionality.
- Do not follow non-vectorizable calculations by vectorizable calculations in the same DO loop.
- Store vector operands contiguously in Central Memory.

THE TEXAS INSTRUMENTS ASC, A VECTOR COMPUTER

The rationale for developing computer codes compatible with array processing may be better appreciated by consideration of a specific system as an example. In some ways, the TI-ASC is a representative vector computer. The vectorizing FORTRAN compiler developed for the ASC recognizes array constructions in standard FORTRAN and generates vector instructions when appropriate. While it requires the programmer's attention to vectorization principles in the code, the compiler does not require special syntax or FORTRAN dialect to generate vector instructions. The ASC system thus illustrates the vectorization principles discussed above.

ASC Architecture

Three architectural features distinguish the Texas Instruments (TI) Advanced Scientific Computer (ASC). It is a pipeline computer; it has a full set of hardware vector instructions in addition to a full set of scalar instructions; and it is a multi-pipe computer.

An ASC arithmetic unit (AU) is logically and physically organized as a twelve-level pipe. Four levels are devoted to instruction decoding and processing, and eight to arithmetic or logical sub-operations. Thus when the AU is operating in scalar mode, up to twelve operations are concurrently at some stage of execution. At each CP clock cycle (80 nanoseconds) each arithmetic or logical operation in progress in the pipe drops to a lower level, and one answer may exit to the memory buffer. Pipe levels unnecessary to a particular instruction are bypassed. Memory buffers are considered part of the pipeline. Operands for calculations and answers are fetched and stored while the calculations are progressing through the pipe.

The most powerful computational capability of the ASC is its ability to run in vector mode. In this situation, a single operation is performed on many pairs of operands. For example, if A, B and C are vectors of length 100, only one vector instruction is needed for computing $c_i = a_i + b_i$; $i = 1, 100$. The A and B values stream continuously into the pipe, additions are performed in discrete steps within the pipe and answers flow back to central memory at the rate of one per clock cycle. The power of the vector instruction is that it guarantees optimum flow of calculations and data through the pipe.

An ASC may have one, two, three, or four pipes. The NRL computer has two pipes and, for fully optimal codes, can provide twice the computing power of a single pipe ASC.

The ASC Vectorizing Complier

A vectorizing/optimizing FORTRAN compiler, known as "NX", is available on the ASC. This compiler transforms ordinary FORTRAN code into vectorized object code which optimally exploits the ASC vector architecture. The NX compiler recognizes vectorizable FORTRAN constructions. When it fails to generate vector instructions, messages to the programmer may suggest how to rearrange or modify the code to achieve vectorization.

The NX compiler has three major levels of optimization. When invoking the compiler, a user specifies either I, J, or K level. An I level compile generates efficient, but unoptimized, scalar code. It is comparable to code generated by the IBM FORTRAN H compiler with OPT=0 or 1. At J level, the NX compiler generates optimized scalar code much like the IBM H compiler with OPT=2. Operating at level K, the NX compiler generates vectorized object code where possible, optimized scalar code elsewhere, and writes vectorization summaries and messages.

VECTORIZATION OF VISCOUS SHOCK-LAYER CODE AND COMPUTING TIME REDUCTIONS

To illustrate the process of vectorizing an existing code and to show the benefits which might be obtained, the vectorization of a moderately large FORTRAN program is discussed.

Description of Viscous Shock-Layer Program

The computer code which was vectorized is a laminar, hypersonic viscous shock-layer code previously developed by the author, (references 2-4). The code was written in FORTRAN and developed on an IBM 370/158. As discussed in references 2-4, the program uses an implicit finite-difference, marching integration procedure to solve the viscous shock-layer equations. Two flow field chemistries were available: dissociating oxygen and multi-component, ionizing air. As in the previous work of Davis (ref. 5), the governing equations are second-order accurate in the inverse Reynolds number parameter ϵ from the body to the shock.

By some criteria, the code might be a poor candidate for vectorization. The program has a significant amount of scalar code and, with 51 grid points used across the viscous layer, the arrays or vectors are much shorter than the vectors of length of 300 or more which have been often suggested for efficient pipeline use. In two ways this code is typical of large computer programs commonly used in solving engineering problems. First it was developed on a scalar processor, and second, efforts were made during its development to write code requiring minimum memory, not to write code that would vectorize. It was also coded in readily transportable FORTRAN. The size of the code, about 3000 FORTRAN statements, is perhaps typical of moderately large programs in use in solving engineering problems.

Computing Time Reductions

Since FORTRAN as implemented on the TI-ASC is very similar to IBM 370 FORTRAN, no changes were needed to run the code on the ASC. Runs were made to verify that the calculations of skin friction and surface heat transfer agreed with previously published results (ref. 2). After verifying the accuracy of the computed results, the program was compiled using the NX compiler at level I (no vectorization), and runs were timed. Other runs were made with the code compiled at J level to determine the gains in computing speed with scalar optimization and at K level to determine how much of the code would vectorize without further modification. Computing times for the viscous shock-layer code on the TI-ASC are given in table 1. The first three lines, for the "scalar" version of the code, give the times for the runs mentioned above. The optimiztion of the object code at J level reduced the computing time by 16%. At K level, enough code with in DO loops vectorized for an additional six percent reduction in the computing time.

As discussed earlier, it is often possible to get statements, which did not originally vectorize, to vectorize with only minor recoding. Recoding segments of the most repetitively used routines reduced computing time by a factor of 4. The computing time for the vectorized code is given in line 4 of table 1.

In vectorizing the code, it was necessary to add additional statements. The vectorized code contained 3497 FORTRAN statements compared with 3176 statements for the scalar version, though some of the additional statements were non-executable (e.g. DIMENSION and COMMENT) statements. The additional statements did not increase computation time when the vectorized version was run in a scalar mode. Line 5 of table 1 gives the computing time for the vectorized version of the code when compiled using the I (scalar) level of the NX compiler. Comparing the times in lines 1 and 5 shows a slight (4%) reduction in computing time for the vectorized version of the code when run in a scalar mode compared with the unvectorized version. The code in the vectorized version is just as transportable as the code in the scalar version and would be expected to run faster on a scalar processor than did the original code.

Table 1 shows the large reduction in time which was obtained by vectorizing the code. Table 2 lists the computing times for the code as various subroutines were modified. Most reductions in computing time were incremental except for subroutines VISCNA and WISUB which gave major reductions. These two subroutines calculate species and mixture properties at each point across the viscous layer. In the original code, the outer loops had the larger range (across the layer) and the inner loops had the smaller range (over the number of species, for example). While loops of length 6 will vectorize, the speed is comparable to scalar code speed. Most of the computing time reduction for these two routines was obtained by rearranging the loops so that the inner loops had the larger range and by eliminating conditionality from the inner loops. This gave typical vector lengths of 51 which run much faster than scalar speeds. It was also necessary to "code around" the exponentiation function ($X**Y$) which is not yet implemented as an ASC vector library function. Other runs were made with the viscous shock-layer code to determine how the number of grid points used in the program affected the computing time. These runs showed that increasing grid resolution is much less costly with a vectorized code running on a vector processor than with a scalar code running on a scalar processor.

We also considered how memory requirements are affected by vectorization and how scalar optimization affects computing time with the vectorized code. In many instances, the code had used scalar temporary variables within loops to conserve memory. In vectorizing the code, the scalar temporary variables caused problems. Either vectorization was inhibited or very inefficient vector code was generated by the NX compiler. By converting the scalar temporary variables to array temporary variables, the problems were overcome; but at the cost of some increase in memory requirements. However, the increase in memory can be minimized by storing the temporary arrays in a scratch common block which can be shared between routines.

A more complete discussion of the reductions in computing times for the viscous shock-layer code has been given by Miner and Brooks (ref. 6). Further information on the TI-ASC architecture and ASC programming considerations is given in references 6 and 7.

VECTORIZATION OF SHOCK TUBE CODE

During the past year, the author has had the opportunity to work with a shock tube code and make some vectorization tests with it. This particular code is a relatively small research code. It had been developed to investigate ways of solving the shock tube equations using a finite-element spatial discretization and various finite-difference techniques for the time integration. The program was developed (in mostly standard FORTRAN) on a minicomputer, a Hewlett-Packard (H-P) 1000-F. During development, the program was coded in standard FORTRAN and the firmware routines of the Vector Instruction Set (VIS) were not used.

The Vector Instruction Set is a group of firmware routines on the H-P 1000-F series computers and a group of equivalent software routines on the other H-P 1000 computers. The appropriate arithmetic operations have corresponding routines, and each routine is equivalent to a "DO" which performs that particular operation. The routines cannot be interrupted by the program logic and thus the programmer cannot include conditionality in these pseudo DO loops. The conditionality might still be coded, but it doesn't inhibit vectorization of neighboring code. A disadvantage of the VIS is that the programmer must vectorize the code explicitly by "commenting out" the old DO loop and inserting the VIS routine calls. This process can be somewhat cumbersome but not overly so. Since the readability of the code is reduced, it is convenient to retain the original code in comment lines. The principal factor motivating the use of the H-P VIS was not the cost of running the code on the H-P 1000 but the long execution time, thirty-five minutes. Fortunately, vectorizing the shock tube code was neither difficult nor time consuming. The vectorization was done in several stages and test runs were made to check results and computing time reductions. After the code had been mostly vectorized, an operations count indicated that about 95% of the candidate arithmetic operations had been replaced by vector routine calls. The computing time was reduced by a factor of six from 2100 seconds to 350 seconds.

It was also of interest to determine the computing times for this code on the TI-ASC. Runs were made with both the vectorizing NX compiler and the non-optimizing, non-vectorizing FX compiler. Since the FX compiler neither optimizes nor vectorizes the object code which it produces, it executes quite rapidly and is normally used for short test runs and code debugging. The resultant object code can be expected to execute slightly slower than the code from the NX compiler in level I. The nominal advantage of the FX compiler can be seen in the fact that it needed only 0.5 seconds to compile the shock tube code while the NX compiler in level K (vectorizing) required 8.0 seconds. The disadvantage of the FX compiler is really only in comparison to the vectorizing NX compiler. The shock tube code compiled using the FX compiler required 103.5 seconds to execute, while the NX-level K (vectorized) object code executed in only 6.25 seconds. For the shock tube code there was a much larger speed increase than for the viscous shock-layer code discussed above primarily because the shock tube code contained less scalar code and vectorized more completely.

Table 3 summarizes the computing times and relative computing speeds for the shock tube code. On the H-P 1000 vectorization increased the relative computing speed by 6, and on the ASC vectorization increased the computing speed by 16. The shock tube code provides an additional example of the potential benefit of designing a computer code so that vectorized object code can easily be generated.

SUMMARY

This paper presents some basic principles for writing FORTRAN code which is compatible with array processing. Many of these principles exact little, if any, penalty in computing time or memory requirements when used on scalar computers. All can be implemented in standard, transportable FORTRAN. When such guidelines are followed, substantial reductions in reprogramming effort will occur if the program is run on an array-oriented computer. Since vector computers, array processors and other array oriented computers are becoming more widely available, easy transportability between scalar and vector computers is a significant, desirable feature of FORTRAN programs.

These principles are illustrated by applying them to a viscous shock-layer code which was written for a scalar computer and then transported to the Texas Instruments Advanced Scientific Computer, a vector machine.

FORTRAN compilers and other software can be expected to recognize vectorizable FORTRAN constructions. The programmer must, however, be responsible for appropriate array-oriented program design, organization, and attention to the details of vectorization.

REFERENCES

1. Feustel, E. A.; Jensen, C. A.; and McMahon, F. H.: Future Trends in Computer Hardware, Proceedings of AIAA Computational Fluid Dynamics Conference, pp. 1-7, Palm Springs, CA., 1973.

2. Miner, E. W. and Lewis, C. H.: Hypersonic Ionizing Air Viscous Shock-Layer Flows over Sphere Cones, AIAA Journal, Vol. 13, January 1975, pp. 80-88.

3. Miner, E. W. and Lewis, C. H.: Computer User's Guide for a Chemically Reacting Viscous Shock-Layer Program, CR-2551, NASA, May 1975.

4. Miner, E. W. and Lewis, C. H.: Viscous Shock-Layer Flows for the Space Shuttle Windward Plane of Symmetry, AIAA Journal, Vol. 14, January 1976, pp. 64-69.

5. Davis, R. T.: Numerical Solution of the Hypersonic Viscous Shock-Layer Equations, AIAA Journal, Vol. 8, May 1970, pp. 843-851.

6. Miner, E. W. and Brooks, B. J.: Comparative Computer Times Between Vectorized and Scalar Versions of a Large Hypersonic Viscous Shock-Layer Code, AIAA Paper 78-1207, Seattle, WA., 1978.

7. Brooks, B.; Brock H.; and Miller, M: Guide to Vectorization on the Naval Research Laboratory's Texas Instruments Advanced Scientific Computer: Volume 1 Vectorization Primer, Memorandum Report 4102, Naval Research Laboratory, Washington, D.C., November, 1979.

TABLE 1.- COMPARISON OF COMPUTING
TIMES ON THE TI-ASC FOR VISCOUS
SHOCK-LAYER PROGRAM[a]

Code Version	Compiler Level[b]	Computing Time; sec
scalar	I	123.6
scalar	J	104.1
scalar	K	98.1
vectorized	K	25.5
vectorized	I	118.7

[a]Central Processor time, test case, 51 grid points across viscous layer.

[b]I level—scalar code only; J level—optimized scalar code; K level—vectorized code with scalar optimization.

TABLE 2.- COMPUTING TIMES ON THE TI-ASC
AS SUBROUTINES VECTORIZED[a]

Subroutine Vectorized	Function of Subroutine	Computing Time; sec
- - -	Base Line Case	98.07
DERIV3	Array Differentiation	96.59
SOLVE	Tridiagonal Solver	96.16
ENERGY	Energy Eq. Coefficients	95.97
SMOMNT	S-Momentum Eq. Coefficients	96.04
NMOMNT	N-Momentum Eq. Coefficients	95.80
SPECIE	Species Eq. Coefficients	95.19
THERM	Thermodynamic Properties	93.84
WISUB	7 Species Production Terms	62.37
VISCNA	7 Species Viscosity and Conductivity	27.52
MASS	Continuity Eq. Integration	27.37
HCPA	Interpolation for H and C_p	25.50

[a]Central Processor time, 51 grid points across viscous layer.

TABLE 3.- COMPUTING TIMES AND SPEEDS
FOR SHOCK TUBE COMPUTER CODE

Computer and Compiler	Computing Time; sec[a]	Relative Computing Speed
H-P, without VIS	2100	1/6
H-P, with VIS	350	1
ASC, FX[b]	103.5	3.4
ASC, NX-K	6.25	56

[a]Central Processor time

[b]CP time on the ASC with FX is estimated to be equivalent to CP time on an IBM 370/168.

EVALUATION OF THE SPAR THERMAL ANALYZER ON THE CYBER-203 COMPUTER

J. C. Robinson
Langley Research Center
Hampton, Virginia

K. M. Riley
Kentron International
Hampton, Virginia

R. T. Haftka
Virginia Polytechnic Institute and State University
Blacksburg, Virginia

CYBER 203

The purpose of this effort is to make the CYBER 203 (fig. 1) vector computer available for thermal calculation and assess the use of such a vector computer for thermal analysis. Strengths of the CYBER 203 include the ability to perform, in vector mode using a 64 bit word, 50 million floating point operations per second (MFLOPS) for addition and subtraction, 25 MFLOPS for multiplication and 12.5 MFLOPS for division. The speed of scalar operation is comparable to that of a CDC 7600 and is some 2 to 3 times faster than Langley's CYBER 175s. The CYBER 203 has 1,048,576 64-bit words of real memory with an 80 nanosecond (nsec) access time. Memory is bit addressable and provides single error correction, double error detection (SECDED) capability. The virtual memory capability handles data in either 512 or 65,536 word pages. The machine has 256 registers with a 40 nsec access time.

The weaknesses of the CYBER 203 include the amount of vector operation overhead and some data storage limitations. In vector operations there is a considerable amount of time before a single result is produced so that vector calculation speed is slower than scalar operation for short vectors. In some cases the vector length at which vector processing becomes faster than scalar may be as large as 70. Also, the terms of a vector must be stored in contiguous locations for vector operations--e.g. terms in a two dimensional array must be used by columns. This last limitation is partially offset by availability of fast routines to "gather" data from non-contiguous locations and store the data in contiguous locations using a vector of indices which indicate which terms are to be collected. Similarly, efficient routines are avilable for the inverse operation (scatter) and transposing a matrix.

CYBER 203

STRENGTHS

- SPEED — VECTOR OPERATION ≈ 30 MFLOPS (±) (64 BIT)[*]
  — SCALAR OPERATION ≈ 7600 ≈ 2 TO 3 * CYBER 175
- MEMORY — 1024 K 64 BIT WORDS, 80 nsec ACCESS
  — SECDED ERROR PROCESSING
- VIRTUAL MEMORY ARCHITECTURE
    SMALL PAGES — 512 WORDS
    LARGE PAGES — 65536 WORDS
- LARGE REGISTER FILE — 256 40 nsec REGISTERS

WEAKNESSES

- VECTOR OPERATION OVERHEAD PENALIZES USE OF SHORT VECTORS
- VECTOR DATA MUST BE IN CONTIGUOUS LOCATIONS PARTIALLY OFFSET BY FAST TRANSPOSE, GATHER/SCATTER

Figure 1

SPAR THERMAL ANALYZER

To provide a general in-house integrated thermal-structural analysis capability the Langley Research Center is having the SPAR Thermal Analyzer (fig. 2) developed under contract by Engineering Information Systems, Inc. The SPAR Thermal Analyzer is a system of finite-element processors for performing steady-state and transient thermal analyses. The processors communicate with each other through the SPAR random access data base. As each processor is executed, all pertinent source data is extracted from the data base and results are stored in the data base.

The tabular input (TAB), element definition (ELD) and arithmetic utility system (AUS) processors are used to describe the finite element model. The data base utility (DCU) processor operates on the data base. The plotting processors (PLTA, PLTB) provide the capability to plot the finite element model for model verification but do not directly plot temperatures. The thermal geometry (TGEO) processor performs geometry checking of the thermal elements and total model. The thermal processors for steady state analysis (SSTA) and transient analysis (TRTA, TRTB and TRTG) are described in References 1 and 2. In addition there are several processors not shown in the figure for extraction of thermal fluxes, system matrices and system operating characteristics.

On a scalar computer the processors may be executed interactively or in a batch mode. A typical analysis is usually performed as a sequence of interactive and batch operations where model development and verification is performed interactively and actual thermal calculations performed in batch mode. The program operates on UNIVAC, CDC, PRIME and VAX computers.

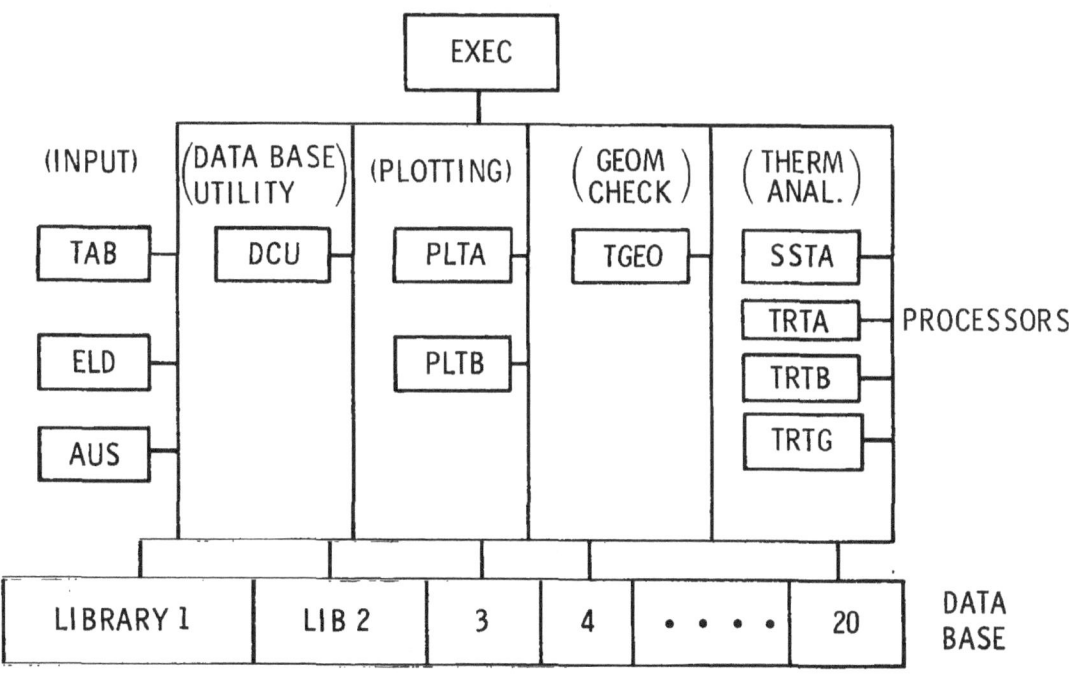

Figure 2

SPAR THERMAL ANALYZER ON THE CYBER 203

The SPAR Thermal Analyzer shown in the last figure was modified to operate on the CYBER 203 in a scalar processing mode (fig. 3). A number of transient thermal analyses were performed with this scalar version to determine the CPU times required and ensure that the program produced correct results. The CPU times were used for comparison with the CYBER-175 and as a basis to evaluate future vectorization. A description of seven of the problems and their scalar mode solution times are presented in subsequent figures.

In addition, six short subroutines were modified so that vector operations could be performed when applicable. The modified subroutines were selected because of their heavy use in implicit solutions where longer vectors are used and the ease with which the modification could be made. No changes were made in the internal data ordering for vector processing. The effect of this simple approach to vectorization will be discussed later.

Several program modifications were required due to differences between the CYBER 203 and other CDC computers at Langley. The virtual memory capability makes it possible to load the complete program without overlaying. It also required changing some data initialization from DATA statements to executable statements since DATA statements are effective only the first time a program segment is placed in memory. The lack of random access to external files required the storage of the data base in dimensioned arrays during execution and the sequential transfer of these arrays to external files upon execution completion for restart capability.

In addition to the changes required by differences in machine architecture several compiler bugs required coding changes to make the program execute properly.

- **CONVERSION EFFORT**

    COMPLETE SCALAR OPERATION
    VECTORIZE A FEW ROUTINES
      HIGH USE, EASY VECTORIZATION

- **CHANGES REQUIRED**

    PARAMETER INITIALIZATION (VIRTUAL MEMORY)
    INTERNAL DATA STRUCTURE (NO RANDOM ACCESS FILES)
    RESTART CAPABILITY

- **COMPILER PROBLEMS**

Figure 3

COMPARISON OF THE CYBER 175 AND CYBER 203 CPU TIME FOR SPAR PROCESSORS

The results presented in figure 4 are discussed with the individual problem slides. In general, the only processor showing appreciable improvement is TRTB which requires the most effort in large problems, and the improvement is based on problem size and probably on the ratio of CPU to I/O effort. Processors that perform large amounts of data input and character manipulation are appreciably slower on the CYBER 203.

PROCESSOR / PROBLEM	TAB	AUS	ELD	TGEO	TRTB	DCU	TOTAL
FRAME	0.40*	0.49	0.33	0.10	6.78	0.40	8.49
	0.96**	1.51	0.96	0.06	4.15	0.39	8.01
ANTENNA	0.24	1.89	0.49	0.14	43.03	0.61	46.41
	0.37	4.33	1.24	0.08	25.55	0.58	32.15
SINGLE BAY	0.25	1.23	0.29	0.10	85.22	1.12	88.21
	0.37	3.49	0.60	0.06	52.45	0.96	57.93
WING	1.87	1.38	3.18	0.59	126.90	10.72	144.64
	4.01	5.78	7.96	0.40	58.77	8.93	85.84
CYLINDER	0.85	0.15	0.99	0.16	156.60	4.30	164.48
	0.82	0.32	1.42	0.63	62.23	2.40	67.82
MULTIWALL	1.14	5.95	2.47	1.27	210.99	1.07	222.89
	1.81	21.91	7.11	0.56	116.31	0.95	148.66
THREE BAYS	1.64	4.97	1.33	1.38	365.23	1.40	375.95
	2.28	18.35	2.54	0.58	184.90	1.25	210.05

* - CYBER 175   ** - CYBER 203 (SCALAR MODE)

Figure 4

SPACE SHUTTLE FRAME

An aluminum space shuttle fuselage frame (Refs. 3 and 4) is shown in figure 5. The finite element model has 190 grid points, 158 thermal elements and is heated by time-dependent surface temperatures. Heat is transferred by conduction in the aluminum and insulation and radiation from the inner insulation surface. Implicit solution times on the CYBER 175 and 203 are shown at the bottom for a temperature history of 1000 seconds with a computational time interval (DT) of 10 seconds. Figure 4 shows the CPU time in seconds (CYBER 175 on upper line, CYBER 203 on lower line) for each of the processors used in the analyses. For the FRAME problem, which is relatively small, the savings in the actual transient analysis (TRTB) is largely offset by the poor relative performance of the CYBER 203 in the other processors where problem input requires a large amount of character manipulation.

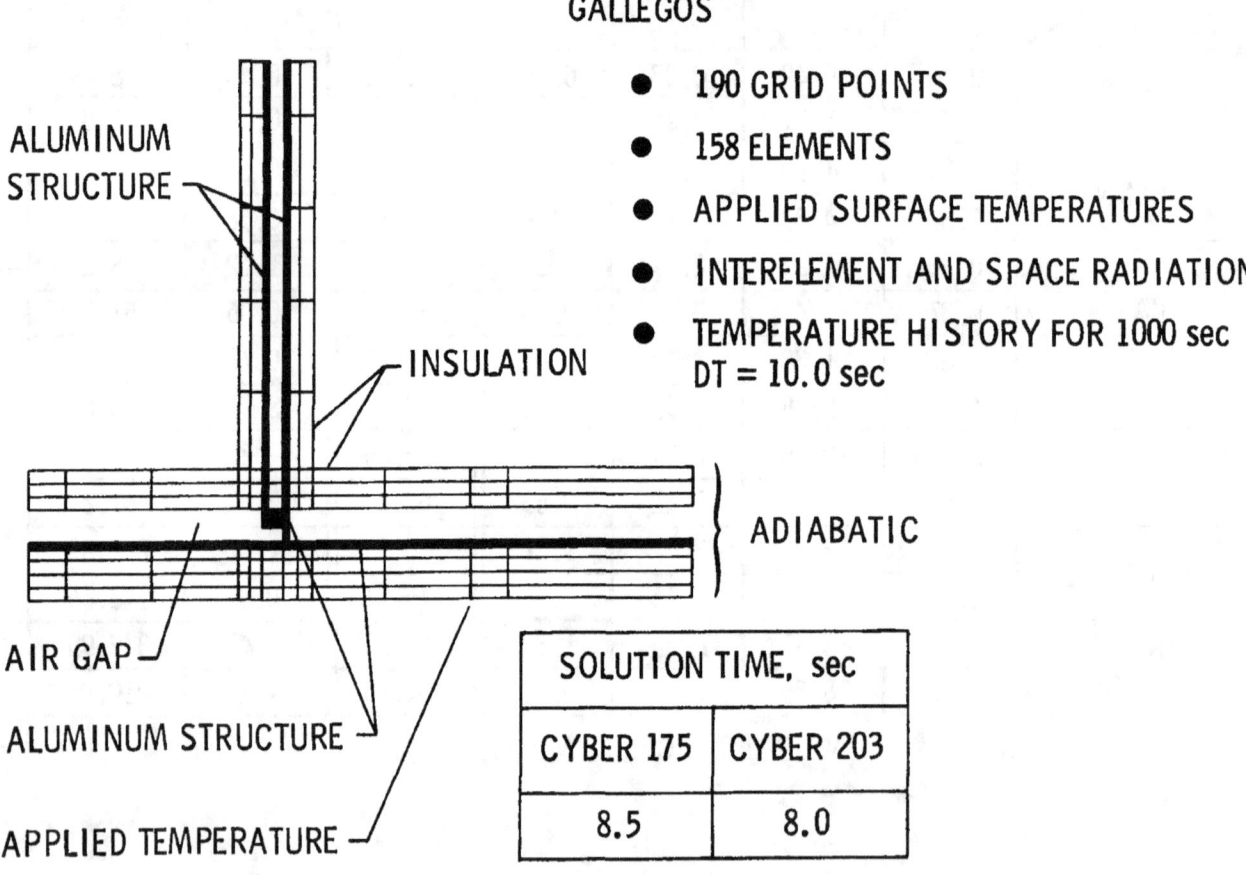

Figure 5

30 METER PRECISION DEPLOYABLE ANTENNA

A model of a 30 meter precision deployable antenna which has 55 grid points and 183 elements is shown in figure 6. Thermal loading is solar irradiation with time-dependent shadowing. Heat transfer includes conduction, interelement radiation and radiation to space. Implicit solution times are shown at the bottom of the figure for a temperature history of 24 hours (one orbit) with a DT of 0.01 hour. The ANTENNA problem CPU time breakdown is shown in figure 4. While this problem is relatively small, the larger amount of effort in TRTB compared to the other processors results in significantly faster operation on the CYBER 203.

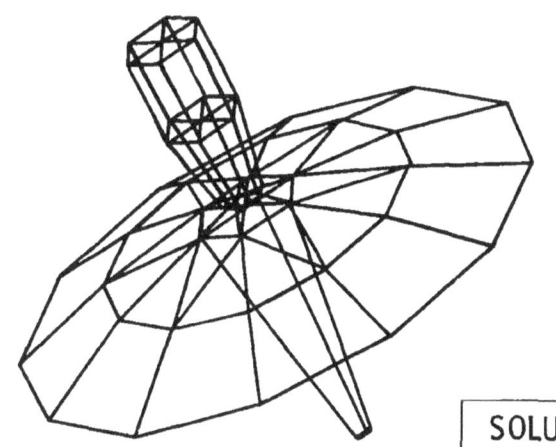

- 55 GRID POINTS
- 183 ELEMENTS
- INTERELEMENT RADIATION
- TIME-DEPENDENT SHADOWING
- TEMPERATURE HISTORY FOR 24 HOURS DT = .01 hr

SOLUTION TIME, sec	
CYBER 175	CYBER 203
46.4	32.1

Figure 6

SINGLE BAY OF SHUTTLE ORBITER WING

A finite element model of a single bay of the space shuttle orbiter wing which has 123 grid points and 151 thermal elements is shown in figure 7. Thermal loading is applied as time-dependent heating on the lower and upper surfaces. Heat transfer is by conduction, internal interelement radiation and surface radiation to space. Implicit solution times for a temperature history of 2500 seconds and a DT of 1.0 sec are shown at the bottom of the figure. The SINGLE BAY problem CPU time breakdown is shown in figure 4. As the problem size increases the relative amount of CPU time spent in TRTB increases and so does the improvement over the CYBER 175.

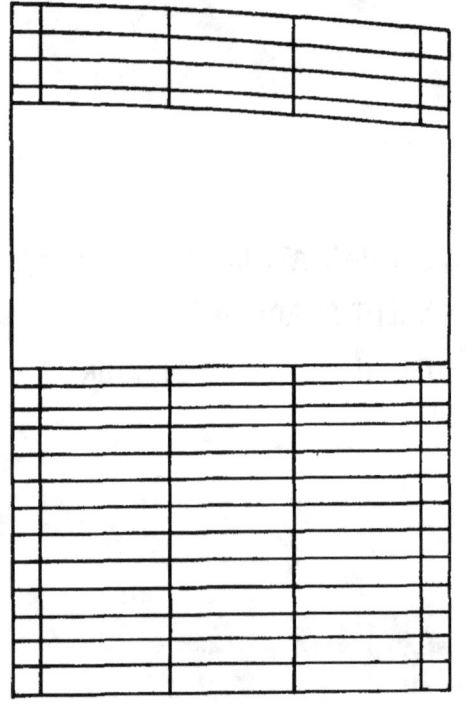

- 123 GRID POINTS
- 151 ELEMENTS
- TIME DEPENDENT SURFACE HEATING
- INTERELEMENT AND SPACE RADIATION
- TEMPERATURE HISTORY FOR 2500 sec DT = 1.0 sec

SOLUTION TIME, sec	
CYBER 175	CYBER 203
88.2	57.9

Figure 7

SPACE SHUTTLE ORBITER WING

A thermal finite element model of the space-shuttle-orbiter primary-wing structure is shown in figure 8 (Ref. 4). The total model including the thermal protection system (TPS) which is not shown has 1542 grid points and 2125 thermal elements. This is a relatively crude model of the wing without the elevons and glove. One dimensional elements were used to model the TPS since solid elements would be much larger in the dimensions parallel to the wing surface than normal to the surface and lateral conduction is much smaller than conduction normal to the surface. Thermal loading is applied as time-dependent surface temperatures. Heat transfer is internal conduction and radiation to space. Implicit solution times for a temperature history of 3000 seconds with a DT of 100 sec are shown at the bottom of the figure. The WING problem CPU time breakdown is shown in figure 4. The larger problem size and the use of the the large number of one dimensional elements for the TPS produces the improved computational efficiency in TRTB such that the CYBER 203 uses approximately half the time of the CYBER 175.

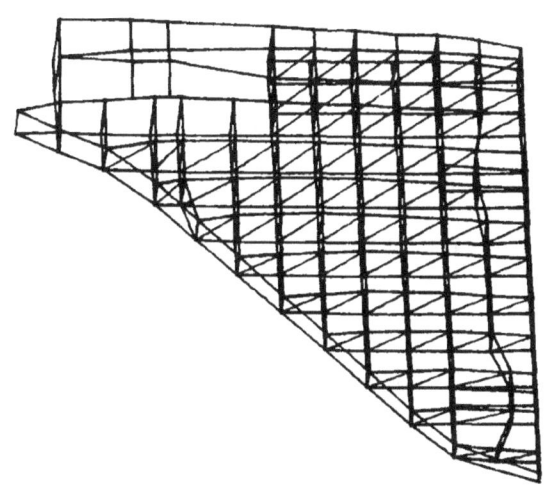

- 1542 GRID POINTS
- 2125 ELEMENTS
- APPLIED SURFACE TEMPERATURES (AERO HEATING)
- 1-D ELEMENTS USED FOR RSI
- TEMPERATURE HISTORY FOR 3000 sec DT = 100.0 sec

SOLUTION TIME, sec	
CYBER 175	CYBER 203
144.6	85.8

Figure 8

CYLINDER

The outer surface of a thermal finite element model of an insulated cylinder developed for solution algorithm testing is shown in figure 9 (Ref. 4). The model has 800 grid points and 650 thermal elements. Thermal loading is applied as time-dependent surface heating. Heat transfer is conduction in the aluminum shell and TPS with radiation to space from the external surface. Implicit solution times for a DT of 10 seconds are shown at the bottom of the figure. The CYLINDER problem CPU time breakdown is shown in figure 4. In this problem, the CYBER 203 takes about 40 percent as much time as the CYBER 175 for the TRTB processor.

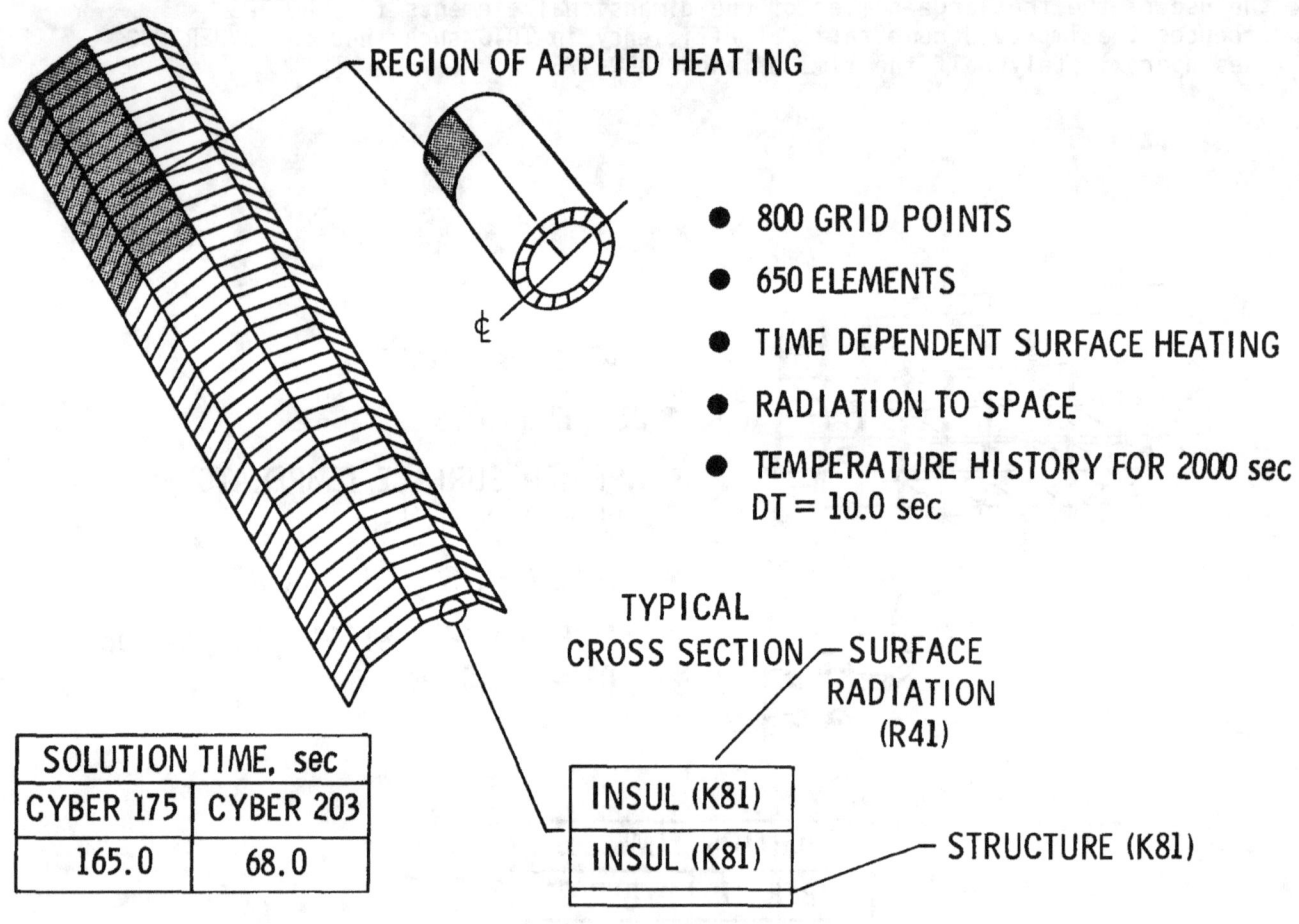

Figure 9

MULTIWALL THERMAL PROTECTION SYSTEM

Several details and the finite element model of a piece of a multiwall thermal protection system (TPS) are shown in figure 10 (Ref. 4). As shown in the upper left sketch, multiwall TPS is made up of alternating flat and dimpled sheets of thin metal welded together at the crests of the dimples. An idealized shape for one of the dimpled sheets is shown in the lower left sketch. The finite element model has 333 grid points and 1096 elements. The thermal loading is a time-dependent temperature on the outer (upper) surface and radiation to a room temperature sink on the lower surface. Heat transfer consists of conduction in the metal sheets, radiation between the sheets and conduction in the air. Solid elements were used to model the heat transfer by conduction in the air between all the sheets but are shown between the lower two sheets only for clarity. Implicit solution times for a temperature history of 2000 seconds and a DT of 1 second are shown at the bottom of the figure. The MULTIWALL problem CPU time breakdown is shown in figure 4. The TRTB time on the CYBER 203 is about 55% of that on the CYBER 175. The increase in the AUS time is due to the input of some 37,000 terms necessary to describe radiation view factors.

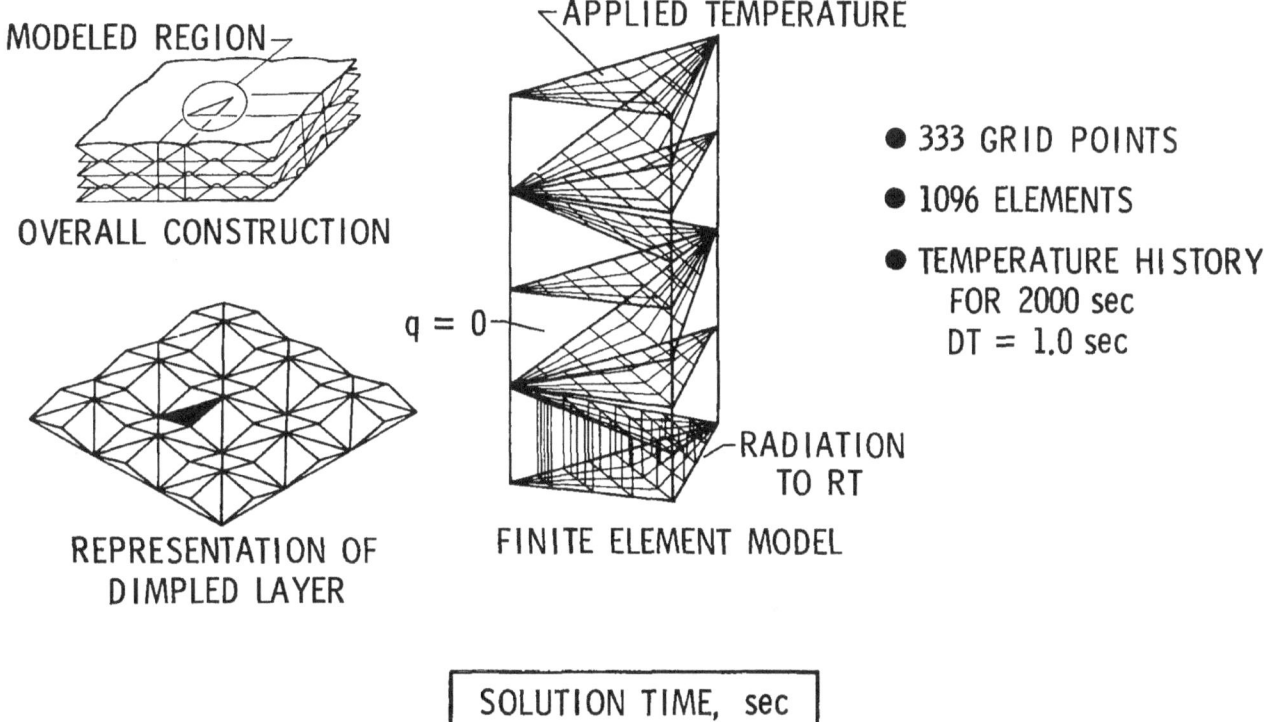

Figure 10

THREE BAYS OF SHUTTLE ORBITER WING

A thermal finite element model of a segment of the space shuttle orbiter wing structure that extends three bays in the chordwise direction and half a bay in the spanwise direction is shown in figure 11 (Ref. 5). Modeling of the upper and lower surface thermal protection systems is shown in the details. The model has 916 grid points and 789 thermal elements. Thermal loading is time-dependent surface heating on both the upper and lower surfaces. Heat transfer consists of conduction in the metal structure and thermal protection system, interelement radiation internally and radiation to space from the outer surfaces. Implicit solution times for a temperature history of 1000 seconds and a DT of 5 seconds are shown at the bottom of the figure. The THREE BAYS problem CPU time breakdown is shown in figure 4. The TRTB time is about half as much on the CYBER 203 as on the CYBER 175.

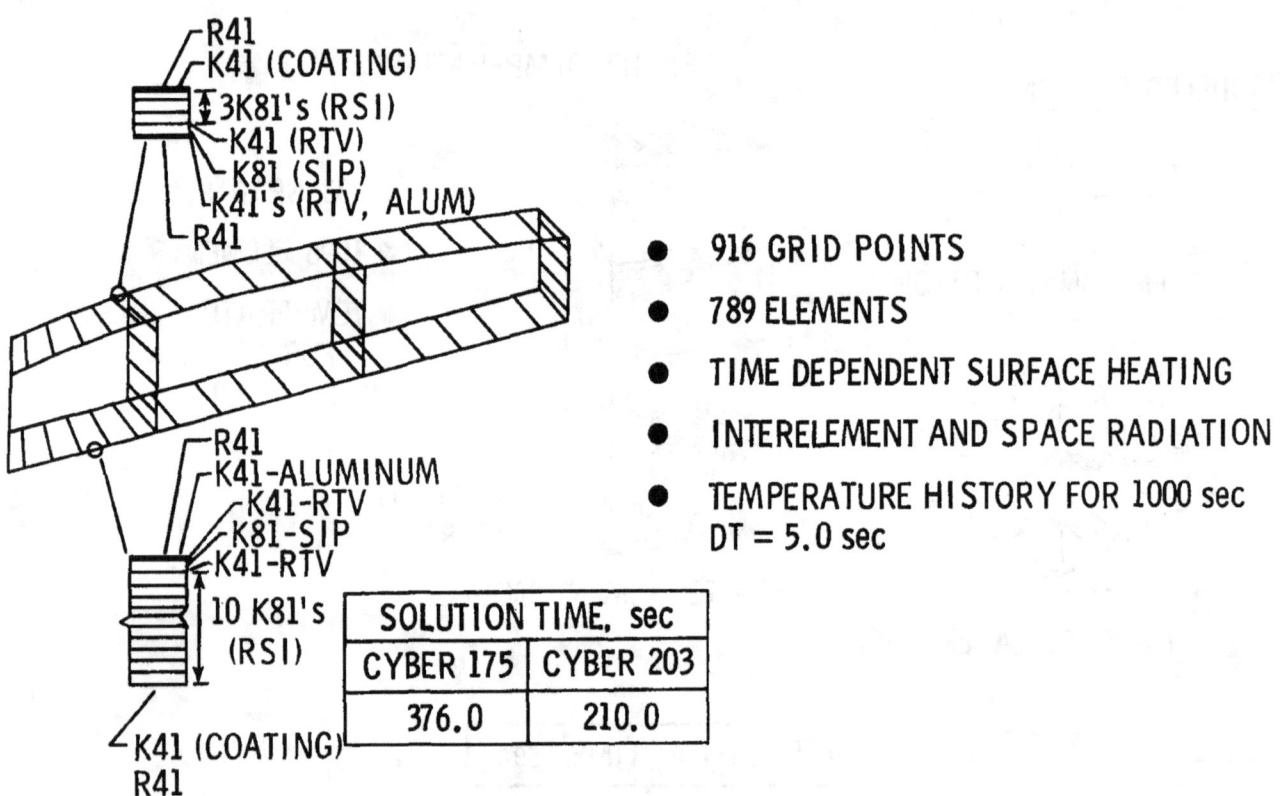

Figure 11

SPAR VECTOR STATISTICS

As stated previously, the CYBER 203 version of SPAR is basically a scalar conversion with some simple vectorization in six small subroutines that may be executed at the user's option. The six subroutines perform operations such as summing vectors and multiplying small matrices. When the seven sample problems, for which scalar results are presented in figure 4, are executed in the optional vector mode, only three of the subroutines are used and there was no decrease in solution time.

To determine why no benefits were achieved in the vector mode, data were collected on the number of times the subroutines were called and the vector lengths involved. The scalar product subroutine was called the most and performs the largest number of operations per call. The vector statistics for the scalar product subroutine are shown in figure 12. This subroutine is used within the inner loops of the implicit method as shown by the number of calls. The column displaying vector calls indicates that vector operations are not always applicable. This is typically the case when the vectors are not stored in contiguous locations. No benefit is received from the vector mode since the vector lengths, on the average, are so small. Redesign is necessary for any significant improvement to be realized.

- **6 SUBROUTINES VECTORIZED**
- **3 CALLED IN THE TEST PROBLEMS**
- **SCALAR PRODUCT SUBROUTINE IS CALLED THE MOST AND PERFORMS MOST OPERATIONS PER CALL**

RESULTS FOR SCALAR PRODUCT					
PROBLEM	NUMBER OF CALLS	NUMBER OF VECTOR CALLS	MIN. VECTOR LENGTH	AVG. VECTOR LENGTH	MAX. VECTOR LENGTH
FRAME	80,275	66,750	1	9	19
ANTENNA	264,967	245,678	1	13	34
SINGLE BAY	768,700	745,662	1	10	105
WING	896,671	693,633	1	13	259
CYLINDER	1,107,200	1,074,794	1	15	25
MULTIWALL	1,641,840	1,588,820	1	40	116
THREE BAYS	3,208,468	3,140,957	1	34	97

Figure 12

CRANKB

CRANKB, a pilot computer program for thermal analysis of an insulated cylinder, is currently being used as a test bed for vectorization techniques. Experience gained from this pilot program will be used in determining if it is worthwhile to vectorize SPAR's thermal analyzer and possible techniques for implementation. The major reason for the selection of CRANKB is that the program, which is both small in size and simple in comparison to SPAR, already exists and has been tested. In addition, since the source code originally came from SPAR, most of the vectorization techniques used can be directly applied to SPAR. CRANKB is designed to model K81 elements and uses an implicit solution technique called CRANK-NICHOLSON. An iterative improvement method is employed in which the conductivity matrix is only updated when the solution does not converge in three iterations. A continuing effort is being applied to CRANKB. The results to date are shown on the following pages.

- TEST BED FOR VECTORIZATION TECHNIQUE

- CODE BASED ON SPAR

- PROGRAM USES K81 ELEMENTS TO MODEL INSULATED CYLINDER

- SOLUTION TECHNIQUE IS CRANK-NICHOLSON (IMPLICIT)

- STUDY NOT COMPLETE

Figure 13

IDENTIFICATION OF TIME CONSUMING OPERATIONS

Two major time consuming operations were identified with the use of a timing utility available on the CYBER 203 (fig. 14). The major time consumer is the multiplication of the conductivity matrix by the temperature vector used in the computation of the temperature derivative. This accounts for 36% of the CPU time. The other major contributor is the factoring and solution subroutines for a symmetric banded system of equations (method LDL^T) which accounts for 40% of the total CPU time. Together, these two operations account for 76% of the CPU time.

OPERATION	PCT OF CPU TIME
SOLUTION OF SYMMETRIC BANDED SYSTEM OF EQUATIONS (METHOD = LDL^T)	40%
$[K]\{T\}$ (AT ELEMENT LEVEL)	36%

Figure 14

IMPACT OF VECTORIZATION ON SOLUTION TIME

Figure 15 shows the vectorization stages that have been completed. The CYBER 203 run time for CRANKB before any modifications is shown in the first line of the table. The next entry displays the benefits from obvious conversions of do loops to explicit vector calls and the vectorization of scaling the element conductivity matrices.

The subroutines which factor and solve the symmetric banded system of equations were replaced by a vectorized subroutine from the CYBER 203 system math library. The answers produced were identical, and the time required for this operation was cut by almost two thirds, saving 20 CPU seconds. The library routine uses a vector length of half the bandwidth plus one. For the insulated cylinder, this turns out to be 26. A larger bandwidth would obviously produce more savings here.

The single most time consuming operation is the multiplication of the conductivity matrix, hereafter referred to as K, by the temperature vector, denoted by T. The original source does this operation at the element level which offers several advantages. The code has already been designed to store the symmetric part of the element K matrices. These are scaled for each change in temperature and then the full element K matrix is built. The corresponding temperature vector is extracted and the multiplication occurs. This is repeated for each element and the results are assembled into the global product. With this method, the global K matrix need not be built. This is advantageous since the assembly is time consuming. The major disadvantage for a vector machine is that multiplication at the element level yields small vector lengths. In the present application, the cylinder is modelled with K81 elements which produce an element K matrix of size 8 x 8.

An alternative is to do this multiplication at the system level. For the cylinder the global K matrix is 800 x 800, which appears ideal for a vector machine. The only problem is that the global K matrix must be reassembled for each multiplication. A single assembly requires 0.06 CPU seconds. For a temperature history of 1000 seconds and a DT of 2.0 seconds, 627 assemblies are required taking a total CPU time of 38 seconds. Even assuming that with vector lengths of 800, the actual multiplication is negligible, no real benefit is found over the element level which takes 33 seconds.

Other less obvious alternatives were found and the actual vectorization applied is described in the next figure.

IMPACT OF VECTORIZATION ON SOLUTION TIME

PROBLEM: 1000 sec TEMPERATURE HISTORY OF 800 NODE CYLINDER, DT = 2.0 sec

LEVEL OF VECTORIZATION	CPU TIME
ORIGINAL - NO VECTORIZATION	92 *
EXPLICIT VECTOR CALLS FOR OBVIOUS LOOPS AND SCALING	85
VECTORIZED ROUTINE FOR SOLUTION OF EQUATIONS (MATH LIBRARY ROUTINE)	65
VECTORIZED $[K]\{T\}$ OPERATION	33

* SPAR TIME FOR SAME PROBLEM   (159)

Figure 15

VECTORIZATION OF [K] {T} OPERATION

The actual vectorization applied is shown in figure 16. It is by no means obvious how this sequence of operations can save time. The available storage includes an EKS matrix, dimensioned NEL by 36 where NEL is the number of elements, which contains the symmetric part of the element K matrices; an index matrix denoted by NODES (not shown) and dimensioned NEL by 8 which stores the 8 node numbers corresponding to each element; and a vector T, dimensioned NOD where NOD is the number of nodes, that contains the temperature at each node. Available on the CYBER 203 are two very efficient functions for gathering and scattering vectors. Both functions require two input vectors; a vector of real numbers representing the values to be used in the operation and a vector of integer numbers which are the array indices of the real terms. For gathering, the index vector determines which elements of the input vector are to be placed in the resultant vector. For scattering, the index vector determines where each element of the input vector is to be placed in the result.

Using the above information, with the assistance of the Langley CYBER 203 consulting office, the vectorization was implemented in the following manner. The temperature for the first node of all the elements is extracted from T using the gather function. The resulting vector (size NEL) is multiplied by the appropriate columns of EKS. Each product vector (size NEL) is scattered into another vector (size NOD) which is then added into the final result. The above is repeated for the extracted temperature vector at each of the 8 nodes. Since the gathers and scatters are efficient operations and the vector lengths are NEL, which for the cylinder application is 650, the run time is greatly reduced. The total CPU time with the vectorized [K] {T} operation is 33 seconds.

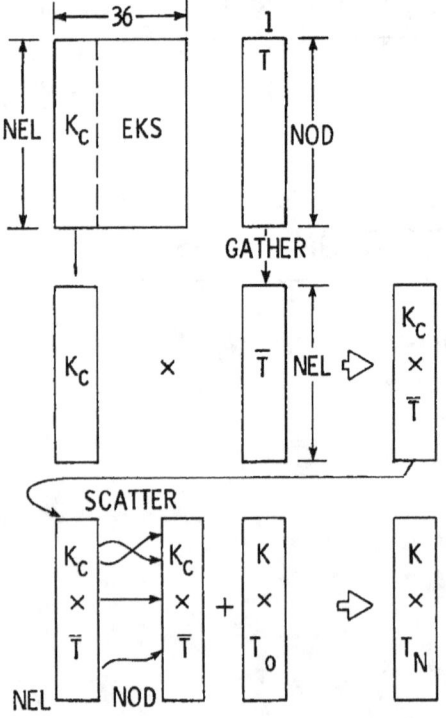

- T (NOD) - TEMPERATURE VECTOR
- EKS (NEL, 36) - DIAG + LOWER PART OF ELEMENT K MATRIX
- LONGEST COLUMNS, NO ZEROES
- "GATHER" TERMS FROM T TO FORM MODIFIED TEMP VECTOR, $\bar{T}$
- DO TERM BY TERM MULTIPLY OF EKS COLUMN (K_c) BY $\bar{T} \Rightarrow K_c \bar{T}$
- "SCATTER" TERMS FROM $K_c \bar{T}$ (NEL) TO $K_c \bar{T}$ (NOD)
- ADD COLUMN PRODUCT TO KT VECTOR
- 36 COLUMNS + 28 COLUMNS FOR SYMMETRIC TERMS

Figure 16

SUMMARY OF CYBER 203 EFFORT

SPAR executes successfully on the CYBER 203 in the scalar mode. A decrease in the scalar mode computation time is realized in the transient thermal analysis processor where most of the CPU time is used. Minimal vectorization was applied with no benefit due to insufficient vector lengths.

Considerable effort was applied to the pilot program CRANKB. The CPU time has been decreased by almost two thirds. The study, although not complete, shows a trade-off between programming effort and time savings for more efficient vectorization of the SPAR Thermal Analyzer for operation on the CYBER 203. (See fig. 17.)

SPAR

- PROGRAM RUNS IN SCALAR MODE

    FOR CALCULATIONS HAVING HIGH CPU/IO SEE SCALAR SPEED ADVANTAGE

- INSUFFICIENT VECTORIZATION TO SHOW ANY ADVANTAGE

    6 ROUTINES VECTORIZED
    AVERAGE VECTOR LENGTH TOO SHORT IN TEST PROBLEMS

PILOT PROGRAM

- CONSIDERABLE VECTORIZATION ACCOMPLISHED
- SHOWS SIGNIFICANT ADVANTAGE (3-D ELEMENT)
- STUDY NOT COMPLETED

SPAR VECTORIZATION

- TRADE-OFF BETWEEN PROGRAMMING EFFORT AND BENEFITS NOT COMPLETE

Figure 17

REFERENCES

1. Marlowe, M. B.; Whetstone, W. D.; and Robinson, J. C.: The SPAR Thermal Analyzer - Present and Future. Computational Aspects of Heat Transfer in Structures, NASA CP-2216, 1982. (Paper no. 3 of this compilation).

2. Marlowe, M. B.; Moore, R. A.; and Whetstone, W. D.: SPAR Thermal Analysis Processors Reference Manual, System Level 16. NASA CR-159162, 1979.

3. Gallegos, J. J.: Thermal Math Model of FRSI Test Article Subjected to Cold Soak and Entry Environment. AIAA Paper 78-1627, 1978.

4. Adelman, H. M.; and Haftka, R. T.: On the Performance of Explicit and Implicit Algorithms for Transient Thermal Analysis of Structures. NASA TM-81880, September 1980.

5. Ko, William L.; Quinn, Robert D; Gong, Leslie; Schuster, Lawrence S.; and Gonzales, David: Reentry Heat Transfer Analysis of the Space Shuttle Orbiter. Computational Aspects of Heat Transfer in Structures, NASA CP-2216, 1982. (Paper no. 18 of this compilation).

DEVELOPMENT OF A CRAY 1 VERSION OF THE
SINDA PROGRAM

Susan M. Juba and Peter E. Fogerson
Lockheed Engineering and Management Services Company, Inc.

SUMMARY

The SINDA thermal analyzer program was transferred from the UNIVAC 1110 computer to a CYBER and then to a CRAY 1. Significant changes to the code of the program were required in order to execute efficiently on the CYBER and CRAY. The program was tested on the CRAY using a thermal math model of the Shuttle which was too large to run on either the UNIVAC or CYBER. An effort was then begun to further modify the code of SINDA in order to make effective use of the vector capabilities of the CRAY.

INTRODUCTION

The computer available for thermal analysis at the National Aeronautics and Space Administration/Johnson Space Center (NASA/JSC) is a UNIVAC 1110 which has a maximum user core of 190,000 words. However, since the 1110 is operated in a time-sharing environment with nonthermal users, its full resources are not available. This computational capability has not proven adequate for all requirements. For example, core storage restraints did not allow construction of a single thermal model of the Space Shuttle. Instead, five separate models were built - three representing the forward, mid, and aft fuselage sections, and one each for the aft propulsion system (APS) and the hydraulic system. A complete transient analysis thus required five interconnected computer runs.

In hopes of producing a combined Space Shuttle model and to provide a backup capability for computation during peak load periods on the UNIVAC 1110, a contract was made with the United Computing System (UCS) for time on their commercial networks. The basic host computer of UCS is a CYBER of the Control Data Corporation (CDC). A CRAY 1 (ref. 1) is also available; however, it requires the CYBER as a front end. The first task was to establish a CDC version of the thermal analysis program, SINDA (ref. 2), and to transfer the five thermal models to CDC files. The UCS CYBER configuration actually offered less core than the NASA/JSC UNIVAC. The CRAY 1, however, has 2 million words available, and additionally offers the prospect of increased speed by virtue of its vector processing capability. The ultimate purpose, then, was to develop a version of SINDA for the CRAY 1.

SYMBOLS

C_i heat capacity of a node i, J/°K (Btu/°R)

G_{ij} thermal conductance between node i and node j, J/hr-°K (Btu/hr-°R)

$GSUM_i$ net heating rate to node i, J/hr (Btu/°R)

i,j   node number indexes

Q_i incident heat to node i, J/hr (Btu/°R)

R_{ij} thermal radiation between node i and node j, J/hr-°K^4 (Btu/hr-°R^4)

T_i current temperature of node i, °K (°R)

T_i' new temperature of node i, °K (°R)

Δt time interval, hr

STRUCTURE OF THE SINDA ANALYZER

SINDA is a general-purpose thermal analyzer, which means that the user can construct a thermal model of anything, unrestricted and unaided by geometry. It uses the finite difference formulation of the thermal diffusion equation, thus requiring a lumped parameter representation of the physical system in a resistor-capacitor (R-C) network. The program has two parts, the preprocessor and the processor. The preprocessor reads the user input data (the definitions of the nodes, heat capacities, temperatures, conductances, etc., which make the thermal model) in an 80-column card input format, and writes:

a. An executable program to perform the analysis

b. Tables listing the thermal parameter actual numbers (assigned by the user) vs. the relative numbers (assigned by the program)

c. A table called the first pseudocompute sequence (PCS1) which tells which nodes are connected to each other and by what type of connection

d. A table called the second pseudocompute sequence (PCS2) which contains the nonlinear thermal parameter information

The processor consists of the program (a) and user-selected subroutines from the SINDA library which, when executed, use as input the data files (b, c, and d) to produce a transient or steady-state simulation.

DEVELOPMENT

UNIVAC-to-CYBER Conversion

NASA was fortunate to have access to a version of SINDA which would run on CDC machines. This version was developed from a UNIVAC source and retained the exact bit configuration in the pseudocompute sequences; i.e., only the first 36 bits of the 60-bit CYBER word were used. The program was transferred to UCS along with an update package which would bring the version to the current UNIVAC level. Successful execution was easily achieved, but Central Processor (CP) second comparisons revealed the CYBER was requiring twice as much time as the UNIVAC. The problem was traced to the unpacking of the PCS data.

In UNIVAC FORTRAN, the unpacking is performed by a special function, FLD, which is fairly efficient. The statement:

NG = FLD(5,16,PCS1(I))

causes 16 consecutive bits, beginning with bit 5, to be taken from the location PCS1(I) and stored, right-adjusted, in the variable NG. The CYBER version had an assembly language subroutine, IFLD, to perform this function. It was of the form:

CALL IFLD(5,16,PCS1(I),NG)

These subroutine calls for unpacking the PCS data were replaced with in-line code using the CYBER SHIFT function and logical AND statements. For example, the call to IFLD shown above was replaced by:

DATA IAND6/0177777000000000000000B/
NG = SHIFT((PCS1(I).AND.IAND6),21)

This causes a logical AND operation between the location PCS1(I) and the variable IAND6, thus picking out bits 5-21 of PCS1(I). The result is then left-circular shifted 21 bits before storing in NG. SUBROUTINE SUBFLD was used to pack the data in the first place. Using similar techniques, its calls were also replaced by in-line code. After making these subroutine call replacements throughout SINDA, the execution time required for some runs was reduced by a factor of 5.

CYBER-to-CRAY Conversion

Having established a working version of SINDA on the CYBER, the CRAY conversion effort was begun by making changes necessitated by differences in byte and word size between CDC and CRAY machines. For the purposes of this discussion, a byte is defined as the sequence of bits required to represent a character. The number of bits in a byte is therefore character-code dependent. Six bits are required to represent a character in Extended Binary Coded Decimal Interchange Code (EBCDIC) which is used on the CYBER, whereas eight bits represent a character in the CRAY version of the American Standard Code for Information Interchange (ASCII). The 60-bit word size of a CYBER machine allows representation of 10 characters per word, opposed to a maximum of 8 characters in the 64-bit word of the CRAY 1. The primary impact of this word and byte size difference is on the preprocessing/decoding processes, where node and conductor-related data are packed into words. The packing and unpacking processes which were implemented to conserve memory became unnecessary with the CRAY 1's 2 million word core memory. However, preprocessor and processor modifications to eliminate these steps would not be cost-effective in terms of programming and checkout time when weighed against the possible increases in execution speed. Therefore, the sections of code where data packing and unpacking were carried out were altered only where alphanumeric data are involved, i.e., where byte size is significant.

Another difference between CDC and CRAY FORTRAN exists in DO-loop handling. CDC FORTRAN will cause loops to be executed at least once, whereas a CRAY DO-loop need never be executed; as in a loop where the initial index value is 1, the final value is 0, and the incrementation value is 1. The correction for this DO-loop handling discrepancy is a simple one: CRAY FORTRAN provides a "J" option on its compiler control command that will ensure CDC-like handling of DO-loops when activated.

One characteristic of CRAY FORTRAN with global ramifications in the SINDA program is the ability to undefine variables. An entity will become undefined if an entity of different type which occupies the same memory location becomes defined. During SINDA processor execution, it is often necessary to access integers and floating-point values from the same array. The preprocessor packs integer information about the number of values in a real array in the first location of that array, and the entire vector is passed to an interpolation subroutine as a floating-point array. What happens upon entry to the interpolation routine is illustrated in the following simplified example:

```
1) SUBROUTINE INTERP(A)
2) DIMENSION A(1)
3) EQUIVALENCE (PN,NP)
4) PN = A(1)
5) LX = NP
```

These five lines of code show the original approach to accessing the integer value contained in A(1). When statement number 4 is executed by the CRAY-1, however, variable NP becomes undefined and statement 5 becomes a meaningless assignment. A quick-fix solution was discovered and is shown below:

```
1)  SUBROUTINE INTERP(A)
2)  DIMENSION A(1)
3)  EQUIVALENCE (PN, NP)
4)  PN = A(1)
4A) NP = NP
5)  LX = NP
```

The addition of statement 4A redefines NP so that the value in A(1) may be accessed as an integer. Statement 4A also has the effect of undefining PN, but since PN is not referenced after statement 4 in this application, further processing is not adversely affected.

As with the CYBER, it was also necessary in the CRAY conversion effort to replace with in-line code the calls to the two bit-manipulation routines, SUBFLD and IFLD. The substitution was accomplished through use of the Boolean selective merge function of CRAY FORTRAN, called CSMG. This function merges two words according to a third mask word, taking bits from word 1 where the mask word bits are set, and from word 2 where the mask word bits are cleared. This approach required that mask words be set up for each subroutine that had previously called SUBFLD or IFLD. In the preprocessor, these mask words could be defined in the driving routine, PREPRO, and accessed through a common block by other routines as necessary. In the processor, however, the driver is created uniquely for each run, so a different solution was needed. All routines that had accessed IFLD and SUBFLD were examined to determine what mask words were required, and DATA statements defining those words were added to each subroutine. A typical example of the changes involved in an IFLD substitution is illustrated below.

Bit manipulation previously effected by

```
VAR2 = IFLD(0,6,VAR1)
```

(right-justify leftmost six bits of VAR1 in VAR2) is now effected by

```
VAR2 = CSMG(SHIFTR(VAR1,(64-(0+6))),0,J6)
```

where J6 is a mask word with the six rightmost bits set.

SUBFLD substitution is slightly more complicated. If the original SUBFLD call was

```
CALL SUBFLD (5,1 VAR1,VAR2)
```

(replace bit 5 in VAR2 with bit 5 from VAR1) the replacement is

```
VAR2 = CSMG(SHIFT(VAR1,(64-(1+5)),VAR2,I5J1)
```

where I5J1 is a mask word containing one set bit in the fifth position from the left, where the leftmost bit is bit 0. The modifications to SUBFLD and IFLD calls in the preprocessor were made on a line-by-line basis, producing a factor of 3 decrease in execution time. Several hundred references to the two routines in the processor library made the use of statement functions more appropriate in the 51 library subroutines in which IFLD and SUBFLD were referenced.

MODEL EXECUTION COMPARISONS

After establishing working versions of SINDA on the CYBER and CRAY 1, some test executions on production size models were made. Selected were the MID model which has 1959 nodes and 15,271 conductors, and a combined Shuttle model which had no external radiation network, leaving it with 6,489 nodes and only 24,445 conductors. Table I shows the comparative performance of SINDA with the MID model on the NASA/JSC UNIVAC and the CYBER and CRAY 1 of UCS. The UNIVAC cost figure is based on $457/hr; the CYBER and CRAY 1 figures are actual costs including the NASA discount. The decrease in execution time in CP seconds going from UNIVAC to CYBER to CRAY 1 was as expected, but there was no significant corresponding decrease in cost. A similar CP second/cost comparison for the UNIVAC and CRAY 1 using the combined Shuttle model is shown in table II. Only preprocessor data is shown because the model is too large to run the processor on the UNIVAC. The CYBER would not handle even the preprocessing of the combined model. The results are about the same as those of the MID preprocessor; the CRAY 1 gives a dramatic reduction in CP seconds but a slightly higher cost.

VECTORIZATION

After establishing a working version of SINDA on the CRAY and recording some comparative execution times for purely scalar code, the next effort was to incorporate some vector code into SINDA. The proportionately large amount of time spent in the network solution routines suggested that a vectorization effort in this area would be the most productive initially. Study led to the conclusion that merely applying vectorizing techniques to the existing code might not provide significant improvement, so a simultaneous attack was begun on the PCS structure.

Execution Routine Code Modification

The first routine to be examined for vectorization potential was the convergent explicit forward differencing routine SNFRWD. This routine was chosen because it produced consistent, reliable predictions of network response, was already one of the most efficient network solution routines, and consequently was heavily used. At this time, there are no cost comparison figures for any of the execution routines, but a brief discussion of the vectorization process for SNFRWD will provide some familiarity with the modifications involved in vectorization.

The ultimate objective when vectorizing code is to significantly reduce array processing time. This reduction can be achieved by accessing array elements so that memory banks are referenced sequentially, and by avoiding statements or constructs that inhibit the pipelining of operands and instructions.

In the SNFRWD routine, some loops cannot be vectorized effectively due to the extensive use of indirect addressing, and the presence of nonvectorizable external references and GO TO's. In loops that appear to be candidates for vectorization, but contain a few nonvectorizable statements, the solution can be the creation of two or more loops so that one or more will vectorize. CRAY FORTRAN offers vectorization aids in the form of vectorizable utility routines to replace conditional assignment statements within loops, as in this example adapted from SNFRWD:

```
C
C  THIS LOOP DOES NOT VECTORIZE
C
      DO 1 I = 1,1000
      LSUM = LSUM + LEN(I)*2
      IF (LEN(I) .NE. 0) LSUM = LSUM + 2
    1 CONTINUE
C
C  THIS CODE VECTORIZES FOR LOOP 1
C
      DO 1 I = 1,1000
      LSUMT(I) = CVMGZ(LSUMT(I) + LEN(I)*2,
             LSUMT(I) + LEN(I)*2 + 2, LEN(I))
    1 CONTINUE
      RSUM = SSUM (1000,LSUMT(1),1)
      LSUM = IFIX (RSUM)
```

Note that scalar summing variable LSUM in the original loop is replaced by temporary array LSUMT in the vector version to avoid scalar register use. With up to four million words of core memory available, and considering the high cost of CRAY processing time, this trade-off becomes economically justifiable in many cases. The vector utility routine CVMGZ performs the conditional test of the original loop and allows the assignment statement that preceded the test to be performed at the same time. CVMGZ tests the third argument, LEN(I), against a zero value, and if the test succeeds, LSUMT(I) is assigned the value of the first calling argument. If the test fails, the value of the second argument is used. LSUMT and LEN are typed real for the second example, so that real library function SSUM can be used to sum vector LSUMT. This real sum is then converted to type integer. The vector version of the loop actually contains more code, plus an external reference, but executes more than 12 times faster than the original. For an iteration count of five or less, vectorization of this loop would not have resulted in any time savings because of the required post-loop processing. Generally, the more times a loop is executed, the greater are the potential savings from vectorizing it.

PCS1 Restructure

The R-C network is contained in PCS1. Each word contains several pieces of information, as illustrated in figure 1. Each node in the network has one of these words for each connection it has to other nodes; i.e., if node 10 is connected to four other nodes, then the portion of PCS1 belonging to node 10 will be four sequential words. Obviously, the length of PCS1 is twice the number of conductors in the network. All current SINDA execution routines perform the new temperature calculations with an outer DO loop with the range of 1 to the number of nodes and an inner loop whose range varies depending upon the number of connections for that node. The basic forward difference formulation is an example:

$$T_i' = T_i + \frac{\Delta t}{C_i}\left[\sum_j G_{ij}(T_j - T_i) + \sum_j R_{ij}(T_j^4 - T_i^4) + Q_i\right] \quad (1)$$

The inner loop, using PCS1, sums the G_{ij} on $R_{ij} \Delta T$ product for each conductor to the node and the outer loop does the calculation of the new temperature, T_i', for each node.

Extracting data from bits of a word is an expensive process in terms of computer time. However, the increase in allowable problem size resulting from packing of the data was judged to be worth the price for the machine on which it was first coded; for the CRAY 1 it is not.

Assuming the inner loop work has been done and stored, i.e.

$$GSUM_i = \sum_j G_{ij}(T_j - T_i) + \sum_j R_{ij}(T_j^4 - T_i^4) \quad (2)$$

the equation resulting from substitution into equation (1):

$$T_i' = T_i + \Delta t/C_i (GSUM_i + Q_i) \quad (3)$$

is well suited for vector processing because all terms are vectors of length i. However, the largest amount of time is spent in forming the $GSUM_i$ term, and the code dictated by the PCS1 structure is not well suited for vectorizing.

One approach taken to make SINDA more efficient on the CRAY 1 was to restructure PCS1 so that a new execution routine could be written which did not require unpacking of data and which contained vectorizable code. (See fig. 2.) The original PCS1 contained four types of information:

a. type of conductor - bits 0-4 and 21
b. conductor number - bits 5-20
c. adjoining node - bits 22-35
d. subject node - implied by current location in PCS1.

The new PCS1 has the conductors sorted by types (instead of by subject node as in the original) and three arrays, one specifying conductor number, and two to list the connected nodes (subject node and adjoining node).

Currently, the new PCS1 is being built from the old PCS1 in a post preprocessor operation, as opposed to recoding the preprocessor to construct it in the new form. It is anticipated that this overhead cost will be more than offset by the elimination of unpacking and the resulting existence of more vectorizable code.

CONCLUDING REMARKS

We have shown that it is possible to build and use for analysis thermal math models on the CRAY 1 computer of UCS that are much larger than those allowed on our local computer, the JSC UNIVAC 1110. We have presented data showing the CRAY 1 executing scalar code to be competitive with the UNIVAC 1110 for the same model. An additional small cost reduction would probably result if the total cost of running the five UNIVAC models were compared to the cost of a single execution of the combined model on the CRAY. However, the largest potential cost savings lie in developing efficient vectorized code for the CRAY. There is no exact method for determining the point at which run time savings gained through vectorization are offset by programming costs accrued in the modification process. Program portability, frequency of use, and life expectancy, along with programmer hours and expected savings should be considered before beginning any vectorization effort.

REFERENCES

1. CRAY-1 Computer System FORTRAN (CFT) Reference Manual. CRAY Research, 2240009, May 1980.

2. SINDA User's Manual. TRW Systems, 14690-H001-R0-00, April 1971.

TABLE I.- COMPARATIVE PERFORMANCE OF SINDA WITH THE MID THERMAL MATH MODEL

	UNIVAC		CYBER		CRAY 1	
	CP seconds	Cost	CP seconds	Cost	CP seconds	Cost
Preprocessor	2059.0	$261.38	1034.3	$304.26	244.1	$279.39
Processor 0.2 - 45.4 hr	5762.0	$731.45	2248.0	$614.38	451.7	$506.33

TABLE II.- COMPARATIVE PERFORMANCE OF SINDA WITH
THE COMBINED SHUTTLE THERMAL MATH MODEL

	UNIVAC		CRAY 1	
	CP seconds	Cost	CP seconds	Cost
Preprocessor	6541.0	$830.30	830.6	$1012.07

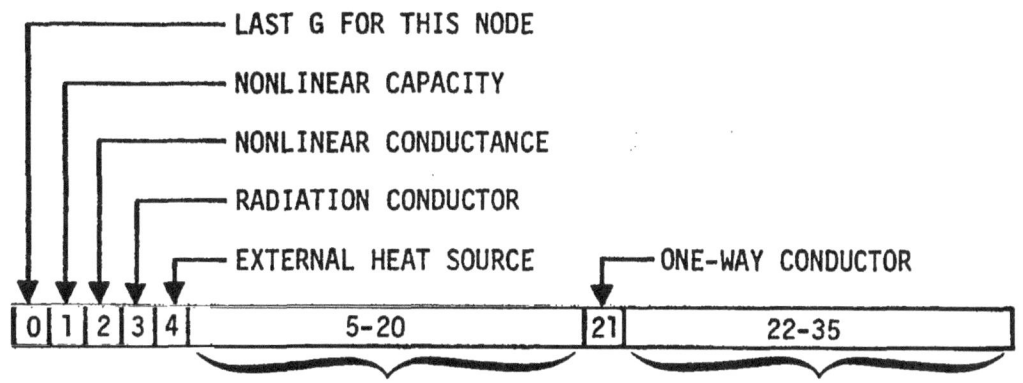

Figure 1.- PCS1 packed data structure.

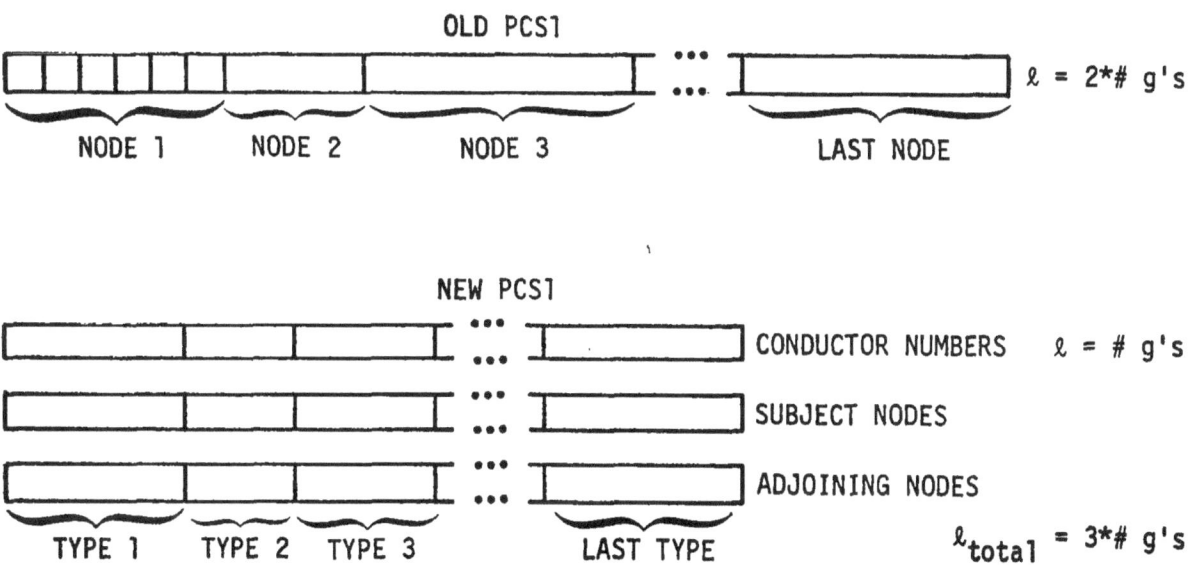

Figure 2.- Old versus new PCS1 organization.

AN EVALUATION OF SUPERMINICOMPUTERS FOR THERMAL ANALYSIS

By

Olaf O. Storaasli[*], James B. Vidal[**], Gary K. Jones[***]

INTRODUCTION

The computer hardware on which thermal analysis is run now and will be run in the future is undergoing significant changes as shown in figure 1. The past and projected market share for mainframes, minicomputers, and microcomputers is shown on the left of figure 1 (refs. 1 and 2). The figure shows a dramatic decrease in market share for mainframe computers as the minicomputers and microcomputers become capable of solving problems formerly solved on mainframes only. Figure 1 shows on the right the most dramatic increase in revenues is projected to be for 32-bit minicomputers to reach 4.3 billion dollars by 1985. In a parallel effort, 32-bit microcomputers (CPU on a chip) with virtual memory and Winchester disk drives are being introduced (micromainframes) which promise to provide mainframe capability in smaller packages at significantly reduced cost (i.e., "VAX on a chip"). The cost vs. capability or "cost-effectiveness" is the driving factor in the choice of future computing capability. The economy of scale criteria used to justify large centralized computer complexes is being challenged by wide-scale use of inexpensive minicomputers which are proliferating in much the same way as hand-held calculators.

Minicomputers of the past were not considered for complex thermal analysis because of insufficient memory, speed, accuracy, and secondary storage. However, two factors have changed the situation: (1) the introduction of virtual memory operating systems, and (2) 32-bit word architecture on minicomputers to produce so-called "superminicomputers." Virtual memory operating systems allow all software and problems (regardless of size) currently running on mainframe computers to run on minicomputers (albeit slower) by using disk memory as an extension of real memory. The 32-bit architecture (with 64-bit double precision) provides the necessary accuracy and compatibility with large computers to simplify software conversion.

The object of this paper is to evaluate the use of superminicomputers for thermal analysis by solving a series of increasingly complex thermal analysis problems on both superminicomputers and large mainframe computers. The approach involved (1) installation and verification of the SPAR thermal analyzer software on superminicomputers at Langley Research Center and Goddard Space Flight Center, (2) solution of six increasingly complex thermal problems on this equipment, and (3) comparison of solution (accuracy, CPU time, turnaround time, and cost) with solutions on large mainframe computers.

---

[*] NASA-Langley Research Center, Hampton, VA 23665.
[**] Digital Equipment Corporation, Richmond, VA 23229.
[***] NASA-Goddard Space Flight Center, Greenbelt, MD 20771.

COMPUTER TERMS

Although minicomputer-related technology is rapidly changing, the following definitions should help to clarify terminology used in the paper.

<u>Microcomputer</u>: 4 to 6 bit computer with the CPU on a single chip.
Cost: $100 to $15,000

<u>Micromainframe</u>: 32-bit computer with the CPU on from 1 to 3 chips.
Cost: $4,000 to $40,000

<u>Minicomputer</u>: 16 to 32 bit computer with the CPU consisting of components on a CPU board.
Cost: $15,000 to $100,000

<u>Superminicomputer</u>: 32-bit minicomputer with virtual memory operating system.
Cost: $50,000 to $500,000

<u>Mainframe Computer</u>: 32 to 64 bit computer with a conventional CPU on many boards.
Cost: Millions of dollars

<u>Supercomputer</u>: 64-bit computer with high-speed CPU.
Cost: Tens of millions of dollars

<u>Virtual Memory</u>: Storage technique in which disk memory is used to augment real memory.

<u>CAD/CAM</u>: Computer-aided design/computer-aided manufacturing.

<u>Turnaround</u>: Wall clock time elapsed from initiation to completion of computations (function of other activity on the computer).

<u>Byte</u>: Unit of computer storage equivalent to one character of 8 bits (i.e., 200 MB = 200 million characters).

<u>DECNET</u>: <u>D</u>igital <u>E</u>quipment Corporation <u>NET</u>work facility to communicate with other DEC computers.

<u>UT200</u>: Control Data Corporation protocol for remote.

<u>PRIMENET</u>: <u>PRIME</u> Computer <u>NET</u>work facility.

SUPERMINICOMPUTER HARDWARE AND SOFTWARE

The configuration of the Langley superminicomputer used in this study is shown in figure 2. It consists of a Digital Equipment Corporation (DEC) 32-bit VAX 11/780 CPU with 1 million bytes (characters) of memory, 0.2 billion bytes of disk memory, a tape drive, remote communications (DECNET, UT200), and 16 terminal ports. The 32-bit architecture provides the engineer with approximately seven or more decimal digits of accuracy. In addition, a second superminicomputer at Langley was used to cross-reference and plot results. It consists of a 32-bit PRIME 750 CPU with 2 million bytes of memory, 0.3 billion bytes of disk memory, a tape drive, printer, remote communications (PRIMENET, UT200), and 16 terminal ports. Both the VAX and PRIME 750 have virtual memory operating systems permitting users to address in excess of 4 billion bytes. This large address space simplifies conversion of thermal analysis software from large mainframe computers.

The configuration of the Goddard superminicomputer (one of fifteen) used in this study is shown in figure 3. It consists of a VAX CPU with four million bytes of memory, 0.1 billion bytes of disk storage, three tape drives, two line printers, card reader, 64-bit array processor, remote communications (DECNET), and 32 terminal ports.

The thermal analysis capability contained in the SPAR finite element structural analysis software (ref. 3) was recently installed and tested using six test problems on both the Control Data Corporation (CDC) CYBER 175 and 203 computers at Langley (ref. 4). Since the basic SPAR software had already been converted to both VAX and PRIME superminicomputers (refs. 5 to 9), the new thermal analysis processor software and six sample problems were transferred from the CDC CYBER to the VAX and PRIME superminicomputers via the UT200 communications link. Several software modifications were made to COMMON block dimensions of the thermal processors to take advantage of virtual memory. Also, slight changes were made to the six input files to reflect the character differences contained in the following table. These are special characters used for SPAR free field input.

Function	CDC 175	CDC 203	VAX	UNIVAC	PRIME
Text/File	"	'	'	'	#
Record Term	;	;	; or :	:	:
Continuation	%	>	>	>	%

The disk input/output subroutines are critical for optimum performance and much could be done to improve the performance of both VAX and PRIME by rewriting these routines as was done on the mainframe. However, for this study, the code was not optimized and the code used was 100 percent FORTRAN. The CYBER 203 results also reflected a minimal use of vectorization.

The following table shows the equivalent CPU charges and accuracy for the computers used in the study based on Langley charging algorithms which have considerable variation on the CYBER due to I/O charges.

Computer	Bits/word	Cost/CPU hour	Significant digits
CDC CYBER 203	64	1492 to 4963[a]	14
CDC CYBER 175	60	735 to 3717[a]	13
PRIME 750 LaRC	32	56	7
DEC VAX 11/780 LaRC	32	33	7

[a] varies according to CPU I/O mix

THERMAL ANALYSIS TEST PROBLEMS

A detailed description of the six thermal analysis test problems is contained in reference 4. Additional descriptions of four of the problems and the algorithms used in transient thermal analysis are contained in reference 10. Figure 4 shows an insulated Space Shuttle orbiter test frame tested under transient heating as described in reference 11. The figure shows an aluminum frame with 190 grid points and 158 elements surrounded by insulation with applied surface temperatures causing conduction in the aluminum structure and insulation, and radiation across the air gap. The material properties are updated every 50 seconds to allow for changes in properties of aluminum and insulation which are functions of temperature and temperature and pressure, respectively. A temperature history at 10 second intervals for 1000 seconds is required.

A model of a 30-meter deployable antenna with 55 grid points and 183 conduction and radiation elements is shown in figure 5. It is subjected to solar radiation heating with interelement time-dependent shadowing considered. A temperature history every 0.01 hour for 24 hours is required.

A model of a single bay of the Shuttle orbiter wing with 123 grid points and 151 elements subject to time-dependent surface heating is shown in figure 6. The model is subject to interelement and space radiation for which a temperature history is required every second for 2500 seconds.

A model of the Space Shuttle orbiter wing with 1542 grid points and 2125 rod triangle and quadrilateral elements subjected to aerodynamic surface heating is shown in figure 7. The external insulation on each surface is modeled by five layers of solid prismatic elements. The material properties were updated in time steps of 100 seconds and the temperature history for 3000 seconds was required.

A half-model of an insulated cylinder with 800 grid points and 650 elements subjected to time-dependent surface heating along the shaded region is shown in figure 8. The material properties were updated every 200 seconds and a temperature history is required every 10 seconds for 2000 seconds.

A model of a multiwall thermal protection system with 333 grid points and 1096 elements located on nine titanium sheets subject to transient temperature imposed at the outer surface of the panel is shown in figure 9. It is assumed that the heat load does not vary in directions parallel to the plane of the panel. A temperature history of 2000 seconds with 1 second intervals is required.

TEST PROCEDURE

The results for the test cases were first obtained on the Langley VAX and then compared with the CYBER 175 and 203 (ref. 4) shown in figure 10. The six test problems were run on the VAX in both heavy use and dedicated environments to evaluate performance. The updated SPAR software with new thermal processors, together with the six sample problems, were transferred to and run on a NASA-Goddard VAX for performance evaluation. Three of the sample problems were also run on the Langley PRIME to evaluate the performance of a superminicomputer from a different manufacturer.

For many cases, the thermal analysis software evaluation was only one of many applications running concurrently on the superminicomputers. Other competing applications included the modification, recompilation, and use of computer-aided design/computer-aided manufacturing (AD-2000, refs. 12 and 13), date base management and finite element modeling, and analysis software.

RESULTS

Figure 10 shows the implicit Frank-Nicholson algorithm solution times (CPU and elapsed) and cost for all six test cases (across the top) for four computers in decreasing size (down the left). The figure shows three entries for each sample problem for each computer. For example, the entries for the Space Shuttle frame on the VAX means the solution took 128 CPU seconds, 278 elapsed seconds, and cost $1.17.

Figure 10 shows that, in general, the Langley VAX was representative of other superminis in that it was somewhat faster than the Goddard VAX and somewhat slower than the PRIME 750. It is expected that the VAX elapsed (and possibly CPU time) results would be significantly faster if fast disk utility calls were used instead of the 100 percent FORTRAN I/O used. The elapsed times are primarily an indication of competing CAD/CAM and analysis applications running concurrently and subject to change as the demand on the CPU changes. Recently announced superminicomputers (i.e., PRIME 850, ref. 14) drastically reduce such elapsed times through the use of dual processors. The results shown in figure 10 are further broken down in figures 11, 12, and 13 for CPU time, elapsed (wall) time, and cost, respectively.

The percent of the VAX CPU time (log scale) of the PRIME 750 and CYBER 175 and 203 computers is shown in figure 11 for each of the six sample problems (represented by bars). The numbers (and corresponding shading) above the bars in figure 11 indicate the actual CPU time taken by the VAX in seconds. The figure shows that the CYBER 175 takes slightly more than 5 percent of the Langley VAX CPU time, the CYBER 203 slightly less than 5 percent, and the PRIME 750 slightly less than the VAX.

The elapsed (wall) time with the maximum value indicated at the top of each bar is shown in figure 12. The figure shows that for the typical loadings (applications) on the computers, the elapsed (wall) time on the superminis was about the same as on mainframe computers. A low showing on this chart may indicate low utilization of a machine.

The cost comparison for the various computers as a function of the maximum cost in dollars is shown at the top of each bar in figure 13. For all problems, the lowest cost was for the VAX with the Langley PRIME 750 about 10 percent higher. The minimum cost to obtain solutions on superminicomputers ranged from 11 to 59 percent of the maximum cost which was accrued on the mainframes. Figure 13 shows the CYBER 203 solutions to be the most costly for the largest and smallest problems, while the CYBER 175 produced the maximum cost for the other four problems.

The general conclusions to be drawn from the results shown in figures 10 to 13 are that superminis under similar loadings with mainframe computers will produce results (elapsed wall time) in approximately the same elapsed time with the superminis taking about 20 times as much CPU time, but costing only 11 to 59 percent of the cost of mainframe solutions. The accuracy of the superminicomputer results from all machines agreed with the CDC mainframe results for all problems to within five significant digits (except for one case where several numbers out of several thousand of one-time history degraded to three significant digits). If increased accuracy is necessary, double precision (64-bit accuracy) is available on both the VAX and PRIME 750 superminicomputers at the expense of additional CPU time. However, it was felt that the five-place accuracy obtained for the superminicomputer results was suitable for this study.

CONCLUDING REMARKS

This paper demonstrates the feasibility and cost-effectiveness of solving thermal analysis problems (both large and small) on interactive superminicomputers. The installation of the SPAR thermal analyzer software on superminicomputers was relatively straightforward since the SPAR data base management was already converted to both PRIME 750 and VAX minicomputers. The interactive features, high speed terminal communications, "user friendly" features of the operating systems, and communications software between the superminicomputers and mainframe computers made the software conversion, evaluation, and testing of the test cases possible in a fraction of the time normally required. The five-place accuracy of the results on the superminicomputers for single precision was better than expected so that the double precision options on the superminicomputers were not required. The time results showing superminicomputers taking about 20 times as much CPU time is about as expected based on the relative CPU speed of the superminicomputer and mainframe processors. The fact that both superminicomputers and mainframes take approximately the same elapsed time for the solutions was surprising. It may be advantageous from an elapsed time standpoint to solve problems on superminicomputers during off-hours since results are likely not to be available on a typically loaded mainframe any faster. The elapsed time on all single processor machines is a function of the number of concurrent computations, and the good performance shown by the CYBER 203 and bad performance of the CYBER 175 are possibly due to low and high utilization, respectively.

The solution cost for superminicomputers ranged from 11 to 59 percent of the mainframe solution costs. One of the important factors leading to the cost-effectiveness shown in the results is the low purchase cost (resulting from high production rates) of superminicomputers. It is felt that the cost-effectiveness and capability demonstrated by the superminicomputers in this paper adds impetus to the trend to produce superminicomputer capability (32-bit architecture with virtual memory) in smaller, less expensive packages.

REFERENCES

1. Micro Mart Growth Seen Threat to IBM. Computerworld, vol. XV, no. 25, June 22, 1981, p. 65.

2. Low-End 32-Bit Minis Join High-End Supermicros in Siege on Traditional 16-Bit Mart. Computerworld, vol. XV, no. 40, Oct. 5, 1981, pp. 67, 71.

3. Marlowe, M. B.; Moore, R. A.; and Whetstone, W. D.: SPAR Thermal Analysis Processors Reference Manual, System Level 16. Volume 1: Program Execution. NASA CR-159162, 1979.

4. Robinson, James C.; Riley, M.; and Haftka, R. T.: Evaluation of the SPAR Thermal Analyzer on the CYBER-203 Computer. Computational Aspects of Heat Transfer in Structures, NASA CP-2216, 1982, pp. 395-414.

5. Storaasli, Olaf O.; and Murphy, Ronald C.: Finite Element Analysis in a Minicomputer/Mainframe Environment. Research in Computerized Structural Analysis and Synthesis, NASA CP-2059, 1978, pp. 77-88.

6. Foster, Edwin P.; and Storaasli, Olaf O.: Using SPAR Structural Analysis on a Minicomputer. Seventh Conference on Electronic Computation, American Soc. Civ. Eng., c.1979, pp. 363-373.

7. Foster, Edwin P.; and Storaasli, Olaf O.: Structural Analysis on a Minicomputer. Engineering Software, R. A. Adey, ed., Pentech Press (London: Plymouth), c.1979, pp. 43-54.

8. Storaasli, Olaf O.; and Foster, Edwin P.: Cost-Effective Use of Minicomputers to Solve Structural Problems. J. Aircr., vol. 16, no. 11, Nov. 1979, pp. 775-779.

9. Storaasli, Olaf O.: On the Role of Minicomputers in Structural Design. Comput. & Struct., vol. 7, no. 1, Feb. 1977, pp. 117-123.

10. Adelman, Howard M.; and Haftka, Raphael T.: On the Performance of Explicit and Implicit Algorithms for Transient Thermal Analysis of Structures. NASA TM-81880, 1980.

11. Gallegos, J. J.: Thermal Math Model Analysis of FRSI Test Article Subjected to Cold Soak and Entry Environments. A Collection of Technical Papers - AIAA/IES/ASTM 10th Space Simulation Conference, Oct. 1978, pp. 131-136. (Available as AIAA Paper 78-1627.)

12. Storaasli, Olaf O.: Integrated Computer-Aided Design Using Minicomputers. Preprint 80-671, American Soc. Civil Eng., Oct. 1980.

13. Blackburn, C. L.; Dovi, A. R.; Kurtze, W. L.; and Storaasli, O. O.: IPAD Applications to the Design, Analysis, and/or Machining of Aerospace Structures. A Collection of Technical Papers, Part 1 - AIAA/ASME/ASCE/AHS 22nd Structures, Structural Dynamics & Materials Conference, Apr. 1981, pp. 96-104. (Available as AIAA-81-0512.)

14. Scannell, Tim: Prime Supermini Most Powerful Yet. Computerworld, vol. XV, no. 16, Apr. 20, 1981, pp. 1, 8.

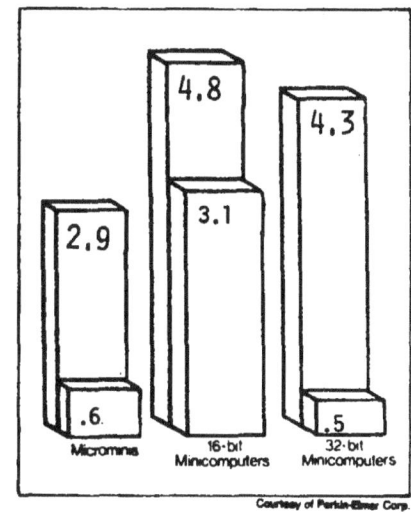

UNIT SHIPMENTS

REVENUE IN BILLIONS
(1980-1985)

Figure 1.- Past and future computer environment.

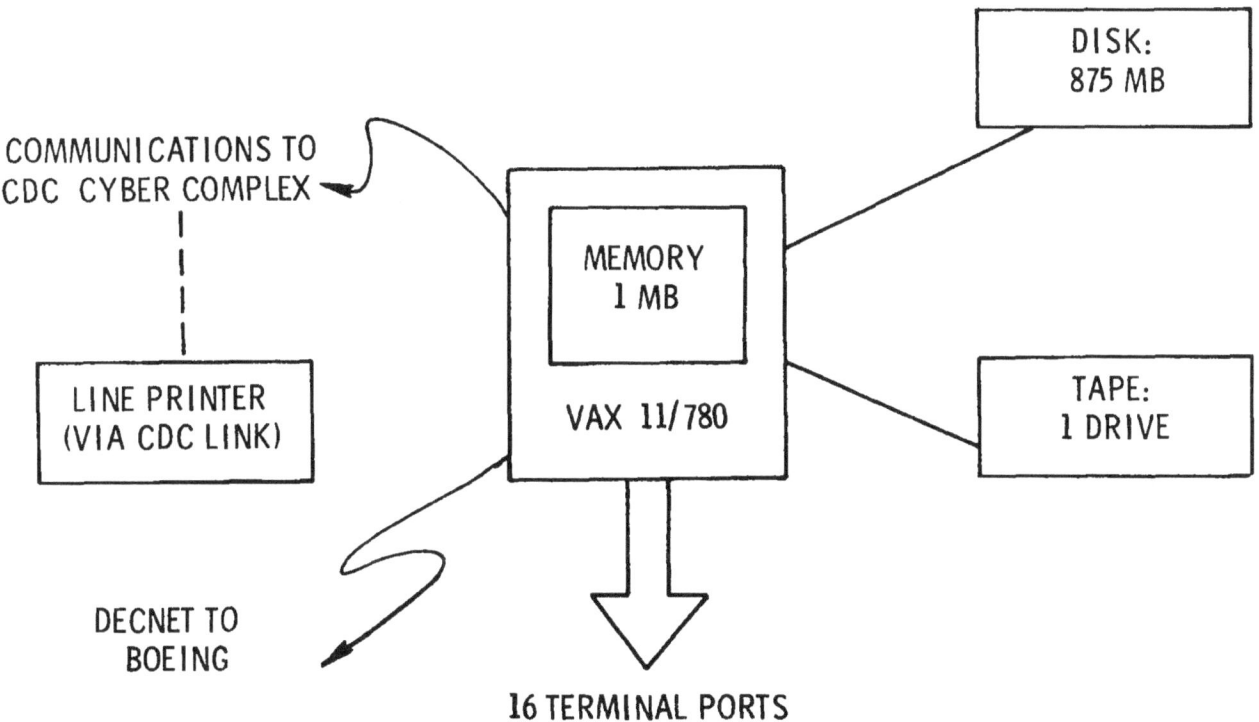

Figure 2.- Langley VAX superminicomputer system.

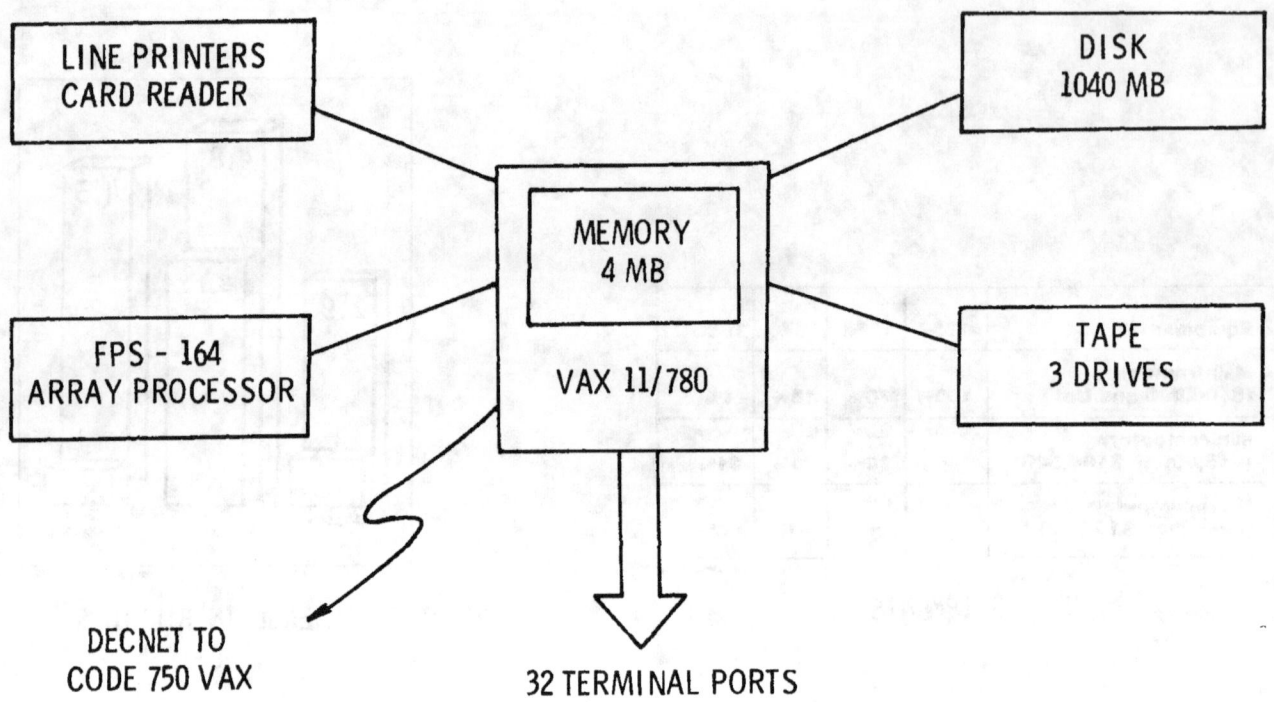

Figure 3.- Goddard Code 730 superminicomputer system.

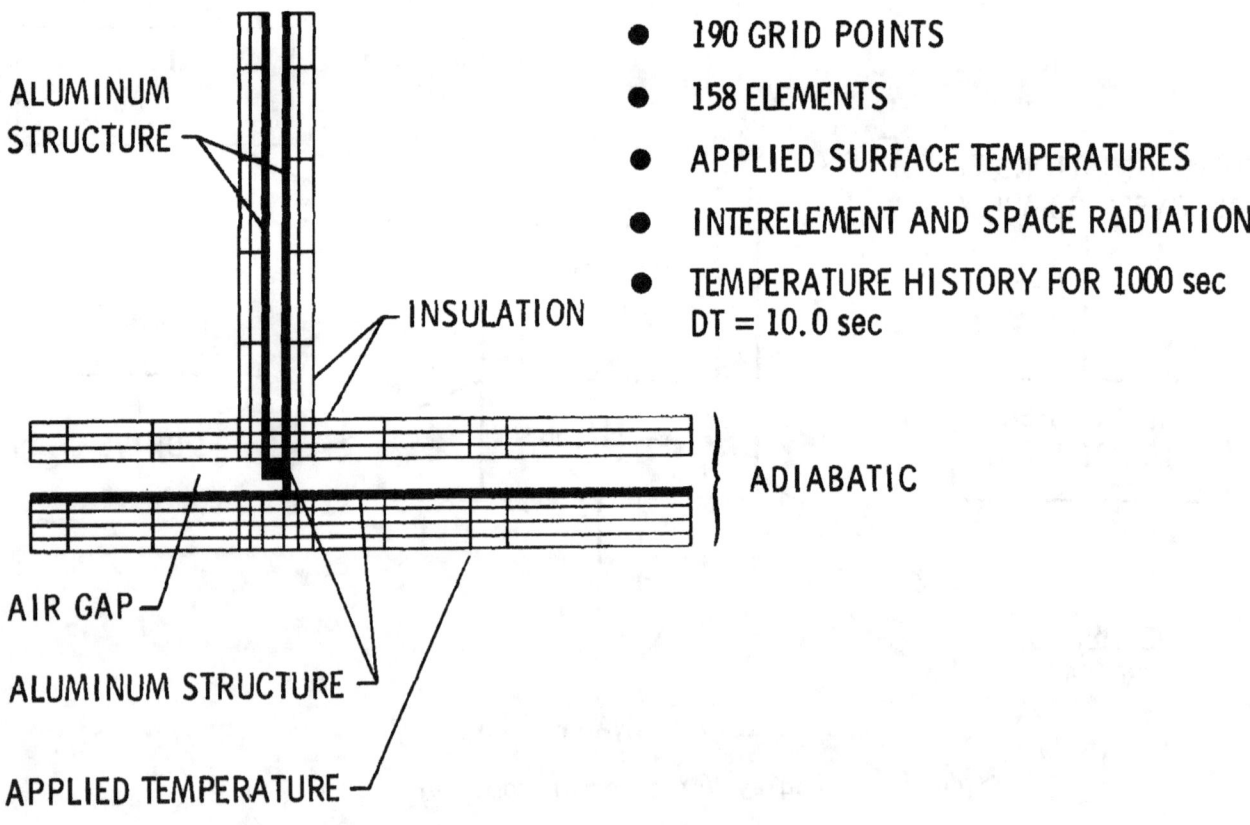

- 190 GRID POINTS
- 158 ELEMENTS
- APPLIED SURFACE TEMPERATURES
- INTERELEMENT AND SPACE RADIATION
- TEMPERATURE HISTORY FOR 1000 sec DT = 10.0 sec

Figure 4.- Insulated Space Shuttle test frame model.

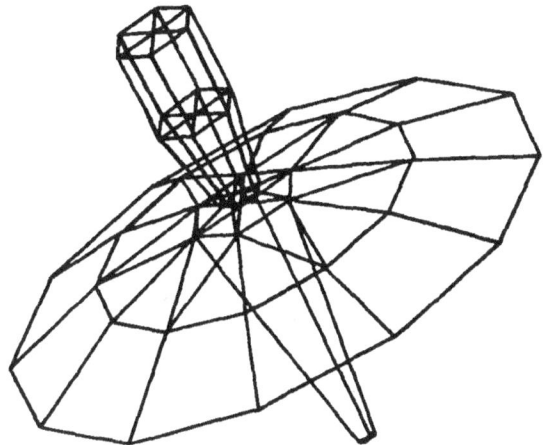

- 55 GRID POINTS
- 183 ELEMENTS
- INTERELEMENT RADIATION
- TIME-DEPENDENT SHADOWING
- TEMPERATURE HISTORY FOR 24 HOURS
  DT = .01 hr

Figure 5.- 30-meter deployable antenna model.

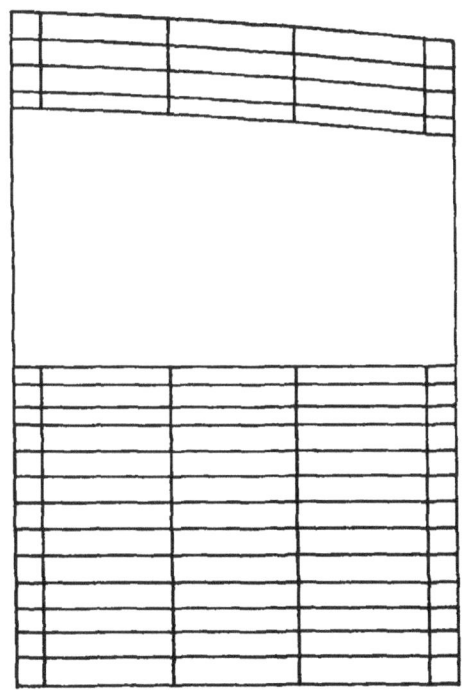

- 123 GRID POINTS
- 151 ELEMENTS
- TIME DEPENDENT SURFACE HEATING
- INTERELEMENT AND SPACE RADIATION
- TEMPERATURE HISTORY FOR 2500 sec
  DT = 1.0 sec

Figure 6.- Space Shuttle orbiter wing-bay model.

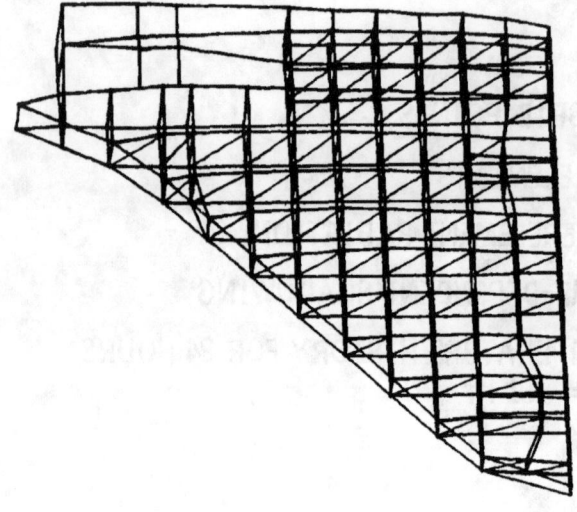

- 1542 GRID POINTS
- 2125 ELEMENTS
- APPLIED SURFACE TEMPERATURES (AERO HEATING)
- 1-D ELEMENTS USED FOR RSI
- TEMPERATURE HISTORY FOR 3000 sec DT = 100.0 sec

Figure 7.- Space Shuttle orbiter wing model.

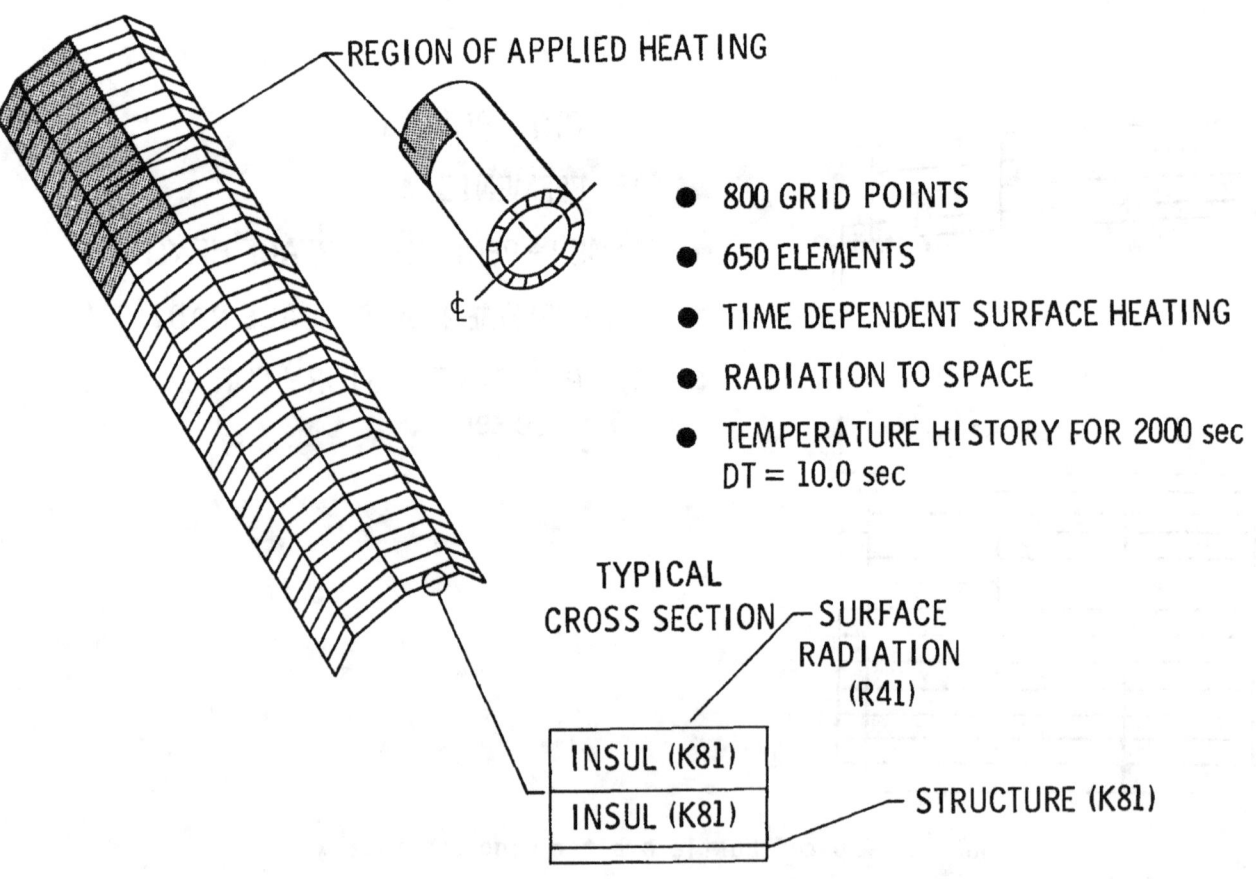

- 800 GRID POINTS
- 650 ELEMENTS
- TIME DEPENDENT SURFACE HEATING
- RADIATION TO SPACE
- TEMPERATURE HISTORY FOR 2000 sec DT = 10.0 sec

Figure 8.- Insulated cylinder model.

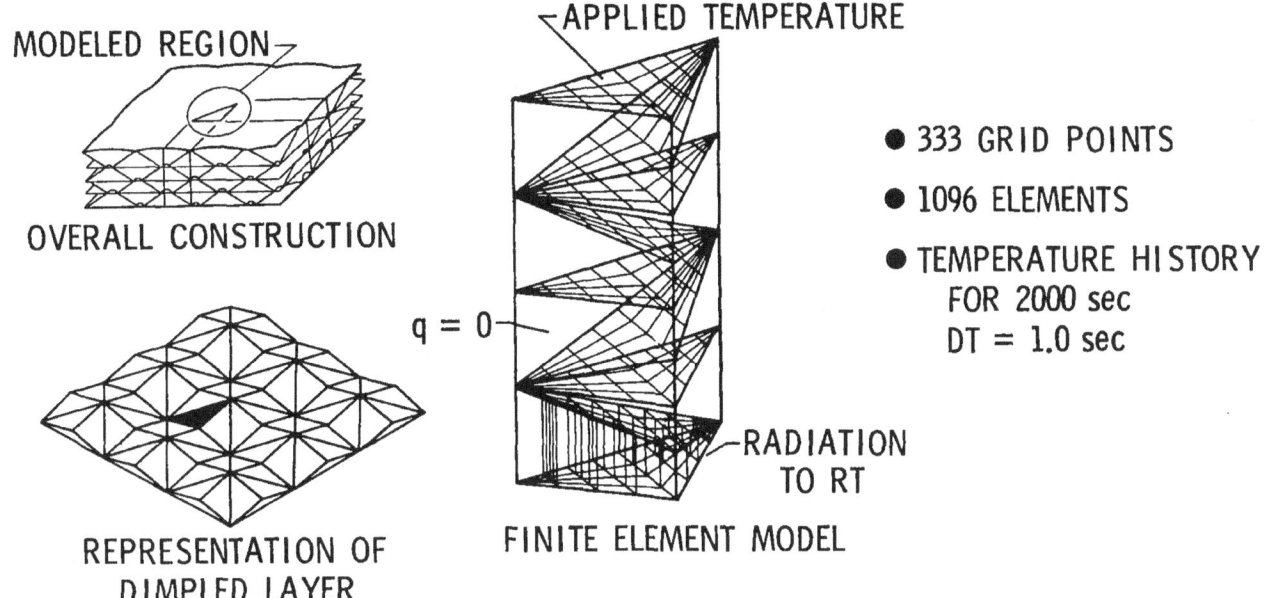

Figure 9.- Multiwall thermal protection system model.

COMPUTER	TEST CASES, CPU (SEC)/ELAPSED (SEC) COST $					
	SHUTTLE FRAME	ANTENNA	SHUTTLE PANEL	SHUTTLE WING	CYLINDER	MULTIWALL
CDC 203	8/13 11.03	32/429 20.80	58/64 34.00	86/120 39.00	68/78 31.85	149/720 61.75
CDC 175	8/91 3.92	46/1306 41.16	88/2199 61.21	145/24000 149.30	164/5708 45.26	223/28949 44.92
PRIME 750	121/170 1.89	672/4025 10.45	1084/2512 16.86	-	-	-
VAX 11/780 (LRC)	128/278 1.17	936/3275 8.58	1264/2494 11.59	1840/9199 16.86	2013/3366 18.45	3060/4357 28.05

Figure 10.- Computer time and cost comparison.

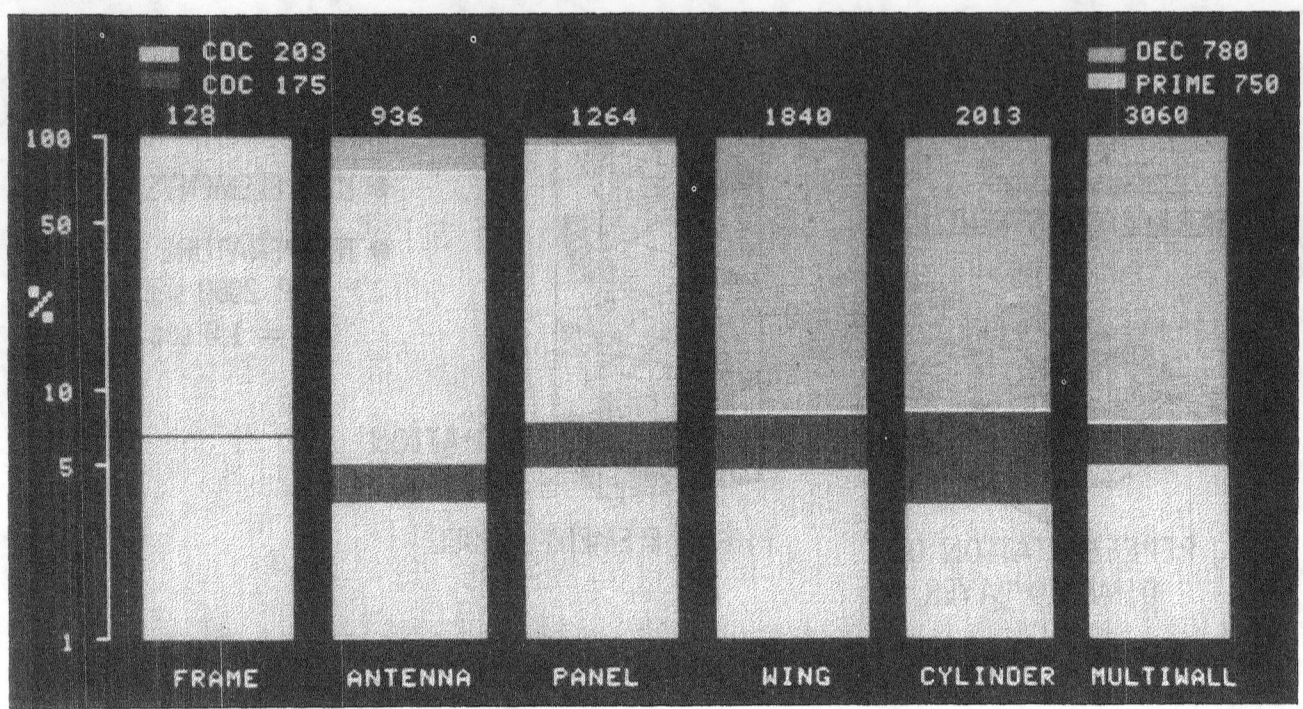

Figure 11.- Computer CPU time comparison. Numbers above bars indicate actual CPU time taken by VAX in seconds.

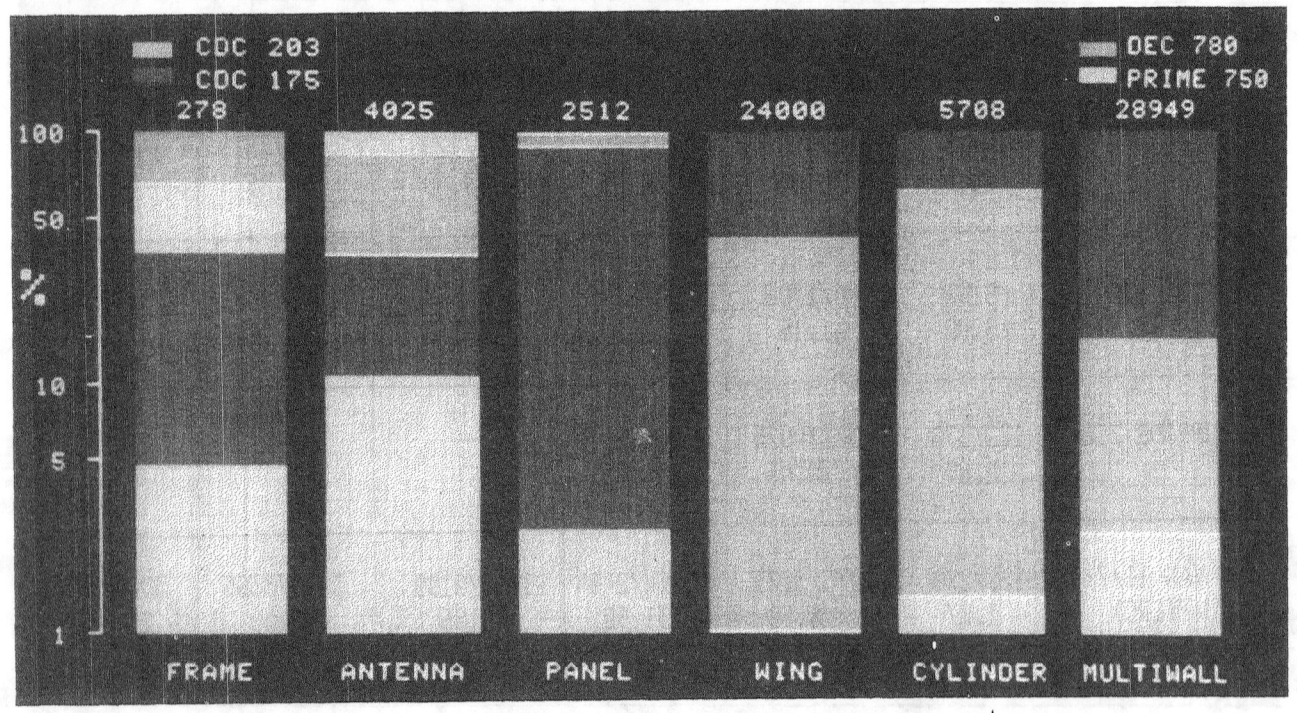

Figure 12.- Computer solution (elapsed) time comparison. Numbers above bars indicate maximum value of elapsed (wall) time.

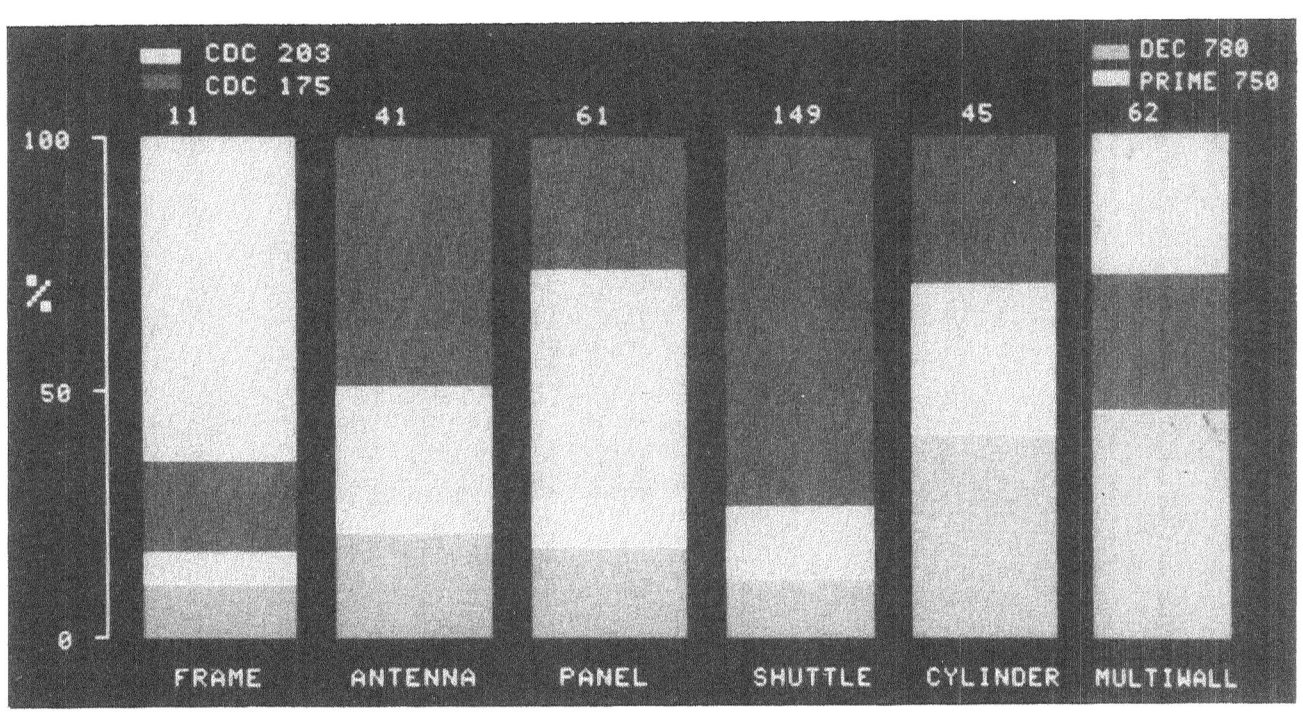

Figure 13.- Computer cost comparison. Numbers above bars indicate maximum cost in dollars.

PANEL DISCUSSION -

CONCERNS, ISSUES, AND FUTURE DIRECTIONS

Sidney Dixon, Moderator
Langley Research Center
Hampton, Virginia

DIXON: We have six panelists and you have probably met them all. Certainly you should know three of them: Rafi Haftka and John Swanson gave papers, and Earl Thornton was chairman of one of our sessions. One of the other three is Ahmed Noor, who is one of the cochairmen. Two who are hopefully benefiting from this but have been sitting quietly in the audience are Al Carter from Dryden and Ed Chimenti from Johnson Space Center. The first area we are going to discuss is computer hardware. Both Rafi and John Swanson had indicated that that was the area they wanted to talk about.

I was looking through a brochure for an SAE meeting in a couple of weeks. One of my people wanted to go to it and I had a travel problem and I found out that both these gentlemen are talking on the same topic there, too, so we are getting a preview of that discussion, and we are going to start with Dr. Haftka.

HAFTKA: Thank you, Sid. We have heard papers today about the impact of computer technology and computer software in our area and I would basically like to sound a note of warning about maybe getting too involved in that area. I have been exposed to some of the temptations of computer technology for some time and I remember a time, maybe 10 or 15 years ago, I found that I could speed up one of my FORTRAN computer programs quite a bit by including in it patches of assembly language. I have done that but it was the first and last computer program where I have taken this approach because I have decided that it is not really the proper usage of the time of a person who is involved with research and development. It may be appropriate for somebody who produces production codes. Probably it is the best usage of the time of a person who writes compilers. Now, I sometimes have similar misgivings about vector machines and minicomputers. In particular, I participated recently in a study of the cost effectiveness of minicomputers versus mainframes for structural analysis problems. I found that the difference in costs is not so much due to mini versus mainframe machines but the difference between owning a computer and renting one or buying time on it. Simply, more people can afford to buy minicomputers. However, the source of the benefit is sometimes due to the fact that when you own a computer you are expected to provide a lot of the services that, when you rent one, you get from your landlord. This is the kind of thing that you have to be careful about if you don't provide in advance or designate the support personnel. What is going to happen is that a lot of engineering or research personnel are going to be converted into computer hardware and software maintenance people. Now it could be that they like it or that management likes it, but it should be a deliberate decision. It should not be something that just happens. Now even though I don't believe that there is much of a cost advantage in minicomputers over mainframes, there is still a lot to be said for having them. Olaf has indicated that there are quite a few other factors involved and, in particular, I found out that in many cases the acquisition of minicomputers has acted as a sort of user liberation from the tyranny of central management that runs the large computer facilities.

DIXON: If you will hold your questions, we will now hear from John Swanson and then we will have questions for both of them.

SWANSON: The slides I am showing are really meant to be illustrative about the sort of things that are going on in the design analysis cycle. Currently in the design analysis cycle we have a large number of computers, finite element software, terminals, preprocessors, data base systems, data centers, array processors, CAD/CAM systems, post-processors, and so on, and that is about the amount of interaction there is among the various systems. If you are going to do design analysis this is the way the modern analysis is done [fig. 1]. You start out with a CAD model, you plug it into some sort of meshing software to make finite element or finite difference models. You get the model, you come into a preprocessor and you put loads and properties on it. You finally get around to running the analysis, then you go into a post-processor, which may be on a minicomputer, maybe a mainframe, then you get some displays and reports. You interpret them, then you go all the way back to the beginning and start all over again. If you are very sophisticated, you do the whole thing on the computer without having punch cards or print outs or something in between, but it is a real mess, because there are so many processes. Ideally, we are looking toward a work station [fig. 2] where you have the engineer working directly with the data base, and I am using this as justification for putting more of the decisions into the computer [fig. 3]. It was pointed out to me the other day that a nodal point is just as much a computed value as a stress result or temperature. Often it is more expensive to get a nodal point than it is to get a result, as far as computer resources go.

So we want to also add loops to the process [figs. 4 and 5]. Currently there are two implied loops in the design analysis--namely an accuracy loop, how accurate is the result (that involves remeshing, looking at different codes, or whatever), and then outside that hopefully there is a design loop. The problem with the design loop, of course, is that you don't get through the process once in time to meet the schedule, to say nothing about coming back, feeding the results into the design process, and benefiting from them. The best you can say is, yes, go ahead and do it. You can go back and say how to do it better. So in the future we want to put all the stuff into a single system, and this sort of thing is going on here at Langley. A single system where we can specify the loads, the desired accuracy (which may vary during the design process) and the design variables, push the button and say spend no more than 5 hrs., or 2 minutes, or whatever; give me the best approximation with that level of investment.

Getting to hardware, the criterion we need for this type of system is how soon can we get the results, and how soon we are talking about is soon enough so that the design can actually impact the process. So what I put into the notes, and I am not going to go through in detail, is how much time is spent in various parts of the analysis task. The analysis task is still the computer-limited part of the problem. Here are numbers [fig. 6] that are sort of similar to numbers we have seen before; the element formulation is 66.9%, the equation solution is 28% (on a vector computer, the CRAY 1), versus 44 and

50% on the scalar computer, the Prime 400. Notice here that the sum of these two is close to 95% of the total time. This is an overlay in space where core memory is the limiting factor. That is no longer true. Now the limiting factor is time [fig. 7], so the question is, what is the concept of an overlay in time? And what we are looking at here is the fact that in finite element work, the element stiffness matrices are completely independent. There is no requirement for any one before or after any other one.

Similarly with matrix triangulation. The row operations involved in Gaussian elimination do not have to be done in any order, so they can be done concurrently. The same can be said for stress recovery. In fact, the only thing that cannot be done concurrently is the back-substitution process. That fortunately is only a few percent of the total time at most. So I have introduced the concept of overlays in time because what we want to do is find out how to get results in a shorter period of time [fig. 8]. Of course, the easiest way of doing it is to increase the CPU speed. If we increase the speed by the factor of 2, we get a factor of 2 improvement in turnaround time. If we increase it by a factor of 10, we get a factor of 10, which is clearly the best we can do. However, there are other ways that are more feasible (possibly). Increasing the vector speed does not help much. The CRAY 1 is pretty much at the limit. The scalar operation now dominates the total time. However, if we can also do the matrix operations, as well as the vector operations, we can show a significant improvement. Hence, we would argue here for matrix hardware as well as vector hardware.

And finally, the concept I want to present is the concept of asynchronous computation. The concept of more than one CPU available to the particular analysis task. The reason I present it is that we can get a much higher return from this than from anything else other than speeding up the CPU itself. In fact, the calculated curve [fig. 9] shows that as the number of CPU's available to run asynchronously on the same task increases, at about 8 we are at the knee of the curve. At the asymptote (at 32 machines) it only takes 2.6% of the time it takes with a single processor, so there is a very large gain because of the finite element task itself, since it is asynchronous. The data structure does not need to be tied in, do this, do this, do this, in fact there are just a few discrete measuring points. The asymptote turns out to be at about 1%, in other words, the factor of 100 improvement with no change in the machine speed at all. So there is an interesting approach. So what we are looking at is to actually design software to take advantage of those concepts with the hope that the concepts will come along [fig. 10]. The control CPU is sort of the big mother hen, as it were, it takes care of all the things like the printers, cardreaders, tape drives, and so on. The user CPU's are the individual terminals each of which can reference sub-CPU's (these might be array processors, for example), and you can carry this one step down--where these have access to matrix operators, these will be hard coded boards. So what this scheme gives us is a way of looking at all the current configurations, be it the HP with its vector processors, the CRAY 1 with vector and matrix capabilities, the floating point FBS 164, which is essentially going to be a task CPU, or the CSPI array-processor which is essentially a matrix operator. What I have tried to do is present here the fact that there are more things that we could look at than we have just talked about today. Thank you.

DIXON: At this time we will take specific questions for Dr. Haftka or Dr. Swanson or any general questions on the area of computing hardware. Do we have any questions?

STORAASLI: I would like to ask Dr. Swanson about his feelings about the impact of the Intel 432.

SWANSON: I would answer it except I am not sure what the Intel 432 is.

STORAASLI: A new Intel micro mainframe.

SWANSON: The impact I think is that this provides the task CPU capability. It would fit in at that level. Task CPU implies FORTRAN capability, so it would fit at the task level and would offer the option of a very low-cost asynchronous CPU unit to be pluged in, so I think that would be the place this would fall in this hierarchy.

ADELMAN: I guess I know what Rafi is trying to get at. I think Rafi's point is, don't spend a lot of engineering time molding your program to a vector computer. I was wondering if Dr. Swanson agrees with this.

SWANSON: Very strongly. The CRAY version of ANSYS has no special code. In other words, we have provided the main part of the program to CRAY research. They looked at it for several weeks and have said there is nothing we can do with the code to get better performance. The CRAY/CFT compiler is, in my opinion, excellent. Now granted the code itself is highly vectorizable because it hasused the Lawrence Livermore STACKLIB routines for years which has forced upon us a very strict vector discipline but with that discipline the CRAY compiler picks up all the vector constructs and that way we do not have to maintain and provide quality assurance on multiple copies of the code which is, of course, extremely expensive.

SOBIESKI: I have a question for Dr. Swanson. I think the curve you showed indicating asymptotic decrease of the cost with the number of CPU's available is very interesting. But I was wondering if it is not missing one ingredient that when you increase the number of processors, even in an asynchronous mode, there is a volume of housekeeping involved that is going up proportionally with the number of processors which would offset the gains from the large number of processors working in parallel. So don't you get, instead of an asymptotic curve, a decrease to a minimum and then an increase?

SWANSON: The nature of the problem is that it will descend to a certain minimum and then there would be no further gains. Basically you cannot do more than all the elements simultaneously. So again with 100 elements, 100 processors would be fine, with 200, 100 would be idle and 100 would be in use. So in fact, that is true; also your point is that there is more overhead. That also is true; but I am leaving that to the hardware developers to make that as small as possible. This is the lower bound of what can be done. What is actually possible, of course, as you point out, is the curve probably has a minimum somewhere.

DIXON: The next two panelists are going to talk about software and methods. The first one will be Earl Thornton. In the first paper of the session on large space structures, the speaker introduced the concept of integrated thermal-structural analysis; you are talking about that Earl, aren't you? OK, I didn't think the full impact was coming out so I added Earl to our panel since he has been working in that area for about six years and I have asked him to spend a couple of minutes talking about that.

THORNTON: Actually, I accused Sid of adding me to the panel to balance the ticket with engineering schools from the state of Virginia since Rafi is here from VPI. Actually, I would like to comment on two points. One, as Sid mentioned, my student colleague, Jack Mahaney, talked about what we call an integrated thermal structural analysis. Later in the program yesterday one of the gentlemen from NASA Goddard talked about an integrated analysis capability. I would like to briefly distinguish these two. The NASA Goddard approach was, as I understand it, talking about interfacing programs to do thermal/structural/controls analysis by writing interfacing programs to transfer data between large existing programs. In our approach at Old Dominion we view the thermal-structural analysis problem as being a problem that since it can be done sequentially, is lending itself very well to elimination of interfacing programs. We view this from the standpoint that finite elements are capable of doing both problems and if properly programmed no interface is necessary. Now early finite element programs which did thermal analysis such as NASTRAN were aimed toward this but fell a little short. One of the speakers (I think Dr. Harder from McNeal-Schwendler) mentioned that frequently the thermal analysis requires a different model. We are working at Old Dominion with our colleagues here at NASA to develop thermal analysis capability through improved elements, improved methods of calculating in the thermal program the thermal forces and thermal moments and so forth, that are exactly needed in the structures program. We have also felt for many years that actually the state of the art in finite element thermal analysis was not quite up to par with what exists in structural analysis. I do not think we have really incorporated in the existing finite element thermal analysis programs all of the knowledge that is known both in heat transfer from the finite difference standpoint or in finite elements from the structures standpoint. So, to summarize briefly, we think there is a great deal of potential in finite element method for doing a truly integrated analysis in thermal stress analysis problems. We are working toward that and my colleague Ahmed Noor is going to mention some specific things that he sees that can be done that I think will complement what I have just said.

DIXON: Thank you, Earl. Ahmed is next. He did one thing also that I asked the panelists to do and that was during the course of the symposium if anything came up that they felt they needed to comment on to do so and he told me he has done that. He is going to be talking about computer programs and some analysis methods.

NOOR: Thank you, Sid. I think my comments here would be quite general, and let me have the first one. I am going to talk about the analysis procedures and computer programs. To start with, when we develop a computational procedure for heat transfer [fig. 11] we are really looking at four disciplines: heat transfer, discretization techniques, numerical analysis, and computer science, and all of these disciplines would very strongly impact our computational algorithm and its effectiveness. This is merely a very list of the analysis techniques that have been used for heat transfer problems [fig. 12]. The first two are the finite element and the finite difference techniques including the lumped-parameter methods, which are by far the most commonly used. Then we have the boundary integral methods and its new version, the boundary element method, is gaining popularity. Then the last three techniques listed here are the more classical ones, the weighted residual, transfer matrices, and the asymptotic perturbation techniques, and these have fields of application but they lack the versatility and the generality of the other techniques. Over the past few months I have attempted to make a literature search in the field of computational heat transfer. I have also looked into a number of programs which have been developed for heat transfer and I came across at least 100 programs. I selected 38 of these to include in a survey paper which will be included in the proceedings [paper no. 27 following this transcript], and what I have here in the middle is just a few comments on what I call the state of the art in computational heat transfer [fig. 13]. With regard to the computer programs, there are a large number of general programs which are user-oriented or, as the term has been used repeatedly here, "user-friendly" programs. I selected 38 of these and a sample is shown in the left slide [fig. 14]. The comments that I have about these is first many of these programs were developed as extensions of the structures programs and in some cases these programs did not represent the state of the art in structures technology. So they did not advance the state of the art in the heat transfer area. With regard to the computational algorithms, as Earl has already mentioned, many of the recent advances in computational structural and fluid mechanics have not been used in heat transfer and I have mentioned some of these here.

This is just what you might call a shopping list of some of the recent advances that could and are impacting computational heat transfer [fig. 15]. The first group is under finite element technology. I put under these subheadings the formulative aspects and element development. There have been a number of papers which discussed situations where you have to have more accurate thermal stresses or more accurate flux components and you can achieve these quite readily by using what we call alternate multifield mixed and hydrid finite element models, which have been widely used in the structures and fluid mechanics areas. Then by taking advantage of the equivalences between several of these finite element models we can cut down a great deal the computational effort in forming the individual elements. There have been a number of special elements developed in the structures and the fluid mechanics areas which did not find their way to the heat transfer area like boundary layer elements and infinite elements for handling infinite or semi-infinite subdomains. Then in the area of mesh design there are the higher order finite elements which have been quite effective in a number of applications, including fracture mechanics, and these are equivalent to the multigrid finite difference methods which have been used in fluid mechanics. Also there is a way now of developing what we call a posteriori error estimates; after you solve the problem you want to get an idea about the error and these computations can be fairly inexpensive. Then in the field of transient analysis, I think we had some papers already on improvements and explicit, implicit, and the mixed explicit/implicit temporal integration techniques. Then there is a whole area of research which I think has contributed much to the field of fluid mechanics in terms of operator splitting and partitioning schemes and there are automatic ways of choosing the time step. I think some of this we heard about in the first two days.

Then in terms of finite difference technology [fig. 16], curvilinear grids have been around for many years in fluid mechanics (we have an example here on the left [fig. 17]) and there are what are known as isoparametric finite differences. Incidentally, these grids on the left were formed automatically. There are grid generators and there are finite difference techniques which would be associated with these grids, and the formation of the finite difference operators is not really any more complicated. This is another example of a curvilinear finite element grid [fig. 18]. Then I mentioned the multigrid finite difference methods which have been used very effectively for adaptive refinement in fluid mechanics. Then we don't have to restrict ourselves to lower-order finite differences, as many people in heat transfer have been using, but we can use what is known as hermitian, or multilocal, finite differences which often give higher-order accuracy without increasing the bandwidth of the equations. In the field of numerical analysis I think we already had words about quasi-Newton methods. The conjugate gradient method has been one of the most effective techniques for solving equations but we can improve the effectiveness a great deal through scaling, which we call preconditioning. We also heard about the incomplete Cholesky factorization. Then in terms of engineering software [fig. 19] there are now definite guidelines which have been given in a number of reports on how to assess and evaluate large general-purpose software, and the assessment should not only be based on analysis capabilities but I think (as many people have mentioned also) the word user-friendly that is adequacy of user-oriented features, the maintainability of the program as well as the adequacy of user-support facilities and the portability of the software. Then there is a lot of development in the area of interactive graphics and (I think we had some of this also in the past two days) pre- and post-processing. The new computing systems we have: on the one end the supercomputers and on the other end the minicomputer. If we form a system of minicomputer arrays this could be a very effective system, as already mentioned by Dr. Swanson (systolic arrays, which use the VLSI technology).

To conclude, I have a slide here which says future directions [fig. 20]. Of course, the driving force would be the need to model large complex hardware systems subject to harsh environments and this means that we have to solve very large-scale thermal problems and we have to have reliable solutions and some way of estimating the error in the solution. Opportunities are provided by the tremendous advances in the computer hardware as well as the software systems, and I have a few comments with regard to computational algorithms. Our experience over the past few years, as well as the experience of others, has shown that trying to improve the efficiency of currently used algorithms, single algorithms, usually results in only marginal reduction in computational cost or effort. However, we feel that the future is for what you call hybridization, which means that you use more than one technique, particularly for large scale problems. The marriage of a number of techniques has resulted in improvements and continues to show very high potential. Examples of these are provided by the mixed explicit/implicit temporal integration scheme, the combined direct/iterative techniques for solution of algebraic equations, for example, in conjunction with something like the multigrid finite difference method. The reduced basis technique has proved to be very effective for steady-state thermal problems and this is nothing but a combined finite element or finite difference Galerkin and perturbation technique. When you combine these techniques in many situations you retain their advantages and alleviate a great deal their drawbacks. Then there is a modified modal superposition for transient problems, which Phil Shore has mentioned and is still working on. Thank You.

DIXON: OK. We will take specific questions for Dr. Thornton or Dr. Noor or any general questions for the panel in the area of computer programs or analysis techniques.

QUESTION: (Inaudible)

NOOR: Well if I understand correctly, first you are asking about an assessment of finite differences versus finite elements. I think a comparison of techniques is very difficult because in order to compare the techniques in any fair way you have to use an optimal finite difference grid for a particular problem and compare that with an optimal finite element grid for the same problem. Many of the comparisons that I have seen in the literature unfortunately did not do that. For example they picked finite elements as very suitable for modeling problems with curved boundaries and they used the very classical finite differences which is not used currently and which does not represent the state of the art. The current state of the art in finite differences is curvilinear grids, higher-order finite differences, so you compare that with the current state of the art in finite elements. It boils down, in my opinion, to a matter of personal preference and the experience of the analyst. There are situations where you can identify some advantages of one method over the other but it is very difficult to make a general statement. I think one has to have lots of qualifications when he makes a statement about one technique being better than the other.

DIXON: Would any of the other panelists want to add to those comments?

THORNTON: I think we had, earlier in the week, a pretty good description of finite differences and finite elements and I am referring to the paper by Ashley Emery from the University of Washington. He made some good points and I think I agree with most of them. He said at the onset of his talk that he did about 90% of his analysis nowadays with finite elements and in the talk he never really said why. So later, in a coffee break, I asked him and he said, because it is easier. I think because it is easier in the long haul and with the availability of computer graphics and the driver of computer-aided design, my personal opinion is that the finite element method, not for highly technical reasons of accuracy and so forth but because it is easier, will show distinct advantages.

DIXON: Any other panelist want to comment?

NOOR: Incidentally, you can use the same grid, in fact, for finite differences and finite elements because you are not restricted in finite differences by a rectangular grid. You can also use curvalinear grids which is the same as the grid used for the structural analysis grid.

THORNTON: I could make a different viewpoint on that also. Ahmed and I are from different universities; we don't always see things the same way. I think he has in mind (I'll speculate a bit here) solving maybe two-dimensional, three-dimensional continuous domains. I know from working here at Langley, with close associations of people like Jim Robinson, that when you get into airframe structures in particular, you are not talking about those kinds of domains. You are talking about highly irregular domains and from a structures standpoint, for example, you are modeling them with completely different kinds of elements structurally than maybe even exist thermally. For example there is no such thing as a thermal beam element. I think that there are distinct problems in transferring data from the finite difference programs to the finite element structural model for those realistic structures.

One of the first slides Sid presented on the opening day was that problem of the scramjet. That was an excellent example and opened my eyes a great deal. In the thermal model of that scramjet strut, we had to model the fluid flow. In the structural model, of course, the fluid flow doesn't enter, so there are distinct differences between the thermal and structural models. I think these are the real problems that have to be faced rather than the continuum type problems in which it probably doesn't make any difference whether you use finite differences or finite elements.

CHIMENTI: I wanted to make one comment concerning the finite element method being maybe easier to use than the finite difference method. We at JSC, in addition to doing what you call structures analysis, have a lot of subsystem analysis related to internal components, tanks, heater systems, components of that sort. I can certainly see for simple structures where the finite element is a pretty straightforward thing to model real quickly and rapidly. I am not so sure that same comment would apply when you get into what I call thermal control subsystem analysis.

NOOR: Well, I don't want to enter into debates here with Earl, but I would say again, being a user of both finite differences and finite elements, it is unfair really to make a strong statement in favor of one, because what Earl presumably was referring to is the classical finite differences where you discretize the equations. You don't have to discretize governing differential equations in finite differences. In fact, I can show (and we have shown it in many situations) that by using what you call, the modern finite difference terminology or methodology you can come up with identical discrete systems as finite elements. So really the discussion of what is better ends up to be a discussion about semantics.

DIXON: Let me ask you one question, Ahmed. Of all these computer programs that you have found in your survey are there any so called finite difference ones that have those techniques today?

NOOR: Most of the programs that I surveyed, particularly the finite difference area, did not include what I call the modern finite difference methodology.

DIXON: I think we need to go on. We'll come back and pick up some comments on this topic later on, but I do want to get to my last two panelists. The next one is Ed Chimenti from JSC, who has been up to his eyeballs for a number of years in the thermal analysis of the Shuttle. You heard several papers about it yesterday but he did not get a chance to say his piece then, so he is going to now.

CHIMENTI: Tuesday, Howard Adelman touched a little bit on the method that was used to perform the thermal-structural analysis of the Shuttle orbiter. From the perspective of somebody that has been on the thermal side of the street for a good number of years with the orbiter program, I wanted to add a few comments to that. This figure [fig. 21] shows the depiction of the thermal finite element surface elements for the Shuttle orbiter. Actually, the model is not analyzed as one piece of structure as shown here. It is broken into some 37 different substructures. There are some 7000 nodes represented by the combined structural model.

The next two figures [figs. 22 and 23] illustrate what the thermal people did in terms of generating temperatures for the structures. I might point out that we are talking here about the thermal analysis that is done specifically for the entry phase and provided to the structures folks so that they can do their thermal stress analysis. When we talk about the on-orbit thermal analysis that is performed, we have integrated models that take into consideration much bigger pieces of the orbiter and, in fact, there are like 4 or 5 pieces that take the forward, the mid, the aft, and the olms pod that include all the various subsystems plus the structure together. The circles [figs. 22 and 23] are the places where we have fairly detailed instrumentation on the vehicle. The thermal analysis and the thermal models were generated a number of years ago so the technology that we have today probably wasn't there. I think that the Rockwell engineers who did it had a number of reasons that the models were broken down like they are shown here. I think number one budget and computing power was probably an overriding reason for doing the job like they did. It is obvious that it requires a great deal of interpolation for those areas where we don't have a model and that is exactly what was done in the procedure.

The next figure [fig. 24] here shows a typical thermal math model of an area in the wing. You will notice that generally the models that were built run from 100 to 300 nodes typically. Incidentally, the finite difference lumped parameter method was used in performing these analyses. The TPS nodes on the top and the bottom, of course, are not shown there but there are a number of nodes that were required to model that sufficiently in order to get the high heat transfer through there and adequate temperatures. You will notice that the vertical truss member, for instance, has a number of nodes. I think the thermal engineer realized when he put this model together that you are going to have high heating from both the bottom and the top essentially, and that putting one node at the interface there was not going to be sufficient because obviously the middle is going to run cooler than the two ends. In contrast to that, at the same location, the picture that just came up [fig. 25] depicts what the finite element model required or was used in that same area. That's really an oversimplification. The little circles there would indicate the node points where the structures person thinks that he might want temperatures. If it were modeled that way and he got temperatures at that location, he certainly would not be getting what he wants because actually for that truss member he is looking for an average temperature of the truss to get what the growth is and if he just had the temperatures at those two end points he certainly would not be getting the right temperatures to use for his thermal stress analysis.

I think that before I go to the conclusions here those last two figures pretty well illustrate that, at least in this case, you could not use the same model for doing your finite element structures analysis and the thermal analysis. And this is generally true in the world of entry heating. I don't believe that the structures people would want to put the number of elements or nodes that the thermal model had in that particular area. In many of the areas when you are talking about on-orbit type analysis there can be some compatibility between the finite element model for the structures analysis and the finite element (if you want to call it that) thermal analysis. If we are to do that same job today, and incidentally, the job that was done probably was initiated seven, eight years ago, we would certainly do it differently. I think you would see a lot more integration of areas so that you would not be dealing with 125 models. I am not convinced that we would be dealing with one model. Of course, I think we are still not there in terms of being able to do something like that with the level of detail that is required to get what I call good temperatures. I have indicated four general areas of importance relative to being able to do a large integrated job [fig. 26]. Computing power: I heard discussions about CRAYs and computers of that sort, so I think we are getting pretty close to where we can do a pretty big job with a number-cruncher like that. Efficient running thermal analyzer: We certainly have programs these days that can analyze a large number of nodes. There is no doubt about that. The last two areas I think are very important. We have to be able to efficiently build the models. If somebody is talking about a 10 000 node model, or some number very large, we can't do it like we used to. It has to be an efficient building technique. Along with that we certainly need an efficient technique to interrogate the data, to check it out and see what we are using. I think we can get overwhelmed by numbers if we don't have an efficient interrogation and check-out procedure.

DIXON: OK. The final panelist is Al Carter who is head of the group out at Dryden that does high temperature work primarily looking at flight loads, but they get into other things from time to time including checking out some of those interpolated temperatures on the Shuttle. Two of the papers we had yesterday were by his people and he will wrap up the prepared comments now.

CARTER: The first thing I would like to do is to say a few words about Howard Adelman's performance on this symposium. I think he has done a hell of a good job. He didn't start on Monday, he didn't quit on Friday, and he kept everybody up to pace so I think he deserves some thanks for doing a good job. I am also surprised to be on this panel since I am a civil engineer and only had one class in thermo, but I would like to tell you a little bit about what we do out at Dryden and some conclusions about our observations from our problems.

As Sid said, our primary mission is to measure loads with strain gages on aircraft structures. The early attempts were very successful on high-aspect-ratio structures. When we got to the advent of low-aspect-ratio plate-type structures, especially with aerodynamic heating, we ran into some difficulty. We instrumented the X-15 and tried to measure loads on that airplane with strain gages and were able to get reasonable load-versus-G slopes, but obviously the thermal stresses pretty much screwed up the absolute values of the loads.

We could see that something had to be done if we were going to use strain gages especially because we anticipated work for the supersonic transport or Dyna-Soar. So a lab was built to conduct heating tests which would help us to do research on how to correct the strain gages for thermal stress. That lab is still operative and we have done a number of projects in it: the heated X-15 wings and tails, we heated the old YF-12 and flew it again. Those experiments were extremely edifying to us and we found out that the thermal and mechanical stresses for our purposes seemed to superimpose quite well. So we found that in flight vehicle testing we could correct the strain gages for temperature rather effectively. We also found a piece of structure that represented current thinking on hypersonic structures put together: Rene 41 beaded panels and spar caps, which we have been heating to generate some data which would help us evaluate thermal stress calculations and also heat transfer calculations.

The current activities [fig. 27] are: we got involved (Sid, Phil Glynn and I) in a discussion on what ought to be done in hot structures and they formed a kind of Gentlemen's Alliance, which is informal, to pursue some of this stuff in as coordinated a way as possible. So we're pursuing the hypersonic wing test structure in our laboratory. We are helping Phil with his orbiter measurements. With the hypersonic wing test structure, we are doing a finite difference thermal analysis. We are also doing a NASTRAN heat transfer analysis that is a fairly complicated structure and we sent the NASTRAN deck up to Ames to run on their 7600 machine. We are also doing some work on the orbiter because of our commitment to help measure loads. We decided we would have to do some heat transfer work to predict the temperatures with enough detail across the cross section, where the strain gages are located, to be able to attempt to make analytical prediction of strains and to make corrections. These diagrams show the cross section of the mid fuselage at station 877. In the middle chart [fig. 28] you can see the temperature distribution along the bottom surface and down the sidewall.

Our first computer run by SPAR uses heat inputs which we have generated the best we could (the kind of things that were discussed by Dr. Ko and Les Gong). We then took the measured temperatures and we tried to modify the theoretical temperatures as best we could. But you can see in the glove area (which is the projection out on the side of the fuselage) that measured temperatures are considerably lower than what we had predicted with our computer. Its quite a complex model, as we saw yesterday, so I don't think we have any concern about the complexity of the heat transfer model but we do have some concern about heating inputs to it. Now looking at the left-hand panel [fig. 29] the thermal stresses are plotted. The solid lines correspond to the solid lines on the temperature chart, the dashed lines correspond with the dashed lines on the temperature chart and the circles are the measured thermal stresses. The way we got those was to read the gages right after landing when it was sitting on the gear hot, read the same gages cold, take the difference and assume that was the thermal stress. This may not be perfectly legitimate because of some drift in the gages. I think the interesting point to be made in this chart is that thermal stresses, being self-equilibrating, are very sensitive to the changes in temperatures around other parts of structure (not just where the thermal stress is being measured). For example, the colder glove kicked the compression stress on the bottom up more than twice. You will also notice that the measured compression

stress is considerably higher than my dashed line. This happens to be a
fairly critical area and it is just an illustration of the fact that we really
need to know the temperature distribution all around the cross section in
order to do the job we want to do. So smearing [interpolation] bothers us.

The next set of charts [figs. 30 and 31] is a similar sort of thing for a wing
cross section. Here the temperatures were quite different (measured versus
predicted) but that was partially due to the fact we didn't account for
convection as discussed yesterday. We didn't have too much data to verify the
gradient between the skin and the spar cap. You will notice that although the
absolute value of the temperature was adjusted, the gradient was not adjusted
very much. You will notice on the stress diagram that the stress predictions
are not too different because we did not change the gradient. Fortunately for
us, the spar cap tensile stresses, which are the things that really hurt us,
seem to be predicted fairly well. The front spar, however, we have no idea
what it is going on. We don't have any measured temperatures. We don't know
how thick the insulation is so we are still pretty lost. That is an
illustration of the kind of problems we are concerned with. I think maybe it
is a good idea for thermal types to get a feeling for the amount of detail
that is required across a cross section to give us something that we can use.

Our future plans [fig. 32] are to continue with this hypersonic wing structure
and to do a SPAR analysis to go along with the NASTRAN and the finite
difference. Not so much to illuminate the world but to give us an idea of
what we would like stick with. Phil Glynn is interested in our proceeding
with this orbiter wing model and putting it on CRAY and trying to do a heat
transfer job on the whole wing. Then, if we ever get that done, he would like
to have us do a combined wing and fuselage. I think I share Ed's feeling that
we ought to be able to get our problem solved without having to go to such
extreme length, but we are now part of Ames. Ames has a CRAY and I think we
may be able to go ahead and do these things as more or less an experimental
investigation.

Our final slide [fig. 33] is some observations on what our feelings were after
doing those cross sections you heard about yesterday. The SPAR system allowed
us to model those very conveniently. In the time that it took for us to model
one 3-D cross section [by finite differences] the finite element system
permitted us to do one fuselage and two 3-D wing cross sections. So the
modeling was really super in SPAR. That was an effective way for us to go.
The view factor generation was a pain. We had a hell of a time. We tried to
use a NASTRAN view factor generator and it took an amazing amount of computer
time probably because we didn't use it right. We finally generated the little
HB something or other, computer program that worked reasonably well but that
was a real work-intensive job. So I am really encouraged to hear about Dr.
Emery's system and the fact that they are going to put it in SPAR because I
think that would be a really big shot in the arm for us. The other problem
was with material property tables. The TPS properties are sensitive to
pressure; therefore, you have to have a different and new set of material
properties for every location on the vehicle for every profile and that is a
mess. And we finally used the Lockheed thermal analyzer (the guts of that
program) to generate material property table for SPAR. That worked out fairly
well once we got it going but that was a real headache. Heat transfer versus
structural analysis: I guess the point I was going to make there is I was

amazed to find that the heat transfer analysis was an order of magnitude more effort than the thermal stress analysis. To think with only one degree of freedom per node that it would be one-third as difficult or one-sixth as difficult, yet it is ten times more difficult. Finally, finite element versus finite difference computer time: I think our conclusion was that finite differences was faster but for our operation we would prefer to use finite elements because we can model so much more quickly. Those are my observations.

CHIMENTI: The use of large models is a big concern of mine. With the large sizes that we are talking about going to are we losing sight maybe of accuracy and losing the feel for the model maybe? Like I say that is a concern that I personally have. I know from experience not with a structural model but with a little experience we had with STS-1, where we had an anomaly in the forward RCS. We had a very large model of that area, something on the order of 2000 nodes. When we got to the problem we immediately dropped down to pulling out the area where we were having the problem and started analyzing that particular area just from the standpoint of turnaround, ease, and understanding what we were doing instead of just number-crunching with an enormous model. I think that kind of thing will be done in the future when you start getting into a specific problem in an area that you can narrow down.

CARTER: I have to agree with Ed Chimenti that we can get really wrapped around the axle with these huge problems. I, however, don't yet have enough feeling for the effect of the adjacent temperatures on the area that I am trying to get thermal stresses. I don't know whether it is very sensitive or not but I am kind of pleased with the 2-D thermal stress analysis results that I showed there because they did not come from a finite element model; they came from a beam model, a Bernoulli kind of 2-D analysis. So it may be that all you got to know is the temperatures at the cross section that you are concerned about but it is pretty obvious from the fuselage that you got to know it pretty well around that cross section.

DIXON: Let me make one comment on that note, Al. When Phil Glynn was contacting us about some of the problems he was having, some of the old timers like Dick Heldenfels did not understand the need for big thermal models. He felt that thermal effects tended to the local. We actually had to send Jim Robinson down there for a couple of days to see what they were talking about and he came back and told us they had a real problem. If you will also remember one of the little jewels that Dick put out yesterday, it was to start with a small model and grow until you are happy with the results and quit. So I think when we are talking about big models, we are not really saying we are going to model the whole orbiter and calculate everything. We are going to try to make them big enough that we can do better than what we are doing with the current setup.

CARTER: I have a feeling that we don't know if the state of the art of the thermal stress analysis is all that mature and we really don't know where we are.

CHIMENTI: There is no doubt in my mind that with regard to the orbiter model we would integrate more than was integrated before in terms of number of small models, less of those and larger models. I don't think we would have a total orbiter [model] by any means with the 300 nodes per model and 115 or 125 models that we are talking about. I think you can see real quick here that with those numbers we are talking about 30 000 or so and you haven't even got many of the areas. So the kind of numbers is mind boggling in terms of total nodes. I think that Rockwell did a little bit of overkill in the level of detail in a lot of these models because they really didn't know at the time how big the gradients were going to be and what was required. I am sure we don't need the level of detail that you saw in that wing model, for instance. But on the other hand you need more than what you saw in the finite element [structural] model.

CARTER: I think Rockwell did an excellent job with the state of the art they had. One thing that came across to me at Downey, it didn't seem to me that thermal guys and structure guys talked a lot together before they started.

CHIMENTI: Well, that's always a problem but nobody asks why were the thermal models where they were. I think basically what dictated where those models cropped up was the structures people had come over to the thermal discipline and said I got a concern in this area and we need some good temperatures here and so the thermal people would go off and build a detailed model in that area and this evolved into the 125 or so thermal math models. So they were doing some talking but maybe not as much as they should have.

CARTER: I kind of also have the impression that the thermal guys speaking to the structures man essentially said, well, here is all of the stuff, you smear it around. So you put guys like me with one class of thermo smearing those temperatures. I would not do it that way if I were doing it.

CHIMENTI: That's a valid comment. As a matter of fact I think the structures people at Rockwell did most of the smearing of temperatures. Maybe the reason was that the structures people knew what they really wanted and they didn't want a temperature here and a temperature there but they wanted integrated average temperatures of this plate or this rod or whatever. So that is the way it was.

DIXON: I would like to make one comment on the comment that Dr. Lee made that out in the real-world environment you have a thermal analyst over here and a structural analyst over there. This was in connection with the integrated analysis capability that Professor Thornton was talking about. Here at Langley our environment is certainly much different than that, but when we got started on that scramjet study our experiences indicated that we needed a new breed of engineer who is a thermal-structural analyst. In our finite element work we are trying to develop thermal finite elements geared to what the structures man is going to want to do when he is calculating the effects of thermal loads. How long it takes before that kind of person begins to find their way into industry I don't know, but here at Langley most of our people that are involved in the thermal work have also had a structures background and are doing the combined analysis to try to get around the communication problem.

CARTER: One of the things that you have to be aware of, as Ed pointed out, is that these guys are doing 125 models; they just cannot all be super experts. So again you eventually need to have something that is practical for the average guy. I don't know what the answer is to that.

SWANSON: I have to open my mouth, I am afraid. The distinction between the thermal analyst and the structural analyst has been around. I was in aerospace back in the mid-60's and we had the thermal group and we had the stress group and in the stress group we redid the thermal to get the temperatures we wanted for the structural. I think this seems to be unique to the aerospace industry. We don't see this in the automobile engine block analyses, transistor analyses, and all the other coupled thermal-structural analyses that go on outside of the aerospace industry. This seems to be a unique phenomenon. I don't know why; maybe it just started that way and then propagated but in types of industries where they said let's start doing detailed analyses, they have not imposed this artificial distinction between thermal and structural analyses. There is a thermal-structures analyst. He does the switching temperature distribution in the transistor. He does the temperatures in the print circuit, read heads, or whatever. It is part of the entire design package. So let us say it looks like it is unique in aerospace and not necessarily in all industries. Perhaps one of the reasons for that is that you got 300 stress men, maybe 50 or 60 thermal men working on this one project, so you know, the order of magnitude of projects is very large.

CHIMENTI: I think there is another point that needs to be brought out concerning that. In the thermal analysis world, I think, the thermal stress part of it is just one small piece of what the thermal analyst is doing. Just like in structures analysis I am sure that the thermal stress is just a small piece of what the structures guys are doing. If all the thermal analyst had to do was thermal stress analysis it would be a different situation.

DIXON: We have time for about one more question or comment.

CHIMENTI: I have one comment and I hope I am not opening up a bag of worms here but I have a concern after listening to a lot of the papers here and maybe interpreting the flavor of the way a lot of people are thinking. It appears that in some instances the concept of a turnkey operation is almost being proposed here where you push the button and out spits a thermal model, a thermal analysis, structures model and a structures analysis. Being an old thermal guy I guess that bothers me. I know that I have enough problems myself with people who have thermal expertise in assuring the temperatures that they generate are correct and I think that problem would really be compounded if you had people that are not thermal people generating temperatures. In some of these programs, specifically with implicit techniques, they're stable, they will give you numbers. They may not be the right number and one of the beauties I have really felt of the explicit technique is that you know when it is stable and when it is not. They do a lot of crunching to give you a number but I have confidence in that number when they get through versus I have always had problems with the implicit technique because you really get into a situation where you think you have a good number but you really don't.

DIXON: I am going to have to turn off the discussion now. I do want to bring up just one final thing. I got several questions last night about were we going to have a conference like this again some time and my initial reaction was, sure it wasn't a whole lot of work and I think we got some things done. Reflecting on it last night, I felt like some members of my staff would have a little bit different answer, but if they have time to recover they probably would have a positive response and I think the thing that will really determine whether we have another one is responses from you attendees. If you felt it was worthwile or in another couple of years it would be worthwhile to do one if you would tell Howard or me or drop us a line or call us some time.

CARTER: Sid, tell them to tell their management.

DIXON: Personally, I think there is a void in the technical societies that they intend to have the thermal people getting together and the structures people getting together and although we tried to aim this at thermal structures people we couldn't avoid bringing in some of the structures things and I think the focus we would like to see in the future is the thermal structural problem as a whole. So that is a question that is up in the air and your actions with comments back to us and your management will go a long way in whether we do this again.

Howard, shall we close it? OK, we are through. Thank You.

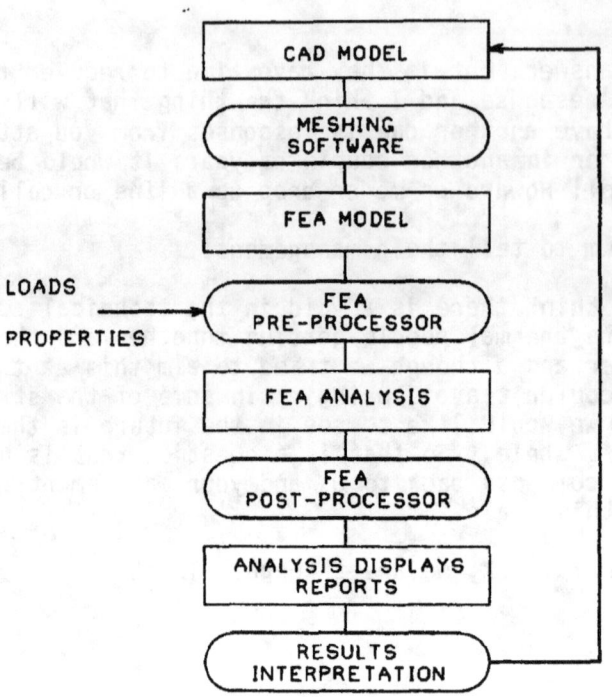

Figure 1.- Present design/analysis system.

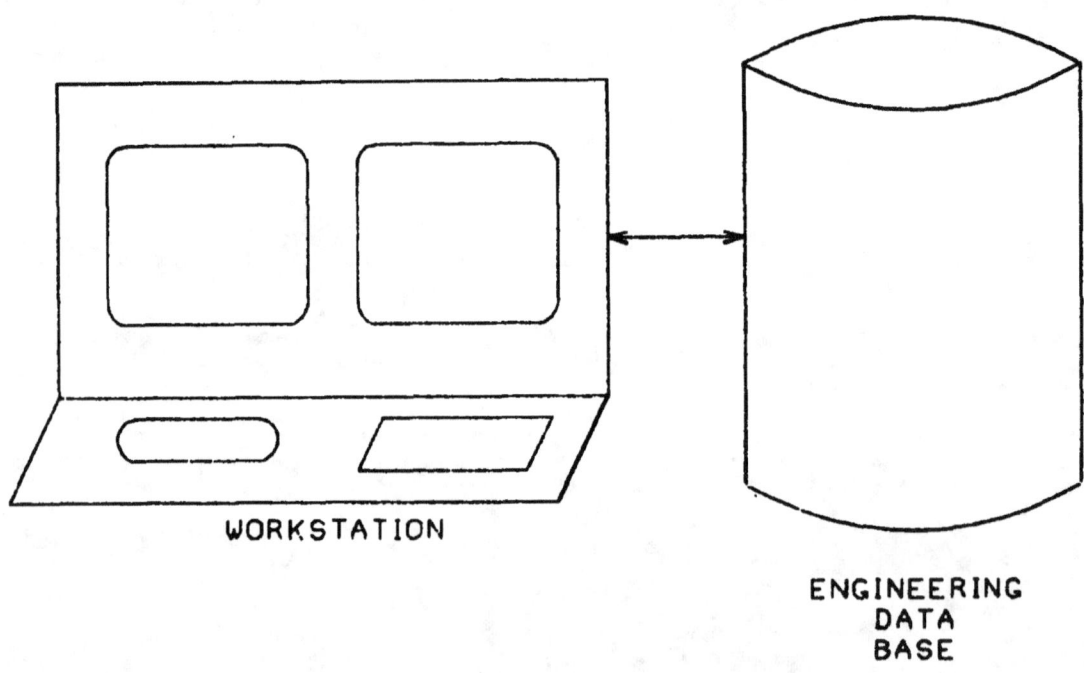

Figure 2.- Future design/analysis system.

MODELING DECISIONS
    FINITE ELEMENT TYPES
    FINITE ELEMENT MESH

SOLUTION
    SOLUTION ALGORITHMS

RESULTS
    INTERPRETATION PROCEDURES

MECHANICS
    DATA GENERATION AND STORAGE OPTIONS
    FILE MANIPULATIONS

Figure 3.- Analysis decisions.

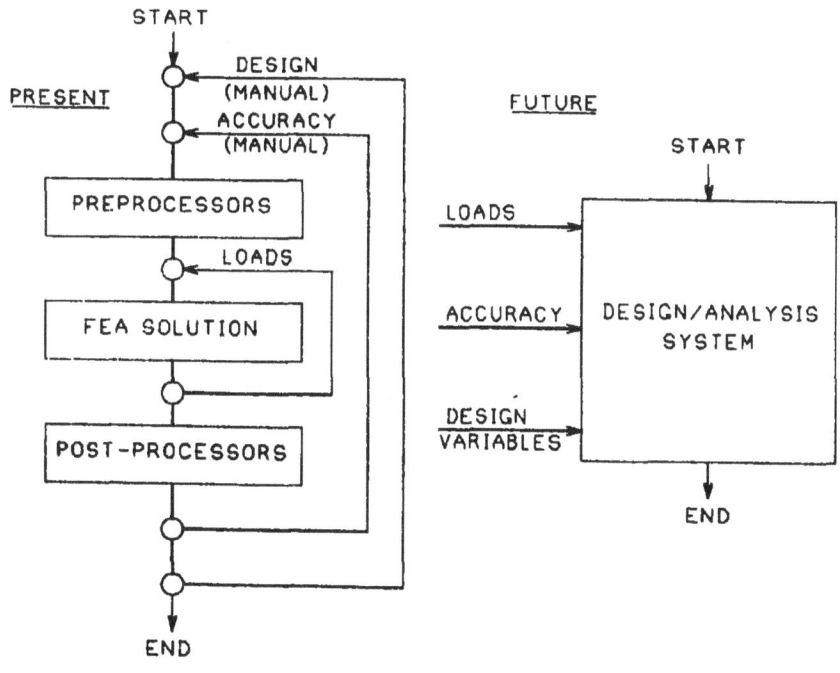

Figure 4.- Design/analysis cycles.

ACCURACY MEASURES

- IDEAL - ON THE MESH
- PROBABLE - ON THE RESULTS
- ACCURACY LEVEL PLOTS

AUTOMATIC MESH GENERATION

- PROBABLE TRIANGLES AND TETRAHEDRONS

GOOD SPATIAL GEOMETRY LANGUAGE

- ALREADY ADVANCED IN CAD SYSTEMS
  - MUST INCLUDE POINTS, INTERSECTIONS, SURFACES AND VOLUMES

INTERPRETATION AND OPTIMIZATION

- STATISTICAL EVALUATION PROCEDURES
- OPTIMIZATION FEEDBACK PROCEDURES

Figure 5.- Software development needs.

VECTOR COMPUTER TIMES (CRAY-1)	1.6%	66.9%	28.0%	0.6%	2.9%
SCALAR COMPUTER TIMES (PRIME 400)	0.5%	44.8%	50.0%	1.2%	3.5%

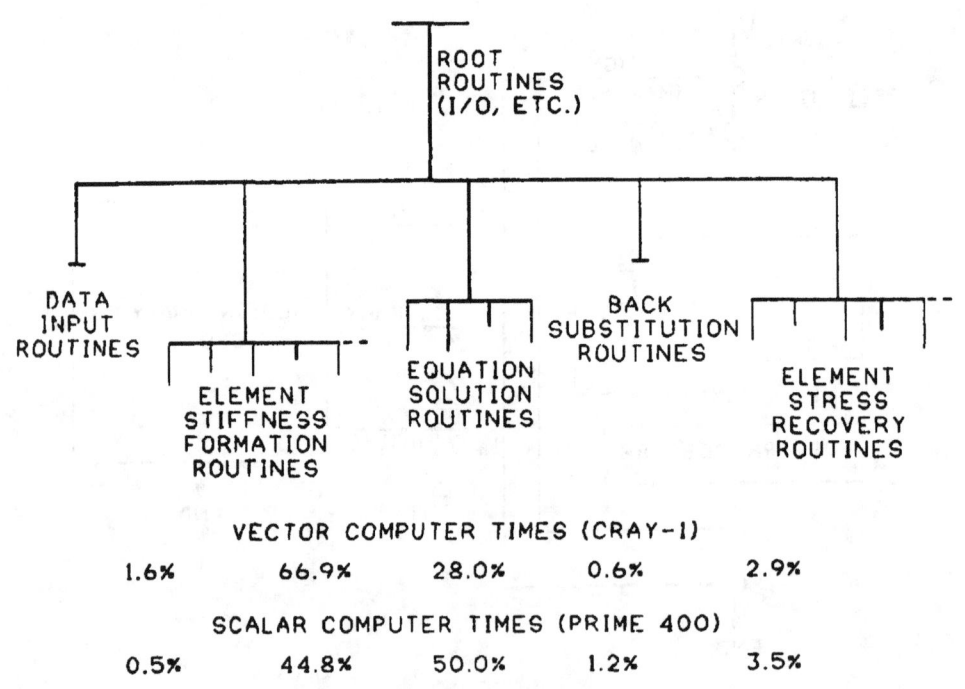

Figure 6.- Finite element overlays in space.

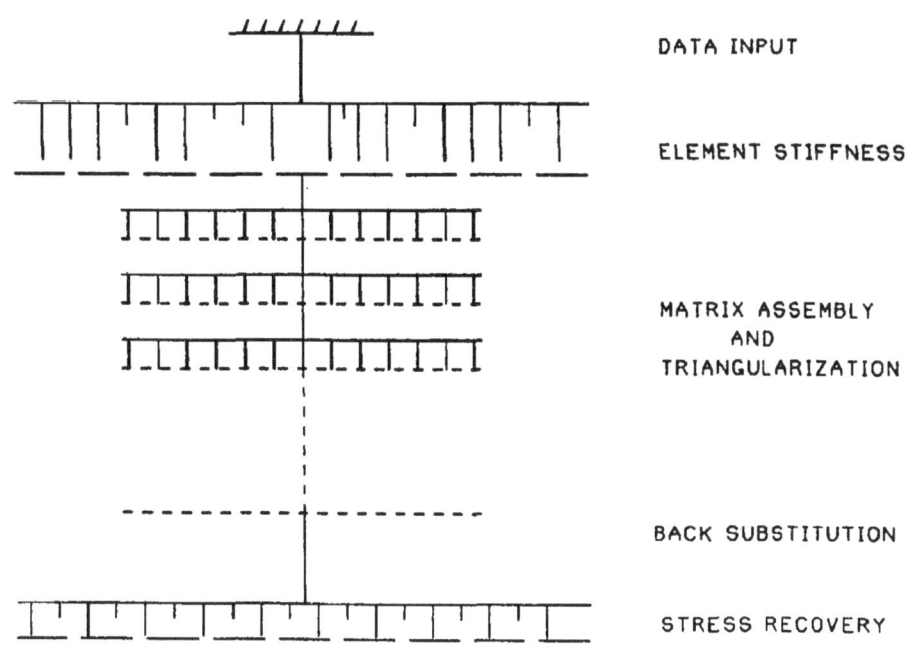

Figure 7.- Finite element overlays in time.

SPEED IMPROVEMENT RELATIVE TO PRESENT COMPUTERS

		2.0 IMPROVEMENT	10.0 IMPROVEMENT
1.	INCREASE CPU SPEED	2.00	10.00
2.	INCREASE VECTOR SPEED	1.16	1.35
3.	ADD MATRIX OPERATIONS	1.68	4.70
4.	ADD ADDITIONAL CPUs TO TASK	1.94	8.70

Figure 8.- Options to decrease analysis time.

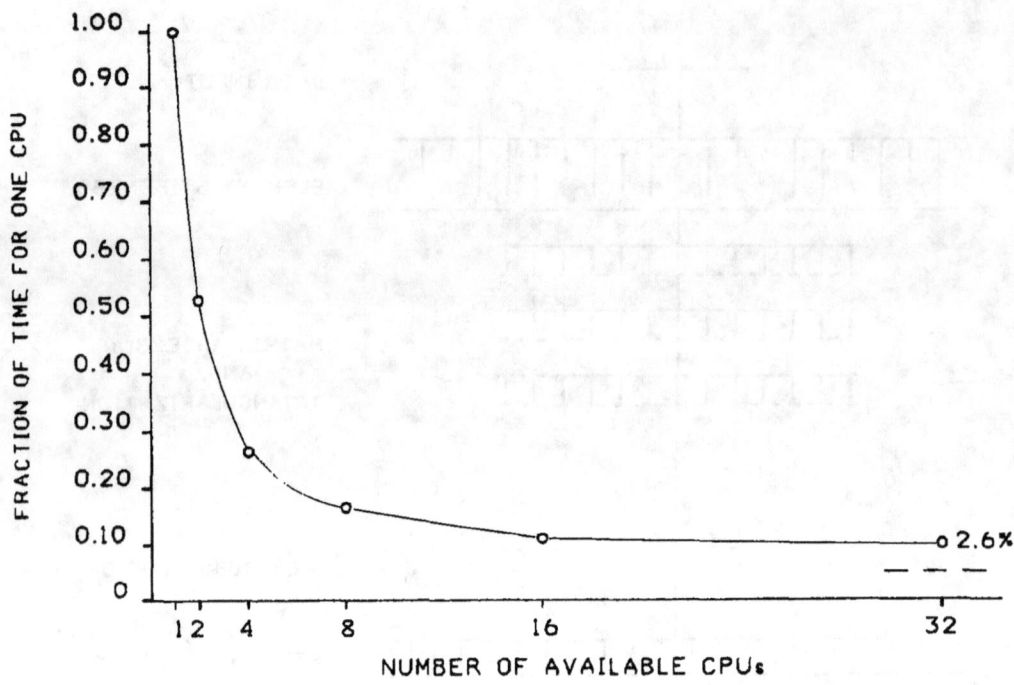

Figure 9.- Finite element performance of multiple CPU system.

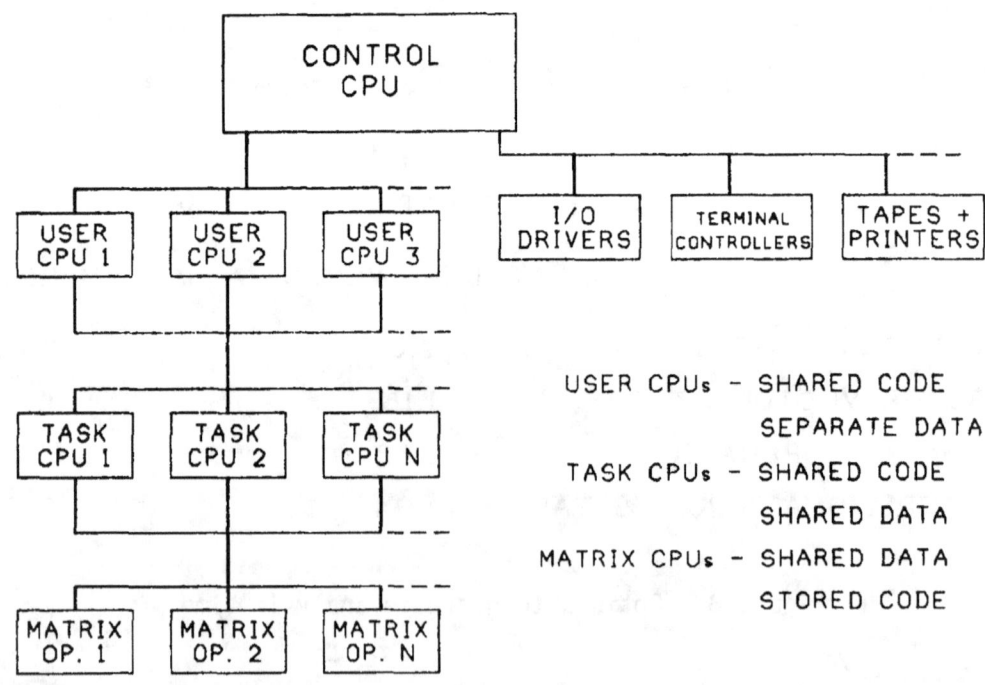

Figure 10.- ANSYS revision 5 design computer.

HEAT TRANSFER PRINCIPLES	DISCRETIZATION TECHNIQUES	NUMERICAL ANALYSIS	COMPUTER SCIENCE
• FORMULATION OF HEAT TRANSFER EQUATIONS	• DEVELOPMENT OF DISCRETE EQUATIONS	• SOLUTION OF EQUATIONS	• USE OF AVAILABLE HARDWARE, FIRMWARE AND SOFTWARE
• CALCULATION OF THERMAL PROPERTIES	• SPATIAL DISCRETIZATION	• CALCULATION OF EIGENSYSTEMS	• EFFICIENT PROGRAM ORGANIZATION
	• TEMPORAL INTEGRATION		• FLEXIBILITY FOR MODIFICATIONS AND EXTENSIONS

Figure 11.- Basic disciplines in the development of computational procedures for heat transfer.

- FINITE ELEMENT TECHNIQUES
- FINITE DIFFERENCE TECHNIQUES
- BOUNDARY ELEMENT METHODS
- WEIGHTED RESIDUAL TECHNIQUES
- TRANSFER MATRIX METHODS
- ASYMPTOTIC AND PERTURBATION TECHNIQUES

Figure 12.- Analysis procedures.

COMPUTER PROGRAMS

- MANY GENERAL-PURPOSE COMPUTER PROGRAMS HAVE BEEN DEVELOPED
  (MAJOR FEATURES AND CAPABILITIES OF 38 PROGRAMS ARE OUTLINED IN A SURVEY PAPER)

- MAJORITY OF FINITE-ELEMENT HEAT TRANSFER PROGRAMS WERE DEVELOPED AS EXTENSIONS OF STRUCTURAL PROGRAMS

COMPUTATIONAL ALGORITHMS

- MANY OF THE RECENT ADVANCES IN COMPUTATIONAL STRUCTURAL AND FLUID MECHANICS HAVE NOT BEEN USED IN HEAT TRANSFER

Figure 13.- State of the art in computational heat transfer.

		ABAQUS	ADINAT	AGTAP	ANDES	ANSYS	ASASHEAT	BERSAFE FLHE	CAVE I	CAVE II	CAVE III	CONFAC	FLUX2	HEATING5	HEATING6	HEATRAN	MARC	MITAS II	MSC/ NASTRAN	NNTB
1.	Goal of Program System																			
	General Purpose	•	•	•	•		•	•	•	•	•	•	•			•	•	•	•	•
	Commercial	•	•				•	•	•								•			
	Research	•	•		•				•	•				•	•				•	
	Educational	•	•				•	•	•								•			
	Others (specify in abstract)																			

		NTEMP	PAFEC	SAHARA	SAMCEF (THERNL)	SESAM-69 (NV-615)	SINDA	SPAR	SSPTA	TACO	TAC2D	TAC3D	TANG	TAU	TEMP	TEPSA	THAC-S1P-3D	THTD	TRUMP	WECAN
1.	Goal of Program System																			
	General Purpose	•	•	•	•	•	•	•	•	•	•	•	•	•	•			•	•	•
	Commercial		•		•	•	•				•	•				•			•	•
	Research		•			•	•				•	•						•	•	
	Educational		•			•	•	•							•			•		•
	Others (specify in abstract)								•											

Figure 14.- Computer programs surveyed.

(A) FINITE ELEMENT TECHNOLOGY

FORMULATIVE ASPECTS AND ELEMENT DEVELOPMENT

- ALTERNATE MULTIFIELD FORMULATIONS (MIXED AND HYBRID FINITE ELEMENT MODELS)
- EQUIVALENCES AND SIMILARITIES BETWEEN ELEMENTS
- SPECIAL ELEMENTS (E.G. BOUNDARY LAYER AND INFINITE ELEMENTS)

MESH DESIGN

- HIERARCHICAL FINITE ELEMENT APPROACHES AND ADAPTIVE REFINEMENT OF MESH (EQUIVALENT TO MULTIGRID FINITE DIFFERENCE METHODS)
- A POSTRIORI ERROR ESTIMATE

TRANSIENT ANALYSIS

- EXPLICIT, IMPLICIT AND MIXED EXPLICIT/IMPLICIT TEMPORAL INTEGRATION TECHNIQUES
- OPERATOR SPLITTING AND PARTITIONING SCHEMES
- AUTOMATIC TIME-STEPPING STRATEGIES

Figure 15.- Some recent advances that will impact computational heat transfer.

(B) FINITE DIFFERENCE TECHNOLOGY

- CURVILINEAR GRIDS (E.G. ISOPARAMETRIC FINITE DIFFERENCES) AND GRID GENERATION
- MULTIGRID FINITE-DIFFERENCE METHODS
- HERMITIAN (MULTILOCAL) FINITE-DIFFERENCE METHODS

(C) NUMERICAL ANALYSIS

- QUASI-NEWTON METHODS (BFGS METHOD)
- PRECONDITIONED CONJUGATE GRADIENT METHOD
- INCOMPLETE CHOLESKY FACTORIZATION

Figure 16.- Additional recent advances that will impact computational heat transfer.

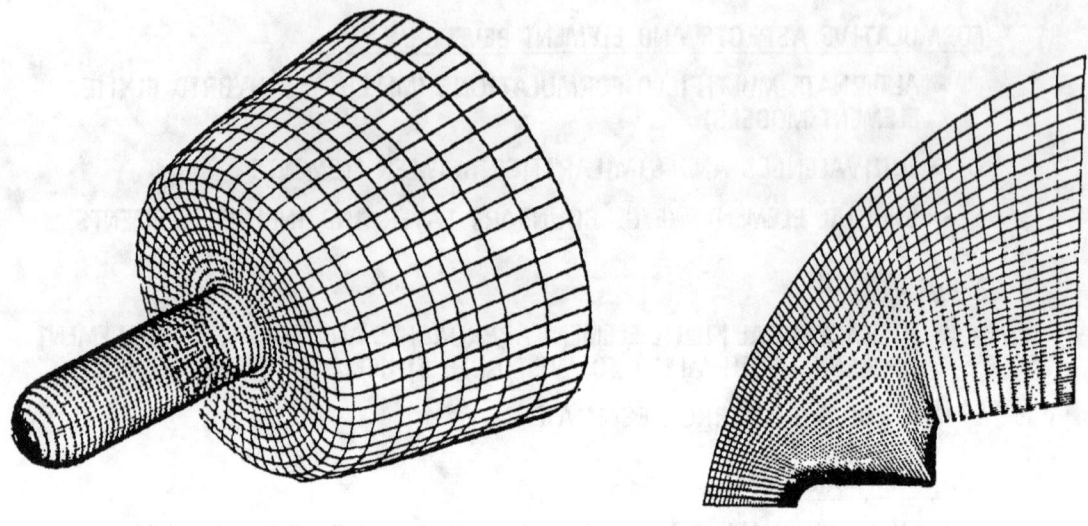

Figure 17.- Curvilinear finite difference grids.

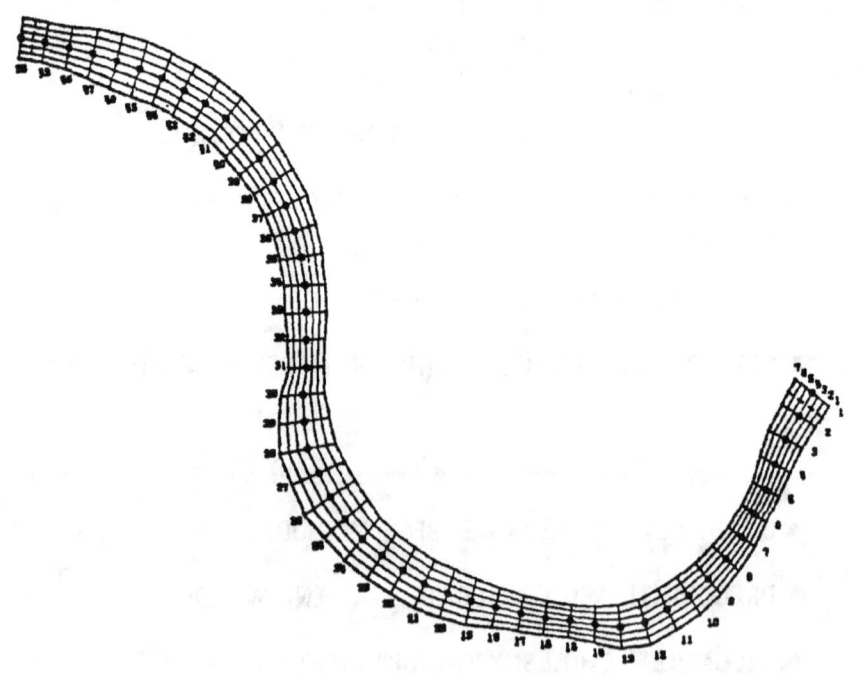

Figure 18.- Curvilinear finite difference grid.

(D) ENGINEERING SOFTWARE

- ASSESSMENT AND EVALUATION OF LARGE GENERAL-PURPOSE SOFTWARE
    - ANALYSIS CAPABILITIES
    - ADEQUACY OF USER-ORIENTED FEATURES
    - MAINTAINABILITY
    - ADEQUACY OF USER-SUPPORT FACILITIES
    - PORTABILITY
- INTERACTIVE GRAPHICS
- PRE- AND POST-PROCESSING

(E) NEW COMPUTING SYSTEMS

- SUPERCOMPUTERS
- MINICOMPUTERS/ARRAY PROCESSORS
- MICROPROCESSORS
- SYSTOLIC ARRAYS

Figure 19.- Recent advances that will impact computational heat transfer.

DRIVING FORCE
- MODELING OF COMPLEX HARDWARE SYSTEMS
    - SOLUTION OF LARGE-SCALE THERMAL PROBLEMS
    - RELIABILITY AND ERROR ESTIMATES OF SOLUTION

OPPORTUNITIES
- NEW ADVANCES IN COMPUTER HARDWARE AND SOFTWARE

COMPUTATIONAL ALGORITHMS
- IMPROVING THE EFFICIENCY OF CURRENTLY-USED ALGORITHMS USUALLY RESULT IN MARGINAL REDUCTIONS IN COMPUTATIONAL EFFORT
- HYBRID TECHNIQUES HAVE HIGH POTENTIAL FOR LARGE-SCALE PROBLEMS
    - MIXED EXPLICIT/IMPLICIT TEMPORAL INTEGRATION SCHEMES
    - COMBINED DIRECT/ITERATIVE TECHNIQUES FOR SOLUTION OF ALGEBRAIC EQUATIONS (IN CONJUNCTION WITH MULTIGRID METHOD)
    - REDUCED BASIS TECHNIQUE FOR STEADY-STATE PROBLEMS - COMBINED FINITE ELEMENT (DIFFERENCE)/BUBNOV-GALERKIN/PERTURBATION TECHNIQUES
    - MODIFIED MODAL SUPERPOSITION FOR TRANSIENT PROBLEMS

Figure 20.- Future directions.

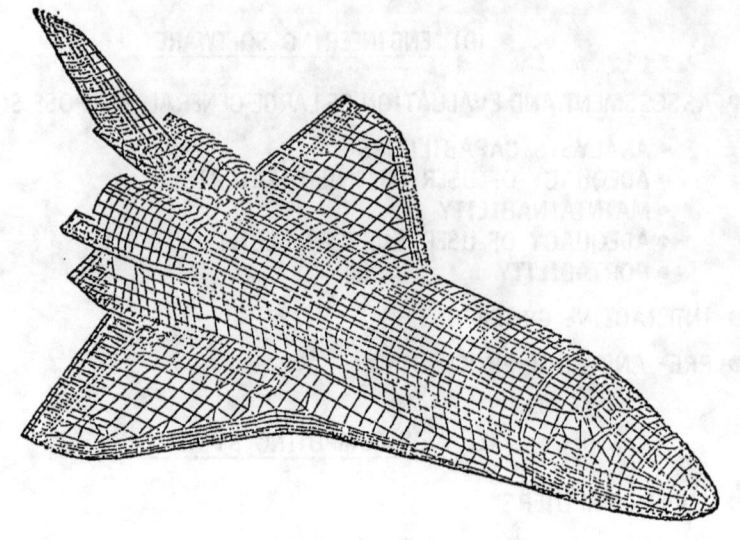

Figure 21.- Overall view of stress/finite element model.

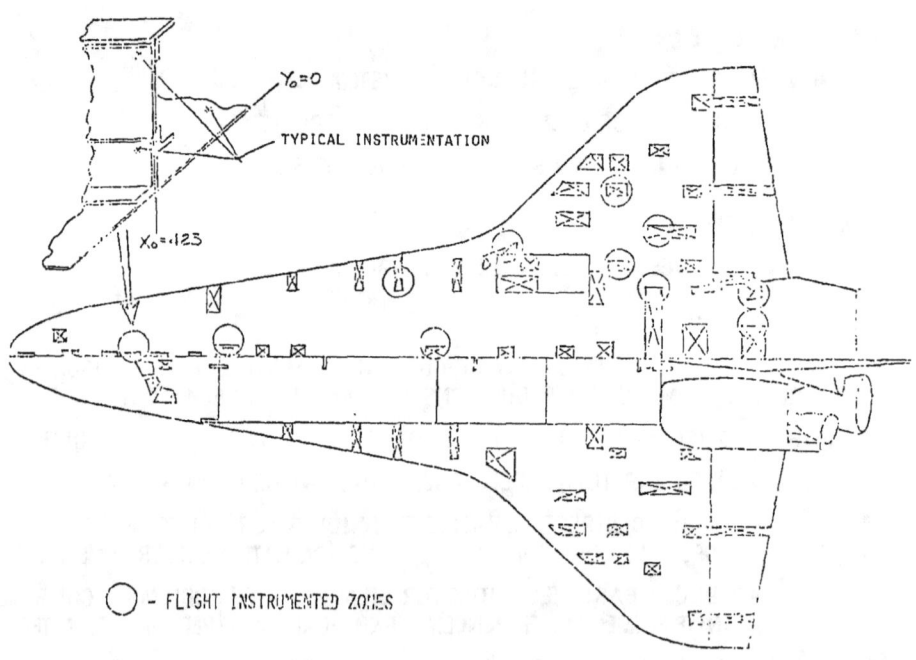

Figure 22.- Locations - detailed thermal gradient models.

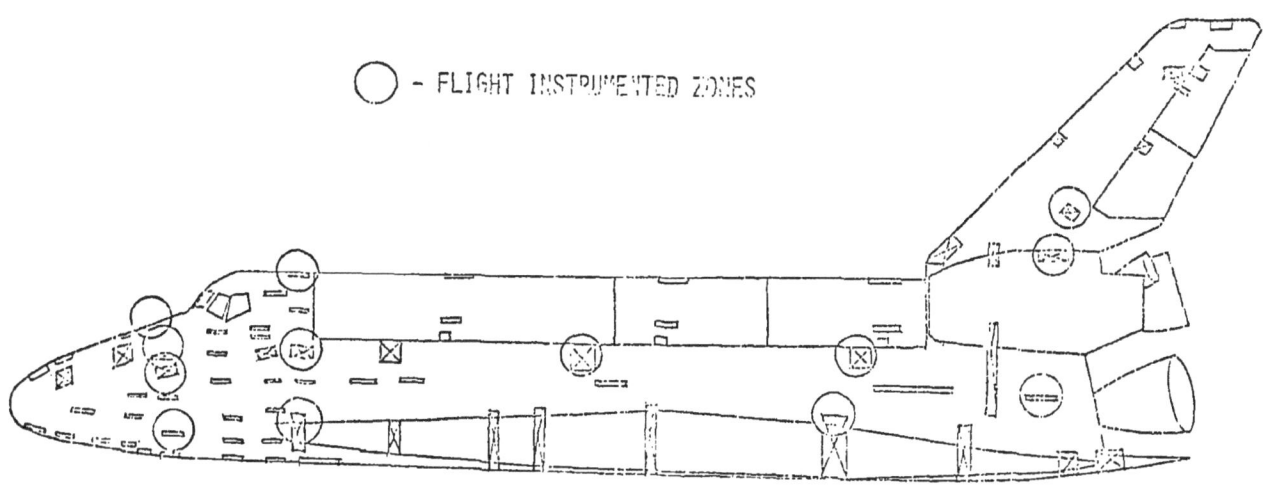

Figure 23.- Locations - detailed thermal gradient models (concluded).

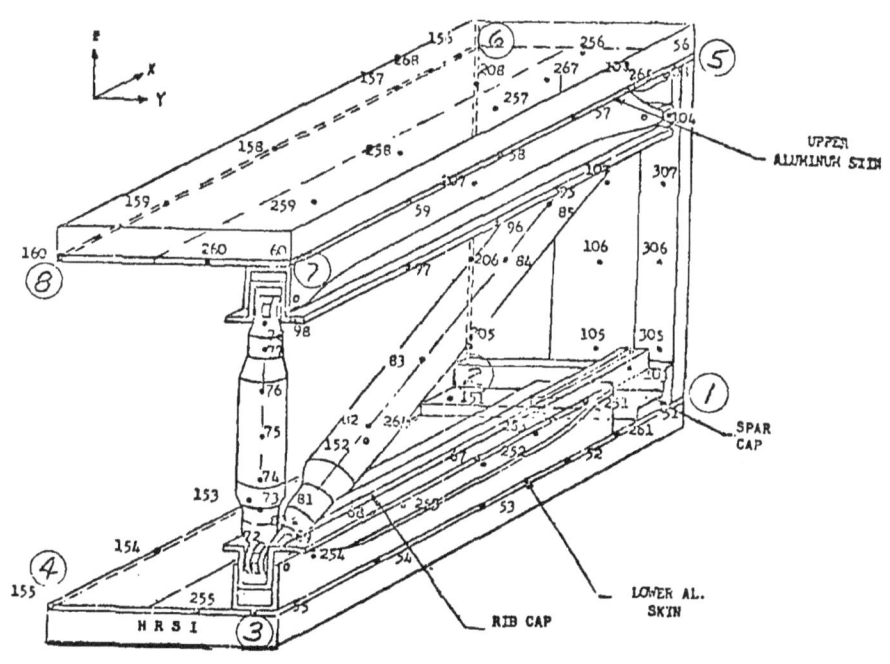

Figure 24.- Wing rib Pratt truss thermal models.

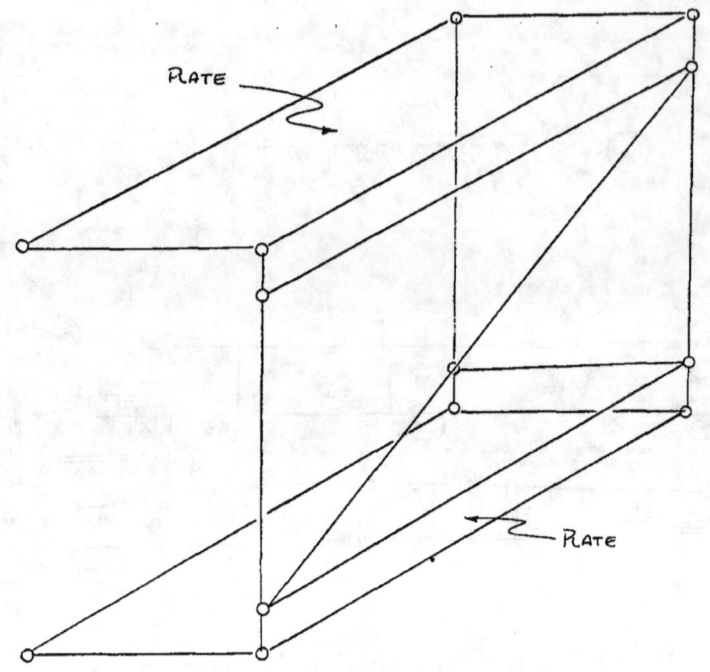

Figure 25.- ASKA structure model nodes.

o COMPUTING POWER

o EFFICIENT RUNNING THERMAL ANALYZER

o EFFICIENT MATH MODEL BUILDING TECHNIQUES

    . INTERACTIVE GRAPHICS, ETC.

o EFFICIENT DATA INTERROGATION & EVALUATION METHODS

    . DATA SATURATION
    . MODEL EXTRACTION TECHNIQUES

Figure 26.- Today's needs for large integrated thermal structures analysis.

- INTERCENTER THERMAL RESEARCH SUPPORT
    - HWTS
    - ORBITER

- HWTS - TESTING
    - FINITE DIFFERENCE ANALYSIS
    - NASTRAN ANALYSIS

- ORBITER - FLIGHT RESULTS STS-1
    - SPAR CROSS SECTION ANALYSIS

- TITANIUM PANEL TESTING
    - THERMAL STRESS
    - THERMAL CREEP

Figure 27.- Dryden thermal structural response research - current activities.

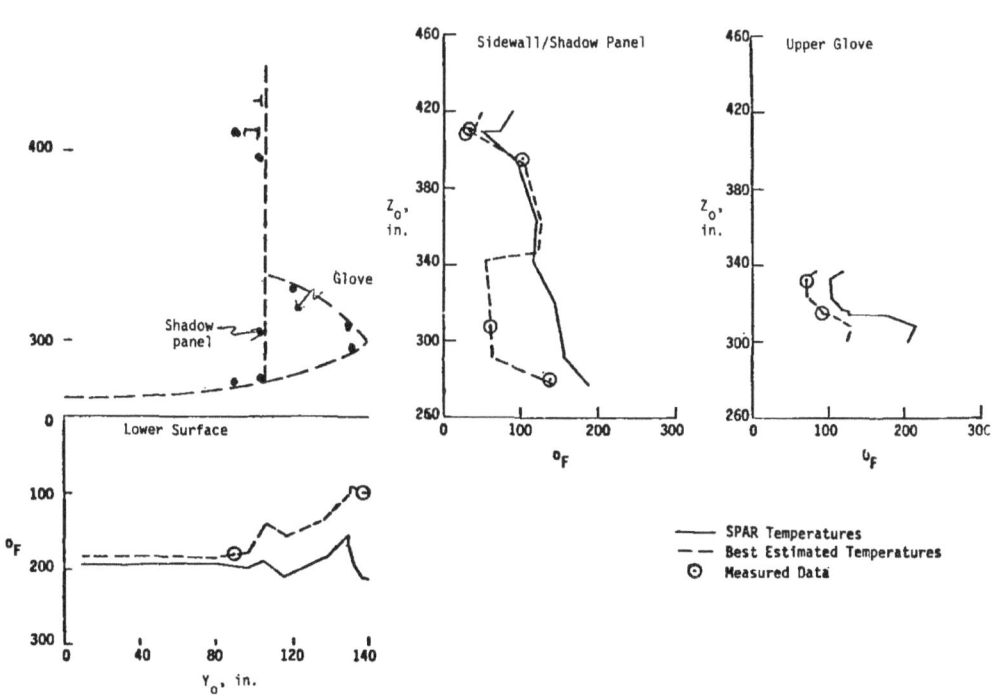

Figure 28.- FSS 877 temperatures.

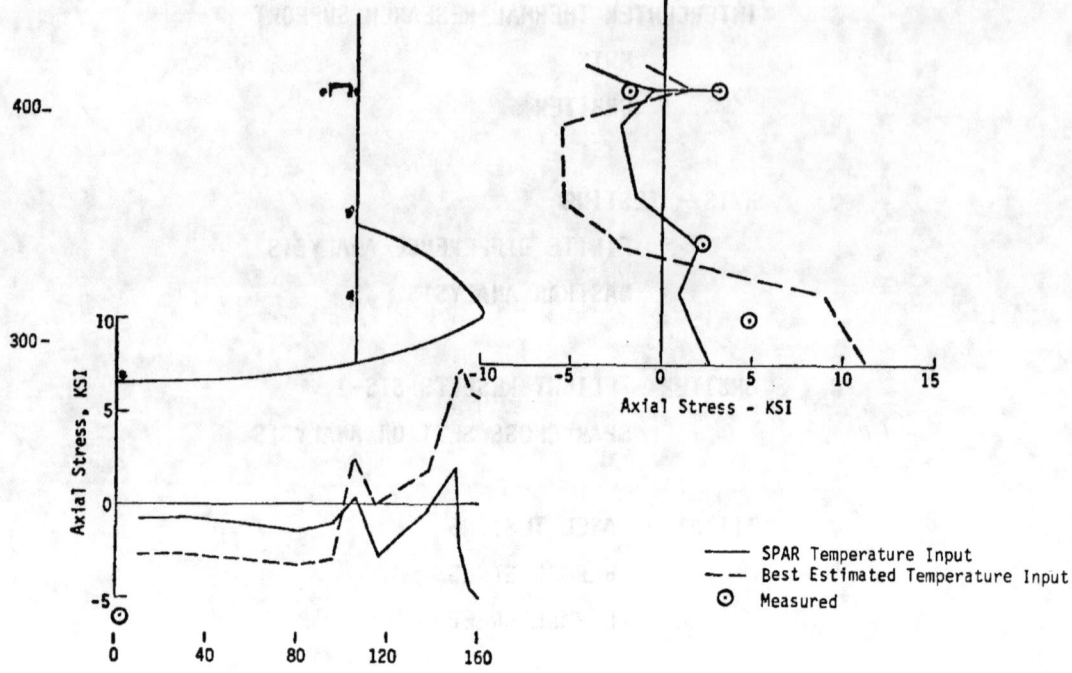

Figure 29.- FSS 877 thermal stress.

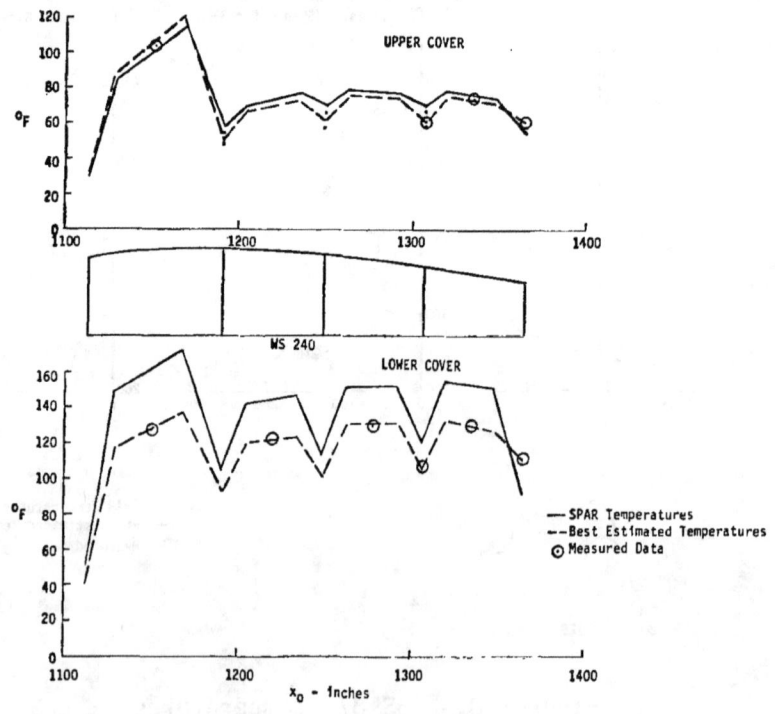

Figure 30.- WS 240 temperature measurements.

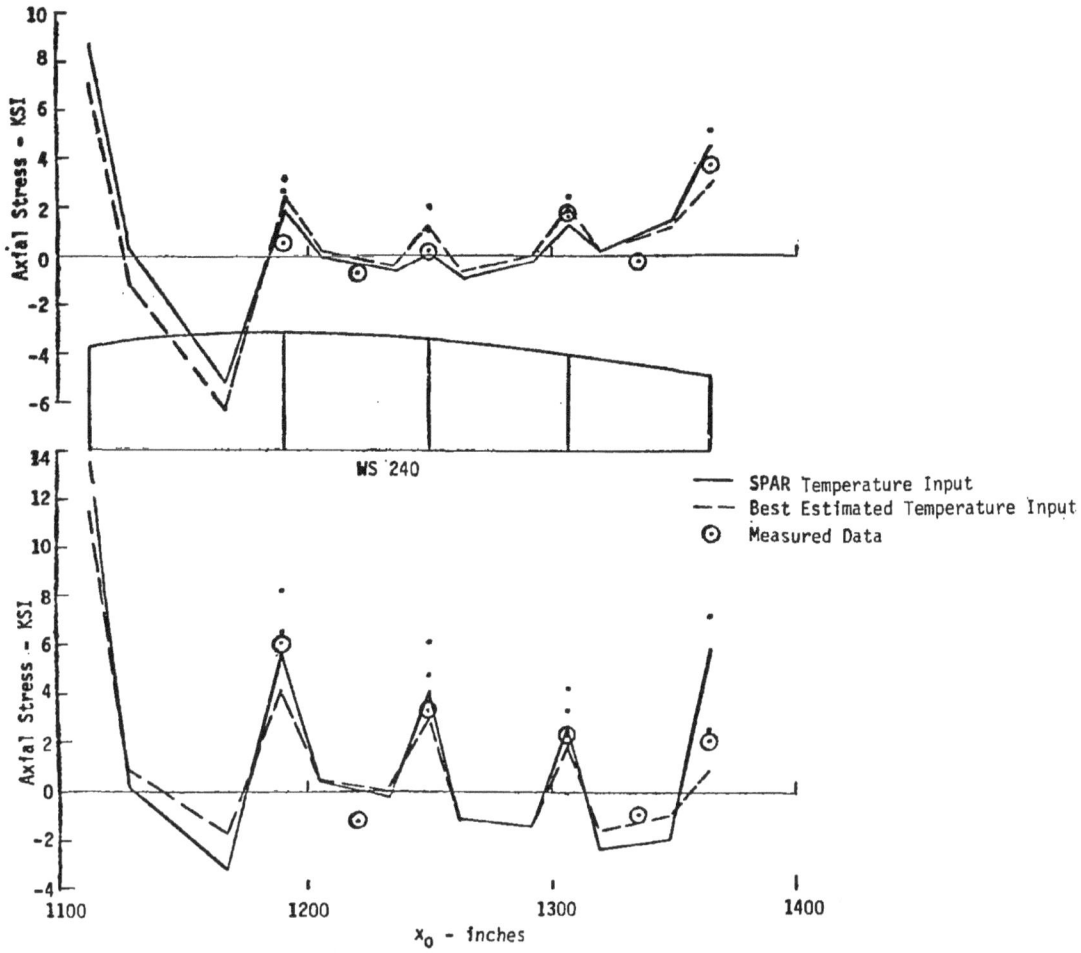

Figure 31.- WS 240 thermal stress.

- HWTS - SPAR ANALYSIS
    FINITE DIFFERENCE, NASTRAN
- ORBITER WING - SPAR/CRAY
- ORBITER WING/FUSELAGE - SPAR/CRAY

Figure 32.- Dryden thermal structural response research - future plans.

- SPAR MODELING VERY EFFECTIVE

- VIEW FACTOR GENERATION

- MATERIAL PROPERTY TABLES

- HEAT TRANSFER VS. STRUCTURAL ANALYSIS

- FINITE ELEMENT VS. FINITE DIFFERENCE COMPUTER TIMES

Figure 33.- Dryden thermal structural response research - observations.

SURVEY OF COMPUTER PROGRAMS FOR HEAT TRANSFER ANALYSIS

Ahmed K. Noor*
Joint Institute for Advancement
of Flight Sciences
NASA Langley Research Center

INTRODUCTION

The significant advances made in numerical discretization techniques, coupled with the rapid developments in computer hardware and software provided the foundation from which general-purpose programs for heat transfer analysis have evolved. After more than two decades of development, a wide variety of these programs are currently being used in government and industry for heat transfer analysis of practical problems. Depending on the criteria used for identifying general-purpose heat transfer analysis programs, estimates of their numbers vary between thirty and seventy. In addition, several special-purpose and research-oriented heat transfer programs are in existence. The potential user of a heat transfer analysis program is now faced with the problems of 1) getting information about, and sorting out, existing heat transfer analysis programs; and 2) identifying the program that is best suited for his particular needs.

While a number of bibliographies, data sheets and tables have been compiled about finite element software (see, for example, Refs. 1, 2 and 3), little has been published on the assessment of programs used for heat transfer analysis. The best known publication on this subject is Ref. 4 which was prepared by the Committee on Computer Technology of the ASME Pressure Vessels and Piping Division and includes information about eleven programs used for thermal analysis. The present paper goes well beyond the scope of Ref. 4. Specifically, the objective of this paper is to give an overview of the current capabilities of thirty-eight computer programs that can be used for solution of heat transfer problems. These programs range from the large, general-purpose codes with a broad spectrum of capabilities, large user community and comprehensive user support (e.g., ANSYS, MARC, MITAS II, MSC/NASTRAN, SESAM-69/NV-615) to the small, special purpose codes with limited user community such as ANDES, NNTB, SAHARA, SSPTA, TACO, TEPSA and TRUMP. The capabilities of the programs surveyed are listed in tabular form followed by a summary of the major features of each program. As with any survey of computer programs, the present one has the following limitations: a) It is useful only in the initial selection of the programs which are most suitable for a particular application. The final selection of the program to be used should, however, be based on a detailed examination of the documentation and the literature about the program; b) Since computer software continually changes, often at a rapid rate, some means must be found for updating this survey and maintaining some degree of currency. Nevertheless, the author feels that the present survey can serve as a focal point for the user community interested in heat transfer analysis.

Before listing the capabilities of the programs, some of the sources of information about computer programs and references on the background material needed for effectively using the programs are listed, and guidelines for selecting the code are discussed.

---

*Professor of Engineering and Applied Science,
George Washington University

SOURCES OF INFORMATION ABOUT COMPUTER PROGRAMS

A partial list of users groups and software dissemination services that provide information about computer programs for heat transfer analysis is given subsequently. A list of cooperative users groups and finite element software dissemination services can be found in Ref. 5.

- ASIAC - Aerospace Structures Information and Analysis Center, AFFDL/FBR, Wright Patterson Air Force Base, Ohio 45433

- COSMIC - Computer Software Management and Information Center, Suite 112, Barrow Hall, University of Georgia, Athens, Georgia 30602

- ICP - International Computer Programs, Inc., 9000 Keystone Crossing, Indianapolis, Indiana 46240

- National Energy Software Center, Argonne National Laboratory, 9700 South Cass Avenue, Argonne, Illinois 60439

- NTIS - National Technical Information Service, U.S. Department of Commerce, 5285 Port Royal Road, Springfield, Virginia 22161.

BACKGROUND MATERIAL NEEDED FOR EFFECTIVE EVALUATION AND USE OF COMPUTER PROGRAMS FOR HEAT TRANSFER ANALYSIS

The user of a computer program for heat transfer analysis is dependent on the detailed knowledge about the principles, algorithms and assumptions behind the program features for the proper selection of models and algorithms as well as for monitoring the solution process. There are also many heat transfer problems whose solution may require modifying (slightly) the program. Therefore, the effective evaluation and use of heat transfer programs, requires some knowledge about the following disciplines:

- Principles of heat transfer and formulation of thermal problems.

- Solution techniques for linear and nonlinear steady-state thermal problems.

- Temporal integration and solution techniques for nonlinear transient thermal problems.

- Considerations for the design of software systems for heat transfer analysis.

GUIDELINES FOR SELECTION OF A COMPUTER PROGRAM

The analysis capabilities and user features vary considerably from one code to the other, and therefore, it is often difficult to identify the proper code that meets a specific need. A number of factors which affect the selection of a code are enumerated in the succeeding paragraphs. The order in which the factors are listed does not necessarily reflect the priority which should be given each factor; this remains the responsibility of the user of the code. For a detailed discussion on the technical, operational and commercial criteria for selecting a code see Refs. 6 and 7.

1. *Analysis Capabilities*

These include the range of applications and limitations of the code. The limitations include both those implied by the formulation aspects and numerical solution procedures adopted by the code as well as the element library available in the code.

2. *Adequacy of User-Oriented Features*

For heat transfer analysis the user's features such as automatic (or semi-automatic) mesh (or model) generation, error checks, displays of original model and of various intermediate results, and mechanism for data transfer to other program packages (e.g., thermal stress analysis program) are essential for the effective use of the analysts' time.

3. *Maintainability*

Because of the rapid advances in computational methods, computer software and hardware technology, the maintenance of heat transfer codes usually includes updating the computational modules, extending the capabilities of the code and improving its performance. There exist well-established formal mechanisms of integration and quality assurance of software extensions. Maintenance of the code by personnel other than the developer (e.g., user's organization) can be quite expensive and time consuming.

4. *Adequacy of User Support Facilities*

In addition to the printed documentation (user manuals, training manuals, programming manuals, sample problems and test cases), the following services are desirable: training courses, users meetings, hotline consulting, assistance by data centers and consulting organizations.

5. *Portability*

Although most of the heat transfer codes are written in standard FORTRAN IV language, a code developed on one computer system may not be entirely compatible with another system due to differences in I/O facilities, operating system, precision of the machine (e.g., UNIVAC versus CDC), etc.

Once a code is acquired and implemented on the user's computer system, it is important to establish its reliability by bench-mark problem runs. For a discussion of verification and qualification procedures see Ref. 8.

6. *Experience of Other Users*

This can be invaluable if the users are objective in their evaluation and are familiar with some of the other software packages.

PROGRAM SURVEY AND DESCRIPTION

This section gives an overview of the capabilities of thirty-eight computer programs for the solution of heat transfer problems. The majority of these programs use either finite elements or finite differences for the spatial discretization. Some of the programs have a much more limited scope than others. The information presented herein is based on a questionnaire sent to the developers of each program. The capabilities of the programs are listed in tabular form followed by a description of each program.

SURVEY OF COMPUTER PROGRAMS FOR HEAT TRANSFER ANALYSIS

Part I - Analytical Capabilities of the Program

		ABAQUS	ADINAT	AGTAP	ANDES	ANSYS	ASASHEAT	BERSAFE FLHE	CAVE I	CAVE II	CAVE III	CONFAC	FLUX2	HEATING5	HEATING6	HEATRAN	MARC	MITAS II	MSC/NASTRAN	NNTB
1.	Goal of Program System																			
	General Purpose	●	●	●	●	●	●	●	●	●	●	●	●			●	●	●	●	●
	Commercial	●	●			●	●	●									●		●	
	Research	●	●			●	●	●						●	●		●		●	
	Educational	●	●					●									●		●	
	Others (see program abstracts)																			
2.	Method of Analysis																			
	Finite Elements	●	●		●	●	●	●				●	●			●	●		●	
	Finite Differences			●																
	Boundary Integral Method																		●	
	Perturbation Technique																			
	Hybrid Analytical - Numerical Technique (see program abstracts)								●	●	●			●	●					●
	Others (see program abstracts)																			
3.	Space Dimensionality																			
	Three-Dimensional	●	●		●	●	●	●	●	●	●	●	●	●	●		●	●	●	●
	Two-Dimensional	●	●		●	●	●	●					●	●	●		●	●	●	●
	One-Dimensional	●	●		●	●	●	●						●	●		●	●	●	●

		NTEMP	PAFEC	SAHARA	SAMCEF (THERNL)	SESAM-69 (NV-615)	SINDA	SPAR	SSPTA	TACO	TAC2D	TAC3D	TANG	TAU	TEMP	TEPSA	THAC-SIP-3D	THTD	TRUMP	WECAN
1.	Goal of Program System																			
	General Purpose	●	●	●	●	●	●	●	●	●	●	●	●	●	●			●	●	●
	Commercial		●		●	●	●				●	●		●		●			●	●
	Research		●		●	●	●				●	●		●		●	●		●	●
	Educational		●			●	●							●		●				●
	Others (see program abstracts)								●											
2.	Method of Analysis																			
	Finite Elements	●	●		●	●	●	●	●	●	●	●		●	●	●				●
	Finite Differences			●									●				●	●	●	
	Boundary Integral Method		●																	
	Perturbation Technique																			
	Hybrid Analytical - Numerical Technique (see program abstracts)						●													
	Others (see program abstracts)																			
3.	Space Dimensionality																			
	Three-Dimensional		●	●	●	●	●	●	●	●		●	●	●			●	●	●	●
	Two-Dimensional	●	●	●	●	●	●	●	●	●	●			●	●	●			●	●
	One-Dimensional		●	●	●		●	●	●	●					●				●	●

Part I - Continued

	ABAQUS	ADINAT	AGTAP	ANDES	ANSYS	ASASHEAT	BERSAFE FLHE	CAVE I	CAVE II	CAVE III	CONFAC	FLUX2	HEATING5	HEATING6	HEATRAN	MARC	MITAS II	MSC/NASTRAN	NNTB
Solids of Revolution	•	•		•	•	•	•								•	•		•	
Boundary Elements	•	•															•	•	
Scalar Elements																•		•	
Point Contact Elements	•				•											•		•	
Others (see program abstracts)																			
4. Range of Applications and Phenomena																			
Linear Steady State	•	•	•	•	•	•	•	•	•	•			•	•	•	•	•	•	•
Nonlinear Steady State	•	•	•	•	•	•	•	•	•				•	•	•	•	•	•	•
Thermal Frequencies and Mode Shapes	•									•									
Linear Transient Response		•	•	•	•	•	•	•	•	•				•	•	•	•	•	•
Nonlinear Transient Response	•	•	•	•	•	•	•							•	•	•	•	•	•
Others (see program abstracts)																			
5. Formulation																			
a) Fundamental Unknowns																			
Temperatures	•	•	•	•	•	•	•	•	•	•			•	•	•	•	•	•	•
Temperatures and Flux																		•	
Enthalpy	•																		
Others (see program abstracts)																			
b) Elemental Matrices																			
Conduction	•	•	•	•	•	•	•	•	•	•			•	•	•	•	•	•	•

492

	NTEMP	PAFEC	SAHARA	SAMCEF (THERNL)	SESAM-69 (NV-615)	SINDA	SPAR	SSPTA	TACO	TAC2D	TAC3D	TANG	TAU	TEMP	TEPSA	THAC-SIP-3D	THTD	TRUMP	WECAN
Solids of Revolution	•	•	•	•					•				•	•	•			•	•
Boundary Elements	•	•	•	•			•											•	•
Scalar Elements		•		•															
Point Contact Elements		•		•														•	
Others (see program abstracts)				•											•				
4. Range of Applications and Phenomena										•	•								
Linear Steady State	•	•	•	•	•	•	•	•	•	•	•		•	•	•			•	•
Nonlinear Steady State		•	•	•	•	•	•	•	•	•	•		•	•	•		•	•	•
Thermal Frequencies and Mode Shapes				•															
Linear Transient Response	•	•	•	•	•	•	•	•	•	•	•	•	•	•	•	•	•	•	•
Nonlinear Transient Response	•	•	•	•	•	•	•	•	•	•	•	•	•	•	•	•	•	•	•
Others (see program abstracts)																			
5. Formulation																			
a) Fundamental Unknowns																			
Temperatures	•	•	•	•	•	•	•	•	•	•	•	•	•	•	•	•	•	•	•
Temperatures and Flux				•			•	•										•	
Enthalpy				•															
Others (see program abstracts)						•													
b) Elemental Matrices																			
Conduction	•	•	•	•	•	•	•	•	•	•	•	•	•	•	•	•	•	•	•

Part I - Continued

		ABAQUS	ADINAT	AGTAP	ANDES	ANSYS	ASASHEAT	BERSAFE/FLHE	CAVE I	CAVE II	CAVE III	CONFAC	FLUX2	HEATING5	HEATING6	HEATRAN	MARC	MITAS II	MSC/NASTRAN	NNTB
Capacitance	a) Consistent	•	•		•	•	•	•								•	•		•	
	b) Lumped	•	•	•	•	•	•		•	•	•			•	•	•	•	•	•	•
Convection	a) Free	•	•		•	•	•	•								•	•		•	
	b) Forced	•	•	•	•	•	•	•	•	•	•					•	•	•	•	•
Radiation		•	•		•		•	•	•	•	•					•	•	•	•	•
Interelement Convection and Radiation		•			•		•				•					•		•	•	
User Supplied Elements (see program abstracts)																			•	
Others (see program abstracts)						•		•												
6. Material Properties and Material Models																				
Isotropic		•	•	•	•	•	•	•	•	•	•	•	•	•	•	•	•	•	•	•
Anisotropic		•	•	•	•	•	•	•						•	•	•	•		•	
Multilayered		•	•																	
Temperature-Dependent Properties	Conductivity	•	•	•	•	•	•	•	•	•	•			•	•	•	•	•	•	•
	Specific Heat	•	•	•	•	•	•	•						•	•	•	•	•	•	
	Density	•	•		•	•	•	•						•	•	•			•	
	Absorptivity (Emissivity Factors)	•	•		•	•	•	•	•	•	•			•	•	•	•		•	•
	Convection Coefficients	•	•	•	•	•	•	•	•	•	•			•	•	•	•	•	•	
Perfect Conductors (Via Multipoint Constraints)			•	•	•	•	•	•						•	•	•	•		•	
Time Dependent Thermal Properties		•	•	•	•	•	•	•	•	•	•			•	•	•	•		•	

		NTEMP	PAFEC	SAHARA	SAMCEF (THERNL)	SESAM-69 (NV-615)	SINDA	SPAR	SSPTA	TACO	TAC2D	TAC3D	TANG	TAU	TEMP	TEPSA	THAC-SIP-3D	THTD	TRUMP	WECAN
Capacitance	a) Consistent	●	●		●	●		●		●				●	●			●		●
	b) Lumped	●	●	●	●	●	●	●	●	●	●	●	●		●	●	●	●	●	●
Convection	a) Free	●	●	●	●	●	●	●	●	●				●	●			●	●	●
	b) Forced	●		●		●	●	●	●	●				●	●			●	●	●
Radiation		●	●	●	●	●	●	●	●	●				●	●			●	●	●
Interelement Convection and Radiation		●		●	●		●	●	●	●			●		●			●	●	●
User Supplied Elements (see program abstracts)			●		●				●											●
Others (see program abstracts)																				
6. Material Properties and Material Models																				
Isotropic		●	●	●	●	●	●	●	●	●	●	●		●	●	●	●	●	●	●
Anisotropic		●	●	●	●	●	●	●	●	●				●	●	●	●			●
Multilayered		●	●						●											
Temperature-Dependent Properties	Conductivity	●	●	●	●	●	●	●		●	●	●		●	●	●	●	●	●	●
	Specific Heat	●	●	●	●	●	●	●		●	●	●		●	●	●	●	●	●	●
	Density	●	●	●	●	●	●	●		●	●	●		●	●	●	●			●
	Absorptivity (Emissivity Factors)						●	●		●				●	●			●		
	Convection Coefficients	●	●	●	●	●	●	●		●	●	●		●	●	●	●	●	●	●
Perfect Conductors (Via Multipoint Constraints)		●	●	●	●	●	●	●						●				●	●	●
Time Dependent Thermal Properties		●						●		●	●	●		●	●	●	●	●	●	

Part I - Continued

	ABAQUS	ADINAT	AGTAP	ANDES	ANSYS	ASASHEAT	BERSAFE FLHE	CAVE I	CAVE II	CAVE III	CONFAC	FLUX2	HEATING5	HEATING6	HEATRAN	MARC	MITAS II	MSC/NASTRAN	NNTB
Latent Heat and Phase Change Effects	•	•	•	•	•		•						•	•		•	•		•
Material Added or Removed During Analysis	•	•														•	•		
User Supplied (see program abstracts)													•			•			
Others (see program abstracts)																•			
7. Initial Conditions																			
Homogeneous	•	•		•	•	•	•						•	•	•	•		•	•
Varying Throughout the Region	•	•		•		•	•						•	•	•	•		•	•
Initial Enthalpy (for Phase Change)																		•	
User-Supplied	•		•				•	•	•	•		•	•	•	•	•	•	•	
Others (see program abstracts)																			
8. Boundary Conditions and Thermal Loads																			
Prescribed Temperatures a) Steady State	•	•	•	•	•	•	•	•	•	•			•	•	•	•	•	•	•
b) Time Dependent	•	•	•	•		•	•			•			•	•	•	•	•	•	•
Thermal Flux Input a) Steady	•	•	•	•		•	•						•	•	•	•	•	•	•
b) Temperature Dependent	•						•						•	•	•	•		•	
c) Time Varying	•		•	•		•	•						•	•	•	•	•	•	•
Convection from a Surface to Its Surroundings a) Steady State	•		•	•	•	•	•	•	•	•			•	•	•	•	•	•	•
b) Time Dependent	•	•	•	•		•	•	•	•	•			•	•	•	•	•	•	•

	NTEMP	PAFEC	SAHARA	SAMCEF	SESAM-69 (THERNL) (NV-615)	SINDA	SPAR	SSPTA	TACO	TAC2D	TAC3D	TANG	TAU	TEMP	TEPSA	THAC- SIP-3D	THTD	TRUMP	WECAN
Latent Heat and Phase Change Effects				●											●		●	●	
Material Added or Removed During Analysis				●	●	●							●						●
User Supplied (see program abstracts)					●	●			●	●	●					●		●	
Others (see program abstracts)				●			●												
7. Initial Conditions																			
Homogeneous	●	●	●		●	●	●	●	●	●	●			●	●	●	●	●	●
Varying Throughout the Region		●	●		●	●	●	●	●	●	●		●	●	●	●	●	●	●
Initial Enthalpy (for Phase Change)					●													●	
User-Supplied		●		●	●	●			●	●	●		●			●			●
Others (see program abstracts)																			
8. Boundary Conditions and Thermal Loads																			
Prescribed Temperatures a) Steady State		●	●		●	●	●	●	●	●			●	●	●	●	●	●	●
b) Time Dependent		●	●		●	●	●	●	●	●	●		●	●	●	●	●	●	●
Thermal Flux Input a) Steady	●	●	●		●	●	●	●	●	●			●	●	●	●		●	●
b) Temperature Dependent	●	●	●		●	●	●	●	●	●			●	●	●	●		●	●
c) Time Varying	●	●	●		●	●	●	●	●	●			●	●	●	●		●	●
Convection from a Surface to Its Surroundings a) Steady State	●	●	●		●	●	●	●	●	●	●		●	●	●	●	●	●	●
b) Time Dependent	●	●	●		●	●	●	●	●	●	●		●	●	●	●	●	●	●

Part I - Continued

		ABAQUS	ADINAT	AGTAP	ANDES	ANSYS	ASASHEAT	BERSAFE FLHE	CAVE I	CAVE II	CAVE III	CONFAC	FLUX2	HEATING5	HEATING6	HEATRAN	MARC	MITAS II	MSC/ NASTRAN	NNTB
Convection from Surroundings to a Surface	a) Steady State	•	•		•	•	•	•						•	•	•	•	•	•	•
	b) Time Dependent	•	•	•	•	•	•	•						•	•	•	•	•	•	
Forced Convection			•	•		•	•	•							•	•	•	•	•	
Radiation from a Surface to Its Surroundings	a) Steady State	•	•	•	•	•	•	•	•	•	•			•	•	•	•	•	•	•
	b) Time Dependent	•	•	•	•	•	•	•	•	•	•			•	•	•	•	•	•	
Radiation from Surroundings to a Surface	a) Steady State	•	•	•	•	•	•	•	•	•	•		•	•	•	•	•	•	•	•
	b) Time Dependent	•	•		•	•	•	•						•	•		•		•	
Radiation Between Narrow Gaps		•												•	•		•			•
Radiation Between n Surfaces with	a) User-Supplied View Factors			•		•		•												
	b) Internally Calculated View Factors							•	•	•	•	•			•					
Prescribed Fluid Flow		•	•			•		•								•			•	
Boundary Layer Convection			•																•	
Volumetric Heat Generation	a) On Element Level	•	•	•	•	•	•	•								•	•		•	
	b) On Node Level		•		•		•	•	•	•	•			•	•	•		•	•	
Gap (Thermal Resistance)			•			•		•									•	•	•	
Boundary Conditions/Loads Added or Removed During Analysis		•	•	•	•	•	•	•						•	•	•	•	•	•	
Others (see program abstracts)															•					

Feature		NTEMP	PAFEC	SAHARA	SAMCEF (THERNL)	SESAM-69 (NV-615)	SINDA	SPAR	SSPTA	TACO	TAC2D	TAC3D	TANG	TAU	TEMP	TEPSA	THAC-SIP-3D	THTD	TRUMP	WECAN
Convection from Surroundings to a Surface	a) Steady State		•	•	•	•	•	•	•	•				•	•	•			•	•
	b) Time Dependent	•	•	•	•	•	•	•	•	•	•	•		•	•	•	•	•	•	•
Forced Convection					•		•	•	•	•	•	•		•		•	•	•	•	•
Radiation from a Surface to Its Surroundings	a) Steady State			•	•		•	•	•	•	•	•		•	•	•	•	•	•	•
	b) Time Dependent	•			•		•		•	•	•	•		•		•	•		•	•
Radiation from Surroundings to a Surface	a) Steady State			•	•		•	•	•	•	•	•		•	•	•	•	•	•	•
	b) Time Dependent	•		•	•		•		•	•	•	•		•		•	•		•	•
Radiation Between Narrow Gaps								•	•	•	•	•		•		•	•	•	•	•
Radiation Between in Surfaces with	a) User-Supplied View Factors	•		•	•			•	•											
	b) Internally Calculated View Factors													•						
Prescribed Fluid Flow		•		•	•		•			•					•				•	
Boundary Layer Convection		•		•	•			•											•	
Volumetric Heat Generation	a) On Element Level	•	•		•		•	•	•	•	•	•		•			•	•	•	•
	b) On Node Level	•		•	•									•	•				•	•
Gap (Thermal Resistance)				•			•	•	•	•	•			•		•		•	•	
Boundary Conditions/Loads Added or Removed During Analysis				•	•		•		•	•	•	•		•					•	•
Others (see program abstracts)																				

Part I - Continued

			ABAQUS	ADINAT	AGTAP	ANDES	ANSYS	ASASHEAT	BERSAFE FLHE	CAVE I	CAVE II	CAVE III	CONFAC	FLUX2	HEATING5	HEATING6	HEATRAN	MARC	MITAS II	MSC/NASTRAN	NNTB
9.	Solution Techniques																				
	Linear Steady State	a) Direct	●	●		●	●	●	●							●	●	●	●	●	●
		b) Iterative				●									●	●			●		
		c) Others (see abstracts)																		●	●
	Nonlinear Steady State	a) Incremental	●	●		●	●											●		●	
		b) Iterative	●	●		●		●	●						●	●		●	●		
		c) Others (see abstracts)																			
	Transient	a) Thermal Mode Superposition																			
		b) Direct Integration i) Explicit		●	●																
		ii) Implicit — User Specified Time Step	●	●		●		●	●				●	●	●	●	●	●	●	●	●
		Automatic Time Step Selection	●			●		●							●	●	●		●		
		iii) Combined Explicit/Implicit													●	●				●	
		c) Finite Elements in the Time Domain																			
		d) Moving - Deforming Grids								●	●	●									
		e) Others (see program abstracts)																			
10.	Other Capabilities																				
	Thermal Stress Analysis Capability	a) Uncoupled	●	●		●	●	●	●									●		●	
		b) Coupled	●	●		●	●											●		●	

		NTEMP	PAFEC	SAHARA	SAMCEF (THERNL)	SESAM-69 (NV-615)	SINDA	SPAR	SSPTA	TACO	TAC2D	TAC3D	TANG	TAU	TEMP	TEPSA	THAC-SIP-3D	THTD	TRUMP	WECAN
9. Solution Techniques																				
Linear Steady State	a) Direct	●	●		●	●	●	●	●	●				●	●	●			●	●
	b) Iterative			●							●	●							●	
	c) Others (see abstracts)				●														●	
Nonlinear Steady State	a) Incremental		●																	
	b) Iterative	●		●	●		●	●	●	●	●	●			●	●		●	●	●
	c) Others (see abstracts)					●													●	
Transient	a) Thermal Mode Superposition																			
	b) Direct Integration i) Explicit																			
	ii) Implicit — User Specified Time Step	●		●	●	●	●	●	●	●				●	●			●	●	●
	— Automatic Time Step Selection						●	●	●	●				●	●	●	●	●	●	●
	iii) Combined Explicit/Implicit		●																	
	c) Finite Elements in the Time Domain				●														●	
	d) Moving — Deforming Grids																		●	
	e) Others (see program abstracts)																		●	
10. Other Capabilities																				
Thermal Stress Analysis Capability	a) Uncoupled		●		●		●	●						●	●	●		●	●	●
	b) Coupled															●				

501

Part I – Continued

		ABAQUS	ADINAT	AGTAP	ANDES	ANSYS	ASAHEAT	BERSAFE FLHE	CAVE I	CAVE II	CAVE III	CONFAC	FLUX2	HEATING5	HEATING6	HEATRAN	MARC	MITAS II	MSC/ NASTRAN	NNTB
	Temperature Field Data Transmitted Directly from Heat Transfer Modules to Thermal Stress Modules	•	•		•	•	•	•								•	•		•	
	Enclosure Radiation with View Factor Calculation											•								
	Heat Input/Output at Constrained Boundaries	•	•			•	•	•								•	•	•	•	•
	Cyclic Symmetry					•													•	
	Substructuring a) Repeated Use of Identical Substructures	•																	•	
	b) Mixing Linear and Nonlinear Substructures						•												•	
	c) Mixing Substructures with Different Types of Nonlinearities						•												•	
	Restart Capability	•	•	•	•	•	•	•				•		•	•	•	•	•	•	•
	Others (see program abstracts)							•												
11.	Program Operational On																			
	CDC		•			•		•	•	•	•	•	•	•	•		•	•	•	
	IBM	•	•	•		•	•	•	•	•	•			•	•	•	•	•	•	•
	UNIVAC	•	•			•													•	
	Honeywell		•																	
	Telefunken		•																	
	AMDAHL					•		•									•		•	
	SIEMENS		•																	
	ICL							•								•				

	NTEMP	PAFEC	SAHARA	SAMCEF (THERNL)	SESAM-69 (NV-615)	SINDA	SPAR	SSPTA	TACO	TAC2D	TAC3D	TANG	TAU	TEMP	TEPSA	THAC-SIP-3D	THTD	TRUMP	WECAN
Temperature Field Data Transmitted Directly from Heat Transfer Modules to Thermal Stress Modules	•	•		•	•		•		•				•		•		•		•
Enclosure Radiation with View Factor Calculation								•	•				•						
Heat Input/Output at Constrained Boundaries	•			•			•	•					•	•				•	
Cyclic Symmetry		•						•					•				•	•	
Substructuring a) Repeated Use of Identical Substructures		•		•	•													•	•
b) Mixing Linear and Nonlinear Substructures		•		•	•			•											
c) Mixing Substructures with Different Types of Nonlinearities				•	•														
Restart Capability	•	•	•	•	•	•	•	•	•	•	•	•	•	•	•	•	•	•	•
Others (see program abstracts)				•															
11. Program Operational On																			
CDC	•	•		•		•	•	•	•					•	•		•	•	•
IBM		•		•		•		•		•	•	•			•	•	•	•	
UNIVAC		•		•	•	•	•			•	•		•		•	•	•	•	
Honeywell		•		•															
Telefunken															•		•		
AMDAHL												•							
SIEMENS				•															
ICL		•											•						

Part I - Concluded

	ABAQUS	ADINAT	AGTAP	ANDES	ANSYS	ASASHEAT	BERSAFE FLHE	CAVE I	CAVE II	CAVE III	CONFAC	FLUX2	HEATING5	HEATING6	HEATRAN	MARC	MITAS II	MSC/NASTRAN	NNTB
Minicomputers (see program abstracts)	●	●			●	●	●									●		●	●
Supercomputers (see program abstracts)	●	●			●											●		●	
Others (see program abstracts)																●			
12. Documentation																			
Programmer's Manual	●	●	●			●	●	●	●	●			●			●		●	
Theoretical Manual	●	●	●	●		●	●	●	●	●			●	●		●	●	●	●
Data Preparation - Users' Manual	●	●	●			●	●	●	●	●			●	●	●	●	●	●	●
Example Problem Manual	●	●	●	●		●		●	●	●						●		●	
Verification/Validation Manual	●	●				●		●	●	●						●		●	
Pre- and Post-Processors' Manual	●		●	●		●										●		●	

504

		NTEMP	PAFEC	SAHARA	SAMCEF (THERNL)	SESAM-69 (NV-615)	SINDA	SPAR	SSPTA	TACO	TAC2D	TAC3D	TANG	TAU	TEMP	TEPSA	THAC-SIP-3D	THTD	TRUMP	WECAN
	Minicomputers (see program abstracts)		•		•		•													
	Supercomputers (see program abstracts)		•	•			•												•	
	Others (see program abstracts)				•				•	•									•	
12.	Documentation																			
	Programmer's Manual	•	•																	
	Theoretical Manual		•	•		•		•				•	•		•			•	•	
	Data Preparation - Users' Manual		•	•	•	•	•	•	•	•	•	•	•	•		•	•	•	•	•
	Example Problem Manual	•	•	•		•		•	•	•		•	•	•	•	•	•	•	•	•
	Verification/Validation Manual		•	•	•	•					•		•	•	•	•			•	•
	Pre- and Post-Processors' Manual		•	•	•	•	•			•		•		•	•			•	•	•

Part II - User Interface and Modeling Capabilities

			ABAQUS	ADINAT	AGTAP	ANDES	ANSYS	ASASHEAT	BERSAFE FLHE	CAVE I	CAVE II	CAVE III	CONFAC	FLUX2	HEATING5	HEATING6	HEATRAN	MARC	MITAS II	MSC/NASTRAN	NNTB
1.	Input Form and Sequence																				
	a) Input Form																				
		Fixed Format	•	•	•				•	•	•	•	•	•	•			•		•	•
		Free Form-List Directed Format	•	•		•	•	•	•							•	•	•	•	•	
		Problem Oriented Language	•						•							•	•		•	•	
		Others (see program abstracts)																				
	b) Input Sequence																				
		User Directed	•	•	•	•	•			•	•	•		•	•	•	•	•	•	•	•
		System Directed						•	•									•		•	
		User Supplied Subroutines (see program abstracts)	•		•																
2.	Model Generation and Checking																				
	a) Automatic or Semi-Automatic Generator for:																				
		Nodal Point Coordinates	•	•	•	•	•	•	•	•	•	•		•	•	•	•	•		•	
		Element Connectivities	•	•	•	•	•	•	•			•			•	•	•	•		•	
		Constraints, Symmetry and Boundary Conditions	•	•	•		•	•	•			•								•	
		Substructure Connectivity						•													
		Repetition of Identical Segments	•	•	•	•	•		•			•		•	•	•		•	•	•	
		Others (see program abstracts)																			

			NTEMP	PAFEC	SAHARA	SAMCEF (THERNL)	SESAM-69 (NV-615)	SINDA	SPAR	SSPTA	TACO	TAC2D	TAC3D	TANG	TAU	TEMP	TEPSA	THAC-SIP-3D	THTD	TRUMP	WECAN
1.	Input Form and Sequence																				
	a) Input Form																				
		Fixed Format	●	●	●	●	●		●		●	●	●	●	●	●	●	●		●	●
		Free Form-List Directed Format			●	●															
		Problem Oriented Language						●		●					●	●			●		
		Others (see program abstracts)																			
	b) Input Sequence																				
		User Directed		●	●	●	●	●	●	●	●			●							
		System Directed	●					●	●			●	●		●	●	●	●	●	●	●
		User Supplied Subroutines (see program abstracts)						●	●	●		●	●								
2.	Model Generation and Checking																				
	a) Automatic or Semi-Automatic Generator for:																				
		Nodal Point Coordinates	●	●		●	●		●		●		●	●	●	●	●	●	●	●	●
		Element Connectivities	●	●		●	●		●		●		●	●	●	●	●	●	●	●	●
		Constraints, Symmetry and Boundary Conditions	●	●		●	●		●		●		●	●	●	●	●	●	●	●	●
		Substructure Connectivity		●		●	●							●			●				●
		Repetition of Identical Segments	●	●					●					●	●	●			●	●	●
		Others (see program abstracts)							●	●										●	

Part II - Continued

	ABAQUS	ADINAT	AGTAP	ANDES	ANSYS	ASASHEAT	BERSAFE / FLHE	CAVE I	CAVE II	CAVE III	CONFAC	FLUX2	HEATING5	HEATING6	HEATRAN	MARC	MITAS II	MSC/NASTRAN	NNTB
b) Automatic or Semi-Automatic Generator for:																			
One-Dimensional Elements	•	•		•	•	•	•									•		•	
Triangular Elements		•		•	•	•	•									•		•	
Quadrilateral Elements	•	•		•	•	•	•			•						•		•	
Body or Shell of Revolution Elements	•	•		•	•	•	•			•						•			
Three-Dimensional Solid Elements	•	•			•	•	•		•	•						•		•	
Two-Dimensional Shell Elements	•				•	•	•			•						•		•	
Curvilinear Finite Difference Grids																			
Others (see program abstracts)								•	•	•									
c) Data Checking Facilities																			
Line Printer	•		•	•	•	•	•			•	•	•	•		•	•	•	•	
Plotter	•			•	•	•	•						•		•	•	•	•	
Interactive Graphics			•	•	•	•	•				•					•		•	
Others (see program abstracts)		•																	
d) Plots and Graphics Display of Mode?																			
Complete Analysis Region	•			•	•	•				•	•	•	•	•	•	•		•	
Part of Analysis Region	•			•	•	•							•	•	•	•		•	
"Blow-Up" Option	•			•	•	•							•	•	•	•		•	

	NTEMP	PAFEC	SAHARA	SAMCEF (THERNL)	SESAM-69 (NV-615)	SINDA	SPAR	SSPTA	TACO	TAC2D	TAC3D	TANG	TAU	TEMP	TEPSA	THAC-SIP-3D	THTD	TRUMP	WECAN
b) Automatic or Semi-Automatic Generator for:																			
One-Dimensional Elements	●	●	●	●										●				●	
Triangular Elements	●	●	●	●			●					●							●
Quadrilateral Elements	●	●	●	●			●					●	●	●				●	●
Body or Shell of Revolution Elements		●	●	●	●		●	●	●						●			●	●
Three-Dimensional Solid Elements		●	●	●	●			●	●			●	●	●	●			●	●
Two-Dimensional Shell Elements		●		●														●	●
Curvilinear Finite Difference Grids																			
Others (see program abstracts)																			
c) Data Checking Facilities																			
Line Printer	●	●	●	●	●	●	●	●	●	●	●		●	●	●	●	●	●	●
Plotter	●	●	●	●	●	●	●	●						●			●	●	●
Interactive Graphics					●							●		●			●	●	●
Others (see program abstracts)													●						
d) Plots and Graphics Display of Model																			
Complete Analysis Region	●	●	●						●			●		●				●	●
Part of Analysis Region	●	●	●				●	●	●					●			●	●	●
"Blow-Up" Option	●	●	●				●	●	●					●			●	●	●

Part II - Continued

	ABAQUS	ADINAT	AGTAP	ANDES	ANSYS	ASASHEAT	BERSAFE FLHE	CAVE I	CAVE II	CAVE III	CONFAC	FLUX2	HEATING5	HEATING6	HEATRAN	MARC	MITAS II	MSC/NASTRAN	NNTB
Hidden Lines or Surfaces				•	•	•	•			•	•					•		•	
Orthographic Views	•			•	•	•	•			•								•	
Perspective and Isometric Views	•			•	•	•	•			•	•					•		•	
Section View on Arbitrary Plane	•			•	•	•				•								•	
Others (see program abstracts)		•					•												
e) Other Facilities																			
Digitizer Input			•		•		•			•	•					•		•	
Automatic Renumbering of Nodes, Elements or Equations					•		•			•					•	•			
Table Lookup of Data	•		•	•				•	•	•		•					•		
Others (see program abstracts)																	•		
3. Results Output Form																			
a) Tabular Output																			
Fixed Set	•	•		•	•	•	•	•	•	•	•								•
User Defined Set and Sequences	•		•	•	•		•												
Maximum and Minimum Quantities				•	•		•						•	•		•	•	•	
Average and Maxima for Blocks of Nodes													•	•		•	•	•	
Temperature or Flux Exceedances					•		•								•		•		
Others (see program abstracts)																		•	

	NTEMP	PAFEC	SAHARA	SAMCEF (THERNL)	SESAM-69 (NV-615)	SINDA	SPAR	SSPTA	TACO	TAC2D	TAC3D	TANG	TAU	TEMP	TEPSA	THAC-SIP-3D	THTD	TRUMP	WECAN
Hidden Lines or Surfaces	•	•										•	•	•					•
Orthographic Views	•			•								•	•	•					•
Perspective and Isometric Views	•	•		•	•		•	•				•	•	•			•		•
Section View on Arbitrary Plane	•	•		•			•	•				•	•	•					•
Others (see program abstracts)		•																	
e) Other Facilities																			
Digitizer Input		•																•	
Automatic Renumbering of Nodes, Elements or Equations	•	•	•	•	•	•	•		•			•	•				•		•
Table Lookup of Data		•				•			•								•	•	
Others (see program abstracts)				•															
3. Results Output Form																			
a) Tabular Output																			
Fixed Set	•	•	•		•	•	•		•			•	•	•	•	•		•	•
User Defined Set and Sequences	•	•	•		•	•	•						•	•	•	•	•	•	
Maximum and Minimum Quantities						•	•			•	•		•	•	•		•	•	•
Average and Maxima for Blocks of Nodes		•					•											•	
Temperature or Flux Exceedances															•		•	•	
Others (see program abstracts)																		•	

Part II - Concluded

	ABAQUS	ADINAT	AGTAP	ANDES	ANSYS	ASASHEAT	BERSAFE/FLHE	CAVE I	CAVE II	CAVE III	CONFAC	FLUX2	HEATING5	HEATING6	HEATRAN	MARC	MITAS II	MSC/NASTRAN	NNTB
b) File Output for User Post-Processing and Plotting	•	•	•	•	•	•	•	•	•	•	•	•	•	•		•	•	•	
c) Plots																			
Isotherm Plots (Contours) of Temperatures/Flux	•		•	•	•	•	•							•	•	•		•	
Surface Functions							•											•	
Selective Output (e.g., by Elements or Regions)	•		•	•	•		•						•	•		•	•	•	
Histories (e.g., Time History)	•		•	•	•		•						•	•		•	•	•	
Others (see program abstracts)															•			•	
4. Interactive Input and Control																			
Parameter Specification (e.g., Flux or Time Steps)					•	•											•		
Singularity Check																•			
Error Correction/Recovery	•		•		•														
User Control of Matrix Decomposition						•													
Others (see program abstracts)																			

Feature	NTEMP	PAFEC	SAHARA	SAMCEF (THERNL)	SESAM-69 (NV-615)	SINDA	SPAR	SSPTA	TACO	TAC2D	TAC3D	TANG	TAU	TEMP	TEPSA	THAC-SIP-3D	THTD	TRUMP	WECAN
b) File Output for User Post-Processing and Plotting		•	•	•	•	•	•	•	•			•	•			•		•	•
c) Plots																			
Isotherm Plots (Contours) of Temperatures/Flux	•	•		•	•				•				•	•		•	•	•	•
Surface Functions																	•		•
Selective Output (e.g., by Elements or Regions)		•		•	•				•				•	•		•		•	•
Histories (e.g., Time History)		•		•		•		•	•					•		•		•	•
Others (see program abstracts)		•	•						•			•	•						
4. Interactive Input and Control																			
Parameter Specification (e.g., Flux or Time Steps)		•		•			•										•	•	
Singularity Check		•		•			•					•							
Error Correction/Recovery		•		•			•								•			•	
User Control of Matrix Decomposition		•					•											•	
Others (see program abstracts)							•											•	

513

ABAQUS

Descriptive Program Title: General Purpose Structural and Heat Transfer Program

Program Developer: Hibbitt, Karlsson and Sorensen, Inc., 35 South Angell Street, Providence, Rhode Island 02906.

Date of First Release and Most Recent Update: 1979 and 1981

General Information:
ABAQUS is a general purpose, structural and heat analysis code developed and maintained by Hibbitt, Karlsson and Sorensen, Inc.

Program Capability and Scope of Analysis:
ABAQUS provides a complete capability for linear and nonlinear analysis. In addition to a general heat transfer capability, static, dynamic, eigenvalue buckling and soil consolidation procedures are included in the code. Also, a procedure for Eully coupled heat/stress analysis is operational.

User Interface and Modeling Capabilities:
- The ABAQUS pre-processor is designed to simplify the task of data specification. Data may be entered in fixed or free format and are identified by leading keyword cards. Extensive data consistency checks are built into the code and clear messages are printed whenever errors are encountered.
- ABAQUS provides a complete range of plotting: mesh plotting, contour and displaced plots at specified points in an analysis may be requested, and time history plots are directly obtained. Each of these plotting capabilities permits detailed 'blow-ups', viewpoints, etc.
- A very general printed/file output is provided as well as a flexible restart capability.

Solution Methods:
- Nonlinear transient response - Backward difference scheme.
- Nonlinear steady state - Newton-Raphson technique.
- Eigenvalue extraction - Subspace iteration procedure.
- Linear equation solver - Wavefront technique.

Notable Items and Limitations:
All solution procedures include automatic time stepping capability. These self-adaptive schemes choose time (loading) increments based on user set tolerances to provide solutions of uniform accuracy. Automatic loading can avoid excessive restarting to obtain convergence and thus generally saves computer costs.

Programming Language: ANSI FORTRAN

Hardware/Operating System: CDC 6600, 7600, CYBER 175, 176, CDC 203, CRAY, IBM 370, 3033, AMDAHL, UNIVAC 1100 series, VAX.

Program Size (Heat Transfer Modules Only): Pre-processors 25,000; Main 50,000 executable statements. Programs load as libraries, so that small problems can be fitted in quite small machine memories.

Documentation: See Ref. 9

Program Availability: The program may be obtained from the developer:
Hibbitt, Karlsson and Sorensen, Inc.
35 South Angell Street
Providence, Rhode Island 02906

It may also be accessed commercially through the CYBERNET System.

ADINAT

Descriptive Program Title: ADINAT - A Finite Element Program for Automatic Dynamic Incremental Nonlinear Analysis of Temperatures

Program Developer: Professor K. J. Bathe, Department of Mechanical Engineering, Massachusetts Institute of Technology, Cambridge, MA 02139

Date of First Release and Most Recent Update: 1977 and 1981

General Information:
ADINAT is part of the ADINA system together with the general purpose code ADINA and the pre- and post-processor ADINA-PLOT. ADINAT is a proprietary code which is maintained and further developed by ADINA Engineering with offices in Västerås, Sweden, and Boston, Massachusetts. The code is available for a fee and members of the users group obtain the source code and all new developments as long as they remain members of the group. The source code is transmitted with sample data cases and their solutions.

Program Capability and Scope of Analysis:
ADINAT is a general purpose linear and nonlinear finite element analysis program for steady-state and transient heat transfer and analogous field problems. The nonlinearities may be due to temperature-dependent material properties including latent heat effects, element birth and death options or boundary convection and radiation conditions. The program can be used to restart at pre-selected time steps. Thermal frequencies and mode shapes can be calculated. Both concentrated and distributed heat flows can be applied.

User Interface and Modeling Capabilities:
The ADINA system includes ADINA-PLOT for pre- and post-processing. At present the capabilities of ADINA-PLOT are oriented towards ADINA but a temperature tape can be output from ADINAT for further processing in ADINA-PLOT with printing and plotting of selected results in the form of tables, curve plots, etc.

Solution Methods:
- Nonlinear transient response - Implicit and explicit time integration, Euler backward and forward method, trapezoidal rule and the α-family method. Equilibrium iteration.
- Nonlinear steady-state problems - Incremental solution, modified Newton-Raphson method.
- Equation solver for linear equations - Compacted out-of-core solver.
- Extraction of frequencies and mode shapes - Determinant search method.

Notable Items and Limitations:
ADINAT offers a very large range of applications in linear and nonlinear analysis with relatively few effective elements, a good library of material models and effective numerical methods. The program can be employed effectively in linear

analysis and then, with only a few input changes, in relatively simple and very complex nonlinear analyses.

Programming Language: FORTRAN IV

Hardware/Operating System: Among mini- and supercomputers, installations are VAX, PRIME, CYBER 203, and CRAY.

Program Size: Approximately 12,000 source statements of the core program; ADINA 45,000 statements.

Documentation: See Refs. 10 to 13

Program Availability: Source program of ADINAT is available by joining the ADINA Users Group (for a fee). Contact:
    ADINA Engineering AB
    Munkgatan 20 D
    S-722 12 Västerås
    Sweden
    Tel. 021-14 40 50
    Telex 40630 ADINA S

AGTAP

Descriptive Program Title: Abbreviated General Thermal Analyzer Program

Program Developer: Grumman Aerospace Corporation, Bethpage, New York 11714

Date of First Release and Most Recent Update: 1965 and 1980

General Information:
AGTAP is a general thermal analyzer designed to solve both simple problems requiring rapid solution and unconventional problems for which extensive supplementary calculations are necessary.

Program Capability and Scope of Analysis:
This program is capable of solving thermal models of up to 1,000 nodes with 2,000 conduction and 2,000 radiation connectivities.

User Interface and Modeling Capabilities:
The preprocessor code TANG provides model generation capability with output formats compatible with AGTAP. A post-processor plotting capability is also available.

Solution Methods:
The solution technique employs a "lumped parameter" approximation of the problem which is solved by a finite difference iterative procedure.

Notable Items and Limitations:
The program features three options to insert specialized calculations into the solution. Evaluation of the maximum critical time step is provided, but not internally controlled.

Programming Language: FORTRAN IV

Hardware/Operating System: IBM 370 (OS/VS)

Program Size (Heat Transfer Modules Only) - Number of Source Statements of Core Program, Pre- and Post-Processors:
The most recent version contains 905 source statements.

Documentation: Informal report describing method of solution, program operation, data preparation and sample problems is available.

Program Availability: For further information contact:
   Dr. John G. Roukis
   Mail Stop B22/35
   Grumman Aerospace Corporation
   Bethpage, New York 11714

ANDES

Descriptive Program Title: Acoustic Non-Destructive Evaluation Stress Analysis

Program Developer: Dr. A. F. Emery, Department of Mechanical Engineering, University of Washington, Seattle, Washington 98195.

Date of First Release and Most Recent Update: September 1980 and September 1981

General Information:
ANDES is designed to calculate stresses and residual stresses for comparison with acoustic experimental tests. Phase changes, temperature-dependent properties, time-dependent properties, and time-dependent boundary conditions are treated. Dynamic core allocation.

Program Capability and Scope of Analysis:
Two-dimensional and axisymmetric problems. Transient with nonlinear boundary conditions.

User Interface and Modeling Capabilities:
Batch processing. Interactive graphic input of mesh. Separate mesh generator or internal generation of simple meshes.

Solution Method:
Direct solution of equations with incremental iteration for nonlinear problems. SOR method if specified by user. User prescribed convergence criterion. Convergence based upon rms or maximum error.

Programming Language: FORTRAN Extended.

Hardware/Operating System: CDC CYBER 175/750, NOS operating system, CDC 6000 series, NOS/BE operating system.

Program Size (Heat Transfer Modules Only):
This program contains approximately 3,000 cards. Pre- and post-processors approximately 3,000 cards each.

Documentation: See Ref. 14

Program Availability: On request from program developer.
Cost: Approximately $200.

ANSYS

Descriptive Program Title: ANSYS

Program Developer: Swanson Analysis Systems, Inc., P.O. Box 65, Johnson Road, Houston, PA 15342.

Date of First Release and Most Recent Update: 1970 and 1981

General Information:
ANSYS is a proprietary finite element program first offered by Swanson Analysis Systems in 1970. ANSYS, Rev. 4, is the most current release. ANSYS is supported by Swanson Analysis Systems, Inc., in Houston, Pennsylvania and by consultants in Los Angeles, California and London, England.

Program Capability and Scope of Analysis:
ANSYS is a general purpose program for steady state and transient heat transfer, as well as structural analyses. Thermal-electrical capabilities and thermal-fluid flow capabilities are also available. Loads include specified temperatures, heat flows, convections and/or internal heat generation. Any thermal solution may be input as a load to a structural analysis. The finite element model is identical; the user need only select structural members from the element library.

User Interface and Modeling Capabilities:
- A powerful preprocessor facilitates complete input data preparation. Model geometry, loads, materials, and analysis options can be described. Many plotting options exist to verify geometry and loads. The preprocessor can be operated in interactive or batch modes.
- Different post-processors aid the user in results evaluation. Isotherms can be plotted, graphs of temperature versus time are available and results can be scanned for user specified temperature and/or heat flow ranges.

Element Library:
A complete library of line, area, shell and solid elements is available. Axisymmetric elements with axisymmetric or nonaxisymmetric loads may be used to perform a three-dimensional analysis with a two-dimensional model. Convection, conduction and radiation element types may be used. Each thermal element has an analogous structural element so that the same model can be used in a structural analysis where the temperature solution is a load.

Solution Methods:
- Transient analysis - Modified Houboldt method.
- Linear equation solver - Wavefront technique, Gaussian elimination on substructures.

Notable Items and Limitations:
Phase change problems can be solved. One of the most powerful features of ANSYS is the ease with which a thermal model can be used in a structural analysis.

Programming Language: ANSI FORTRAN

Hardware/Operating System: CRAY, CDC, IBM, AMDAHL, UNIVAC, Honeywell, PRIME, DEC VAX, Harris.

Program Size: 100,000 lines of code.

Documentation: See Refs. 15 and 16

Program Availability: ANSYS is available at most data centers in the United States and Europe. ANSYS can be leased on an in-house basis. Charges are cents/CP second based on machine speed or fixed cost/month for some in-house leases. Contact the developer for further information.

ASASHEAT

Descriptive Program Title: Linear/Nonlinear Thermal Analyzer of the ASAS Range of Finite Element Programs

Program Developer: Atkins Research and Development, Parkside House, Woodcote Grove, Ashley Road, Epsom, Surrey, England

Date of First Release and Most Recent Update: 1973 and 1981

General Information:
- ASASHEAT is a proprietary code developed and maintained by Atkins Research and Development. Development started in 1972 with the first release in 1973. Nonlinear capabilities were incorporated in 1981 which is also the date of the latest update. The program is supported from the headquarters in Epsom, England, and from Houston, and H. G. Engineering, Ltd., in Ontario.
- The program is modular in design with free format-list directed format input and extensive data checking, model creation and solution modules. Complete saving and restart facilities are incorporated.
- The nonlinear capabilities include: temperature dependent material properties, surface radiation to and from surroundings, temperature dependent free and boundary layer convection, temperature dependent internal/nodal heat generation and thermal flux for both steady state and transient analysis.

User Interface and Modeling Capabilities:
As a result of developments in the field of interactive mesh generation, the ASAS system incorporating ASASHEAT is designed to interface with proprietary pre- and post-processors.

Element Library:
The linear/nonlinear elements include: uniaxial, two-dimensional Cartesian, axisymmetric and three-dimensional isoparametric elements. For coupled thermal-structural analysis, structural elements default automatically to thermal ones.

Solution Methods for Nonlinear Problems:
Nonlinear transient response: Implicit integration; coupled Crank-Nicholson and corrective iterative scheme.
Nonlinear steady state response: Corrective iterative scheme applied to steady state.
Equation solvers for linear problems: Out-of-core, in-core modified frontal solver.

Notable Items and Limitations:
The program is under continual development. Greater flexibility in heat flux output, intersurface radiation, forced convection and phase change effects are being developed and included.

Programming Language: Portable ANSI FORTRAN 66

Hardware/Operating System: UNIVAC 1100 series, SIGMA, PRIME, VAX 11/780, IBM 360 series.

Program Size (Heat Transfer Modules Only):
Core Program: 35,000 statements

Documentation: See Ref. 17

Program Availability: The program is available at several bureaus and computer installations. For further information the developer's Support Manager should be contacted. Absolute versions only are distributed. Program fees are negotiable with developer.

BERSAFE (FLHE)

Descriptive Program Title: FLHE - FLow of Heat by Finite Elements

Program Developer: The overall system was developed by Dr. T. K. Hellen and colleagues, Central Electricity Generating Board, Berkeley Nuclear Laboratories, Berkeley, Gloucestershire GL13 9PB, England. FLHE was developed originally by Mr. K. Fullard and is maintained by Dr. M. A. Keavey at the above address.

Date of First Release and Most Recent Update: 1971. Level 3 released in 1981.

General Information:
BERSAFE is a general purpose finite element system started in 1968. It has been developed for the Central Electricity Generating Board at Berkeley Nuclear Laboratories, and has been available for purchase since 1970. FLHE is the component dealing with thermal analysis within the overall BERSAFE system.

Program Capability and Scope of Analysis:
General purpose program linking the functions of thermal analysis, stress analysis (elasticity, plasticity, creep, large displacements), and linear dynamics. Extensive pre- and post-processor aids are available. A wide range of finite elements exist for two-dimensional and three-dimensional beams, plates and shells.

Solution Methods:
- Transient analysis - Crank-Nicholson scheme.
- Linear equation solver - Wavefront technique.

Notable Items and Limitations:
Storage use is dynamic so the limitations on most variables are imposed by available core - on our system this is very large. Semibandwidth is the only notable limitation (currently 1,000 for stress analysis, but can easily be increased). The stress analysis package uses substructuring techniques and is particularly powerful for

fracture mechanics, as is the nonlinear version. Plasticity, creep and cycling are also well used, often coupled to previous transient temperature analyses.

Programming Language: FORTRAN IV

Hardware/Operating System: IBM (MVS, MVT, DOS), UNIVAC, AMDAHL (MVS), ICL, VAX (MVS), Burroughs, PRIME.

Program Size (Heat Transfer Modules Only) - Number of Source Statements of Core Program, Pre- and Post-Processors:
Size of core is approximately 420 K bytes on IBM. Source statements for FLHE is approximately 10,000. The stress analysis and pre- and post-processor programs are much larger - the whole system being well over 100,000 statements.

Documentation: See Refs. 18 to 20.

Program Availability: The programs are available from the BERSAFE Advisory Group at the above address (Mr. G. Marshall). The price for a five year license for source, including support and maintenance, is from $10,000 for the thermal analysis program including relevant pre- and post-processors.

CAVE I, II, III

Descriptive Program Title: CAVE (Conductive Analysis Via Eigenvalues)
A General Transient Heat Transfer Computer Code Utilizing Eigenvectors and Eigenvalues.

Program Developer: Grumman Aerospace Corporation, Bethpage, New York, under a contract for NASA Langley Research Center, Hampton, Virginia.

Date of First Release and Most Recent Update: November 1977 through October 1979

Brief Summary of the Major Capabilities of the Program:
- The computer code CAVE III (Conduction Analysis Via Eigenvalues for Three-Dimensional Geometries) provides a convenient and economical tool for predicting the transient temperature response of structures. This code is an extension of the work done under contract NAS1-13655 for two-dimensional geometries. CAVE III is written in FORTRAN IV and is operational on both the IBM 370/165 and CDC 6600 computers.
- The method of solution is a hybrid analytical-numerical technique which utilizes eigenvalues (thermal frequencies) and eigenvectors (thermal mode vectors). The method is inherently stable, permitting large time steps even with the best of conductors with the finest of mesh sizes which can provide a factor-of-five reduction in machine time compared to conventional explicit finite difference methods when structures with small time constants are analyzed over long time periods. This code will find utility in analyzing hypersonic missile and aircraft structures which fall naturally into this class.
- The code is a completely general one in that problems involving any geometry, boundary layer conditions and materials can be analyzed. This is made possible by requiring the user to establish the thermal network, e.g., node capacitances, conductances between nodes, etc. Dynamic storage allocation is used to minimize core storage requirements.
- The report is primarily a user's manual for the CAVE III code. Input and output formats are presented and explained. Sample problems are included which illustrate the usage of the code as well as establish the validity and accuracy of the method.

Program Size (Heat Transfer Modules Only) - Number of Source Statements of Core Program, Pre- and Post-Processors:
The CAVE programs contain approximately 1,600 source statements. The pre-processor network generator and graphics package for CAVE III contains an added 1,400 lines of code.

Documentation: See Refs. 21 to 23.

Program Availability: This program is available from Grumman Aerospace Corporation. For further information contact:
  Dr. John G. Roukis
  Mail Stop B22/35
  Grumman Aerospace Corporation
  Bethpage, New York 11714

Descriptive Program Title: Geometric Configuration Factor Program

Program Developer: Grumman Aerospace Corporation, Bethpage, New York 11714

Date of First Release and Most Recent Update: 1964 and 1975

General Information:
This program was originally developed by Grumman to determine geometric configuration factors between surfaces for the LUNAR Module and Orbiting Astronomical Observatory Projects.

Program Capability and Scope of Analysis:
The program determines geometric configuration factors between convex planar polygons. Thermal radiative behavior is in accordance with Lambert's Law.

User Interface and Modeling Capabilities:
Versions have been developed for up to two hundred surface geometries. Graphics pre-processor packages will provide model generation and model verification capability.

Solution Method:
CONFAC uses the method of contour integration and includes the effects of intervening surfaces in the results. Surfaces are divided into subareas based on a user selected mesh size.

Notable Items and Limitations:
This program is limited to planar convex polygons that are described by a maximum of ten vertices. Intersecting surfaces (other than a common edge) cannot be treated without subdivision to eliminate the intersection.

Programming Language: FORTRAN IV

Hardware/Operating System: IBM 370 (OS/VS)

Program Size (Heat Transfer Modules Only) - Number of Source Statements of Core Program, Pre- and Post-Processors:
This program contains approximately 1,250 source statements.

Documentation: Informal users guide documentation is available.

Program Availability: For further information contact:
Dr. John G. Roukis
Grumman Aerospace Corporation
Mail Stop B22/35
Bethpage, New York 11714

FLUX2

Descriptive Program Title: Grumman Orbital Heat Flux Program

Program Developer: Grumman Aerospace Corporation, Bethpage, New York 11714

Date of First Release and Most Recent Update: 1965 and 1980

General Information:
This program was developed by Grumman to evaluate the orbital environments of the Orbiting Astronomical Observatory and the Lunar Module. Its verification was based on extensive correlation with flight test data from these programs.

Program Capability and Scope of Analysis:
This program calculates solar, planetary albedo and planetary emission fluxes for up to one hundred surfaces of any orbiting vehicle. Blockage of the environment by intervening surfaces can be accomplished by tabular input. Six different vehicle orientation modes are available.

Solution Methods:
Computation of albedo and IR fluxes can be accomplished with options of either numerical integration or an approximate technique for determining the form factors between the planet and the vehicle.

Notable Items and Limitations:
Blockage effects for albedo and IR fluxes are only available if the numerical integration option is selected.

Programming Language: FORTRAN IV

Hardware/Operating System: IBM 370 (OS/VS)

Program Size (Heat Transfer Modules Only) - Number of Source Statements of Core Program, Pre- and Post-Processors:
This program contains 770 source statements.

Documentation: Informal users guide documentation is available.

Program Availability: For further information contact:
Dr. John G. Roukis
Grumman Aerospace Corporation
Mail Stop B22/35
Bethpage, New York 11714

HEATING5

Descriptive Program Title: Finite Difference Heat Conduction Program

Program Developer: W. D. Turner, D. C. Elrod, I. I. Siman-Tov, Union Carbide Corporation, Nuclear Division, Oak Ridge, Tennessee 37830

Date of First Release and Most Recent Update: March 1977 and July 1979

General Information:
HEATING5 was written over a period of years by personnel at Union Carbide Corporation, Nuclear Division, Oak Ridge, Tennessee, and was funded by various departments of the Department of Defense and the U.S. Nuclear Regulatory Commission.

Program Capability and Scope of Analysis:
HEATING5 is a general purpose heat conduction code designed to solve steady-state and/or transient heat conduction problems in one-, two- or three-dimensional Cartesian or cylindrical coordinates or one-dimensional spherical coordinates.

User Interface and Modeling Capabilities:
The user defines the problem by a series of regions having common characteristics. HEATING5 generates the nodal configuration from this information. Parameters may be defined by built-in functions or by user-supplied functions.

Solution Methods:
Steady-state problems are solved by SOR with Aitken's extrapolation. Transient problems may be solved by either implicit schemes ranging from Crank-Nicholson to fully implicit or by forward difference technique or Levy's extrapolation procedure.

Notable Items and Limitations:
Variable-dimensioned with respect to maximum number of nodes.

Programming Language: FORTRAN IV

Hardware/Operating System: IBM 360, IBM 370, IBM 3033, CDC 6600, CDC 7600.

Documentation: See Ref. 24

Program Availability:
    Radiation Shielding Information Center
    Oak Ridge National Laboratory
    Oak Ridge, Tennessee 37830

or:

    National Energy Software Center
    Argonne National Laboratory
    9700 South Cass Avenue
    Argonne, Illinois 60439

HEATING6

Descriptive Program Title: Finite Difference Heat Conduction Program

Program Developer: W. D. Turner, D. C. Elrod, G. E. Giles, Union Carbide Corporation, Nuclear Division, Oak Ridge, Tennessee 37830

Date of First Release and Most Recent Update: October 1981

General Information:
HEATING6, an extensive revision of HEATING5, was written by personnel at Union Carbide Corporation, Nuclear Division, Oak Ridge, Tennessee, and was primarily funded by the Transportation Branch, Nuclear Materials Safety and Safeguards, U.S. Nuclear Regulatory Commission.

Program Capability and Scope of Analysis:
HEATING6 is a general purpose heat conduction code designed to solve steady-state and/or transient heat conduction problems in one-, two-, or three-dimensional Cartesian or cylindrical coordinates or one-dimensional spherical coordinates.

User Interface and Modeling Capabilities:
The user defines the problem by a series of regions having common characteristics. HEATING6 generates the nodal configuration from this information. Parameters may be defined by built-in functions or by user-supplied functions.

Solution Methods:
Steady state problems are solved by direct solution techniques or by SOR with Aitken's extrapolation. Transient problems may be solved by either implicit schemes ranging from Crank-Nicholson to fully implicit or by forward difference technique or Levy's extrapolation procedure.

Notable Items and Limitations:
All arrays whose length is a function of the input parameters are variable-dimensioned. Extensive error checking facilities are incorporated into the code.

Programming Language: FORTRAN IV

Hardware/Operating System: IBM 360, IBM 370, IBM 3033

Program Size (Heat Transfer Modules Only) - Number of Source Statements of Core Program, Pre- and Post-Processors:
Core Program 15,000

Documentation: See Ref. 25

Program Availability:
    Radiation Shielding Information Center
    Oak Ridge National Laboratory
    Oak Ridge, Tennessee 37830

HEATRAN

Descriptive Program Title: A Two-Dimensional Finite Element Program for HEAT TRANsfer Analysis

Program Developer: Dr. W. D. Collier, United Kingdom Atomic Energy Authority (UKAEA), Risley Nuclear Power Development Establishment, Warrington, Cheshire WA3 6AT, England.

Date of First Release and Most Recent Update: 1969 and 1979

General Information:
HEATRAN was developed to meet the need of users in the UKAEA. The intention was to provide a simple but concise and natural means of inputting the data and to provide a wide range of boundary conditions. It runs about 4,000 jobs a year for UKAEA and other companies associated with the nuclear industry, but work is being phased over gradually to TAU.

Program Capability:
HEATRAN deals with conduction in materials of varying composition with material data varying with position, time or temperature. Radiation may be specified between arbitrary surfaces. Boundary conditions include natural and forced convection, convection to a fluid whose temperature is to be found, radiation to ambient, fixed temperatures, fixed flux. Calculations are performed in one- and two-dimensional (slab and axisymmetric) for steady state and transient situations.

User Interface and Modeling Capabilities:
There is a simple and natural system for inputting nodes and connections. Various shorthands are provided. Mesh and result plots are provided and a tabular output of temperatures is available. Temperatures may be saved for transfer to a stress analysis program.

Element Library: Linear triangular elements only.

Solution Methods:
- A sparse equation solver using a variant of Gauss elimination is used. Many solutions stay in core using work space provided by the program at run time. Data is moved out of core automatically in an efficient systematic manner if in-core storage is insufficient.
- Nonlinear cases use a Newton-Raphson iterative scheme.
- Transients use a fully implicit method (backward difference) with time steps adjusted automatically for accuracy.

Programming Language: FORTRAN IV

Hardware/Operating System: IBM 370, IBM 3033 (OS/VS); ICL 2900 (VME/B)

Program Size (Heat Transfer Modules Only) - Number of Source Statements of Core Program, Pre- and Post-Processors:
9,500 lines

Documentation: See Ref. 26

Program Availability: Available for use under contract on the Risley, Harwell and Winfrith computers of the UKAEA.

Source (1976 vintage) available from:
   Nuclear Energy Agency - Computer Program Library
   Gif-sur-Yvette
   Paris, France

MARC

Descriptive Program Title: General Purpose Finite Element Code

Program Developer: MARC Analysis Research Corporation, 260 Sheridan Avenue, Suite 200, Palo Alto, California 94306.

Date of First Release and Most Recent Update: 1970; Version J.2 - September 1981

General Information:
MARC is a proprietary code supported by MARC Analysis Research Corporation with offices in Palo Alto, California, Tokyo, Japan, and The Hague, Holland. The current state-of-the-art of finite element technology is adapted and incorporated into the program. New releases of the program are generated at the rate of about one per year.

Program Capability and Scope of Analysis:
MARC is designed for the linear and nonlinear analysis of structures in the transient heat transfer, static and dynamic regimes. Anisotropic heat conduction, and latent heat effects are included. The heat transfer model uses element types of which a structural analog exists, making a decoupled thermo-mechanical analysis using the identical mesh possible. All properties may be thermally dependent. Numerous user interfaces for specification of user selected parameters make the program extremely flexible. Properties may be specified as a function of other state variables through user subroutines.

User Interface and Modeling Capabilities:
An interactive pre- and post-processor, MENTAT, assists in the two- and three-dimensional mesh generation and other data preparation areas. Post-processing includes displaced geometries and contours of element quantities. MENTAT interfaces to MARC, NASTRAN and other FEM programs. MARC also contains complete pre- and post-processing capabilities.

Solution Methods:
Backward difference (modified Crank-Nicholson).

Notable Items and Limitations:
Most of the different options can be used simultaneously to cover an extremely wide range of nonlinear applications: tying degrees of freedom, joining shell and solid elements, fine and coarse mesh, user supplied constraints, friction and gap element.

Programming Language: FORTRAN IV

Hardware/Operating System: CDC 6600, 7600, CYBER 175, 176, AMDAHL, IBM, UNIVAC, CRAY, HITAC, ACOS, FUJITSU, MAGNUSON, PRIME, VAX.

Program Size (Heat Transfer Modules Only):
40,000 core programs, 20,000 pre-processors; 20,000 post-processors.

Documentation:
    MARC User Information Manual, Version J.2
      Volume A - User Information Manual
      Volume B - MARC Element Library
      Volume C - Program Input Manual
      Volume D - User Subroutine and System Description
      Volume E - Demonstration Problems
      Volume F - Structural Analysis with MARC Course Notes
      Volume G - MARC Background Papers
    Published by MARC Analysis Research Corporation
For background material see Refs. 27 to 29.

Program Availability: MARC is available internationally at the following data centers: CDC Cybernet, McAuto, Information Systems Design, Boeing Computer Services, Westinghouse, and Babcock and Wilcox. The program is available from MARC Analysis Research Corporation on a lease basis in either a binary or a binary and source form.

MITAS II

Descriptive Program Title: Martin Marietta Interactive Thermal Analysis System, Version 2.0 (MITAS II), Lumped Parameter Finite Difference Thermal Analysis Program

Program Developer: R. E. Kannady, Jr., R. J. Connor, C. E. Shirley, Martin Marietta Corporation, P.O. Box 179, Denver, Colorado 80201

Date of First Release and Most Recent Update: 1969 and 1981

General Information:
MITAS II is a computer code developed and maintained by Martin Marietta Corporation, Aerospace Division. Development started in 1969 and was an offshot from the CINDA-3G code developed by Chrysler-Aerospace for NASA.

Program Capability and Scope of Analysis:
MITAS II is a general purpose program system that provides a solution to a lumped parameter representation of the diffusion equation. The boundary conditions can be conduction, convection, radiation or some user-defined heat flow function. There is a large library of subroutines and functions that the user can employ in setting up boundary functions or determining values during the course of problem solution.

User Interface and Modeling Capabilities:
- The problem is defined in terms of nodes which represent the lumped mass of the system. The specific heat of these nodes can be allowed to change as a function of temperature or some user-defined function. Nodes with a very small thermal mass can be considered to have a zero mass. Boundary nodes' temperatures are specified by the user and can be altered during the course of the solution.
- Heat flow paths (conductors) can be linear for conduction and convection, non-linear for radiation heat transfer or defined by the user to simulate some process. Conductors can be generated by using short hand input statements and can be constants, temperature varying or varied by some user definition.
- During the solution process the user has access to the temperature solution vector and the matrix of conductors. The contents can be tested and altered if the user desires.

Solution Methods:
The program provides for a steady state temperature solution and transient solutions. The transient solutions available are explicit (forward differencing) and implicit (backward differencing and forward-backward differencing).

Notable Items and Limitations:
Due to a word packing technique and code which adjusts to the size of the problem, MITAS II has very few practical limitations as to the size of the problem to be solved. The structure of MITAS II allows the user to code in logic which will be executed prior to and after the temperature solution is obtained. This feature allows the user to describe conditions or printouts as he desires.

Programming Language: CDC FORTRAN 2.4

Hardware/Operating System: CDC 6000 series, CDC CYBER series/Operating systems NOS, NOS/BE1, SCOPE3

Program Size: Preprocessor = 50,000 statements
Library = 100,000 statements

Documentation: See Ref. 30

Program Availability: MITAS II is available through the Cyber Service network executing on a CYBER 750 computer. For further information as to this service, contact Martin Marietta Corporation, Data Systems, 104 Inverness Circle East, Suite 310, Englewood, Colorado 80112, Alan R. Cheuvront or John W. Davis (303) 740-3012.

The MITAS II source code is available through Martin Marietta Denver Aerospace, P.O. Box 179, Denver, Colorado 80201; Roy E. Kannady, Jr., (303) 977-3075.

MSC/NASTRAN

Descriptive Program Title: The MacNeal-Schwendler Corporation, "NASA STRuctural ANalysis"

Program Developer: The MacNeal-Schwendler Corporation, 7442 North Figueroa Street, Los Angeles, California 90041.

Date of First Release and Most Recent Update:
The initial release of the NASA funded program NASTRAN occurred in 1969 (NASTRAN is a registered trademark of NASA). The MacNeal-Schwendler Corporation has marketed and serviced an advanced version of that code since 1972. MSC releases a new, updated version of MSC/NASTRAN twice yearly. The latest version was released in March 1981.

General Information:
MSC/NASTRAN is a large-scale, general purpose computer program which solves a wide variety of engineering problems by the finite element method. The program is an advanced, proprietary version of the NASA-funded structural analysis program NASTRAN. MSC/NASTRAN is marketed and serviced from MSC's offices in the United States, Europe and Japan, and is available at most major public data processing centers. Customer hotline service is available to users in need of assistance. The engineering community is also aided by a wide variety of MSC/NASTRAN instructional courses.

Program Capability:
- The program's capabilities include static and dynamic structural analysis, both geometric and material nonlinear static analyses, thermal analysis, acoustics, aeroelasticity, electromagnetism and other types of field problems. Many substructuring options are also available with the above capabilities.
- MSC/NASTRAN's thermal analysis capabilities include linear steady state, nonlinear steady state, and transient heat transfer. As outlined in the tables, these basic capabilities are available in scalar, one, two and three dimensions with various material properties and various boundary, loading and initial conditions. These basic capabilities can be further enhanced through many options available to the analyst. Some options not included in the tables are, for example:
- The analyst may input his own elemental matrices, thereby defining his own elements (these matrices may be symmetric or unsymmetric);
- He may input transfer functions for use with active thermal mechanisms such as heat pipes and thermostatic controls;
- He may use matrix order reduction methods such as Guyan reduction;
- He may use the thermal analysis capabilities to solve analogous electrostatic problems.

User Interface and Modeling Capabilities:
MSC/NASTRAN contains many special features to support and enhance its "user-friendliness" and modeling capabilities. These features include:
- Pre- and Post-Processors:

MSC/NASTRAN's internal pre-processor MSGMESH automatically generates finite element models from analyst supplied descriptions of one, two and three dimensional regions. Since MSGMESH is an integral part of MSC/NASTRAN, finite element model generation and solution may be accomplished in a single execution. Post-processing may be performed by either of two post-processors. MSGVIEW enables the user to graphically display undeformed structural models in both batch and interactive modes. MSC's newest post-processor, MSC/GRASP (scheduled for release later this year) provides extensive interactive post-processing capabilities including: many model viewing and manipulation functions, deformed and modal plots, element stress contours, output scanning, x-y plots and keyframe animation.
- Data Checking and Error Analysis:

Special aids help to detect errors in such input data as geometry, element connections, elastic properties, mass properties, constraints, loads and temperatures. These aids include editing of input card formats, verification that all required cards are present, and verification that data for specified elements are geometrically compatible. In addition, formatted tables present summaries of grid point geometry, coordinate systems, constrained degrees of freedom, and element connections. This information can be assessed prior to initiating the problem's matrix operations.
- Matrix Operations:
  - Automatic Resequencing - The user can request automatic internal resequencing of grid points in order to minimize the time for equation solution. The sequencing processor also provides time estimates for the generation and decomposition of the stiffness matrix.
  - Automatic Singularity Suppression - At user option, singular or nearly singular degrees of freedom will be suppressed. Output data are supplied which indicate the degrees of freedom which have been removed.
  - Sparse Matrix Routines - All of MSC/NASTRAN's matrix routines have been designed for the efficient solution of very large problems. Detailed patterns of nonzero terms are recognized and processed in condensed form. Efficient automatic spill logic is provided for matrices which are too large to be kept in high speed memory. The availability of several methods of matrix multiplication provides optimum efficiency for wide ranges of matrix size and density.

- External Post-Processing:

The program contains a number of methods by which the analyst may output intermediate or final results (on punched cards or FORTRAN readable files) for easy input to external post-processing.

Solution Methods:

- MSC/NASTRAN's solution methods are directed toward large-order complex structural models. The linear static solution methods provide for multiple loading conditions and boundary conditions in a single run. Superelement/substructure and modal methods are also available to reduce the size of the final solution matrices for thermal as well as structural analysis.

- Heat transfer solutions are performed by straightforward matrix methods similar to the structural analysis methods. For the nonlinear statics solution, a modified Newton/Raphson iteration is performed, starting from a user-selected initial estimate of temperatures. Conduction, convection, and surface coefficients may be functions of temperature.

- Both linear and nonlinear transient solutions are performed using a user-specified parameter for controlling the time steps and forward-backward difference ratio. The default provides an implicit integration with an optimum balance between stability and drift errors.

Notable Items and Limitations:

- Large Problem Capabilities:

MSC/NASTRAN employs many features that make possible the solution of large problems in an accurate and efficient manner. These features include: multilevel superelement substructuring, cyclic symmetry substructuring, and matrix reduction methods.

- Problem Oriented Language:

MSC/NASTRAN's DMAP (Direct Matrix Abstraction Program) allows the user to specify his own series of matrix operations in order to perform a specific type of problem solution that is not contained in a solution sequence supplied with the program.

Programming Language: FORTRAN IV, with isolated machine dependent assembly language routines.

Hardware/Operating System: MSC/NASTRAN is available at most major public data centers and is currently operational on more than 200 computers, including the IBM 360/370 series, the AMDAHL series, the ITEL AS, the Fujitsu M series, the CDC 7600, and the CDC CYBER series, the UNIVAC 1100 series, Digital's VAX 11/700 series and the CRAY-1.

Program Size (Heat Transfer Modules Only) - Number of Source Statements of Core Program, Pre- and Post-Processors:
The delivered system contains approximately 430,000 source statements.

Documentation: The following manuals are available from MSC:
    NASTRAN Theoretical Manual
    MSC/NASTRAN User's Manual (2 volumes)
    MSC/NASTRAN Programmer's Manual (3 volumes)
    MSC/NASTRAN Application Manual (2 volumes)
    MSC/NASTRAN Demonstration Problem Manual
    MSC/NASTRAN Aeroelastic Supplement
    MSGMESH Analyst's Guide
    MSC/NASTRAN Handbook for Linear Static Analysis
    MSC/NASTRAN Primer, Statics and Normal Modes Analysis
    MSC/NASTRAN Handbook for Superelement Analysis
(See Refs. 31 to 34).

In addition, NASTRAN user's conferences have been sponsored by MSC and NASA. Many papers have been published in these proceedings.

A manual describing MSC/NASTRAN's thermal capabilities, with how-to descriptions and many example problems is in preparation.

Program Availability:
An executable version of the program is available under lease agreement from MSC through any one of our regional sales offices in the United States, Europe and Japan. Prices vary with computer type, therefore, interested persons should contact one of the sales offices listed below:

The MacNeal-Schwendler Corporation
7442 North Figueroa Street
Los Angeles, CA 90041 U.S.A.
Tel. (213) 254-3456
TWX: (910) 321-2492 MACN SCHW LSA

MSC Southwest Regional Office
P.O. Box 1606
Grapevine, TX 76051 U.S.A.
Tel. (817) 481-4812

MSC Midwest Regional Office
5745 Oxford Drive
New Berlin, WI 53151 U.S.A.
Tel. (414) 542-5747

MSC Eastern Regional Office
P.O. Box 504
Oakdale, NY 11769 U.S.A.
Tel. (516) 589-8316

MacNeal-Schwendler GmbH
8000 Munchen 80
Prinzregentenstrasse 78
West Germany
Tel. (089) 47 02 068
Telex: (41) 523784 MSG D

MacNeal-Schwendler Representative Office
Kyodo Building (Kodenma-cho)
16-8 Kodenma-cho, Nihonbashi
Chuo-ku, Tokyo 103
Japan
Tel. (03) 661-0133
Telex: (781) J23363 MSGWATA

Descriptive Program Title: Nodal Network Thermal Balance Program

Program Developer: J. T. Skladany, NASA-Goddard Space Flight Center, Code 732, Beltsville, Maryland 20771

Date of First Release: 1967

General Information:
Performs thermal analysis of lumped parameter networks consisting of both radiation and conduction. Nodal performs a thermal analysis of a spacecraft during all stages from launch to orbital dynamic steady state. The program is designed to solve integro-differential equations taken from a thermal network composed of radiatively and conductively coupled modes. A finite difference form of the equation is used but they are solved using a matrix inversion technique known as Gauss elimination. Nodal is designed to handle both steady state and transient problems.

User Interface and Modeling Capabilities:
The program contains no automatic modeling capabilities.

Solution Methods:
A finite difference form of the thermal equations is used but is solved implicitly by the Gauss elimination method.

Notable Items and Limitations:
The program was written to handle thermal analysis of spacecraft. Hence, great emphasis was placed on radiation between nodes including temperature varying radiation. The program can handle up to 300 nodes.

Programming Language: FORTRAN IV

Hardware/Operating System: OS-360, VAX 11/780

Program Size (Heat Transfer Modules Only) - Number of Source Statements of Core Program, Pre- and Post-Processors:
Total program includes 2,000 statements.

Documentation: See Ref. 35

Program Availability:
   Computer Library
   Goddard Space Flight Center
   Beltsville, Maryland 20771

   Computer Software Management
    and Information Center (COSMIC)
   Computing and Information Services
   Suite 112, Barrow Hall
   The University of Georgia
   Athens, Georgia 30602

See TEMP

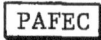

Descriptive Program Title: Program for Automatic Finite Element Calculations

Program Developer: PAFEC, Ltd., Strelley Hall, Strelley, Nottingham NG8 6PE, England.

Date of First Release and Most Recent Update: 1970 and November 1981

General Information:
PAFEC is an independent company which grew from a research group active in finite element methods during the 1960's. First copies of the system were sold while members of the group were still working at the University of Nottingham, but, subsequently, the extent of commercial involvement became too great and PAFEC, Ltd. was formed. Since then, PAFEC has continued to increase its share of the finite element market and the system is now supported in the United States, Europe, Australia and South Africa.

Program Capability and Scope of Analysis:
- Many advanced facilities are available with PAFEC in the fields of statics and dynamics as well as heat transfer. The basic system includes linear and nonlinear elastic stress analysis, large displacements, modes and frequencies and dynamic response, creep and plasticity.
- Problems may be solved involving steady state and transient heat transfer with conduction and convection. Nodal temperatures and heat flux are user definable with time as are all material and boundary layer properties. Temperature dependent properties are available in both isotropic and anisotropic elements. Stress analyses can be performed directly from the temperatures calculated in a thermal analysis or in combinations with thermal and static loads applied from another data file. PAFEC's development of the Boundary Integral Method for heat transfer is currently under test and is expected to be released in the very near future.

User Interface and Modeling Capabilities:
PAFEC's greatest selling feature is its extreme versatility and ease of use. Data preparation is the simplest yet devised for any major finite element system. PAFEC employs free format input with engineering key words and modular layout. The use of system defaults, abbreviated headings, constant properties and data generation reduces to a minimum the amount of user effort required for data preparation. Data may also be prepared interactively using the APES (nongraphics) package or the PAFEC Interactive Graphics System (PIGS). PIGS allows the user to generate many elements and shapes automatically and then replicate these shapes to form other parts of the structure. Nodes and elements may be added and removed with ease. Post-processing options allow temperature contours and many graph options including temperature, stress and displacement. Sophisticated hard copy plots of all of the above options are available without PIGS.

Element Library:
PAFEC has over eighty element types including four two-dimensional and two three-dimensional temperature distribution elements for isotropic applications and duplicates of these for anisotropic calculations. In addition to these there are four two- and three-dimensional boundary layer elements.

Solution Methods:
Three system solution methods are available: front solution, blocked front solution, and partitioned banded solution. Each has some advantages. Steady state temperature problems are solved directly while incremental methods are used for nonlinear steady state. Transient analyses use a modified Crank-Nicholson technique.

Additional Information:
There is no limitation on problem size other than those imposed by machine capacity. PAFEC always supplies source code.

Programming Language: A subset of ANSI FORTRAN.

Hardware/Operating System:
PAFEC is currently running on the following computers: DEC (10, 20 and VAX), IBM, UNIVAC, CRAY, Burroughs, GEC, SEL, PRIME, CDC, Perkin-Elmer, Honeywell, ICL (1900 and 2900), Sigma and Itel.

Documentation: See Refs. 36 to 38

Program Availability:

PAFEC Engineering Consultants, Inc.
5401 Kingston Pike
Knoxville, TN 37919
Tel. (615) 584-2117

PAFEC, Ltd.
Strelley Hall, Strelley
Nottingham NG8 6PE
England
Tel. 0602-292291

PAFEC, Ltd.
3 Marsh Street
Bristol BS1 1SS
England
Tel. 0272-213914

Matrix Computing Services Pty, Ltd.
64 Randhill Building
106 Jan Smuts Avenue
Randburg 2194
South Africa
Tel. (011) 481084-5

Megadata, Pty, Ltd.
24 Falcon Street
Crows Nest
New South Wales 2065
Australia
Tel. 438-1233

Value Engineering (WA), Pty, Ltd.
32 Kings Park Road
West Perth
Western Australia 6005
Tel. 322-2211

Cost: The current (1981) prices of the FORTRAN source and installation are: PAFEC - $41,900; Interactive Graphics - $9,600; Substructures - $7,200. Six-month trial installations are currently $7,500.

SAHARA

Descriptive Program Title: General Purpose Conduction, Convection and Radiation Heat Transfer Code Used at Sandia Laboratories, Livermore, California.

Program Developer: V. K. Gabrielson, Organization 8331, Sandia National Laboratories, Livermore, California 94550.

Date of First Release and Most Recent Update: 1972 and 1981

General Information:
The SAHARA set of codes contains a very general finite difference pre-processor and several interactive post-processor codes. The SAHARA codes have developed over the past fifteen years for adaptations to Sandia Laboratories' special needs and computer hardware. The work is funded by the United States government.

Program Capability:
SAHARA is a general purpose conduction, convection and radiation heat transfer code applicable to nonlinear steady-state and transient analysis. Model sizes can be easily extended to capacity of each computer. A variety of boundary and initial conditions are available with special emphasis on radiation heat transport. Boundary and fluid elements may be defined within the model.

User Interface and Modeling Capabilities:
SAHARA uses the HEATMESH pre-processor for developing the finite difference mesh of two-dimensional axisymmetric solids with some capability for three-dimensional modeling.

Solution Methods:
Successive overrelaxation iteration used for most transient problems. An adaptive conjugate gradient method used for steady-state problems.

Notable Items:
- Input for both HEATMESH and SAHARA use mnemonic keywords for all data sets.
- Several interactive post-processors are available providing time history and mesh plots.

Programming Language: CRAY-CFT and CDC-FTN FORTRAN

Hardware/Operating System: CRAY-COS and CDC-NOS-BE systems

Program Size (Heat Transfer Modules Only) - Number of Source Statements of Core Program, Pre- and Post-Processors:
Pre-processors: 2,000 statements
SAHARA: 4,500 statements
Post-processors: 2,000 statements

Documentation: See Refs. 39 and 40

Program Availability: For CRAY or CDC versions - contact author. Little or no experience on machines other than CRAY and CDC.

SAMCEF (THERNL)

Descriptive Program Title: Systeme d'Analyse des Milieux Continus par Elements Finis (Thermique Non Lineaire)

Program Developer: L.T.A.S., Aerospace Laboratory, University of Liege, Rue Ernest Solvay 21, B-4000 Liege, Belgium

Date of First Release and Most Recent Update: Both 1981

General Information:
SAMCEF is a university code. The package was first developed as a research tool in the structural mechanics area. Progressively, further developments were financed by industries and since 1970 a general purpose code is operational. It is being used in an industrial as well as research environment. The LTAS Group provides the users with the service they require for using the program, which is available for various computers. The price includes the source code, installation, documentation and training. A yearly contribution allows the users to obtain updated versions.

Program Capability and Scope of Analysis:
THERNL is a member of SAMCEF Systems dedicated to linear and nonlinear, steady-state and transient heat transfer analysis. The finite element library covers one-dimensional, two-dimensional, and three-dimensional geometries including shells and transition-to-volume elements. Boundary elements for forced convection and radiation are included. Specialized elements for forced convection in ducts are available. Thermal conductivity of the material can be anisotropic and all thermal characteristics are temperature dependent. Step-by-step analysis is allowed: data preparation - element generation, steady-state and transient responses and restarts.

User Interface and Scope of Analysis:
THERNL allows to use the general pre- and post-processors of SAMCEF, including graphic displays of the data and results.

Solution Methods:
The nonlinear algebraic systems are solved via a secant method performed block-by-block.

Notable Items and Limitations:
- Explicit mesh deformation for modeling steep gradients and implicit mesh deformation for phase change problems are included for one-dimensional situations.
- User supplied elements are easily introduced. The front width is limited to one thousand degrees of freedom.

Programming Language: Subset of FORTRAN IV

Hardware/Operating System: IBM (OS, DOS, VS, CMS) - UNIVAC (double precision), CDC (6400, 6600, 7600), DEC 2040, VAX, SIEMENS.

Program Size (Heat Transfer Modules Only) - Number of Source Statements of Core Program, Pre- and Post-Processing:
40,000 source statements

Documentation: See Refs. 41 to 45

Program Availability and Cost: Contact developer. Cost negotiable depending on source availability, maintenance, assistance, etc. Special rate for universities.

```
SESAM-69
NV615
```

Descriptive Program Title: Analysis of Heat Conduction in Three-Dimensional Solids, Stationary and Transient

Program Developer: A. S. Computas, Data Division of Det Norske Veritas, Veritasveien 1, P.O. Box 310, 1322 Høvik, Norway.

Date of First Release and Most Recent Update: 1976 and 1981

General Information:
The general heat conduction program NV615 is part of SESAM-69 (see Ref. 46) which is a proprietary code developed and maintained by A. S. Computas which is a subsidiary and the Data Division of the Norwegian classification society Det Norske Veritas. Development of NV615 started in 1974 and was released for external use in 1976. The program is supported from the headquarters in Oslo, Norway, and from four European branch offices (London, Paris, Rotterdam and Hamburg).

Program Capability and Scope of Analysis:
NV615 is a general purpose batch program for calculation of stationary- and transient temperature distributions in three-dimensional solids. Both linear and nonlinear heat conduction problems may be analyzed. Isotropic material properties are assumed, but the properties may be temperature dependent. NV615 is based on the finite element method. The multilevel superelement technique (see Ref. 47) may be used.

Considerable advantages are obtained with this technique both for linear and nonlinear problems. The program has complete saving and restart facilities.
- Boundary Conditions/Thermal Loads:
    1) Prescribed zero or nonzero temperature
    2) Given heat flow
    3) Given heat flux
    4) Given ambient temperature (convective heat transfer)

Given heat flux, heat flow and ambient temperature may be time-dependent in transient analyses.
- The convective heat transfer coefficients used in connection with ambient temperatures may be temperature and/or time-dependent. In stationary analyses the heat transfer coefficients are, of course, limited to temperature dependency. The initial temperature of the structure may be at zero temperature or a nonzero value varying throughout the region.

User Interface and Modeling Capabilities:
NV615 utilizes a highly efficient data generator (see Ref. 46) both for the geometric modeling and specification of thermal loads, boundary conditions, etc. The input is fixed format batch input specifications. The input data generator has extensive checking and visualization facilities (plots).
- Element Library:
    1) Isoparametric hexahedral solid element, eight nodes
    2) Isoparametric hexahedral solid element, twenty nodes

In connection with eight node basic elements, diagonal or consistent capacitance matrix may be requested. Twenty node basic elements always require the use of consistent capacitance matrix.
- A separate post-processor (NV340) (see Ref. 48) performs print and plot of temperatures in the form of tables, isoplots, etc.

Solution Methods:
- In the case of stationary heat conduction the system equation is solved by Choleski factorization accompanied by forward and backward substitution (see Ref. 49).
- For the solution of transient heat conduction problems the time integration is carried out by means of the trapezoidal formula, i.e., a step-by-step time integration (see Ref. 49).

Notable Items and Limitations:
- Adding some supplementary input data, the output (i.e., nodal temperatures) may be used as input for a corresponding thermoelastic static stress analysis using the same finite element mesh. The communication is via magnetic tape (see Refs. 50, 51).
- Due to extensive use of superelements, NV615 gives very few limitations with respect to problem size (total number of elements, nodes, etc.). In direct analysis, i.e., not using the superelement technique, the limits are: 1,000 nodes, 175 elements, 400 boundary condition nodes.
- When the superelement technique is employed the above limitations relate to each first level superelement. For higher levels the following limitations have to be observed: 5,000 nodes, 100 elements (superelements), 10 levels.

Programming Language: Simplified ANSI 1966 FORTRAN

Hardware/Operating System: UNIVAC 1100 Series/Exec 8

Program Size (Heat Transfer Modules Only) - Number of Source Statements of Core Program, Pre- and Post-Processors:
Core program abt. 80,000; post-processor abt. 30,000.

Documentation: See Refs. 46 to 52

Program Availability:
- The program is available at several European service Bureaus and computer installations. For further information the developer should be contacted. Normally absolute versions only are distributed. However, program sales will also include source code. Program fee to be negotiated with the developer.
- Contact A. S. Computas, Section for Technical Software Series, P.O. Box 310, 1322 Høvik, Norway.

SINDA

Descriptive Program Title: Systems Improved Numerical Differencing Analyzer

Program Developer: Chrysler Corporation, Space Division, New Orleans, Louisiana; TRW Systems, Redondo Beach, California; TRW Systems, Houston, Texas; LTV Aerospace Corporation, Dallas, Texas; Lockheed, Houston, Texas.

Date of First Release and Most Recent Update: 1967 and 1975

General Information:
- SINDA has been in use at NASA and throughout industry since 1967. Several modifications have been made to the original code to adapt it to various computers. Pre- and post-processor codes have been adapted to run with SINDA.
- The SINDA system consists of two main pieces: (1) the preprocessor and (2) the library. The SINDA preprocessor is a program which accepts problems written in the SINDA language and converts them to the FORTRAN language. The preprocessor also accepts 'program-like' logic statements and subroutine calls (requesting some particular routine from the library) as data, which permits the user to tailor the program to suit his particular problem. The SINDA library consists of many pre-written FORTRAN subroutines which perform a large variety of commonly needed actions and which reduce the programming effort which might have been required to solve a given problem. These routines are fully compatible with the FORTRAN routines produced by the preprocessor. It should be recognized that the use of a preprocessor provides a system with a large capability and considerable flexibility, but because of the numerous options that are generally offered, user instructions are more difficult than other thermal analyzer-type programs which have less flexibility.

Program Capability and Scope of Analysis:
- SINDA, the Systems Improved Numerical Differencing Analyzer, is a software system which possesses capabilities which make it well suited for solving lumped parameter representation of physical problems governed by diffusion-type equations, such as Fourier, Poissons, or Laplace equations. The system was originally designed as a general thermal analyzer accepting resistor-capacitor (R-C) network representations of thermal systems, although, with due attention to units and thermally oriented peculiarities, SINDA will accept R-C networks representing other types of systems (e.g., electrical networks).
- As a thermal analyzer, SINDA can handle such interrelated complex phenomena as sublimation, diffuse radiation within enclosures, transport delay effects, sensitivity analysis, and thermal network error correction methods. The thermal analysis is performed on thermal analog modes presented in network format. The network represents a one-to-one correspondence with both the physical and mathematical models. SINDA has been used in the analysis of networks containing about 2,000 nodes without

requiring unreasonable amounts of computer time. The thermal network can be coupled to an iterative solution of a lumped parameter fluid network. Nonlinear material properties and boundary conditions may be calculated simultaneously as a function of one or more independent variables.

• The general fluid flow solution capabilities include extensive valve characterization and ability to match pump curves and system pressure-flow characteristics. The valves have been formulated so that either cooling or heating situations may be controlled with any of the valve types. Pump options included are pressure rise as a tabulated function of system flow rate and pressure rise as a polynomial function of flow rate. Special subroutines are included in the SINDA library to facilitate the thermal analysis of systems containing counter flow heat exchangers, parallel flow heat exchangers, cross flow heat exchangers, condensing heat exchangers, and any heat exchanger with an input effectiveness. The Flow-Hybrid method is incorporated for calculating fluid temperatures, with improved calculation accuracy obtained by using fluid enthalpy rather than specific heat for the convective term of the fluid temperature equation. To facilitate the speedy analysis of a general flow problem, provisions have been made for the user to divide the flow system network into sub-network elements.

User Interface and Modeling Capabilities:
Software using DISPLA produces temperature history plots through the use of a post-processor in batch or demand mode.

Solution Methods:
The use of SINDA is based on a lumped parameter representation of a physical system. Thus, SINDA solves numerically a set of ordinary (in general nonlinear) differential equations that represent the transient behavior of a lumped parameter system or a set of nonlinear algebraic equations representing steady state conditions. The numerical techniques used by SINDA are based on finite difference algorithms as opposed to finite element methods. For user decision flexibility, SINDA provides a number of implicit and explicit numerical solution methods for both steady-state and transient solutions.

Notable Items and Limitations:
The generality of the SINDA code is largely accomplished by being able to program the driver code for each particular problem. This flexibility often requires more tedious input to somewhat standard types of problems.

Programming Language: FORTRAN

Hardware/Operating System: CDC 7600, UNIVAC 1100, VAX, IBM, CRAY

Program Size (Heat Transfer Modules Only) - Number of Source Statements of Core Program, Pre- and Post-Processors:
There are approximately 30,000 cards for the SINDA pre-processor and library.

Documentation: See Refs. 53 to 55

Program Availability: Computer Software Management
and Information Center (COSMIC)
Computing and Information Services
Suite 112, Barrow Hall
The University of Georgia
Athens, Georgia 30602

SPAR

Descriptive Program Title: SPAR Thermal Analysis Processors - SSTA, TRTA, TRTB and TRTG
- SSTA - Steady State Thermal Analyzer
- TRTA - Transient Thermal Analyzer, Explicit Method
- TRTB - Transient Thermal Analyzer, Implicit (Galerkin) Method
- TRTG - Transient Thermal Analyzer, Implicit (Gear) Method

Program Developer: Engineering Information Systems, Inc., 5120 West Campbell Avenue, Suite 240, San Jose, California 95130.

Date of First Release and Most Recent Update: Level 12 - 1977; Level 20 - 1981

General Information:
The SPAR thermal analyzer is a system of finite element processors that perform steady-state and transient thermal analyses. The processors communicate with each other through a random-access data base. As each processor is executed, all pertinent source data is automatically extracted from the data base, and computed results are stored in the data base. Each processor may be executed in interactive or batch mode.

Program Capability and Scope of Analysis:
- The steady-state processor SSTA performs linear and nonlinear analyses. The user may exercise complete control over the solution process through a variety of commands. The processor may be restarted, as required, from any initial temperature state.
- The transient processors TRTA, TRTB and TRTG perform linear and nonlinear transient analyses. Each processor may be started from any point in time using any stored temperature vector.
- The thermal element repertoire consists of a complete set of conduction, forced convection, fluid-surface convective-exchange, mass-transport, and radiation-exchange elements.
- All properties may be functions of temperature, pressure and time. Properties may be isotropic or anisotropic.
- Thermal loading may be of any combination of the following excitation types:
  - Time-dependent volumetric heat generation
  - Temperature-dependent volumetric heat generation
  - Time-dependent surface heat fluxes
  - Time-dependent convective-exchange temperatures
  - Fluid-surface convective-exchange
  - Time-dependent prescribed nodal temperatures
  - Radiation-exchange.

User Interface and Modeling Considerations:
The SPAR thermal analysis processors utilize data generated by the SPAR or EAL processors TAB (geometry), ELD (element connectivities), AUS (tables of properties and thermal excitation), and SEQ (minimization of the system K matrix rms bandwidth). Source data and results may be scanned and displayed using processor DCU.

Solution Methods:
Processors SSTA, TRTB and TRTG use a skyline method to solve systems of equations. A modified Newton method is used to perform nonlinear analyses. Processor TRTA uses an explicit method to compute solutions.

Notable Items and Limitations:
- An experimental element capability is provided for those users who wish to insert their own element formulation into any of the processors.
- An automated verification package is used to assure proper functioning of the thermal analysis processors each time new capabilities are added to the system or the system is installed on a new host machine.

Programming Language: FORTRAN V

Hardware/Operating System: UNIVAC 1100, CDC CYBER 175 - NOS BE1.3

Program Size (Heat Transfer Modules Only) - Number of Source Statements of Core Program, Pre- and Post-Processors:
Approximately 18,000 statements (thermal analysis processors only).

Documentation: See Ref. 56

Program Availability: Contact: James C. Robinson
Loads and Aeroelasticity Division
Mail Stop 243
NASA Langley Research Center
Hampton, Virginia 23665

Engineering Information Systems, Inc.
5120 West Campbell Avenue, Suite 240
San Jose, California 95130

SSPTA

Descriptive Program Title: The Simplified Shuttle Payload Thermal Analyzer

Program Developer: Arthur D. Little, Inc., Acorn Park, Cambridge, Massachusetts 02140

Date of First Release and Most Recent Update: First release - November 1977. Second release - September 1979. Current update in process.

General Information:
SSPTA was developed by Arthur D. Little, Inc., under contract to NASA Goddard Space Flight Center, to simplify the computational procedures involved in defining the thermal design of shuttle payloads. It comprises a number of subroutines which have been in use for ten to fifteen years. The input to the program has been designed to simplify its use through the use of engineering terminology and stored Job Control Language (JCL) procedures. The program has the capability to easily store and keep a record of data for use with subsequent runs.

Program Capability and Scope of Analysis:
Transient or steady-state thermal analysis, in general, with shuttle payloads as a specific example. Programs can generate and automatically store view areas, orbital fluxes, thermal models, etc., so that in subsequent runs only new data has to be inputted or generated. The data file management system has the capability to handle up to fifty stored data sets with automatic backup. The program was designed primarily for transient, orbital, thermal design analysis of shuttle payloads.

User Interface and Modeling Capabilities:
A user language has been defined, based on simple thermal engineering terms, to input data and to define files for the storage of data. Geometric modeling of a cargo bay full of payloads is enhanced by an internally stored model of the shuttle cargo bay. Geometric models of individual payloads can also be stored in the program file storage system and placed within the geometric model of the cargo bay using coordinate transformations. Radiative view areas and orbital fluxes (with shadowing and multiple diffuse reflections) are then computed by the program using the unified geometric models.

Solution Methods:
Contour integration is used to compute black body view areas. The thermal analyzer uses the Newton-Raphson iterative procedure on each nonlinear energy balance equation and the modified Gauss-Seidel procedure for solving the set of energy balance equations.

Notable Items and Limitations:
SSPTA greatly simplifies the engineer-computer interface associated with developing a thermal design for a shuttle payload. This is done by a system of user oriented input data formats and data storage schemes. In order to keep the program as simple as possible, the analytical capabilities do not automatically handle variable properties.

Programming Language: FORTRAN

Hardware/Operating System: IBM System/370 Model 4341, VM370 System Product Release 1, TI ASC, VAX 11, VMS

Program Size (Heat Transfer Modules Only) - Number of Source Statements of Core Program, Pre- and Post-Processors:
Core Program: 5,000 statements

Documentation: See Refs. 57 and 58

Program Availability:

> Computer Software Management
> and Information Center (COSMIC)
> Computing and Information Services
> Suite 112, Barrow Hall
> The University of Georgia
> Athens, Georgia 30602
>
> Dr. David W. Almgren
> Arthur D. Little, Inc.
> Room 20-531
> Acorn Park
> Cambridge, Massachusetts 02140

TACO

Descriptive Program Title: A Finite Element Heat Transfer Code

Program Developer: W. E. Mason, Applied Mechanics Department, Sandia National Laboratories, Livermore, California 94550

Date of First Release and Most Recent Update: 1979 and 1980

General Information:
TACO is a two-dimensional implicit finite element code for heat transfer analysis. It can perform both linear and nonlinear analyses and can be used to solve either transient or steady state problems. Either plane or axisymmetric geometries can be analyzed.

Program Capability and Scope of Analysis:
- TACO has the capability to handle time or temperature dependent material properties and materials may be either isotropic or orthotropic. A variety of time and temperature dependent loadings and boundary conditions are available including temperature, flux, convection and radiation boundary conditions and internal heat generation.
- Additionally, TACO has some specialized features such as internal surface conditions (e.g., contact resistance), bulk nodes, enclosure radiation with view factor calculations, and chemical reactive kinetics. A user subprogram feature allows for any type of functional representation of any independent variable. A bandwidth and profile minimization option is also available in the code.

User Interface and Modeling Capabilities:
- TACO has some limited capability for generation of nodal, element and boundary data. However, it relies on separate mesh generation codes for complex models.
- Graphical representation of results in the form of time histories, isoplots, and profile plots are provided by a companion post-processor named POSTACO.
- Temperature states calculated by TACO are written to a file which can be read by mechanical codes for uncoupled thermal stress calculations.

Solution Methods:
- Time integration: Generalized trapezoidal method (Ref. 59) varying from forward explicit to backward implicit.
- Nonlinear solution scheme: Modified direct iteration and BFGS method.
- Equation solver for linearized equations: Compact out-of-core skyline.

Programming Language: ANSI FORTRAN

Hardware/Operating System: CDC 6600/NOS BE, CDC 7600/LTSS, CRAY 1-S/COS and CTSS

Program Size (Heat Transfer Modules Only) - Number of Source Statements of Core Program, Pre- and Post-Processors:
Core Program - 10,000; Post-Processor - 2,000

Documentation: See Refs. 59 to 61

Program Availability: Presently available from the developer at no cost.

TAC2D

Descriptive Program Title: Thermal Analysis Code - Two-Dimensional

Program Developer: General Atomic Company, P.O. Box 81608, San Diego, California 92138

Date of First Release and Most Recent Update: 1969 and 1976

General Information:
TAC2D is a code for calculating steady state and transient temperatures in two-dimensional problems by the finite difference method.

Program Capability:
Linear and nonlinear problems may be treated with TAC2D. Internal and external flowing coolants may be used, and there may be internal and external thermal radiation. Thermal expansion of materials may also be accounted for.

User Interface and Modeling Capabilities:
The configuration to be analyzed is described in the rectangular, cylindrical or circular (polar) coordinate system. The input of thermal properties is by FORTRAN statement functions. This permits flexibility as many of the calculation variables (time, local temperature, local position, etc.) are available for use in these functions. There is a wide selection of optional output.

Solution Methods:
Alternating direction implicit method for two-dimensional problems (Peaceman-Rachford).

Notable Items and Limitations:
The grid lines must be orthogonal, and the entire problem must be bounded by four grid lines in one of the coordinate systems. All radiation is one-dimensional. There is no provision for phase change.

Programming Language: FORTRAN V

Hardware/Operating System: UNIVAC 1110/Exec 8

Program Size (Heat Transfer Modules Only) - Number of Source Statements of Core Program, Pre- and Post-Processors:
65,000 word storage; 120,000 statements total

Documentation: See Refs. 62 to 64

Program Availability: Contact developer. Cost is negotiable.

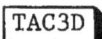

Descriptive Program Title: Thermal Analysis Code - Three-Dimensional

Program Developer: General Atomic Company, P.O. Box 81608, San Diego, California 92138

Date of First Release and Most Recent Update: 1969

General Information:
TAC3D is a code for calculating steady state and transient temperatures in three-dimensional problems by the finite difference method.

Program Capability:
Linear and nonlinear problems may be treated with TAC3D. Internal and external flowing coolants may be used, and there may be internal and external thermal radiation.

User Interface and Modeling Capabilities:
The configuration to be analyzed is described in the rectangular or cylindrical coordinate system. The input of thermal properties is by FORTRAN statement functions. This permits flexibility as many of the calculation variables (time, local temperature, local position, etc.) are available for use in these functions. There is a wide selection of optional output.

Solution Methods:
Alternating direction implicit method for three-dimensional problems (Douglas).

Notable Items and Limitations:
The grid planes must be orthogonal, and the entire problem must be bounded by six grid planes in one of the coordinate systems. All radiation is one-dimensional. There is no provision for phase change or thermal expansion of materials.

Programming Language: FORTRAN V

Hardware/Operating System: UNIVAC 1110/Exec 8

Program Size (Heat Transfer Modules Only) - Number of Source Statements of Core Program, Pre- and Post-Processors:
65,000 word storage; 150,000 statements total

Documentation: See Refs. 65 and 66

Program Availability: Contact developer. Cost is negotiable.

TANG

Descriptive Program Title: Thermal Analyzer Network Generator

Program Developer: Grumman Aerospace Corporation, Bethpage, New York 11714

Date of First Release and Most Recent Update: December 1976

General Information:
The Thermal Analyzer Network Generator software has been developed to automate the creation of finite difference thermal models. The program is a proprietary code belonging to Grumman Aerospace Corporation.

Program Capability and Scope of Analysis:
The program is capable of generating two- and three-dimensional thermal models of general geometries.

User Interface and Modeling Capabilities:
The program makes use of the interactive capabilities of the IBM operating system and user graphics hardware (digitizers and CRT displays) to expedite data display and verification.

Solution Methods:
The code employs standard conduction coupling formulation and surface area evaluation for convection and radiation coupling determination.

Programming Language: FORTRAN IV

Hardware/Operating System: IBM 370 (OS/VS)

Program Size (Heat Transfer Modules Only) - Number of Source Statements of Core Program, Pre- and Post-Processors:
The pre-processor code contains 384 source statements and performs digitizing and input data graphics functions. The core program for model generation contains 573 lines of code. Post-processor plotting - 173 lines of code.

Documentation: See Ref. 67

Program Availability: For further information contact:
  Dr. John G. Roukis
  Grumman Aerospace Corporation
  Mail Stop B22/35
  Bethpage, New York 11714

TAU

Descriptive Program Title: <u>T</u>hermal <u>A</u>nalysis of <u>U</u>ncle

Program Developer: J. A. Enderby, United Kingdom Atomic Energy Authority (UKAEA), Risley Nuclear Power Development Establishment, Warrington, Cheshire WA3 6AT, England.

Date of First Release and Most Recent Update: 1975 and 1981

General Information:
TAU is the heat transfer module of a general finite element system UNCLE which provides all the facilities needed by a general purpose finite element program. It is used extensively within the UKAEA and by a number of companies associated with the nuclear industry.

Program Capability:
TAU deals with conduction in materials of varying composition with material data optionally a function of temperature, time, or space and with internal heat generation. Radiation is allowed between all surfaces with a direct line of sight. Surface conditions include natural and forced convection, fixed flux, radiation to ambient, convection to an internal fluid whose temperature is to be found, and fixed temperatures. Calculations are performed for two-dimensional (slab and axisymmetric) and three-dimensional models in steady state and transient modes.

User Interface and Modeling Capabilities:
A unique element-cell-array-structure hierarchy gives an extremely concise and powerful yet natural way of inputting finite element models. The difficult task of providing radiation between surfaces, calculating viewfactors, and allowing for "shadows" can be accomplished with minimal effort on the part of the user. Mesh plots (including hidden line plots of three-dimensional structures) are provided. On-line interaction with the output routines allows selection of type of plot, size, viewpoint at run time.

Printed and plotted presentations of results are available (contour and line plots of temperature against position) and dump files may be used for restart purposes, for providing additional output, and for transfer of temperatures to the associated stressing program.

Element Library:
Two- and three-dimensional isoparametric elements are provided including triangles, quadrilaterals, tetrahedra, bricks and prisms with and without mid-side nodes. Boundary conditions are attached using compatible surface elements whose dimensionality is one less than that of the whole problem.

Solution Methods:
A sparse equation solver using a variant of Gauss elimination is provided. Many solutions stay in core using work space provided by the program at run time. Data is moved out of core automatically in an efficient systematic manner if in-core storage is insufficient. Nonlinear cases use a Newtonian iteration, or a modified Newton method in which the matrix of equations is set up and eliminated only once.

Transients use a backward difference (implicit) technique with time steps adjusted automatically for accuracy.

Programming Language: FORTRAN IV

Hardware/Operating System: IBM 370, 3033 (OS/VS), ICL 2900 (VME/B)

Program Size (Heat Transfer Modules Only) - Number of Source Statements of Core Program, Pre- and Post-Processors:
UNCLE (input and solution - 18,000 statements; UNCLE (output) - 8,500 statements; TAU - 4,300 statements; LISTIN - 3,300 statements.

Documentation: See Refs. 68 to 70

Program Availability: Available for use under contract on the Risley and Harwell computers of the UKAEA. Expected to be available in the United Kingdom and the United States on bureau by early 1982.

TEMP and NTEMP

Descriptive Program Title: TEMP - Temperature Analyzer; NTEMP - New Temperature Analyzer

Program Developer: Dr. A. F. Emery, Department of Mechanical Engineering, University of Washington, Seattle, Washington 98195

Date of First Release and Most Recent Update: TEMP - 1970 and 1981; NTEMP - 1980

General Information:
- TEMP handles temperature dependent, time dependent properties and boundary conditions. Dynamic core allocation.
  - NTEMP - Same as TEMP with the following additions:
    - Algebraic formula input for time and temperature dependent variables.
    - Direct solution on modification and resolution using R. Young's method - choice determined on basis of execution times.
    - Convergence based upon dT/dt or dF/dt using rms or mat. value.

Scope of Analysis:
Two-dimensional and axisymmetric. Radiation and convection. Nonlinear problems. Used for input to SAAS.

User Interface and Modeling Capabilities:
Batch processing. Interactive graphic input of mesh. Separate mesh generator or internal generation of simple meshes.

Solution Methods:
Direct solution of equations combined with iteration for nonlinear problems. Young's method used to reduce solution time for nonlinear problems. User prescribed convergence criterion. Convergence based upon rms or maximum error.

Notable Items and Limitations:
NTEMP has the following two mesh generators:
- Batch processor based on solution of Laplace's equation. Arbitrary mesh capability.
- Interactive generator. Triangles or quadrilaterals. Library of different surfaces. Spheres, cylinders, cones, etc.

The post-processor has the following facilities:
- Contour plotter
- Three-dimensional perspective hidden line plotter
- Determined mesh plotter
- Temperature-time plots
- Heat flux - time plots

Programming Language: FORTRAN Extended

Hardware/Operating System: CDC 6000 series/NOS BE; CDC CYBER 175/750/NOS

Program Size (Heat Transfer Modules Only) - Number of Source Statements of Core Program, Pre- and Post-Processors:
1,500 cards. Pre- and post-processors - 3,000 cards each.

Documentation: See Refs. 71 and 72

Program Availability: Contact developer. Cost: $250

TEPSA

Descriptive Program Title: Thermal Elasto-Plastic Stress Analysis

Program Developer: T. R. Hsu, Department of Mechanical Engineering, University of Manitoba, Winnipeg, Manitoba, Canada R3T 2N2

Date of First Release and Most Recent Update: September 1973 and November 1980

General Information:
TEPSA is a proprietary finite element computer code developed and maintained by Professor Tai-Ran Hsu. The code was initially developed to predict the thermo-mechanical behavior of reactor core components especially the fuel elements. However, it can be used equally well for other structures which comply, either three-dimensional axisymmetric or two-dimensional plane geometries. This code can

be used to handle thermal only, mechanical only, or coupled thermomechanical problems.

Program Capability:
The thermal analysis part of the TEPSA code is fully capable of handling both steady state and transient heat conduction of solids involving temperature-dependent material properties. Nonlinear convective and radiative boundary conditions can be applied, as well as heat generating elements. The analysis applies to the instantaneous geometry of the solid in the transient cases. Phase changes of the structure can also be handled. Extensive experimental verification of the code predictions has been made and a comprehensive user's manual is available.

Solution Methods:
Incremental process was used in the code for the nonlinear loads, material properties and geometries involved in the solution. Both two-level implicit and three-level explicit time difference schemes have been used. The "averaging enthalpy" algorithm was used for the evaluation of latent heat involved in the phase change process.

Programming Language: FORTRAN IV

Hardware/Operating System: AMDAHL, IBM

Program Size (Heat Transfer Modules Only) - Number of Source Statements of Core Program, Pre- and Post-Processors:
Preprocessor: approximately 100 statements.
Core program: approximately 1,684 statements.

Documentation: See Refs. 73 and 74

Program Availability: Contact developer. Cost is negotiable.

THAC-SIP-3D

Descriptive Program Title: A Three-Dimensional, Transient Heat Analysis Code Using the Strongly Implicit Procedure

Program Developer: W. D. Turner, Union Carbide Corporation, Nuclear Division, Oak Ridge, Tennessee 37830

Date of First Release and Most Recent Update: September 1978

General Information:
THAC-SIP-3D was written at Union Carbide Corporation, Nuclear Division, Oak Ridge, Tennessee, and was funded through the Office of Waste Isolation as part of the National Waste Thermal Storage Program.

Program Capability and Scope of Analysis:
THAC-SIP-3D is a transient analysis code designed to calculate temperature distributions for problems that can be modeled in the three-dimensional Cartesian coordinate system.

User Interface and Modeling Capabilities:
The user defines the problem by a series of regions having common characteristics. THAC-SIP-3D generates the nodal configuration from this information. Parameters

may be defined by built-in functions or by user-supplied functions.

Solution Methods:
Uses Stone's strongly implicit procedure for three-dimensional, transient problems.

Notable Items and Limitations:
Variable-dimensioned with respect to maximum number of nodes.

Programming Language: FORTRAN IV

Hardware/Operating System: IBM 360, IBM 370, IBM 3033

Program Size (Heat Transfer Modules Only) - Number of Source Statements of Core Program, Pre- and Post-Processors:
Core program - 5000

Documentation: See Ref. 75

Program Availability:
    Radiation Shielding Information Center
    Oak Ridge National Laboratory
    Oak Ridge, Tennessee 37830

Descriptive Program Title: Transient Heat Transfer Program Version D

Program Developer: General Electric Company, AEBG, Cincinnati, Ohio

General Information:
THTD is a proprietary code developed and maintained by the General Electric Company. Development started in 1956 with the first version released for use in 1957. Since then the program has undergone numerous changes and updates with continuing ongoing work to expand program and pre- and post-processor capabilities.

Program Capability and Scope of Analysis:
THTD is a general heat transfer code based on a finite difference implicit formulation of the partial differential equation for heat conduction. It computes transient and steady-state temperatures for three-dimensional geometries with a large variety of optional boundary, interface and internal conditions including:
- Internal heat generation, volumetric, time dependent
- Surface flux, time dependent
- Contact coefficient, constant
- Node to node or node to boundary radiation
- Convective boundaries with and without fluid flow, natural and forced convection, time and temperature dependent
- Temperature dependent physical properties
- Phase change (melting)

User Interface and Modeling Capabilities:
- THTD includes options for output data generation for direct interfacing with finite element stress programs through appropriate file manipulation.
- Extensive pre- and post-processor plotting capabilities for input and output

data analyses and documentation including isotherm plots, geometry plots and time and space dependent temperature plots are part of the THTD system.

• Input to the program is greatly simplified by the use of pre-processors in the preparation of geometry and boundary condition input data and a conversational input generation program. Geometry input data processing is accomplished by use of digitized nodal coordinate data generated by use of digitizer or mesh generator programs.

Solution Methods:
The solution method used is based on the Gauss-Seidel procedure with solution obtained by iterative solution of simultaneous algebraic equations for node temperatures derived from finite difference analysis. Convergence is recognized by user specified successive sets of maximum temperature changes (tolerance) permitted between iteration sweeps. The fully implicit formulation of the finite difference solution precludes any stability limitations on time increments and permits a direct steady state solution.

Notable Items and Limitations:
• THTD is a strongly user-oriented program with extensive data checks and edited output features and capabilities to accumulate solution results on binary tapes for problem restart, stress program interface and for graphical display.

• Although the program includes node to node and node to boundary radiation among its boundary condition options, it is generally not recommended for use in node to node radiation dominant problems.

• The program is currently limited to 2047 nodal elements, but can be readily expanded to 6,999 nodes.

Programming Language: Basic program language is FORTRAN IV. However, several service type subroutines are encoded in the GMAP assembly language for the Honeywell 6000 computers.

Hardware/Operating System: THTD is operational on the Honeywell 6000 computers at several locations within the General Electric Company. Earlier versions of the program are operational on the CDC 6400 at Battelle Memorial Institute and the UNIVAC computer at the NASA Manned Spacecraft Center, Houston, Texas.

Program Size (Heat Transfer Modules Only) - Number of Source Statements of Core Program, Pre- and Post-Processors:
• Core Program: 45K, additional memory required for loading of input data and depends on the number of nodes and tables used in the input.
• Source Statements: 22,000

Documentation: See Refs. 76 to 83

Program Availability: Inquiries should be addressed to:
W. K. Koffel
General Electric Company AEBG
Mail Drop K70
Neumann Way
Cincinnati, Ohio 45215

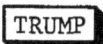

Descriptive Program Title: A Computer Program for Transient and Steady State Temperature Distributions in Multidimensional Systems

Program Developer: Arthur L. Edwards, Lawrence Livermore National Laboratory, P.O. Box 808, Livermore, California 94550

Date of First Release and Most Recent Update: 1967 and 1974

General Information:
TRUMP is a computer program in the public domain available from the Argonne Code Center, Argonne, Illinois. TRUMP is currently in use at Lawrence Livermore National Laboratory, and at a number of other locations in the United States, Canada and several other countries.

Program Capability and Scope of Analysis:
- TRUMP is a general purpose program for solving linear or nonlinear, steady-state or transient potential flow problems, including heat flow in temperature fields, Darcy flow in pressure fields, and fluxes in electrical and magnetic fields. In addition, TRUMP solves two additional equations representing, in thermal problems, heat production by decomposition of two reactants having rate constants with a general Arrhenius temperature dependence.
- Geometrical configurations may consist of complex three-dimensional structures of many materials which are described by specifying the dimensions or volumes of volume elements, and the dimensions or areas of their connections or surfaces.
- Material properties (e.g., thermal conductivity, specific heat) may be tabulated functions of the field variable (e.g., temperature) or time. Initial conditions may be specified for each volume element. Sources (e.g., heat) may be specified for each volume element, and may be tabulated functions of the field variable (i.e., temperature) or time, or may be given an exponential time dependence. Boundary conditions may be specified for each surface and each volume element, and may consist of a time-dependent field variable (e.g., temperature) or flux (e.g., heat), or a combination of a time-dependent external field variable and an interface conductance which may also be a tabulated function of time or the surface field variable (e.g., temperature). In thermal problems both convective and radiative transport may be represented at boundaries and between volume elements, and the surface conductance may be made proportional to a specified power of the difference between surface and external temperature.
- On thermal problems, a mass flow field may be specified, which may be either time or temperature dependent, constrained only by the requirement that all mass flow connections are between volume elements of the same material, and that inflow equals outflow for each volume element. In problems of Darcy flow in a pressure field, this field may be used to model the effects of gravity.
- Special elements may be specified that measure linear combinations (sums, averages, differences) of the field variable (e.g., temperature), or its rate of change. Any property that may be a tabulated function of the field variable (e.g., temperature) in one volume element may be made to depend on the field variable in another volume element, including the special elements specified for measurement purposes. In thermal-reactive problems, these properties include specific heat, thermal conductivity, heat of reaction, collision frequency, activation energy, heat generation rate, mass flow rate, and surface convection coefficient. This capability allows the solution of problems involving remote or automatic control.

• The solution method and accuracy may be determined by the program or specified by the user. Save-restart capability is provided by the program.

User Interface and Modeling Capability:
TRUMP geometric input can be produced directly by the user or by a pre-processor such as FED (Dale Schauer, LLNL). At LLNL the user may interact to determine the progress of the calculation, change output intervals, interrupt and restart, or end the problem. The user controls output intervals and quantity from minimum (e.g., temperature and global heat balance values) to maximum (e.g., detailed heat balance data for each volume element and connection, phase concentrations, chemical reactant concentrations, flow totals and rates, etc.). Plots include snapshots, time histories and contour plots, produced either directly or by a post-processor.

Physical Property Library:
A collection of critically evaluated thermal properties of over 1,000 materials in the required input format for TRUMP is available (see UCRL-50589) as part of the TRUMP package at Argonne Code Center.

Solution Methods:
TRUMP uses a combination of explicit and implicit methods to solve the algebraic set of difference equations for each time increment, or the user can choose a particular method, such as explicit, or two forms of implicit (backwards time-step or Crank-Nicholson) methods. In the combination explicit-implicit method, the zones done implicitly are determined by the program, but others may be added by the user. The particular mix varies as the problem proceeds, to optimize the use of computer time for the accuracy specified. The implicit method uses a one-point iterative scheme, with an extrapolated first estimate and local and global convergence criteria.

Notable Items and Limitations:
TRUMP is a very general and powerful solver of the general nonlinear parabolic partial differential equation describing flow in various kinds of potential fields in complex geometries. Geometry is specified independently of any global coordinate system, which limits the types of plots which can be made. Coordinate data is easily added when a geometric pre-processor is used. The number of volume elements and their interconnections and boundary connectors is limited by the memory size of the computer.

Programming Language: LRLTRAN at Lawrence Livermore National Laboratory, and various versions of FORTRAN at other locations.

Hardware/Operating System: CDC 7600, 6000 and 3000 series, IBM 360, UNIVAC 1100 series and GE 200

Program Size (Heat Transfer Modules Only):
Pre-processors - 5,000 statements, variable
Core Program - 5,000 statements
Post-processors - variable.

Documentation: See Refs. 84 to 86

Program Availability: National Energy Software Center
Argonne National Laboratory
9700 South Cass Avenue
Argonne, Illinois 60439

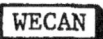

Descriptive Program Title: <u>W</u>estinghouse <u>E</u>lectric <u>C</u>omputer <u>An</u>alysis

Program Developer: Analytical Mechanics, Westinghouse Research and Development Center, Pittsburgh, Pennsylvania 15235

Date of First Release and Most Recent Update: 1973 and 1981

General Information:
WECAN is a proprietary code. It was and is developed jointly by Westinghouse Research and Development and other Westinghouse user divisions to be efficient, capable and easy to use. WECAN and its close relatives WAPPP (WECAN Pre- and Post-Processors) and FIGURES II (Finite Element Interactive Graphics User Routines) were designed to provide a complementary system of computer programs for structural analysis. Maintenance is funded by a surcharge. New developments are funded by user organizations desiring new development and the surcharge.

Program Capability:
WECAN is a general purpose static, dynamic transient linear buckling, and heat transfer and analogous field problems. Isotropic, orthotropic, and anisotropic materials are permitted. Material properties are defined as a fifth order polynomial of temperature. Substructures are linear but can be combined with nonlinear elements in the solution phase. Multilevel substructures are permitted. Substructures may be rotated, reflected or scaled. WECAN may be restarted at preselected time steps.

User Interface and Modeling Capabilities:
WAPPP is a collection of batch pre- and post-processors for WECAN. The pre-processors generate meshes and loads, check isoparametric element shapes, reduce wave fronts, prepare input for general matrix input and for composite materials. The post-processors edit heat transfer results, edit mode shapes and frequencies, combine results, plot contours, deformed shapes, transients and general xy curves, process seismic data, calculate J-integrals and calculate Fourier coefficients.

FIGURES II is a collection of interactive pre-processors that prepare input for WECAN. It can interactively plot what is being generated.

Solution Methods for Nonlinear Problems:
- WECAN solves transient heat transfer problems using the Crank-Nicholson-Galerkin integration scheme with $\alpha = 2/3$ or 1 or else uses a quadratic integration scheme.
- WECAN uses the wave front equation solver with as much in core as possible.

Notable Items and Limitations:
WECAN's heat transfer capabilities offer a wide range of applications. By analogy it has been used to solve electromagnetic field problems, fluid-structure interaction, torsion of prismatic bars, incompressible inviscid fluid flow, corrosion, seepage, acoustics, and electrical conductance problems. Interactive post-processing of results is under development. Users may specify any element conductivity or specific heat matrix through the general matrix input element. Basic workshops and advanced training sessions are offered periodically to train inexperienced and experienced WECAN users. An annual user's colloquium is held each Fall where users present papers in competition for prizes.

Programming Language: WECAN and WAPPP are over 99 percent FORTRAN IV, and less than 1 percent COMPASS. FIGURES - 100 percent FORTRAN IV

Hardware/Operating System: WECAN and WAPPP - CDC 7600 (SCOPE);
FIGURES II - PRIME 750 (PRIMOS)

Program Size (Heat Transfer Related):
WECAN 35,000; WAPPP 20,000; FIGURES II 100,000.

Documentation: See Refs. 87 to 89

Program Availability: Program can be used on Westinghouse PSCC Engineering Computer System or CDC Cybernet System for a surcharge on each run of WECAN, WAPPP or FIGURES II. The object tapes are available with terms negotiable. For further information please contact William Kunkel, Advance Systems Technology, Westinghouse Electric Corporation, 777 Penn Center Boulevard, Pittsburgh, Pennsylvania 15235 (412) 824-9100.

REFERENCES

1. Pilkey, W. D.; Saczalski, K. J.; and Schaeffer, H. G. (editors): Structural Mechanics Computer Programs. University Press of Virginia, Charlottesville, Virginia, 1974.

2. Fredriksson, B.; and Mackerle, J.: Structural Mechanics Finite Element Computer Programs - Surveys and Availability. Advanced Engineering Corporation, Box 3044, S-580 03, Linköping, Sweden.

3. Fredriksson, B.; and Mackerle, J.: Overview and Evaluation of Some Versatile General Purpose Finite Element Computer Programs. In Finite Element Methods in the Commercial Environment, vol. 2, pp. 390-419. Robinson and Associates 1978.

4. Thermal Structural Analysis Programs - A Survey and Evaluation. Sponsored by the Computer Technology Committee of the Pressure Vessels and Piping Division of the American Society of Mechanical Engineers, 1972.

5. Pilkey, B. F.: Computerized Sources of Abstracts of the Engineering Literature. In Structural Mechanics Software Series, vol. 3, pp. 29-39. University Press of Virginia, Charlottesville, Virginia, 1980.

6. Taig, I. C.: Selection Criteria for Structural Analysis Programs. AGARD Report No. 670, NATO-AGARD, pp. 11-20. Neuilly Sur Seine, France, Jan. 1979.

7. Sollogoob, P.; Wahl, L.; and Dreyer, F.: Main Computer Systems Programs, Presentation, Criteria for Selection. CNEXO/CTICM, 1978 (to be obtained from: CNEXO, B.P. 337, F-29273, Brest, or: CTICM, 20 Rue Jean Jaures, F-92807, Puteaux).

8. Berman, I. (editor): Engineering Computer Software: Verification, Qualification and Certification. ASME Publication, 1971.

9. ABAQUS User's Manual, Theory Manual, Example Problems Manual and Systems Manuel, Hibbitt, Karlsson and Sorensen, Inc., Providence, Rhode Island, Sept. 1981.

10. ADINAT User's Manual, 1981 Version. AE Report 2-81, ADINA Engineering AB, Västerås, Sweden, 1981.

11. Bathe, K. J.; and Khosgoftaar, M. R.: Finite Element Formulation and Solution of Nonlinear Heat Transfer. J. Nuc. Eng. and Design, vol. 51, 1979, pp. 389-401.

12. Bathe, K. J. (editor): Applications Using ADINA. Acoustics and Vibration Laboratory, Mechanical Engineering Dept., Report 82448-6, Massachusetts Institute of Technology, Cambridge, MA., Aug. 1977.

13. Bathe, K. J. (editor): Nonlinear Finite Element Analysis and ADINA. Acoustics and Vibration Laboratory, Mechanical Engineering Dept., Report 82448-9, Massachusetts Institute of Technology, Cambridge, MA., Aug. 1979.

14. Emery, A. F.: ANDES - A Computer Program for the Computation of Temperature, Stress and NDE Acoustic Signal Response. Sandia Laboratories, Livermore, CA (to be published in 1982).

15. Kohnke, P. C.: ANSYS Theoretical Manual. Swanson Analysis Systems, Inc., Houston, PA., 1977.

16. DeSalvo, G. J.; and Swanson, J. A.: ANSYS User's Manual. Swanson Analysis Systems, Inc., Houston, PA., July 1979.

17. ASASHEAT User's Manual, Verification Manual and Programmer's Manual, Parts 1-5. Atkins Research and Development, Surrey, England, 1981.

18. Hellen, T. K.: The BERSAFE System. BM/SM/SAM201, Bersafe Advisory Group, Berkeley Nuclear Laboratories, Berkeley, Glos., England.

19. Hellen, T. K.: A User's Guide to BERSAFE Phase II (Level 3). To be published. Berkeley Nuclear Laboratories, Berkeley, Glos., England.

20. Keavey, M. A.: A User's Guide to FLHE Phase II (Level 3). To be published. Berkeley Nuclear Laboratories, Berkeley, Glos., England.

21. Rathjen, K.: CAVE: A Computer Code for Two-Dimensional Transient Heating Analysis of Conceptual Thermal Protection Systems for Hypersonic Vehicles. NASA CR-2897, Nov. 1977.

22. Rathjen, K.; Palmieri, J.: CAVE3 - A General Transient Heat Transfer Computer Code Utilizing Eigenvectors and Eigenvalues. NASA CR-145290, Feb. 1978.

23. Rathjen, K.; Palmieri, J.: Nodal Network Generator for CAVE3. NASA CR-TBD, Oct. 1979.

24. Turner, W. D.; Elrod, D. C.; Sieman-Tov, I. I.: Heating5 - An IBM 360 Heat Conduction Program. ORNL/CSD/TM-15, Mar. 1977.

25. Elrod, D. C.; Giles, G. E.; and Turner, W. D.: Heating6 - A Multidimensional Heat Conduction Analysis with the Finite Difference Formulation. Nov. 1981.

26. Collier, W. D.: The Solution of Heat Transfer Problems Using HEATRAN. TR9 Report 2512(R). United Kingdom Atomic Energy Authority, Risley, Warrington, England.

27. Marcal, P. V.: Nonlinear Analysis with a General Purpose Code in a Commercial Environment. Second World Congress on Finite Element Methods. Bournemouth, England, Oct. 1970.

28. Hibbitt, H. D.; and Marcal, P. V.: A Numerical Thermomechanical Model for the Welding and Subsequent Loading of a Fabricated Structure. Brown University, 1973.

29. Hughes, T. J. R.; Belytschko, T.; Nagtegaal, J. C.; and Burke, M. A.: Advanced Topics in Nonlinear Finite Element Analysis. MARC Analysis Research Corporation, 1981.

30. Martin Marietta Interactive Thermal Analysis System, Version 2.0 (MITAS II) User's Manual. Engineering Department Technical Manual, Report No. M-76-2, Martin Marietta Corporation, Denver, CO., May 1976.

31. MacNeal, R. H. (editor): NASTRAN Theoretical Manual. MacNeal-Schwendler Corporation, Los Angeles, CA., (MSR-40), Dec. 1972.

32. McCormick, C. W. (editor): MSC/NASTRAN User's Manual. MacNeal-Schwendler Corporation (MSR-34), Los Angeles, CA., Feb. 1981.

33. Joseph, J. A. (editor): MSC/NASTRAN Application Manual. MacNeal-Schwendler Corporation (MSR-59), monthly updates, Los Angeles, CA.

34. McLean, D. M. (editor): MSC/NASTRAN Programmer's Manual (three volumes). Sept. 1981.

35. Nodal Network Thermal Balance Program. Temperature Control Section, Thermophysics Branch, Goddard Space Flight Center, Greenbelt, MD., Feb. 1968.

36. Henshell, R. D.: PAFEC Data Preparation. PAFEF, Ltd., Nottingham, England, 1978.

37. Akin, J. E.: PAFEC for the Engineer. PAFEC Engineering Consultants, Inc., Knoxville, TN., 1980.

38. Henshell, R. D.: PAFEC Theory, Results. PAFEC, Ltd., Nottingham, England, 1975.

39. Gabrielson, V. K.: SAHARA: A Multidimensional Heat Transfer Code (User's Manual). SCL-DR-720024, Sandia National Laboratories, Livermore, CA., 1972.

40. Gabrielson, V. K.: HEATMESH-71: A Computer Code for Generating Geometrical Data Required for Studies of Heat Transfer in Axisymmetric Structures. SCL-DR-720004, Sandia National Laboratories, Livermore, CA., 1972.

41. Hogge, M.; Gerrekens, P.; and Laschet, G.: Module de Reponse Thermique Non Lineaire THERNL. User's Manual. LTAS Report SF-100. University of Liege, Belgium, 1981.

42. L.T.A.S. (editor): SAMCEF - Modules de Reponse Thermique. User's and Theoretical Manual No. 7. University of Liege, Belgium, 1981.

43. Hogge, M.: Secant Versus Tangent Methods in Nonlinear Heat Transfer Analysis. Int. Journal for Num. Meth. Eng., vol. 16, 1980, pp. 51-64.

44. Hogge, M.: A Comparison of Two- and Three-Level Integration Schemes for Nonlinear Heat Conduction. In Numerical Methods in Heat Transfer, R. W. Lewis, et al. (eds.), John Wiley, 1981, pp. 75-90.

45. Hogge, M.; and Gerrekens, P.: One-Dimensional Finite Element Analysis of Thermal Ablation with Pyrolysis. Proceedings of FENOMECH '81, University of Stuttgart, West Germany, Aug. 1981.

46. Klem, H. F., et al.: SESAM-69 General Description. Computas Report No. 78-921. Veritas, Høvik, Norway, 1978.

47. Klem, H. F., et al.: SESAM-69/NV336 General Superelement Program, User's Manual. Computas Report No. 78-926. Veritas, Høvik, Norway, 1978.

48. Pahle, E.: SESAM-69/NV340 General Plotter Program, User's Manual. Computas Report No. 78-929. Veritas, Høvik, Norway, 1978.

49. Sandsmark, N.: Analysis of Stationary and Transient Heat Conduction by the Use of Finite Element Method. Report UR-80-06. Marine Technology Centre, Trondheim, Norway, 1979.

50. Kråkeland, B., et al.: SESAM-69/NV333 Analysis of Solids, User's Manual. Computas Report No. 78-925. Veritas, Høvik, Norway, 1978.

51. Hagen, K., et al.: SESAM-69/NV615 Analysis of Heat Conduction in Three-Dimensional Solids (Linear and Nonlinear, Stationary and Transient), User's Manual. Computas Report No. 79-947. Veritas, Høvik, Norway, 1978.

52. Hagen, K., et al.: SESAM-69/NV615 Analysis of Heat Conduction in Three-Dimensional Solids, Maintenance Manual. Computas Report No. 78-948. Veritas, Høvik, Norway, 1978.

53. Smith, J. P.: SINDA User's Manual. TRW Systems, Houston, Texas, April 1971.

54. Ishimoto, T.; and Fink, L. C.: SINDA Engineering-Program Manual. TRW Systems, Redondo Beach, CA., 1971.

55. Oren, J. A.; and Williams, D. R.: SINDA/SINFLO Computer Routine, Rev. A. Vols. 1 and 2. LTV Aerospace Corporation, Dallas, TX, Feb. 1975.

56. Marlowe, M. B.; Moore, R. A.; and Whetstone, W. D.: SPAR Thermal Analysis Processors - Reference Manual - System Level 16. NASA CR-159162, 1979.

57. Program Manual for the Simplified Shuttle Payload Thermal Analyzer (Version 2.0/IBM). Prepared by Arthur D. Little, Inc., for NASA Goddard Space Flight Center under Contract NAS5-23392, Sept. 1979.

58. Bartoszek, J. T.; and Huckins, B.: A Simplified Shuttle Payload Thermal Analyzer (SSPTA) Program. AIAA Paper 79-1052 presented at the AIAA Fourteenth Thermophysics Conference, Orlando, FL., June 4-6, 1979.

59. Hughes, T. J. R.: Stability of One-Step Methods in Transient Nonlinear Heat Conduction. Transactions of the Fourth International Conference on Structural Mechanics in Nuc. Reactor Tech., San Francisco, CA., No. B2/10, Aug. 1977.

60. Mason, W. E.: TACO – A Finite Element Heat Transfer Code. UCID-17980, Lawrence Livermore National Laboratory, 1978.

61. Mason, W. E.: POSTACO – A Post-Processor for Scalar Finite Element Codes. UCID-17980, Rev. 1, Lawrence Livermore National Laboratory, 1980.

62. Clark, S. S.; and Petersen, J. F.: TAC2D – Mathematical Formulations and Programmer's Guide. GA-9262. General Atomic Company, San Diego, CA., Sept. 1969.

63. Boonstra, R. H.: TAC2D – User's Manual. GA-A14032. General Atomic Company, San Diego, CA., July 1976.

64. Morcos, S. M.; and Williams, K. A.: The TAC2D Code Verification and Benchmark Problems. GA-A13415. General Atomic Company, San Diego, CA., June 1975.

65. Clark, S. S.; Del Bene, J. V.; and Petersen, J. F.: TAC3D – Mathematical Formulations and Programmer's Guide. GA-9264. General Atomic Company, San Diego, CA., Sept. 1969.

66. Petersen, J. F.: TAC3D – User's Manual. GA-9263. General Atomic Company, San Diego, CA., Sept. 1969.

67. Palmieri, J. V.; and Bocchicchio, R.: Thermal Analyzer Network Generator. Proceedings of the Symposium on Applications of Computer Methods in Engineering, held in Los Angeles, CA., Aug. 23-26, 1977, vol. 2, pp. 1175-1183.

68. Johnson, D.: TAU – A Computer Program for the Analysis of Temperature in Two- and Three-Dimensional Structures Using the UNCLE Finite Element Scheme. ND-R-218(R). United Kingdom Atomic Energy Authority, Risley, Warrington, England.

69. Enderby, J. A.: An Introduction to the UNCLE Finite Element Scheme. ND-R-225(R). United Kingdom Atomic Energy Authority, Risley, Warrington, England.

70. Collier, W. D.: Radiation of Heat in the Heat Transfer Program TAU. ND-R-555(R). United Kingdom Atomic Energy Authority, Risley, Warrington, England.

71. Emery, A. F.: TEMP – A Nonlinear Thermal Analyzer. Report 6. Dept. of Mech. Engineering, University of Washington, Seattle, WA., 1964.

72. Emery, A. F.; and Bicler, T. R.: NTEMP – A Finite Element Temperature Code. Report 80-2. Dept. of Mech. Engineering, University of Washington, Seattle, WA., Sept. 1980.

73. Hsu, T. R.; Bertels, A. W. M.; Banerjee, S.; and Harrison, W. C.: Theoretical Basis for a Transient Thermal Elastic-Plastic Stress Analysis of Nuclear Reactor Fuel Elements. AECL-5233. Whiteshell Nuclear Research Establishment, Pinawa, Manitoba, Canada, July 1976.

74. Wu, R. Y.: User's Manual for TEPSA Code, Version 2. Thermomechanics Laboratories, University of Manitoba, Winnipeg, Manitoba, Canada R3T 2N2, Report 79-8-65, 1979.

75. Turner, W. D.: THAC-SIP-3D - A Three-Dimensional Transient Heat Analysis Code Using the Strongly Implicit Procedure. Union Carbide Corporation, Nuclear Division, Oak Ridge, TN., Sept. 1978.

76. Anderson, J. T.; Botje, J. M.; and Koffel, W. K.: Digital Computer Solution of Complex Transient Heat Transfer Problems. Tech. Bulletin No. 62, West Virginia University Bulletin, Engineering Experiment Station.

77. Koffel, W. K.: Transient Heat Transfer IBM 704 Program Experience and Extensions. R59FPD768, General Electric Company, 1959.

78. Kaplan, B.; and Clark, N.: Accelerating the Convergence of the Generalized Transient Heat Transfer Program (THT). ANPD, XDC 60-5-8, General Electric Company, 1960.

79. Skirvin, S. C.: Partial Text for a Programmer's Handbook for THTD Computer Program (Transient Heat Transfer - Version D). S-69-1105.

80. Skirvin, S. C.: User's Manual for the Standard THTD Computer Program on GE-600 Series Computers (Transient Heat Transfer - Version D). General Electric Research and Development Center Report 69-C-205, May 1969.

81. Brown, D. L.: Dynamic Memory Allocation During Execution From an H* Program File on the GE-625/635 Computer for Programs THTD and THTE. General Electric Irradiation Processing Operation Report NEDO-12018, July 1969.

82. Brown, D. L.; and Skirvin, S. C.: Extensions to the THTD Program (Transient Heat Transfer - Version D) for Binary Gas Mixture Properties and Radial Thermal Expansion. General Electric Nuclear Energy Division Group Report NEDE-12717.

83. Wilton, M. E.; and Pope, M. L.: User's Manual for the THTD Computer Program on GE635 Computer (Transient Heat Transfer - Version D). General Electric TIS Report R71AEG155, April 1971.

84. Edwards, A. L.: TRUMP: A Computer Program for Transient and Steady State Temperature Distributions in Multidimensional Systems. UCRL-14754, Rev. 3, Dept. 1, 1972.

85. Edwards, A. L.: A Compilation of Thermal Property Data for Computer Heat-Conduction Calculations. UCRL-50589, Feb. 24, 1969.

86. Schauer, D. A.: FED: A Computer Program to Generate Geometric Input for the Heat Transfer Code TRUMP. UCRL-50816, Rev. 1, Jan. 12, 1973.

87. Filstrup, A. W. (editor): WECAN User's Manual. Advanced Systems Technology, Westinghouse Electric Corp., Pittsburgh, PA., March 1981.

88. Gabrielse, S. E.: WAPPP User's Manual. Advanced Systems Technology, Westinghouse Electric Corp., Pittsburgh, PA., 1980.

89. Morris, J. W.; Newman, M. B.; and Snyder, J. R.: FIGURES II User's Guide. Advanced Systems Technology, Westinghouse Electric Corp., Pittsburgh, PA., June 1981.

ATTENDEES

Dr. Howard M. Adelman
NASA/Langley Research Center
Mail Stop 243
Hampton, VA 23665

Mr. John Allred
Mercury Engineers
17 Research Drive
Hampton, VA 23666

Mr. James E. Akin
PAFEC Engineering Incorporated
Knoxville, TN 37919

Dr. Carl M. Andersen
College of William & Mary
Williamsburg, VA 23185

Mr. James P. Bailey
Kentron International Incorporated
Hampton Technical Center
Hampton, VA 23665

Mr. Chad Balch
George Washington University
NASA/Langley Research Center
Mail Stop 246
Hampton, VA 23665

Professor K. Jurgen Bathe
Massachusetts Institute of Technology
Department of Mechanical Engineering
Boston, MA 02139

Mr. William M. Berrios
NASA/Langley Research Center
Mail Stop 434
Hampton, VA 23665

Mr. Max L. Blosser
NASA/Langley Research Center
Mail Stop 395
Hampton, VA 23665

Mr. Obie H. Bradley, Jr.
NASA/Langley Research Center
Mail Stop 434
Hampton, VA 23665

Mr. Elliot W. Brogren
Boeing Aerospace Company
P. O. Box 3999
Seattle, WA 98124

Mr. Charles J. Camarda
NASA/Langley Research Center
Mail Stop 243
Hampton, VA 23665

Mr. Alan L. Carter
NASA/Dryden Flight Research
 Facility
Mail E-EAS
Edwards, CA 93523

Mr. David J. Carter, Jr.
NASA/Langley Research Center
Mail Stop 258
Hampton, VA 23665

Mr. John J. Catherines
NASA/Langley Research Center
Mail Stop 431
Hampton, VA 23665

Mr. T. F. Chen
Kaiser Aluminum -
 Center for Technology
P. O. Box 877
Pleasanton, CA 94566

Mr. Edward T. Chimenti
NASA/Johnson Space Center
Mail Code ES3
Houston, TX 77058

Mr. Charles Class
Martin Marietta - Denver Aerospace
P. O. Box 179
Mail Stop 50484
Denver, CO 80201

Mr. Marvin C. Clemmons
NASA/Langley Research Center
Mail Stop 125
Hampton, VA 23665

Mr. O. R. Cluraman
Mercury Engineers
17 Research Drive
Hampton, VA 23666

Mr. Howard Collicott
Bendix Corporation
P. O. Box 5029
Southfield, MI 48037

Mr. Robert B. Davis
NASA/Langley Research Center
Mail Stop 431
Hampton, VA 23665

Mr. Pramote Dechaumphai
Old Dominion University
Norfolk, VA 23508

Dr. Sidney C. Dixon
NASA/Langley Research Center
Mail Stop 398
Hampton, VA 23665

Mr. Augustine Dovi
Kentron International Incorporated
Hampton Technical Center
Hampton, VA 23665

Dr. Ashley F. Emery
University of Washington
Seattle, WA 98185

Mr. Edward A. Eiswirth
McDonnell Douglas Aircraft Company-
 St. Louis
P. O. Box 516
E242/HQ/45
St Louis, MO 63166

Mr. Richard W. Faison
NASA/Langley Research Center
Mail Stop 431
Hampton, VA 23665

Mr. David S. Fine
General Motors Corporation
C-2-232
30003 Van Dyke
Warren, MI 48090

Mr. Darrell Fletcher
Garrett Turbine Engine Company
111 South 34th Street
P. O. Box 5217
Phoenix, AZ 85010

Mr. Richard A. Foss
NASA/Langley Research Center
Mail Stop 431
Hampton, VA 23665

Mr. Laurence E. Frank
Northrop Aircraft Company
6124 Queenridge Drive
R. Palos Verdes, CA 90274

Mr. V. K. Gabrielson
Sandia National Laboratories
Livermore, CA 94550

Dr. Bernard L. Garrett
NASA/Langley Research Center
Mail Stop 364
Hampton, VA 23665

Mr. J. D. Gaski
Sinda Industries Incorporated
P. O. Box 8007
Fountain Valley, CA 92708

Mr. Larry Gaudreau
Hughes Aircraft Company
Box 91961
Los Angeles, CA 90009

Mr. Stuart Glazer
Jet Propulsion Laboratory
4800 Oak Grove Drive
Pasadena, CA 91103

Mr. Leslie Gong
NASA/Dryden Flight Research
 Facility
Mail E-EAS
Edwards, CA 93523

Mr. R. Grandhi
Virginia Polytechnic Institute and
 State University
Blacksburg, VA 24061

Mr. Robert F. Greene, Jr.
NASA/Langley Research Center
Mail Stop 434
Hampton, VA 23665

Mr. William H. Greene
NASA/Langley Research Center
Mail Stop 190
Hampton, VA 23665

Mr. F. H. Gregory
Army Ballistics Research Laboratory
Aberdeen, MD 21005

Mr. R. L. Haddock
PDA Engineering
1560 Brookhollow Drive
Santa Anna, CA 92705

Dr. Raphael T. Haftka
Dept. of Aerospace & Ocean Engineering
Virginia Polytechnic Institute and
 State University
Blacksburg, VA 24061

Mr. T. W. E. Hankinson
NASA/Langley Research Center
Mail Stop 431
Hampton, VA 23665

Mr. Perry W. Hanson
NASA/Langley Research Center
Mail Stop 398
Hampton, VA 23665

Mr. Robert L. Harder
MacNeal-Schwendler Corporation
7442 North Figueroa Street
Los Angeles, CA 90041

Mr. Richard R. Heldenfels, DRA
NASA/Langley Research Center
Mail Stop 109
Hampton, VA 23665

Mr. H. P. Henderson
United Technology - Chemical Systems
1050 East Arques
P. O. Box 358
Sunnyvale, CA 94088

Mr. Leroy M. Herold
TRW Systems and Energy
Building M3, Room 2542
One Space Park
Redondo Beach, CA 90277

Mr. Joseph L. Heywood
Thiokol Corporation
P. O. Box 524
Brigham City, UT 84302

Mr. F. T. Hung
Aerospace Corporation
Los Angeles, CA 90041

Mr. L. Roane Hunt
NASA/Langley Research Center
Mail Stop 395
Hampton, VA 23665

Dr. William F. Hunter
NASA/Langley Research Center
Mail Stop 431
Hampton, VA 23665

Mr. Ben B. James
Kentron International Inc.
Hampton Technical Center
Hampton, VA 23665

Mr. Carl Jensen
Martin Marietta-Denver Aerospace
P. O. Box 179
Mail Stop 50484
Denver, CO 80201

Mr. Gary Jones
NASA/Goddard Space Flight Center
Mail Code 731
Greenbelt, MD 20771

Mr. Harry E. Jones
Hercules Incorporated
Bacchus Works
P. O. Box 98
Magna, UT 84044

Ms. Susan M. Juba
Lockheed Engineering and Management
 Services Incorporated
Houston, TX 77058

Dr. O. S. Kandil
Old Dominion University
Norfolk, VA 23508

Mr. H. Neale Kelly
NASA/Langley Research Center
Mail Stop 395
Hampton, VA 23665

Mr. Charles E. Knight
Department of Mechanical Engineering
Virginia Polytechnic Institute and
 State University
Blacksburg, VA 24061

Mr. Norman F. Knight, Jr.
NASA/Langley Research Center
Mail Stop 190
Hampton, VA 23665

Dr. William L. Ko
NASA/Dryden Flight Research
 Facility
Mail E-EAS
Edwards, CA 93523

Dr. Edwin T. Kruszewski
Old Dominion University
Norfolk, VA 23508

Mr. A. Kwas
TRW Systems and Energy
Building M3, Room 2542
One Space Park
Redondo Beach, CA 90277

Dr. Jules J. Lambiotte, Jr.
NASA/Langley Research Center
Mail Stop 125
Hampton, VA 23665

Dr. H. P. Lee
NASA/Goddard Space Flight Center
Mail Stop
Greenbelt, MD 20771

Mr. Wendell H. Lee
NASA/Langley Research Center
Mail Stop 434
Hampton, VA 23665

Dr. Wing K. Liu
Northwestern University
Evanstown, IL 60201

Mr. Duncan MacDonald
AVCO Systems Division
201 Lowell Street
Room 3108
Wilmington, MA 01887

Mr. Paul E. McGowan
NASA/Langley Research Center
Mail Stop 395
Hampton, VA 23665

Mr. James M. McKee
Naval Ship R&D Center
Code 1844
Bethesda, MD 20084

Mr. Jack Mahaney
Old Dominion University
Norfolk, VA 23508

Mr. James A. Martin
NASA/Langley Research Center
Mail Stop 365
Hampton, VA 23665

Dr. William E. Mason
Sandia National Laboratories
Livermore, CA 94550

Ms. Maryellen E. Maxson
Mercury Engineers
17 Research Drive
Hampton, VA 23666

Mr. E. Wade Miner
Naval Research Laboratory
Washington, DC 20375

Mr. E. Leon Morrisette
NASA/Langley Research Center
Mail Stop 164
Hampton, VA 23665

Mr. Tom Nakamura
Hughes Aircraft Company
Building 313, Mail Stop R124
Centinela and Trale
Culver City, CA 90230

Mr. R. W. Newman
Johns Hopkins University
Applied Physics Laboratory
Laurel, MD 20810

Mr. Gim Shek Ng
NASA/Langley Research Center
Mail Stop 431
Hampton, VA 23665

Dr. Ahmed K. Noor
George Washington University
NASA/Langley Research Center
Mail Stop 246
Hampton, VA 23665

Mr. N. D. Osella
General Dynamics Corporation
Eastern Point Road
Groton, CT 06340

Ms. Ann B. Patten
NASA/Langley Research Center
Mail Stop 431
Hampton, VA 23665

Mr. S. B. Paul
AFWAL/FIBE
Wright-Patterson AFB, OH 45433

Mr. Richard L. Peterson
Hercules Incorporated
P. O. Box 98
Magna, UT 84044

Mr. Dennis H. Petley
NASA/Langley Research Center
Mail Stop 431
Hampton, VA 23665

Mr. Claud M. Pittman
NASA/Langley Research Center
Mail Stop 396
Hampton, VA 23665

Mr. Anthony S. Pototzky
Kentron International Incorporated
Hampton Technical Center
Hampton, VA 23665

Dr. Robert C. Ried
NASA/Johnson Space Center
Mail Code ES3
Houston, TX 77058

Ms. Kathleen M. Riley
Kentron International Incorporated
Hampton Technical Center
Hampton, VA 23665

Mr. James C. Robinson
NASA/Langley Research Center
Mail Stop 243
Hampton, VA 23665

Dr. Aileen Rogers
Drexel University
Philadelphia, PA 19104

Mr. A. L. Rosenblatt
Grumman Aerospace Company
Bethpage, NY 11714

Mr. H. Rosenetti
Rohr Industries Incorporated
P. O. Box 878
Chula Vista, CA 92012

Mr. Herman A. Rosenthal
Rohr Industries Incorporated
P. O. Box 878
Chula Vista, CA 92012

Mr. E. J. Shemenski
Newport News Shipbuilding and
 Dry Dock Company
Newport News, VA 23601

Mr. John L. Shideler
NASA/Langley Research Center
Mail Stop 395
Hampton, VA 23665

Mr. Charles P. Shore
NASA/Langley Research Center
Mail Stop 243
Hampton, VA 23665

Mr. Joseph T. Skladany
NASA/Goddard Space Flight Center
Mail Stop
Greenbelt, MD 20771

Mr. Dewey M. Smith
NASA/Langley Research Center
Mail Stop 431
Hampton, VA 23665

Dr. Jaroslaw Sobieski
NASA/Langley Research Center
Mail Stop 243
Hampton, VA 23665

Mr. J. A. Stone
McDonnell Douglas Aircraft Company
St. Louis, MO 63166

Dr. Olaf O. Storaasli
NASA/Langley Research Center
Mail Stop 246
Hampton, VA 23665

Dr. John A. Swanson
Swanson Analysis Systems
P. O. Box 65
Houston, PA 15342

Mr. Doyle P. Swofford
NASA/Langley Research Center
Mail Stop 431
Hampton, VA 23665

Mr. Kumar Tamma
Old Dominion University
Norfolk, VA 23508

Mr. D. W. Taylor
Naval Ship R&D Center
Code 1844
Bethesda, MD 20084

Mr. Robert L. Thompson
NASA/Lewis Research Center
Mail Stop 49-6
Cleveland, OH 44135

Dr. Earl A. Thornton
Old Dominion University
Norfolk, VA 23508

Mr. Robert A. Vogt
NASA/Johnson Space Center
Mail Code ES3
Houston, TX 77058

Mr. Woodrow W. Wagner, Jr.
NASA/Langley Research Center
Mail Stop 431
Hampton, VA 23665

Ms. Dolores R. Wallace
Naval Ship R&D Center
Code 1844
Bethesda, MD 20084

Ms. Joanne L. Walsh
NASA/Langley Research Center
Mail Stop 243
Hampton, VA 23665

Mr. Joseph E. Walz
NASA/Langley Research Center
Mail Stop 230
Hampton, VA 23665

Mr. Irving Weinstein
NASA/Langley Research Center
Mail Stop 395
Hampton, VA 23665

Mr. Larry Weisstein
George Washington University
NASA/Langley Research Center
Mail Stop 246
Hampton, VA 23665

Mr. Allan R. Wieting
NASA/Langley Research Center
Mail Stop 395
Hampton, VA 23665

Mr. Willy Wolter
Grumman Aerospace Company
Bethpage, NY 11714

Mr. Robert L. Wright
NASA/Langley Research Center
Mail Stop 364
Hampton, VA 23665

Ms. Kathryn E. Wurster
NASA/Langley Research Center
Mail Stop 365
Hampton, VA 23665

Mr. Y. X. Yang
Old Dominion University
Norfolk, VA 23508

Dr. Clarence P. Young, Jr.
NASA/Langley Research Center
Mail Stop 431
Hampton, VA 23665

Mr. Joseph P. Young
NASA/Goddard Space Flight Center
Mail Code 731
Greenbelt, MD 20771

www.ingramcontent.com/pod-product-compliance
Lightning Source LLC
Chambersburg PA
CBHW081713170526
45167CB00009B/3566